T0271206

CAMBRIDGE STUDIES IN
ADVANCED MATHEMATICS 106

LINEAR OPERATORS AND THEIR SPECTRA

This wide ranging but self-contained account of the spectral theory of non-self-adjoint linear operators is ideal for postgraduate students and researchers, and contains many illustrative examples and exercises.

Fredholm theory, Hilbert-Schmidt and trace class operators are discussed, as are one-parameter semigroups and perturbations of their generators. Two chapters are devoted to using these tools to analyze Markov semigroups.

The text also provides a thorough account of the new theory of pseudospectra, and presents the recent analysis by the author and Barry Simon of the form of the pseudospectra at the boundary of the numerical range. This was a key ingredient in the determination of properties of the zeros of certain orthogonal polynomials on the unit circle.

Finally, two methods, both very recent, for obtaining bounds on the eigenvalues of non-self-adjoint Schrödinger operators are described. The text concludes with a description of the surprising spectral properties of the non-self-adjoint harmonic oscillator.

CAMBRIDGE STUDIES IN ADVANCED MATHEMATICS

All the titles listed below can be obtained from good booksellers or from Cambridge University Press. For a complete series listing visit:
http://www.cambridge.org/series/sSeries.asp?code=CSAM

Already published

58   J. McCleary *A user's guide to spectral sequences II*
59   P. Taylor *Practical foundations of mathematics*
60   M. P. Brodmann & R. Y. Sharp *Local cohomology*
61   J. D. Dixon *et al. Analytic pro-P groups*
62   R. Stanley *Enumerative combinatorics II*
63   R. M. Dudley *Uniform central limit theorems*
64   J. Jost & X. Li-Jost *Calculus of variations*
65   A. J. Berrick & M. E. Keating *An introduction to rings and modules*
66   S. Morosawa *Holomorphic dynamics*
67   A. J. Berrick & M. E. Keating *Categories and modules with K-theory in view*
68   K. Sato *Levy processes and infinitely divisible distributions*
69   H. Hida *Modular forms and Galois cohomology*
70   R. Iorio & V. Iorio *Fourier analysis and partial differential equations*
71   R. Blei *Analysis in integer and fractional dimensions*
72   F. Borceaux & G. Janelidze *Galois theories*
73   B. Bollobás *Random graphs*
74   R. M. Dudley *Real analysis and probability*
75   T. Sheil-Small *Complex polynomials*
76   C. Voisin *Hodge theory and complex algebraic geometry, I*
77   C. Voisin *Hodge theory and complex algebraic geometry, II*
78   V. Paulsen *Completely bounded maps and operator algebras*
79   F. Gesztesy & H. Holden *Soliton equations and their algebro-geometric solutions, I*
81   S. Mukai *An Introduction to invariants and moduli*
82   G. Tourlakis *Lectures in logic and set theory, I*
83   G. Tourlakis *Lectures in logic and set theory, II*
84   R. A. Bailey *Association schemes*
85   J. Carlson, S. Müller-Stach & C. Peters *Period mappings and period domains*
86   J. J. Duistermaat & J. A. C. Kolk *Multidimensional real analysis I*
87   J. J. Duistermaat & J. A. C. Kolk *Multidimensional real analysis II*
89   M. Golumbic & A. Trenk *Tolerance graphs*
90   L. Harper *Global methods for combinatorial isoperimetric problems*
91   I. Moerdijk & J. Mrcun *Introduction to foliations and Lie groupoids*
92   J. Kollar, K. E. Smith & A. Corti *Rational and nearly rational varieties*
93   D. Applebaum *Levy processes and stochastic calculus*
94   B. Conrad *Modular forms and the Ramanujan conjecture*
95   M. Schecter *An introduction to nonlinear analysis*
96   R. Carter *Lie algebras of finite and affine type*
97   H. L. Montgomery, R. C. Vaughan & M. Schechter *Multiplicative number theory I*
98   I. Chavel *Riemannian geometry*
99   D. Goldfeld *Automorphic forms and L-functions for the group GL(n,R)*
100 M. Marcus & J. Rosen *Markov processes, Gaussian processes, and local times*
101 P. Gille & T. Szamuely *Central simple algebras and Galois cohomology*
102 J. Bertoin *Random fragmentation and coagulation processes*
104 A. Ambrosetti & A. Malchiodi *Nonlinear analysis and semilinear elliptic problems*
105 T. Tao & V. H. Vu *Additive combinatorics*

LINEAR OPERATORS AND
THEIR SPECTRA

E. BRIAN DAVIES

CAMBRIDGE
UNIVERSITY PRESS

CAMBRIDGE
UNIVERSITY PRESS

University Printing House, Cambridge CB2 8BS, United Kingdom

One Liberty Plaza, 20th Floor, New York, NY 10006, USA

477 Williamstown Road, Port Melbourne, VIC 3207, Australia

4843/24, 2nd Floor, Ansari Road, Daryaganj, Delhi - 110002, India

79 Anson Road, #06-04/06, Singapore 079906

Cambridge University Press is part of the University of Cambridge.

It furthers the University's mission by disseminating knowledge in the pursuit of education, learning and research at the highest international levels of excellence.

www.cambridge.org
Information on this title: www.cambridge.org/9780521866293

First published 2007

A catalogue record for this publication is available from the British Library

ISBN 978-0-521-86629-3 Hardback

Contents

Preface   *page* ix

1 Elementary operator theory   1
    1.1   Banach spaces   1
    1.2   Bounded linear operators   12
    1.3   Topologies on vector spaces   19
    1.4   Differentiation of vector-valued functions   23
    1.5   The holomorphic functional calculus   27

2 Function spaces   35
    2.1 L^p spaces   35
    2.2   Operators acting on L^p spaces   45
    2.3   Approximation and regularization   54
    2.4   Absolutely convergent Fourier series   60

3 Fourier transforms and bases   67
    3.1   The Fourier transform   67
    3.2   Sobolev spaces   77
    3.3   Bases of Banach spaces   80
    3.4   Unconditional bases   90

4 Intermediate operator theory   99
    4.1   The spectral radius   99
    4.2   Compact linear operators   102
    4.3   Fredholm operators   116
    4.4   Finding the essential spectrum   124

5   **Operators on Hilbert space**   135

     5.1   Bounded operators   135

     5.2   Polar decompositions   137

     5.3   Orthogonal projections   140

     5.4   The spectral theorem   143

     5.5   Hilbert-Schmidt operators   151

     5.6   Trace class operators   153

     5.7   The compactness of $f(Q)g(P)$ 160

6   **One-parameter semigroups**   163

     6.1   Basic properties of semigroups   163

     6.2   Other continuity conditions   177

     6.3   Some standard examples   182

7   **Special classes of semigroup**   190

     7.1   Norm continuity   190

     7.2   Trace class semigroups   194

     7.3   Semigroups on dual spaces   197

     7.4   Differentiable and analytic vectors   201

     7.5   Subordinated semigroups   205

8   **Resolvents and generators**   210

     8.1   Elementary properties of resolvents   210

     8.2   Resolvents and semigroups   218

     8.3   Classification of generators   227

     8.4   Bounded holomorphic semigroups   237

9   **Quantitative bounds on operators**   245

     9.1   Pseudospectra   245

     9.2   Generalized spectra and pseudospectra   251

     9.3   The numerical range   264

     9.4   Higher order hulls and ranges   276

     9.5   Von Neumann's theorem   285

     9.6   Peripheral point spectrum   287

10   **Quantitative bounds on semigroups**   296

     10.1   Long time growth bounds   296

     10.2   Short time growth bounds   300

     10.3   Contractions and dilations   307

     10.4   The Cayley transform   310

10.5	One-parameter groups	315
10.6	Resolvent bounds in Hilbert space	321
11	**Perturbation theory**	**325**
11.1	Perturbations of unbounded operators	325
11.2	Relatively compact perturbations	330
11.3	Constant coefficient differential operators on the half-line	335
11.4	Perturbations: semigroup based methods	339
11.5	Perturbations: resolvent based methods	350
12	**Markov chains and graphs**	**355**
12.1	Definition of Markov operators	355
12.2	Irreducibility and spectrum	359
12.3	Continuous time Markov chains	362
12.4	Reversible Markov semigroups	366
12.5	Recurrence and transience	369
12.6	Spectral theory of graphs	374
13	**Positive semigroups**	**380**
13.1	Aspects of positivity	380
13.2	Invariant subsets	386
13.3	Irreducibility	390
13.4	Renormalization	393
13.5	Ergodic theory	395
13.6	Positive semigroups on $C(X)$	399
14	**NSA Schrödinger operators**	**408**
14.1	Introduction	408
14.2	Bounds on the numerical range	409
14.3	Bounds in one space dimension	412
14.4	The essential spectrum of Schrödinger operators	420
14.5	The NSA harmonic oscillator	424
14.6	Semi-classical analysis	427
	References	436
	Index	446

Preface

This volume is halfway between being a textbook and a monograph. It describes a wide variety of ideas, some classical and others at the cutting edge of current research. Because it is directed at graduate students and young researchers, it often provides the simplest version of a theorem rather than the deepest one. It contains a variety of examples and problems that might be used in lecture courses on the subject.

It is frequently said that over the last few decades there has been a decisive shift in mathematics from the linear to the non-linear. Even if this is the case it is easy to justify writing a book on the theory of *linear* operators. The range of applications of the subject continues to grow rapidly, and young researchers need to have an accessible account of its main lines of development, together with references to further sources for more detailed reading.

Probability theory and quantum theory are two absolutely fundamental fields of science. In terms of their technological impact they have been far more important than Einstein's relativity theory. Both are entirely linear. In the first case this is in the nature of the subject. Many sustained attempts have been made to introduce non-linearities into quantum theory, but none has yet been successful, while the linear theory has gone from triumph to triumph. Nobody can predict what the future will hold, but it seems likely that quantum theory will be used for a long time yet, even if a non-linear successor is found.

The fundamental equations of quantum mechanics involve self-adjoint and unitary operators. However, once one comes to applications, the situation changes. Non-self-adjoint operators play an important role in topics as diverse as the optical model of nuclear scattering, the analysis of resonances using complex scaling, the behaviour of unstable lasers and the scattering of atoms by periodic electric fields.[1]

---

[1] A short list of references to such problems may be found in [Berry, website].

There are many routes into the theory of non-linear partial differential equations. One approach depends in a fundamental way on perturbing linear equations. Another idea is to use comparison theorems to show that certain non-linear equations retain desired properties of linear cousins. In the case of the Kortweg-de Vries equation, the exact solution of a highly non-linear equation depends on reducing it to a linear inverse problem. In all these cases progress depends upon a deep technical knowledge of what is, and is not, possible in the linear theory. A standard technique for studying the non-linear stochastic Navier-Stokes equation involves reformulating it as a Markov process acting on an infinite-dimensional configuration space X. This process is closely associated with a *linear* Markov semigroup acting on a space of observables, i.e. bounded functions $f : X \to \mathbf{C}$. The decay properties of this semigroup give valuable information about the behaviour of the original non-linear equation. The material in Section 13.6 is related to this issue.

There is a vast number of applications of spectral theory to problems in engineering, and I mention just three. The unexpected oscillations of the London Millennium Bridge when it opened in 2000 were due to inadequate eigenvalue analysis. There is a considerable literature analyzing the characteristic timbres of musical instruments in terms of the complex eigenvalues of the differential equations that govern their vibrations. Of more practical importance are resonances in turbines, which can destroy them if not taken seriously.

As a final example of the importance of spectral theory I select the work of Babenko, Mayer and others on the Gauss-Kuzmin theorem about the distribution of continued fractions, which has many connections with modular curves and other topics; see [Manin and Marcolli 2002]. This profound work involves many different ideas, but a theorem about the dominant eigenvalue of a certain compact operator having an invariant closed cone is at the centre of the theory. This theorem is close to ideas in Chapter 13, and in particular to Theorem 13.1.9.

Once one has decided to study linear operators, a fundamental choice needs to be made. Self-adjoint operators on Hilbert spaces have an extremely detailed theory, and are of great importance for many applications. We have carefully avoided trying to compete with the many books on this subject and have concentrated on the non-self-adjoint theory. This is much more diverse – indeed it can hardly be called a theory. Studying non-self-adjoint operators is like being a vet rather than a doctor: one has to acquire a much wider range of knowledge, and to accept that one cannot expect to have as high a rate of success when confronted with particular cases. It comprises a collection of methods, each of which is useful for some class of such operators. Some of these are described in the recent monograph of Trefethen and Embree on pseudospectra, Haase's monograph based on the holomorphic functional calculus, Ouhabaz's detailed theory of the L^p semigroups associated with NSA second order elliptic operators, and the much older work of Sz.-Nagy and Foias, still being actively developed by Naboko and others. If there is a common thread in all of these it is the idea of using theorems from analytic function theory to understand NSA operators.

One of the few methods with some degree of general application is the theory of one-parameter semigroups. Many of the older monographs on this subject (particularly my own) make rather little reference to the wide range of applications of the subject. In this book I have presented a much larger number of examples and problems here in order to demonstrate the value of the general theory. I have also tried to make it more user-friendly by including motivating comments.

The present book has a slight philosophical bias towards explicit bounds and away from abstract existence theorems. I have not gone so far as to insist that every result should be presented in the language of constructive analysis, but I have sometimes chosen more constructive proofs, even when they are less general. Such proofs often provide new insights, but at the very least they may be more useful for numerical analysts than proofs which merely assert the existence of a constant or some other entity.

There are, however, many entirely non-constructive proofs in the book. The fact that the spectrum of a bounded linear operator is always non-empty depends upon Liouville's theorem and a contradiction argument. It does not suggest a procedure for finding even one point in the spectrum. It should therefore come as no surprise that the spectrum can be highly unstable under small perturbations of the operator. The pseudospectra are more stable, and because of that arguably more important for non-self-adjoint operators.

It is particularly hard to give precise historical credit for many theorems in analysis. The most general version of a theorem often emerges several decades after the first one, with a proof which may be completely different from the original one. I have made no attempt to give references to the original literature for results discovered before 1950, and have attached the conventional names to theorems of that era. The books of Dunford and Schwartz should be consulted for more detailed information; see [Dunford and Schwartz 1966, Dunford and Schwartz 1963]. I only assign credit on a systematic basis for results proved since 1980, which is already a quarter of a century ago. I may not even have succeeded in doing that correctly, and hope that those who feel slighted will forgive my failings, and let me know, so that the situation can be rectified on my website and in future editions.

I conclude by thanking the large number of people who have influenced me, particularly in relation to the contents of this book. The most important of these have been Barry Simon and, more recently, Nick Trefethen, to both of whom I owe a great debt. I have also benefited greatly from many discussions with Wolfgang Arendt, Anna Aslanyan, Charles Batty, Albrecht Böttcher, Lyonell Boulton, Ilya Goldsheid, Markus Haase, Evans Harrell, Paul Incani, Boris Khoruzhenko, Michael Levitin, Terry Lyons, Reiner Nagel, Leonid Parnovski, Michael Plum, Yuri Safarov, Eugene Shargorodsky, Stanislav Shkarin, Johannes Sjöstrand, Dan Stroock, John Weir, Hans Zwart, Maciej Zworski and many other good friends and colleagues. Finally I want to record my thanks to my wife Jane, whose practical and moral support over many years has meant so much to me. She has also helped me to remember that there is more to life than proving theorems!

1

Elementary operator theory

1.1 Banach spaces

In this chapter we collect together material which should be covered in an introductory course of functional analysis and operator theory. We do not always include proofs, since there are many excellent textbooks on the subject.[1] The theorems provide a list of results which we use throughout the book.

We start at the obvious point. A normed space is a vector space $\mathcal{B}$ (assumed to be over the complex number field $\mathbf{C}$) provided with a norm $\|\cdot\|$ satisfying

$$\|f\| \geq 0,$$

$$\|f\| = 0 \text{ implies } f = 0,$$

$$\|\alpha f\| = |\alpha| \|f\|,$$

$$\|f + g\| \leq \|f\| + \|g\|,$$

for all $\alpha \in \mathbf{C}$ and all $f, g \in \mathcal{B}$. Many of our definitions and theorems also apply to real normed spaces, but we will not keep pointing this out. We say that $\|\cdot\|$ is a seminorm if it satisfies all of the axioms except the second.

A Banach space is defined to be a normed space $\mathcal{B}$ which is complete in the sense that every Cauchy sequence in $\mathcal{B}$ converges to a limit in $\mathcal{B}$. Every normed space $\mathcal{B}$ has a completion $\overline{\mathcal{B}}$, which is a Banach space in which $\mathcal{B}$ is embedded isometrically and densely. (An isometric embedding is a linear, norm-preserving (and hence one-one) map of one normed space into another in which every element of the first space is identified with its image in the second.)

---

[1] One of the most systematic is [Dunford and Schwartz 1966].

Problem 1.1.1 Prove that the following conditions on a normed space $\mathcal{B}$ are equivalent:

(i) $\mathcal{B}$ is complete.
(ii) Every series $\sum_{n=1}^{\infty} f_n$ in $\mathcal{B}$ such that $\sum_{n=1}^{\infty} \|f_n\| < \infty$ is norm convergent.
(iii) Every series $\sum_{n=1}^{\infty} f_n$ in $\mathcal{B}$ such that $\|f_n\| \leq 2^{-n}$ for all n is norm convergent.

Prove also that any two completions of a normed space $\mathcal{B}$ are isometrically isomorphic. $\quad\square$

The following results from point set topology are rarely used below, but they provide worthwhile background knowledge. We say that a topological space X is normal if given any pair of disjoint closed subsets A, B of X there exists a pair of disjoint open sets U, V such that $A \subseteq U$ and $B \subseteq V$. All metric spaces and all compact Hausdorff spaces are normal. The size of the space of continuous functions on a normal space is revealed by Urysohn's lemma.

Lemma 1.1.2 *(Urysohn)*[2] *If A, B are disjoint closed sets in the normal topological space X, then there exists a continuous function $f : X \to [0, 1]$ such that $f(x) = 0$ for all $x \in A$ and $f(x) = 1$ for all $x \in B$.*

Problem 1.1.3 Use the continuity of the distance function $x \to \text{dist}(x, A)$ to provide a direct proof of Urysohn's lemma when X is a metric space. $\quad\square$

Theorem 1.1.4 *(Tietze) Let S be a closed subset of the normal topological space X and let $f : S \to [0, 1]$ be a continuous function. Then there exists a continuous extension of f to X, i.e. a continuous function $g : X \to [0, 1]$ which coincides with f on S.[3]*

Problem 1.1.5 Prove the Tietze extension theorem by using Urysohn's lemma to construct a sequence of functions $g_n : X \to [0, 1]$ which converge uniformly on X and also uniformly on S to f. $\quad\square$

If K is a compact Hausdorff space then $C(K)$ stands for the space of all continuous complex-valued functions on K with the supremum norm

$$\|f\|_{\infty} := \sup\{|f(x)| : x \in K\}.$$

$C(K)$ is a Banach space with this norm, and the supremum is actually a maximum. We also use the notation $C_{\mathbf{R}}(K)$ to stand for the real Banach space of all continuous, real-valued functions on K.

---

[2] See [Bollobas 1999], [Simmons 1963, p. 135] or [Kelley 1955, p. 115].
[3] See [Bollobas 1999].

The following theorem is of interest in spite of the fact that it is rarely useful: in most applications it is equally evident that all four statements are true (or false).

Theorem 1.1.6 *(Urysohn) If K is a compact Hausdorff space then the following statements are equivalent.*

(i) *K is metrizable;*
(ii) *the topology of K has a countable base;*
(iii) *K can be homeomorphically embedded in the unit cube* $\Omega := \prod_{n=1}^{\infty} [0, 1]$ *of countable dimension;*
(iv) *the space $C_{\mathbf{R}}(K)$ is separable in the sense that it contains a countable norm dense subset.*

The equivalence of the first three statements uses methods of point-set topology, for which we refer to [Kelley 1955, p. 125]. The equivalence of the fourth statement uses the Stone-Weierstrass theorem 2.3.17.

Problem 1.1.7 Without using Theorem 1.1.6, prove that the topological product of a countable number of compact metrizable spaces is also compact metrizable. □

We say that $\mathcal{H}$ is a Hilbert space if it is a Banach space with respect to a norm associated with an inner product $f, g \to \langle f, g \rangle$ according to the formula

$$\|f\| := \sqrt{\langle f, f \rangle}.$$

We always assume that an inner product is linear in the first variable and conjugate linear in the second variable. We assume familiarity with the basic theory of Hilbert spaces. Although we do not restrict the statements of many theorems in the book to separable Hilbert spaces, we frequently only give the proof in that case. The proof in the non-separable context can usually be obtained by either of two devices: one may replace the word sequence by generalized sequence, or one may show that if the result is true on every separable subspace then it is true in general.

Example 1.1.8 If X is a finite or countable set then $l^2(X)$ is defined to be the space of all functions $f : X \to \mathbf{C}$ such that

$$\|f\|_2 := \sqrt{\sum_{x \in X} |f(x)|^2} < \infty.$$

This is the norm associated with the inner product

$$\langle f, g \rangle := \sum_{x \in X} f(x)\overline{g(x)},$$

the sum being absolutely convergent for all $f, g \in l^2(X)$. $\square$

A sequence $\{\phi_n\}_{n=1}^{\infty}$ in a Hilbert space $\mathcal{H}$ is said to be an *orthonormal sequence* if

$$\langle \phi_m, \phi_n \rangle = \begin{cases} 1 & \text{if } m = n, \\ 0 & \text{otherwise.} \end{cases}$$

It is said to be a *complete orthonormal sequence* or an *orthonormal basis*, if it satisfies the conditions of the following theorem.

Theorem 1.1.9 *The following conditions on an orthonormal sequence $\{\phi_n\}_{n=1}^{\infty}$ in a Hilbert space $\mathcal{H}$ are equivalent.*

(i) *The linear span of $\{\phi_n\}_{n=1}^{\infty}$ is a dense linear subspace of $\mathcal{H}$.*
(ii) *The identity*

$$f = \sum_{n=1}^{\infty} \langle f, \phi_n \rangle \phi_n \tag{1.1}$$

holds for all $f \in \mathcal{H}$.
(iii) *The identity*

$$\|f\|^2 = \sum_{n=1}^{\infty} |\langle f, \phi_n \rangle|^2$$

holds for all $f \in \mathcal{H}$.
(iv) *The identity*

$$\langle f, g \rangle = \sum_{n=1}^{\infty} \langle f, \phi_n \rangle \langle \phi_n, g \rangle$$

holds for all $f, g \in \mathcal{H}$, the series being absolutely convergent.

The formula (1.1) is sometimes called a *generalized Fourier expansion* and $\langle f, \phi_n \rangle$ are then called the *Fourier coefficients* of f. The rate of convergence in (1.1) depends on f, and is discussed further in Theorem 5.4.12.

Problem 1.1.10 (Haar) Let $\{v_n\}_{n=0}^{\infty}$ be a dense sequence of distinct numbers in $[0, 1]$ such that $v_0 = 0$ and $v_1 = 1$. Put $e_1(x) := 1$ for all $x \in (0, 1)$ and define $e_n \in L^2(0, 1)$ for $n = 2, 3, \ldots$ by

$$e_n(x) := \begin{cases} 0 & \text{if } x < u_n, \\ \alpha_n & \text{if } u_n < x < v_n, \\ -\beta_n & \text{if } v_n < x < w_n, \\ 0 & \text{if } x > w_n, \end{cases}$$

where

$$u_n := \max\{v_r : r < n \text{ and } v_r < v_n\},$$

$$w_n := \min\{v_r : r < n \text{ and } v_r > v_n\},$$

and $\alpha_n > 0$, $\beta_n > 0$ are the solutions of

$$\alpha_n(v_n - u_n) - \beta_n(w_n - v_n) = 0,$$

$$(v_n - u_n)\alpha_n^2 + (w_n - v_n)\beta_n^2 = 1.$$

Prove that $\{e_n\}_{n=1}^\infty$ is an orthonormal basis in $L^2(0, 1)$. If $\{v_n\}_{n=0}^\infty$ is the sequence $\{0, 1, 1/2, 1/4, 3/4, 1/8, 3/8, 5/8, 7/8, 1/16, \ldots\}$ one obtains the standard Haar basis of $L^2(0, 1)$, discussed in all texts on wavelets and of importance in image processing. If $\{m_r\}_{r=1}^\infty$ is a sequence of integers such that $m_1 \geq 2$ and m_r is a proper factor of m_{r+1} for all r, then one may define a generalized Haar basis of $L^2(0, 1)$ by concatenating 0, 1, $\{r/m_1\}_{r=1}^{m_1}$, $\{r/m_2\}_{r=1}^{m_2}$, $\{r/m_3\}_{r=1}^{m_3}, \ldots$ and removing duplicated numbers as they arise. □

If X is a set with a σ-algebra Σ of subsets, and $\mathrm{d}x$ is a countably additive σ-finite measure on Σ, then the formula

$$\|f\|_2 := \sqrt{\int_X |f(x)|^2 \, \mathrm{d}x}$$

defines a norm on the space $L^2(X, \mathrm{d}x)$ of all functions f for which the integral is finite. The norm is associated with the inner product

$$\langle f, g \rangle := \int_X f(x)\overline{g(x)} \, \mathrm{d}x.$$

Strictly speaking one only gets a norm by identifying two functions which are equal almost everywhere. If the integral used is that of Lebesgue, then $L^2(X, \mathrm{d}x)$ is complete.[4]

Notation If $\mathcal{B}$ is a Banach space of functions on a locally compact, Hausdorff space X, then we will always use the notation $\mathcal{B}_c$ to stand for all those

---

[4] See [Lieb and Loss 1997] for one among many more complete accounts of Lebesgue integration. See also Section 2.1.

functions in $\mathcal{B}$ which have compact support, and $\mathcal{B}_0$ to stand for the closure of $\mathcal{B}_c$ in $\mathcal{B}$. Also $C_0(X)$ stands for the closure of $C_c(X)$ with respect to the supremum norm; equivalently $C_0(X)$ is the space of continuous functions on X that vanish at infinity. If X is a region in $\mathbf{R}^N$ then $C^n(X)$ will stand for the space of n times continuously differentiable functions on X.

Problem 1.1.11 The space $L^1(a, b)$ may be defined as the abstract completion of the space $\mathcal{P}$ of piecewise continuous functions on $[a, b]$, with respect to the norm

$$\|f\|_1 := \int_a^b |f(x)| \, dx.$$

Without using any properties of Lebesgue integration prove that $C^k[a, b]$ is dense in $L^1(a, b)$ for every $k \geq 0$. $\quad\square$

Lemma 1.1.12 *A finite-dimensional normed space V is necessarily complete. Any two norms $\|\cdot\|_1$ and $\|\cdot\|_2$ on V are equivalent in the sense that there exist positive constants a and b such that*

$$a\|f\|_1 \leq \|f\|_2 \leq b\|f\|_1 \tag{1.2}$$

for all $f \in V$.

Problem 1.1.13 Find the optimal values of the constants a and b in (1.2) for the norms on $\mathbf{C}^n$ given by

$$\|f\|_1 := \sum_{r=1}^n |f_r|, \qquad \|f\|_2 := \left\{ \sum_{r=1}^n |f_r|^2 \right\}^{1/2}. \qquad\square$$

A bounded linear functional $\phi : \mathcal{B} \to \mathbf{C}$ is a linear map for which

$$\|\phi\| := \sup\{|\phi(f)| : \|f\| \leq 1\}$$

is finite. The dual space $\mathcal{B}^*$ of $\mathcal{B}$ is defined to be the space of all bounded linear functionals on $\mathcal{B}$, and is itself a Banach space for the norm given above. The Hahn-Banach theorem states that if L is any linear subspace of $\mathcal{B}$, then any bounded linear functional ϕ on L has a linear extension ψ to $\mathcal{B}$ which has the same norm:

$$\sup\{|\phi(f)|/\|f\| : 0 \neq f \in L\} = \sup\{|\psi(f)|/\|f\| : 0 \neq f \in \mathcal{B}\}.$$

It is not always easy to find a useful representation of the dual space of a Banach space, but the Hilbert space is particularly simple.

Theorem 1.1.14 *(Fréchet-Riesz)*[5] *If $\mathcal{H}$ is a Hilbert space then the formula*

$$\phi(f) := \langle f, g \rangle$$

defines a one-one correspondence between all $g \in \mathcal{H}$ and all $\phi \in \mathcal{H}^*$. Moreover $\|\phi\| = \|g\|$.

<u>Note</u> The correspondence $\phi \leftrightarrow g$ is conjugate linear rather than linear, and this can cause some confusion if forgotten.

Problem 1.1.15 Prove that if ϕ is a bounded linear functional on the closed linear subspace $\mathcal{L}$ of a Hilbert space $\mathcal{H}$, then there is only one linear extension of ϕ to $\mathcal{H}$ with the same norm. $\quad\square$

The following theorem is not elementary, and we will not use it until Chapter 13.1. The notation $C_{\mathbf{R}}(K)$ refers to the real Banach space of continuous functions $f : K \to \mathbf{R}$ with the supremum norm.[6]

Theorem 1.1.16 *(Riesz-Kakutani) Let K be a compact Hausdorff space and let $\phi \in C_{\mathbf{R}}(K)^*$. If ϕ is non-negative in the sense that $\phi(f) \geq 0$ for all non-negative $f \in C_{\mathbf{R}}(K)$ then there exists a non-negative countably additive measure μ on K such that*

$$\phi(f) = \int_X f(x)\, \mu(\mathrm{d}x)$$

for all $f \in C_{\mathbf{R}}(K)$. Moreover $\|\phi\| = \phi(1) = \mu(K)$.

One may reduce the representation of more general bounded linear functionals to the above special case by means of the following theorem. Given $\phi, \psi \in C_{\mathbf{R}}(K)^*$, we write $\phi \geq \psi$ if $\phi(f) \geq \psi(f)$ for all non-negative $f \in C_{\mathbf{R}}(K)$.

Theorem 1.1.17 *If K is a compact Hausdorff space and $\phi \in C_{\mathbf{R}}(K)^*$ then one may write $\phi := \phi_+ - \phi_-$ where $\phi_\pm$ are canonically defined, non-negative, bounded linear functionals. If $|\phi| := \phi_+ + \phi_-$ then $|\phi| \geq \pm\phi$. If $\psi \geq \pm\phi \in C_{\mathbf{R}}(K)^*$ then $\psi \geq |\phi|$. Finally $\|\,|\phi|\,\| = \|\phi\|$.*

---

[5] See [Dunford and Schwartz 1966, Theorem IV.4.5] for the proof.

[6] A combination of the next two theorems is usually called the Riesz representation theorem. According to [Dunford and Schwartz 1966, p. 380] Riesz provided an explicit representation of $C[0, 1]^*$. The corresponding theorem for $C_{\mathbf{R}}(K)^*$, where K is a general compact Hausdorff space, was obtained some years later by Kakutani. The formula $\phi := \phi_+ - \phi_-$ is called the Jordan decomposition. For the proof of the theorem see [Dunford and Schwartz 1966, Theorem IV.6.3]. A more abstract formulation, in terms of Banach lattices and AM-spaces, is given in [Schaefer 1974, Proposition II.5.5 and Section II.7].

Proof. The proof is straightforward but lengthy. Let $\mathcal{B} := C_{\mathbf{R}}(K)$, let $\mathcal{B}_+$ denote the convex cone of all non-negative continuous functions on K, and let $\mathcal{B}_+^*$ denote the convex cone of all non-negative functionals $\psi \in \mathcal{B}^*$.

Given $\phi \in \mathcal{B}^*$, we define $\phi_+ : \mathcal{B}_+ \to \mathbf{R}_+$ by

$$\phi_+(f) := \sup\{\phi(f_0) : 0 \le f_0 \le f\}.$$

If $0 \le f_0 \le f$ and $0 \le g_0 \le g$ then

$$\phi(f_0) + \phi(g_0) = \phi(f_0 + g_0) \le \phi_+(f + g).$$

Letting f_0 and g_0 vary subject to the stated constraints, we deduce that

$$\phi_+(f) + \phi_+(g) \le \phi_+(f + g)$$

for all $f, g \in \mathcal{B}_+$.

The reverse inequality is harder to prove. If $f, g \in \mathcal{B}_+$ and $0 \le h \le f + g$ then one puts $f_0 := \min\{h, f\}$ and $g_0 := h - f_0$. By considering each point $x \in K$ separately one sees that $0 \le f_0 \le f$ and $0 \le g_0 \le g$. hence

$$\phi(h) = \phi(f_0) + \phi(g_0) \le \phi_+(f) + \phi_+(g).$$

Since h is arbitrary subject to the stated constraints one obtains

$$\phi_+(f + g) \le \phi_+(f) + \phi_+(g)$$

for all $f, g \in \mathcal{B}_+$.

We are now in a position to extend ϕ_+ to the whole of $\mathcal{B}$. If $f \in \mathcal{B}$ we put

$$\phi_+(f) := \phi_+(f + \alpha 1) - \alpha \phi_+(1)$$

where $\alpha \in \mathbf{R}$ is chosen so that $f + \alpha 1 \ge 0$. The linearity of ϕ_+ on $\mathcal{B}_+$ implies that the particular choice of α does not matter subject to the stated constraint.

Our next task is to prove that the extended ϕ_+ is a linear functional on $\mathcal{B}_+$. If $f, g \in \mathcal{B}$, $f + \alpha 1 \ge 0$ and $g + \beta 1 \ge 0$, then

$$\begin{aligned}
\phi_+(f + g) &= \phi_+(f + g + \alpha 1 + \beta 1) - (\alpha + \beta)\phi_+(1) \\
&= \phi_+(f + \alpha 1) + \phi_+(g + \beta 1) - (\alpha + \beta)\phi_+(1) \\
&= \phi_+(f) + \phi_+(g).
\end{aligned}$$

It follows immediately from the definition that $\phi_+(\lambda h) = \lambda \phi_+(h)$ for all $\lambda \ge 0$ and $h \in \mathcal{B}_+$. Hence $f \in \mathcal{B}$ implies

$$\phi_+(\lambda f) = \phi(\lambda f + \lambda \alpha 1) - \lambda \alpha \phi_+(1) = \lambda \phi(f + \alpha 1) - \lambda \alpha \phi_+(1) = \lambda \phi_+(f).$$

If $\lambda < 0$ then

$$0 = \phi_+(\lambda f + |\lambda| f) = \phi_+(\lambda f) + \phi_+(|\lambda| f) = \phi_+(\lambda f) + |\lambda| \phi_+(f).$$

Therefore

$$\phi_+(\lambda f) = -|\lambda| \phi_+(f) = \lambda \phi_+(f).$$

Therefore ϕ_+ is a linear functional on $\mathcal{B}$. It is non-negative in the sense defined above.

We define ϕ_- by $\phi_- := \phi_+ - \phi$, and deduce immediately that it is linear. Since $f \in \mathcal{B}_+$ implies that $\phi_+(f) \geq \phi(f)$, we see that ϕ_- is non-negative. The boundedness of $\phi_\pm$ will be a consequence of the boundedness of $|\phi|$ and the formulae

$$\phi_+ = \tfrac{1}{2}(|\phi| + \phi), \qquad\qquad \phi_- = \tfrac{1}{2}(|\phi| - \phi).$$

We will need the following formula for $|\phi|$. If $f \in \mathcal{B}_+$ then the identity $|\phi| = 2\phi_+ - \phi$ implies

$$\begin{aligned} |\phi|(f) &= 2\sup\{\phi(f_0) : 0 \leq f_0 \leq f\} - \phi(f) \\ &= \sup\{\phi(2f_0 - f) : 0 \leq f_0 \leq f\} \\ &= \sup\{\phi(f_1) : -f \leq f_1 \leq f\}. \end{aligned} \tag{1.3}$$

The inequality $|\phi| \geq \pm\phi$ of the theorem follows from

$$\begin{aligned} |\phi| = \phi + 2\phi_- &\geq \phi \\ |\phi| = 2\phi_+ - \phi &\geq -\phi. \end{aligned}$$

If $\psi \geq \pm\phi$, $f \geq 0$ and $-f \leq f_1 \leq f$ then adding the two inequalities $(\psi + \phi)(f - f_1) \geq 0$ and $(\psi - \phi)(f + f_1) \geq 0$ yields $\psi(f) \geq \phi(f_1)$. Letting f_1 vary subject to the stated constraint we obtain $\psi(f) \geq |\phi|(f)$ by using (1.3). Therefore $\psi \geq |\phi|$.

We finally have to evaluate $\| |\phi| \|$. If $f \in \mathcal{B}$ and $\phi \in \mathcal{B}^*$ then

$$\begin{aligned} |\phi(f)| &= |\phi_+(f_+) - \phi_+(f_-) - \phi_-(f_+) + \phi_-(f_-)| \\ &\leq \phi_+(f_+) + \phi_+(f_-) + \phi_-(f_+) + \phi_-(f_-) \\ &= |\phi|(|f|) \\ &\leq \| |\phi| \| \, \| |f| \| \\ &= \| |\phi| \| \, \|f\|. \end{aligned}$$

Since f is arbitrary we deduce that $\|\phi\| \leq \| |\phi| \|$.

Conversely suppose that $f \in \mathcal{B}$. The inequality $-|f| \leq f \leq |f|$ implies

$$-|\phi|(|f|) \leq |\phi|(f) \leq |\phi|(|f|).$$

Therefore

$$| \, |\phi|(f)| \leq |\phi|(|f|)$$
$$= \sup\{\phi(f_1) : -|f| \leq f_1 \leq |f|\}$$
$$\leq \|\phi\| \sup\{\|f_1\| : -|f| \leq f_1 \leq |f|\}$$
$$= \|\phi\| \, \|f\|.$$

Hence $\| \, |\phi| \, \| \leq \|\phi\|$. $\square$

If L is a closed linear subspace of the normed space $\mathcal{B}$, then the quotient space $\mathcal{B}/L$ is defined to be the algebraic quotient, provided with the quotient norm

$$\|f + L\| := \inf\{\|f + g\| : g \in L\}.$$

It is known that if $\mathcal{B}$ is a Banach space then so is $\mathcal{B}/L$.

Problem 1.1.18 If $\mathcal{B} = C[a, b]$ and L is the subspace of all functions in $\mathcal{B}$ which vanish on the closed subset K of $[a, b]$, find an explicit representation of $\mathcal{B}/L$ and of its norm. $\square$

The Hahn-Banach theorem implies immediately that there is a canonical and isometric embedding j from $\mathcal{B}$ into the second dual space $\mathcal{B}^{**} = (\mathcal{B}^*)^*$, given by

$$(jx)(\phi) := \phi(x)$$

for all $x \in \mathcal{B}$ and all $\phi \in \mathcal{B}^*$. The space $\mathcal{B}$ is said to be reflexive if j maps $\mathcal{B}$ one-one onto $\mathcal{B}^{**}$.

We will often use the more symmetrical notation $\langle x, \phi \rangle$ in place of $\phi(x)$, and regard $\mathcal{B}$ as a subset of $\mathcal{B}^{**}$, suppressing mention of its natural embedding.

Problem 1.1.19 Prove that $\mathcal{B}$ is reflexive if and only if $\mathcal{B}^*$ is reflexive. $\square$

Example 1.1.20 The dual $\mathcal{B}^*$ of a Banach space $\mathcal{B}$ is usually not isometrically isomorphic to $\mathcal{B}$ even if $\mathcal{B}$ is reflexive. The following provides a large number of spaces for which they are isometrically isomorphic. We simply choose any reflexive Banach space $\mathcal{C}$ and consider $\mathcal{B} := \mathcal{C} \oplus \mathcal{C}^*$ with the norm

$$\|(x, \phi)\| := (\|x\|^2 + \|\phi\|^2)^{1/2}. \qquad \square$$

If X is an infinite set, $c_0(X)$ is defined to be the vector space of functions f which converge to 0 at infinity; more precisely we assume that for all $\varepsilon > 0$ there exists a finite set $F \subset X$ depending upon f and ε such that $x \notin F$ implies $|f(x)| < \varepsilon$.

Problem 1.1.21 Prove that $c_0(X)$ is a Banach space with respect to the supremum norm. $\square$

Problem 1.1.22 Prove that $c_0(X)$ is separable if and only if X is countable. $\square$

Problem 1.1.23 Prove that the dual space of $c_0(X)$ may be identified naturally with $l^1(X)$, the pairing being given by

$$\langle f, g \rangle := \sum_{x \in X} f(x)g(x)$$

where $f \in c_0(X)$ and $g \in l^1(X)$. $\square$

Problem 1.1.24 Prove that the dual space of $l^1(X)$ may be identified with the space $l^\infty(X)$ of all bounded functions $f : X \to \mathbf{C}$ with the supremum norm. Prove also that if X is infinite, $l^1(X)$ is not reflexive. $\square$

Problem 1.1.25 Use the Hahn-Banach theorem to prove that if $\mathcal{L}$ is a finite-dimensional subspace of the Banach space $\mathcal{B}$ then there exists a closed linear subspace $\mathcal{M}$ of $\mathcal{B}$ such that $\mathcal{L} \cap \mathcal{M} = \{0\}$ and $\mathcal{L} + \mathcal{M} = \mathcal{B}$. Moreover there exists a constant $c > 0$ such that

$$c^{-1}(\|l\| + \|m\|) \leq \|l + m\| \leq c(\|l\| + \|m\|)$$

for all $l \in \mathcal{L}$ and $m \in \mathcal{M}$. $\square$

We will frequently use the concept of integration[7] for functions which take their values in a Banach space $\mathcal{B}$. If $f : [a, b] \to \mathcal{B}$ is a piecewise continuous function, there is an element of $\mathcal{B}$, denoted by

$$\int_a^b f(x)\,dx$$

which is defined by approximating f by piecewise constant functions, for which the definition of the integral is evident. It is easy to show that the integral depends linearly on f and that

$$\left\| \int_a^b f(x)\,dx \right\| \leq \int_a^b \|f(x)\|\,dx.$$

Moreover

$$\left\langle \int_a^b f(x)\,dx, \phi \right\rangle = \int_a^b \langle f(x), \phi \rangle\,dx$$

for all $\phi \in \mathcal{B}^*$, where $\langle f, \phi \rangle$ denotes $\phi(f)$ as explained above. Both of these relations are proved first for piecewise constant functions. The integral may also be defined for functions $f : \mathbf{R} \to \mathcal{B}$ which decay rapidly enough at infinity. Many other familiar results, such as the fundamental theorem of calculus, and the possibility of taking a bounded linear operator under the integral sign, may be proved by the same method as is used for complex-valued functions.

1.2 Bounded linear operators

A bounded linear operator $A : \mathcal{B} \to \mathcal{C}$ between two Banach spaces is defined to be a linear map for which the norm

$$\|A\| := \sup\{\|Af\| : \|f\| \leq 1\}$$

is finite. In this chapter we will use the term 'operator' to stand for 'bounded linear operator' unless the context makes this inappropriate. The set $\mathcal{L}(\mathcal{B}, \mathcal{C})$ of all such operators itself forms a Banach space under the obvious operations and the above norm.

The set $\mathcal{L}(\mathcal{B})$ of all operators from $\mathcal{B}$ to itself is an algebra, the multiplication being defined by

$$(AB)(f) := A(B(f))$$

---

[7] We treat this at a very elementary level. A more sophisticated treatment is given in [Dunford and Schwartz 1966, Chap. 3], but we will not need to use this.

for all $f \in \mathcal{B}$. In fact $\mathcal{L}(\mathcal{B})$ is called a Banach algebra by virtue of being a Banach space and an algebra satisfying

$$\|AB\| \leq \|A\| \|B\|$$

for all A, $B \in \mathcal{L}(\mathcal{B})$. The identity operator I satisfies $\|I\| = 1$ and $AI = IA = A$ for all $A \in \mathcal{L}(\mathcal{B})$, so $\mathcal{L}(\mathcal{B})$ is a Banach algebra with identity.

Problem 1.2.1 Prove that $\mathcal{L}(\mathcal{B})$ is only commutative as a Banach algebra if $\mathcal{B} = \mathbf{C}$, and that $\mathcal{L}(\mathcal{B})$ is only finite-dimensional if $\mathcal{B}$ is finite-dimensional. □

Every operator A on $\mathcal{B}$ has a dual operator A^* acting on $\mathcal{B}^*$, satisfying the identity

$$\langle Af, \phi \rangle = \langle f, A^* \phi \rangle$$

for all $f \in \mathcal{B}$ and all $\phi \in \mathcal{B}^*$. The map $A \to A^*$ from $\mathcal{L}(\mathcal{B})$ to $\mathcal{L}(\mathcal{B}^*)$ is linear and isometric, but reverses the order of multiplication.

For every bounded operator A on a Hilbert space $\mathcal{H}$ there is a unique bounded operator A^*, also acting on $\mathcal{H}$, called its adjoint, such that

$$\langle Af, g \rangle = \langle f, A^* g \rangle,$$

for all f, $g \in \mathcal{H}$. This is not totally compatible with the notion of dual operator in the Banach space context, because the adjoint map is conjugate linear in the sense that

$$(\alpha A + \beta B)^* = \overline{\alpha} A^* + \overline{\beta} B^*$$

for all operators A, B and all complex numbers α, β. However, almost every other result is the same for the two concepts. In particular $A^{**} = A$. The concept of self-adjointness, $A = A^*$, is peculiar to Hilbert spaces, and is of great importance. We say that an operator U is unitary if it satisfies the conditions of the problem below.

Problem 1.2.2 Let U be a bounded operator on a Hilbert space $\mathcal{H}$. Use the polarization identity

$$4\langle x, y \rangle = \|x + y\|^2 - \|x - y\|^2 + i\|x + iy\|^2 - i\|x - iy\|^2$$

to prove that the following three conditions are equivalent.

(i) $U^*U = UU^* = I$;
(ii) U is one-one onto and isometric in the sense that $\|Ux\| = \|x\|$ for all $x \in \mathcal{H}$;
(iii) U is one-one onto and $\langle Uf, Ug \rangle = \langle f, g \rangle$ for all f, $g \in \mathcal{H}$. □

The inverse mapping theorem below establishes that algebraic invertibility of a bounded linear operator between Banach spaces is equivalent to invertibility in the category of bounded operators.[8]

Theorem 1.2.3 *(Banach) If the bounded linear operator A from the Banach space $\mathcal{B}_1$ to the Banach space $\mathcal{B}_2$ is one-one and onto, then the inverse operator is also bounded.*

Let $\mathcal{A}$ be an associative algebra over the complex field with identity element e. The number $\lambda \in \mathbf{C}$ is said to lie in the resolvent set of $a \in \mathcal{A}$ if $\lambda e - a$ has an inverse in $\mathcal{A}$. We call $R(\lambda, a) := (\lambda e - a)^{-1}$ the resolvent operators of a. The Spec(a) of a is by definition the complement of the resolvent set. If A is a bounded linear operator on a Banach space $\mathcal{B}$ we assume that the spectrum and resolvent are calculated with respect to $\mathcal{A} = \mathcal{L}(\mathcal{B})$, unless stated otherwise.

The appearance of the spectrum and resolvent at such an early stage in the book is no accident. They are the key concepts on which everything else is based. An enormous amount of effort has been devoted to their study for over a hundred years, and sophisticated software exists for computing both in a wide variety of fields. No book could aspire to treating all of this in a comprehensive manner, but we can describe the foundations on which this vast subject has been built. One of these is the resolvent identity.

Problem 1.2.4 Prove the resolvent identity

$$R(z, a) - R(w, a) = (w - z)R(z, a)R(w, a)$$

for all $z, w \notin$ Spec(a). □

Problem 1.2.5 Let a, b lie in the associative algebra $\mathcal{A}$ with identity e and let $0 \neq z \in \mathbf{C}$. Prove that $ab - ze$ is invertible if and only if $ba - ze$ is invertible. □

Problem 1.2.6 Let A, B be linear maps on the vector space $\mathcal{V}$ and let $0 \neq z \in \mathbf{C}$. Prove that the eigenspaces

$$\mathcal{M} := \{f \in \mathcal{V} : ABf = zf\}, \qquad \mathcal{N} := \{g \in \mathcal{V} : BAg = zg\}$$

have the same dimension. □

---

[8] See [Dunford and Schwartz 1966, Theorem II.2.2].

Problem 1.2.7 Let a be an element of the associative algebra $\mathcal{A}$ with identity e. Prove that

$$\mathrm{Spec}(a) = \mathrm{Spec}(L_a)$$

where $L_a : \mathcal{A} \to \mathcal{A}$ is defined by $L_a(x) := ax$. $\quad\square$

Problem 1.2.8 Let A be an operator on the Banach space $\mathcal{B}$ satisfying $\|A\| < 1$. Prove that $(I - A)$ is invertible and that

$$(I - A)^{-1} = \sum_{n=0}^{\infty} A^n, \tag{1.4}$$

the sum being norm convergent. $\quad\square$

Theorem 1.2.9 *The set $\mathcal{G}$ of all bounded invertible operators on a Banach space $\mathcal{B}$ is open. More precisely, if $A \in \mathcal{G}$ and $\|B - A\| < \|A^{-1}\|^{-1}$ then $B \in \mathcal{G}$.*

Proof. If $C := I - BA^{-1}$ then under the stated conditions

$$\|C\| = \|(A - B)A^{-1}\| \leq \|A^{-1}\|^{-1} \|A^{-1}\| < 1.$$

Therefore $(I - C)$ is invertible by Problem 1.2.8. But $B = (BA^{-1})A = (I - C)A$, so B is invertible with

$$B^{-1} = A^{-1} \sum_{n=0}^{\infty} C^n. \tag{1.5}$$

$\square$

Theorem 1.2.10 *The resolvent operator $R(z, A)$ satisfies*

$$\|R(z, A)\| \geq \mathrm{dist}(z, \mathrm{Spec}(A))^{-1} \tag{1.6}$$

for all $z \notin \mathrm{Spec}(A)$, where $\mathrm{dist}(z, S)$ denotes the distance of z from the set S.

Proof. If $z \notin \mathrm{Spec}(A)$ and $|w - z| < \|R(z, A)\|^{-1}$ then

$$D := R(z, A) \{I - (z - w)R(z, A)\}^{-1}$$
$$= \sum_{n=0}^{\infty} (z - w)^n R(z, A)^{n+1}$$

is a bounded invertible operator on $\mathcal{B}$; the inverse involved exists by Problem 1.2.8. It satisfies

$$D \{I - (z - w)R(z, A)\} = R(z, A)$$

and hence

$$D\{zI - A - (z - w)I\} = I.$$

We deduce that $D(wI - A) = I$, and similarly that $(wI - A)D = I$. Hence $w \notin \mathrm{Spec}(A)$. The statement of the theorem follows immediately. $\square$

Our next theorem uses the concept of an analytic operator-valued function. This is developed in more detail in Section 1.4.

Theorem 1.2.11 *Every bounded linear operator A on a Banach space has a non-empty, closed, bounded spectrum, which satisfies*

$$\mathrm{Spec}(A) \subseteq \{z \in \mathbf{C} : |z| \leq \|A\|\}. \tag{1.7}$$

If $|z| > \|A\|$ then

$$\|(zI - A)^{-1}\| \leq (|z| - \|A\|)^{-1}. \tag{1.8}$$

The resolvent operator $R(z, A)$ is a norm analytic function of z on $\mathbf{C}\backslash\mathrm{Spec}(A)$.

Proof. If $|z| > \|A\|$ then $zI - A = z(I - z^{-1}A)$ and this is invertible, with inverse

$$(zI - A)^{-1} = z^{-1} \sum_{n=1}^{\infty} (z^{-1}A)^n.$$

The bound (1.8) follows by estimating each of the terms in the geometric series. This implies (1.7). Theorem 1.2.10 implies that $\mathrm{Spec}(A)$ is closed. An examination of the proof of Theorem 1.2.10 leads to the conclusion that $R(z, A)$ is a norm analytic function of z in some neighbourhood of every $z \notin \mathrm{Spec}(A)$. It remains only to prove that $\mathrm{Spec}(A)$ is non-empty.

   Since

$$(zI - A)^{-1} = \sum_{n=0}^{\infty} z^{-n-1} A^n$$

if $|z| > \|A\|$, we see that $\|(zI - A)^{-1}\| \to 0$ as $|z| \to \infty$. The Banach space version of Liouville's theorem given in Problem 1.4.9 now implies that if $R(z, A)$ is entire, it vanishes identically. The contradiction establishes that $\mathrm{Spec}(A)$ must be non-empty. $\square$

We note that this proof is highly non-constructive: it does not give any clues about how to find even a single point in $\mathrm{Spec}(A)$. We will show in Section 9.1 that computing the spectrum may pose fundamental difficulties.

Problem 1.2.12 Let a be an element of the Banach algebra $\mathcal{A}$, whose multiplicative identity 1 satisfies $\|1\| = 1$. Prove that a has a non-empty, closed, bounded spectrum, which satisfies

$$\mathrm{Spec}(a) \subseteq \{z \in \mathbf{C} : |z| \leq \|a\|\}. \qquad \square$$

Our definition of the spectrum of an operator A was algebraic in that it only refers to properties of A as an element of the algebra $\mathcal{L}(\mathcal{B})$. One can also give a characterization that is geometric in the sense that it refers to vectors in the Banach space.

Lemma 1.2.13 *The number λ lies in the spectrum of the bounded operator A on the Banach space $\mathcal{B}$ if and only if at least one of the following occurs:*

(i) λ *is an eigenvalue of A. That is $Af = \lambda f$ for some non-zero $f \in \mathcal{B}$.*

(ii) λ *is an eigenvalue of A^*. Equivalently the range of the operator $\lambda I - A$ is not dense in $\mathcal{B}$.*

(iii) *There exists a sequence $f_n \in \mathcal{B}$ such that $\|f_n\| = 1$ for all n and*

$$\lim_{n \to \infty} \|Af_n - \lambda f_n\| = 0.$$

Proof. The operator $B := \lambda I - A$ may fail to be invertible because it is not one-one or because it is not onto. In the second case it may have closed range not equal to $\mathcal{B}$ or it may have range which is not closed. If it has closed range L not equal to $\mathcal{B}$, then there exists a non-zero $\phi \in \mathcal{B}^*$ which vanishes on L by the Hahn-Banach theorem. Therefore 0 is an eigenvalue of $B^* = \lambda I - A^*$, with eigenvector ϕ. If B is one-one with range which is not closed, then B^{-1} is unbounded; equivalently there exists a sequence f_n such that $\|f_n\| = 1$ for all n and $\lim_{n \to \infty} \|Bf_n\| = 0$. $\square$

In case (iii) we say that λ lies in the approximate point spectrum of A.

$\quad$ Note$\quad$ In the Hilbert space context we must replace (ii) by the statement that $\overline{\lambda}$ is an eigenvalue of A^*.

Problem 1.2.14 Prove that

$$\mathrm{Spec}(A) = \mathrm{Spec}(A^*)$$

for every bounded operator $A : \mathcal{B} \to \mathcal{B}$. $\quad \square$

Problem 1.2.15 Prove that if λ lies on the topological boundary of the spectrum of A, then it is also in its approximate point spectrum. $\quad \square$

Problem 1.2.16 Find the spectrum and the approximate point spectrum of the shift operator

$$Af(x) := f(x+1)$$

acting on $L^2(0, \infty)$, and of its adjoint operator. □

Problem 1.2.17 Let $a_1, \ldots, a_n$ be elements of an associative algebra $\mathcal{A}$ with identity. Prove that if the elements commute then the product $a_1 \ldots a_n$ is invertible if and only if every a_i is invertible. Prove also that this statement is not always true if the a_i do not commute. Finally prove that if $a_1 \ldots a_n$ and $a_n \ldots a_1$ are both invertible then a_r is invertible for all $r \in \{1, \ldots, n\}$. □

The following is the most elementary of a series of spectral mapping theorems in this book.

Theorem 1.2.18 *If p is a polynomial and a is an element of the associative algebra $\mathcal{A}$ with identity e then*

$$\mathrm{Spec}(p(a)) = p(\mathrm{Spec}(a)).$$

Proof. We assume that p is monic and of degree n. Given $w \in \mathbf{C}$ we have to prove that $w \in \mathrm{Spec}(p(a))$ if and only if there exists $z \in \mathrm{Spec}(a)$ such that $w = p(z)$. Putting $q(z) := p(z) - w$ this is equivalent to the statement that $0 \in \mathrm{Spec}(q(a))$ if and only if there exists $z \in \mathrm{Spec}(a)$ such that $q(z) = 0$. We now write

$$q(z) = \prod_{r=1}^{n} (z - z_r)$$

where z_r are the zeros of q, so that

$$q(a) = \prod_{r=1}^{n} (a - z_r e).$$

The theorem reduces to the statement that $q(a)$ is invertible if and only if $(a - z_r e)$ is invertible for all r. This follows from Problem 1.2.17. □

Problem 1.2.19 Let $A : \mathcal{B} \to \mathcal{B}$ be a bounded operator. We say that the closed linear subspace $\mathcal{L}$ of $\mathcal{B}$ is invariant under A if $A(\mathcal{L}) \subseteq \mathcal{L}$. Prove that this implies that $\mathcal{L}$ is also invariant under $R(z, A)$ for all z in the unbounded component of $\mathbf{C} \backslash \mathrm{Spec}(A)$. Give an example in which $\mathcal{L}$ is not invariant under $R(z, A)$ for some other $z \notin \mathrm{Spec}(A)$. □

1.3 Topologies on vector spaces

We define a topological vector space (TVS) to be a complex vector space V provided with a topology $\mathcal{T}$ such that the map $\{\alpha, \beta, u, v\} \to \alpha u + \beta v$ is jointly continuous from $\mathbf{C} \times \mathbf{C} \times V \times V$ to V. All of the TVSs in this book are generated by a family of seminorms $\{p_a\}_{a \in A}$ in the sense that every open set $U \in \mathcal{T}$ is the union of basic open neighbourhoods

$$\bigcap_{r=1}^{n} \{v : p_{a(r)}(v - u) < \varepsilon_r\}$$

of some central point $u \in V$. In addition we will assume that if $p_a(u) = 0$ for all $a \in A$ then $u = 0$.[9]

Problem 1.3.1 Prove that the topology generated by a family of seminorms turns V into a TVS as defined above. $\square$

Problem 1.3.2 Prove that the topology on V generated by a countable family of seminorms $\{p_n\}_{n=1}^{\infty}$ coincides with the topology for the metric

$$d(u, v) := \sum_{n=1}^{\infty} 2^{-n} \frac{p_n(u - v)}{1 + p_n(u - v)}. \qquad \square$$

One says that a TVS V is a Fréchet space if $\mathcal{T}$ is generated by a countable family of seminorms and the metric d above is complete.

Every Banach space $\mathcal{B}$ has a weak topology in addition to its norm topology. This is defined as the smallest topology on $\mathcal{B}$ for which the bounded linear functionals $\phi \in \mathcal{B}^*$ are all continuous. It is generated by the family of seminorms $p_\phi(f) := |\phi(f)|$. We will write

$$\text{w-}\lim_{n \to \infty} f_n = f \qquad \text{or} \qquad f_n \xrightarrow{w} f$$

to indicate that the sequence $f_n \in \mathcal{B}$ converges weakly to $f \in \mathcal{B}$, that is

$$\lim_{n \to \infty} \langle f_n, \phi \rangle = \langle f, \phi \rangle$$

for all $\phi \in \mathcal{B}^*$.

Problem 1.3.3 Use the Hahn-Banach theorem to prove that a linear subspace L of a Banach space $\mathcal{B}$ is norm closed if and only if it is weakly closed. $\square$

---

[9] Systematic accounts of the theory of TVSs are given in [Narici and Beckenstein 1985, Treves 1967, Wilansky 1978].

Our next result is called the uniform boundedness theorem and also the Banach-Steinhaus theorem.[10]

Theorem 1.3.4 *Let* $\mathcal{B}$, $\mathcal{C}$ *be two Banach spaces and let* $\{A_\lambda\}_{\lambda \in \Lambda}$ *be a family of bounded linear operators from* $\mathcal{B}$ *to* $\mathcal{C}$. *Then the following are equivalent.*

(i) $\sup_{\lambda \in \Lambda} \|A_\lambda\| < \infty$;

(ii) $\sup_{\lambda \in \Lambda} \|A_\lambda x\| < \infty$ *for every* $x \in \mathcal{B}$;

(iii) $\sup_{\lambda \in \Lambda} |\phi(A_\lambda x)| < \infty$ *for every* $x \in \mathcal{B}$ *and* $\phi \in \mathcal{C}^*$.

Proof. Clearly (i)$\Rightarrow$(ii)$\Rightarrow$(iii). Suppose that (ii) holds but (i) does not. We construct sequences $x_n \in \mathcal{B}$ and $\lambda(n) \in \Lambda$ as follows. Let x_1 be any vector satisfying $\|x_1\| = 1/4$. Given $x_1, \ldots, x_{n-1} \in \mathcal{B}$ satisfying $\|x_r\| = 4^{-r}$ for all $r \in \{1, \ldots, n-1\}$, let

$$c_{n-1} := \sup_{\lambda \in \Lambda} \|A_\lambda(x_1 + \cdots + x_{n-1})\|.$$

Since (i) is false there exists $\lambda(n)$ such that

$$\|A_{\lambda(n)}\| \geq 4^{n+1}(n + c_{n-1}).$$

There also exists $x_n \in \mathcal{B}$ such that $\|x_n\| = 4^{-n}$ and

$$\|A_{\lambda(n)} x_n\| \geq \frac{2}{3} \|A_{\lambda(n)}\| \, \|x_n\|.$$

The series $x := \sum_{n=1}^{\infty} x_n$ is norm convergent and

$$\|A_{\lambda(n)} x\| \geq \|A_{\lambda(n)} x_n\| - \|A_{\lambda(n)}(x_1 + \cdots + x_{n-1})\| - \|A_{\lambda(n)}\| \sum_{r=n+1}^{\infty} \|x_r\|$$

$$\geq \frac{2}{3} \|A_{\lambda(n)}\| 4^{-n} - c_{n-1} - \frac{1}{3} \|A_{\lambda(n)}\| 4^{-n}$$

$$\geq \|A_{\lambda(n)}\| 4^{-n-1} - c_{n-1}$$

$$\geq n.$$

The contradiction implies (i).

The proof of (iii)$\Rightarrow$(ii) uses (ii)$\Rightarrow$(i) twice, with appropriate choices of $\mathcal{B}$ and $\mathcal{C}$. $\square$

---

[10] According to [Carothers 2005, p. 53], whom we follow, the proof below was first published by Hausdorff in 1932, but the 'sliding hump' idea was already well-known. Most texts give a longer proof based on the Baire category theorem. The sliding hump argument is also used in Theorem 3.3.11.

Corollary 1.3.5 *If the sequence $f_n \in \mathcal{B}$ converges weakly to $f \in \mathcal{B}$ as $n \to \infty$, then there exists a constant c such that $\|f_n\| \leq c$ for all n.*

In applications, the hypothesis of the corollary is usually harder to prove than the conclusion. Indeed the boundedness of a sequence of vectors or operators is often one of the ingredients used when proving its convergence, as in the following problem.

Problem 1.3.6 Let A_t be a bounded operator on the Banach space $\mathcal{B}$ for every $t \in [a, b]$, and let $\mathcal{D}$ be a dense linear subspace of $\mathcal{B}$. If $\|A_t\| \leq c < \infty$ for all $t \in [a, b]$ and $t \to A_t f$ is norm continuous for all $f \in \mathcal{D}$, prove that $(t, f) \to A_t f$ is a jointly continuous function from $[a, b] \times \mathcal{B}$ to $\mathcal{B}$. $\square$

We define the weak* topology of $\mathcal{B}^*$ to be the smallest topology for which all of the functionals $\phi \to \langle f, \phi \rangle$ are continuous, where $f \in \mathcal{B}$. It is generated by the family of seminorms $p_f(\phi) := |\phi(f)|$ where $f \in \mathcal{B}$. If $\mathcal{B}$ is reflexive the weak and weak* topologies on $\mathcal{B}^*$ coincide, but generally they do not.

Theorem 1.3.7 *(Banach-Alaoglu) Every norm bounded set in $\mathcal{B}^*$ is relatively compact for the weak* topology, in the sense that its weak* closure is weak* compact.*

Proof. It is sufficient to prove that the ball

$$B_1^* := \{\phi \in \mathcal{B}^* : \|\phi\| \leq 1\}$$

is compact. We first note that the topological product

$$S := \prod_{f \in \mathcal{B}} \{z \in \mathbf{C} : |z| \leq \|f\|\}$$

is a compact Hausdorff space. It is routine to prove that the map $\tau : B_1^* \to S$ defined by

$$\{\tau(\phi)\}(f) := \langle f, \phi \rangle$$

is a homeomorphism of B_1^* onto a closed subset of S. $\square$

Problem 1.3.8 Prove that the unit ball B_1^* of B^* provided with the weak* topology is metrizable if and only if $\mathcal{B}$ is separable. $\square$

Problem 1.3.9 Suppose that $f_n \in l^p(\mathbf{Z})$ and that $\|f_n\|_p \leq 1$ for all $n = 1, 2, \dots$. Prove that if $1 < p < \infty$ then the sequence f_n converges weakly to 0 if and only if the functions converge pointwise to 0, but that if $p = 1$ this is not always the case. Deduce that the unit ball in $l^1(\mathbf{Z})$ is not weakly compact, so $l^1(\mathbf{Z})$ cannot be reflexive. $\square$

Bounded operators between two Banach spaces $\mathcal{B}$ and $\mathcal{C}$ can converge in three different senses. Given a sequence of operators $A_n : \mathcal{B} \to \mathcal{C}$, we will write $A_n \overset{n}{\to} A$, $A_n \overset{s}{\to} A$ and $A_n \overset{w}{\to} A$ respectively in place of

$$\lim_{n \to \infty} \|A_n - A\| = 0,$$

$$\lim_{n \to \infty} \|A_n f - A f\| = 0 \text{ for all } f \in \mathcal{B},$$

$$\lim_{n \to \infty} \langle A_n f, \phi \rangle = \langle A f, \phi \rangle \text{ for all } f \in \mathcal{B}, \phi \in \mathcal{C}^*.$$

Another notation is $\lim_{n \to \infty} A_n = A$, s-$\lim_{n \to \infty} A_n = A$, w-$\lim_{n \to \infty} A_n = A$.

Problem 1.3.10 Let A, A_n be bounded operators on the Banach space $\mathcal{B}$ and let $\mathcal{D}$ be a dense linear subspace of $\mathcal{B}$. Use the uniform boundedness theorem to prove that $A_n \overset{s}{\to} A$ if and only if there exists a constant c such that $\|A_n\| \leq c$ for all n and $\lim_{n \to \infty} A_n f = A f$ for all $f \in \mathcal{D}$. $\square$

Problem 1.3.11 Given two sequences of operators $A_n : \mathcal{B} \to \mathcal{C}$ and $B_n : \mathcal{C} \to \mathcal{D}$, prove the following results:
(a) If $A_n \overset{s}{\to} A$ and $B_n \overset{s}{\to} B$ then $B_n A_n \overset{s}{\to} BA$.
(b) If $A_n \overset{s}{\to} A$ and $B_n \overset{w}{\to} B$ then $B_n A_n \overset{w}{\to} BA$.
(c) If $A_n \overset{w}{\to} A$ and $B_n \overset{w}{\to} B$ then $B_n A_n \overset{w}{\to} BA$ may be false.
Prove or give counterexamples to all other combinations of these types of convergence. $\square$

From the point of view of applications, norm convergence is the best, but it is too strong to be true in many situations; weak convergence is the easiest to prove, but it does not have good enough properties to prove many theorems. One is left with strong convergence as the most useful concept.

Problem 1.3.12 Let P_n be a sequence of projections on $\mathcal{B}$, i.e. operators such that $P_n^2 = P_n$ for all n. Prove that if $P_n \overset{s}{\to} P$ then P is a projection, and give a counterexample to this statement if one replaces strong convergence by weak convergence. $\square$

Problem 1.3.13 Let A, A_n be operators on the Hilbert space $\mathcal{H}$. Prove that if $A_n \overset{s}{\to} A$ then $A_n^* \overset{w}{\to} A^*$, and give an example in which A_n^* does not converge strongly to A^*. $\square$

One sometimes says that A_n converges in the strong* sense to A if $A_n \overset{s}{\to} A$ and $A_n^* \overset{s}{\to} A^*$.

1.4 Differentiation of vector-valued functions

We discuss various notions of differentiability for two functions $f : [a, b] \to \mathcal{B}$ and $\phi : [a, b] \to \mathcal{B}^*$. We write C^n to denote the space of n times continuously differentiable functions if $n \geq 1$, and the space of continuous functions if $n = 0$.

Lemma 1.4.1 *If $\langle f(t), \psi \rangle$ is C^1 for all $\psi \in \mathcal{B}^*$ then $f(t)$ is C^0. Similarly, if $\langle g, \phi(t) \rangle$ is C^1 for all $g \in \mathcal{B}$ then $\phi(t)$ is C^0.*

Proof. By the uniform boundedness theorem there is a constant N such that $\| f(t) \| \leq N$ for all $t \in [a, b]$. If $a \leq c \leq b$ then

$$\lim_{\delta \to 0} \langle \delta^{-1} \{ f(c + \delta) - f(c) \}, \psi \rangle = \frac{\mathrm{d}}{\mathrm{d}c} \langle f(c), \psi \rangle,$$

so using the uniform boundedness theorem again there exists a constant M such that

$$\| \delta^{-1} \{ f(c + \delta) - f(c) \} \| \leq M$$

for all small enough $\delta \neq 0$. This implies that

$$\lim_{\delta \to 0} \| f(c + \delta) - f(c) \| = 0.$$

The other part of the lemma has a similar proof. $\square$

Lemma 1.4.2 *If $\langle f(t), \psi \rangle$ is C^2 for all $\psi \in \mathcal{B}^*$ then $f(t)$ is C^1. Similarly, if $\langle g, \phi(t) \rangle$ is C^2 for all $g \in \mathcal{B}$ then $\phi(t)$ is C^1.*

Proof. By the uniform boundedness theorem there exist $g(t) \in \mathcal{B}^{**}$ for each $t \in [a, b]$ such that

$$\frac{\mathrm{d}}{\mathrm{d}t} \langle f(t), \psi \rangle = \langle g(t), \psi \rangle.$$

Moreover $\langle g(t), \psi \rangle$ is C^1 for all $\psi \in \mathcal{B}^*$, so by Lemma 1.4.1 $g(t)$ depends norm continuously on t. Therefore

$$\int_a^t g(s) \, \mathrm{d}s$$

is defined as an element of $\mathcal{B}^{**}$, and

$$\frac{\mathrm{d}}{\mathrm{d}t} \left\langle f(t) - f(a) - \int_a^t g(s) \, \mathrm{d}s, \psi \right\rangle = 0$$

for all $t \in [a, b]$ and $\psi \in \mathcal{B}^*$. It follows that

$$f(t) - f(a) = \int_a^t g(s) \, \mathrm{d}s$$

for all $t \in [a, b]$. We deduce that

$$g(t) = \lim_{h \to 0} h^{-1}\{f(t+h) - f(t)\}$$

the limit being taken in the norm sense. Therefore $g(t) \in \mathcal{B}$, and $f(t)$ is C^1. The proof for $\phi(t)$ is similar. $\square$

Corollary 1.4.3 *If $\langle f(t), \psi \rangle$ is C^∞ for all $\psi \in \mathcal{B}^*$ then $f(t)$ is C^∞. Similarly, if $\langle g, \phi(t) \rangle$ is C^∞ for all $g \in \mathcal{B}$ then $\phi(t)$ is C^∞.*

Proof. One shows inductively that if $\langle f(t), \psi \rangle$ is C^{n+1} for all $\psi \in \mathcal{B}^*$ then $f(t)$ is C^n. $\square$

We will need the following technical lemma later in the book.

Lemma 1.4.4 *(i) If $f : [0, \infty) \to \mathbf{R}$ is continuous and for all $x \geq 0$ there exists a strictly monotonic decreasing sequence x_n such that*

$$\lim_{n \to \infty} x_n = x, \qquad \limsup_{n \to \infty} \frac{f(x_n) - f(x)}{x_n - x} \leq 0$$

then f is non-increasing on $[0, \infty)$.

(ii) If $f : [0, \infty) \to \mathcal{B}$ is norm continuous and for all $x \geq 0$ there exists a strictly monotonic decreasing sequence x_n such that

$$\lim_{n \to \infty} x_n = x, \qquad \lim_{n \to \infty} \left\langle \frac{f(x_n) - f(x)}{x_n - x}, \phi \right\rangle = 0$$

for all $\phi \in \mathcal{B}^*$ then f is constant on $[0, \infty)$.

Proof. (i) If $\alpha > 0$, $a \geq 0$ and

$$S_{\alpha, a} := \{x \geq a : f(x) \leq f(a) + \alpha(x - a)\}$$

then $S_{\alpha, a}$ is closed, contains a, and for all $x \in S_{\alpha, a}$ and $\varepsilon > 0$ there exists $t \in S_{\alpha, a}$ such that $x < t < x + \varepsilon$. If $u > a$ then there exists a largest number $s \in S_{\alpha, a}$ satisfying $s \leq u$. The above property of $S_{\alpha, a}$ implies that $s = u$. We deduce that $S_{\alpha, a} = [a, \infty)$ for every $\alpha > 0$, and then that $f(x) \leq f(a)$ for all $x \geq a$.

(ii) We apply part (i) to $\mathrm{Re}\{e^{i\theta} \langle f(x), \phi \rangle\}$ for every $\phi \in \mathcal{B}^*$ and every $\theta \in \mathbf{R}$ to deduce that $\langle f(x), \phi \rangle = 0$ for all $x \geq 0$. Since $\phi \in \mathcal{B}^*$ is arbitrary we deduce that $f(x)$ is constant. $\square$

All of the above ideas can be extended to operator-valued functions. We omit a systematic treatment of the various topologies for which one can define differentiability, but mention three results.

Problem 1.4.5 Prove that if A, $B : [a, b] \to \mathcal{L}(\mathcal{B})$ are continuously differentiable in the strong operator topology then they are norm continuous. Moreover $A(t)B(t)$ is continuously differentiable in the same sense and

$$\frac{\mathrm{d}}{\mathrm{d}t}\{A(t)B(t)\} = A(t)'B(t) + A(t)B(t)'$$

for all $t \in [a, b]$. $\square$

Problem 1.4.6 Prove that if $A : [a, b] \to \mathcal{L}(\mathcal{B})$ is differentiable in the strong operator topology then $A(t)^{-1}$ is strongly differentiable and

$$\frac{\mathrm{d}}{\mathrm{d}t}A(t)^{-1} = -A(t)^{-1}A(t)'A(t)^{-1}$$

for all $t \in [a, b]$. $\square$

Problem 1.4.7 Prove that if $A(t)$ is a differentiable family of $m \times m$ matrices for $t \in [a, b]$ then

$$\frac{\mathrm{d}}{\mathrm{d}t}A(t)^n \neq nA(t)'A(t)^{n-1}$$

in general, but nevertheless

$$\frac{\mathrm{d}}{\mathrm{d}t}\mathrm{tr}[A(t)^n] = n\,\mathrm{tr}[A(t)'A(t)^{n-1}]. \qquad \square$$

We now turn to the study of analytic functions. Let $f(z)$ be a function from the region (connected open subset) U of the complex plane $\mathbf{C}$ taking values in the complex Banach space $\mathcal{B}$. We say that f is analytic on U if it is infinitely differentiable in the norm topology at every point of U.

Lemma 1.4.8 *If $\langle f(z), \phi \rangle$ is analytic on U for all $\phi \in \mathcal{B}^*$ then $f(z)$ is analytic on U.*

Proof. We first note that by a complex variables version of Lemma 1.4.1, $z \to f(z)$ is norm continuous. If γ is the boundary of a disc inside U then

$$\langle f(z), \phi \rangle = \frac{1}{2\pi i}\int_\gamma \frac{\langle f(w), \phi \rangle}{w - z}\,\mathrm{d}w$$

$$= \left\langle \frac{1}{2\pi i}\int_\gamma \frac{f(w)}{w - z}\,\mathrm{d}w, \phi \right\rangle$$

for all $\phi \in \mathcal{B}^*$. This implies the vector-valued Cauchy's integral formula

$$f(z) = \frac{1}{2\pi i} \int_\gamma \frac{f(w)}{w - z} \, dw, \tag{1.9}$$

the right-hand side of which is clearly an analytic function of z. □

Problem 1.4.9 Prove a vector-valued Liouville's theorem: namely if $f : \mathbf{C} \to \mathcal{B}$ is uniformly bounded in norm and analytic then it is constant. □

Lemma 1.4.10 *Let* $f_n \in \mathcal{B}$ *and suppose that*

$$\sum_{n=0}^{\infty} \langle f_n, \phi \rangle z^n$$

converges for all $\phi \in \mathcal{B}^*$ *and all* $|z| < R$. *Then the power series*

$$\sum_{n=0}^{\infty} f_n z^n \tag{1.10}$$

is norm convergent for all $|z| < R$, *and the limit is a* $\mathcal{B}$*-valued analytic function.*

Proof. We define the linear functional $f(z)$ on $\mathcal{B}^*$ by

$$\langle f(z), \phi \rangle := \sum_{n=0}^{\infty} \langle f_n, \phi \rangle z^n.$$

The uniform boundedness theorem implies that $f(z) \in \mathcal{B}^{**}$ for all $|z| < R$. An argument similar to that of Lemma 1.4.1 establishes that $z \to f(z)$ is norm continuous, and an application of the Cauchy integral formula as in Lemma 1.4.8 shows that $f(z)$ is norm analytic. A routine modification of the usual proof for the case $\mathcal{B} = \mathbf{C}$ now establishes that the series (1.10) is norm convergent, so we finally see that $f(z) \in \mathcal{B}$ for all $|z| < R$. □

If $a_n \in \mathcal{B}$ for $n = 0, 1, 2, \ldots$ then the power series $\sum_{n=0}^{\infty} a_n z^n$ defines a $\mathcal{B}$-valued analytic function for all z for which the series converges. The radius of convergence R is defined as the radius of the largest circle with centre at 0 within which the series converges. As in the scalar case $R = 0$ and $R = +\infty$ are allowed.

Problem 1.4.11 Prove that

$$R = \sup\{\rho : \{\|a_n\|\rho^n\}_n \text{ is a bounded sequence}\}.$$

Alternatively

$$R^{-1} = \limsup_{n \to \infty} \|a_n\|^{1/n}. \qquad \square$$

The following theorem establishes that the powers series of an analytic function converges on the maximal possible ball $B(0, r) := \{z : |z| < r\}$.

Theorem 1.4.12 *Let* $f : B(0, r) \to \mathcal{B}$ *be an analytic function which cannot be analytically continued to a larger ball. Then the power series of* f *has radius of convergence* r.

Proof. If we denote the radius of convergence by R, then it follows immediately from Problem 1.4.11 that $R \leq r$. If $|z| < r$ and $t = (r + |z|)/2$ then by adapting the classical proof (which depends on using (1.9)) we obtain

$$f(z) = f(0) + f'(0)\frac{z}{1!} + \cdots + f^{(n)}(0)\frac{z^n}{n!} + \mathrm{Rem}(n)$$

where

$$\mathrm{Rem}(n) := \frac{1}{2\pi i} \int_{|w|=t} \frac{f(w)}{w-z} \left(\frac{z}{w}\right)^{n+1} \mathrm{d}w.$$

This implies that

$$\|\mathrm{Rem}(n)\| \leq c_{z,t}(z/t)^{n+1}$$

which converges to 0 as $n \to \infty$. Therefore the power series converges for every z such that $|z| < r$. This implies that $R \geq r$. $\square$

All of the results above can be extended to operator-valued analytic functions. Since the space $\mathcal{L}(\mathcal{B})$ is itself a Banach space with respect to the operator norm, the only new issue is dealing with weaker topologies.

Problem 1.4.13 Prove that if $A(z)$ is an operator-valued function on $U \subseteq \mathbf{C}$, and $z \to \langle A(z)f, \phi \rangle$ is analytic for all $f \in \mathcal{B}$ and $\phi \in \mathcal{B}^*$, then $A(z)$ is an analytic function of z. $\square$

1.5 The holomorphic functional calculus

The material in this section was developed by Hilbert, E. H. Moore, F. Riesz and others early in the twentieth century. A functional calculus is a procedure for defining an operator $f(A)$, given an operator A and some class of complex-valued functions f defined on the spectrum of A. One requires $f(A)$ to satisfy certain properties, including (1.11) below. The following theorem defines a holomorphic functional calculus for bounded linear operators. Several of the proofs in this section apply with minimal changes to unbounded operators, and we will take advantage of that fact later in the book.

Theorem 1.5.1 *Let S be a compact component of the spectrum* $\mathrm{Spec}(A)$ *of the operator A acting on $\mathcal{B}$, and let $f(\cdot)$ be a function which is analytic on a neighbourhood U of S. Let γ be a closed curve in U such that S is inside γ and $\mathrm{Spec}(A)\backslash S$ is outside γ. Then*

$$B := \frac{1}{2\pi i} \int_{\gamma} f(z) R(z, A) \, dz$$

is a bounded operator commuting with A. It is independent of the choice of γ, subject to the above conditions. Writing B in the form $f(A)$ we have

$$f(A)g(A) = (fg)(A) \tag{1.11}$$

for any two functions f, g of the stated type. The map $f \to f(A)$ is norm continuous from the stated class of functions with $\|f\| := \max\{|f(z)| : z \in \gamma\}$ to $\mathcal{L}(\mathcal{B})$.

Proof. It is immediate from its definition that

$$BR(w, A) = R(w, A)B$$

for all $w \notin \mathrm{Spec}(A)$. This implies that B commutes with A. In the following argument we label B according to the contour used to define it. If σ is a second contour with the same properties as γ, and we put $\delta := \gamma - \sigma$, then

$$B_{\gamma} - B_{\sigma} = B_{\delta} = 0$$

by the operator version of Cauchy's Theorem.

To prove (1.11), let γ, σ be two curves satisfying the stated conditions, with σ inside γ. Then

$$f(A)g(A) = -\frac{1}{4\pi^2} \int_{\sigma} \int_{\gamma} f(z)g(w)R(z, A)R(w, A) \, dz \, dw$$

$$= -\frac{1}{4\pi^2} \int_{\sigma} \int_{\gamma} \frac{f(z)g(w)}{z - w} (R(w, A) - R(z, A)) \, dz \, dw$$

$$= \frac{1}{2\pi i} \int_{\sigma} f(w)g(w)R(w, A) \, dw$$

$$= (fg)(A). \qquad \square$$

Problem 1.5.2 Let A be a bounded operator on $\mathcal{B}$ and let γ be the closed curve $\theta \to re^{i\theta}$ where $r > \|A\|$. Prove that if $p(z) := \sum_{m=0}^{n} a_m z^m$ then

$$p(A) = \sum_{m=0}^{n} a_m A^m. \qquad \square$$

Example 1.5.3 Let A be a bounded operator on $\mathcal{B}$ and suppose that $\mathrm{Spec}(A)$ does not intersect $(-\infty, 0]$. Then there exists a closed contour γ that winds around $\mathrm{Spec}(A)$ and which does not intersect $(-\infty, 0]$. If $t > 0$ then the function z^t is holomorphic on and inside γ, so one may use the holomorphic functional calculus to define A^t. However, one should not suppose that $\|A^t\|$ must be of the same order of magnitude as $\|A\|$ for $0 < t < 1$. Figure 1.1 displays the norms of A^t for $n := 100$, $c := 0.6$ and $0 < t < 2$, where A is the $n \times n$ matrix

$$A_{r,s} := \begin{cases} r/n & \text{if } s = r+1, \\ c & \text{if } r = s, \\ 0 & \text{otherwise.} \end{cases}$$

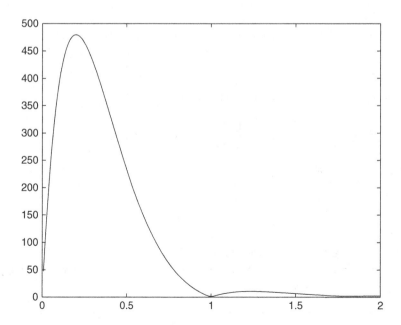

Figure 1.1: Norms of fractional powers in Example 1.5.3

Note that $\|A^t\|$ is of order 1 for $t = 0$, 1, 2. It can be much larger for other t because the resolvent norm must be extremely large on a portion of the contour γ, for any contour satisfying the stated conditions. See also Example 10.2.1. $\square$

Theorem 1.5.4 *(Riesz) Let γ be a closed contour enclosing the compact component S of the spectrum of the bounded operator A acting in $\mathcal{B}$, and suppose that $T = \mathrm{Spec}(A)\backslash S$ is outside γ. Then*

$$P := \frac{1}{2\pi i} \int_\gamma R(z, A)\, dz$$

is a bounded projection commuting with A. The restriction of A to $P\mathcal{B}$ has spectrum S and the restriction of A to $(I - P)\mathcal{B}$ has spectrum T. P is said to be the spectral projection of A associated with S.

Proof. It follows from Theorem 1.5.1 with $f = g = 1$ that $P^2 = P$. If we put $\mathcal{B}_0 = \mathrm{Ran}(P)$ and $\mathcal{B}_1 = \mathrm{Ker}(P)$ then $\mathcal{B} = \mathcal{B}_0 \oplus \mathcal{B}_1$ and $A(\mathcal{B}_i) \subseteq \mathcal{B}_i$ for $i = 0$, 1. If A_i denotes the restriction of A to $\mathcal{B}_i$ then

$$\mathrm{Spec}(A) = \mathrm{Spec}(A_0) \cup \mathrm{Spec}(A_1).$$

The proof is completed by showing that

$$\mathrm{Spec}(A_0) \cap T = \emptyset, \qquad \mathrm{Spec}(A_1) \cap S = \emptyset.$$

If w is in T, then it is outside γ, and we put

$$C_w := \frac{1}{2\pi i} \int_\gamma \frac{1}{w - z} R(z, A)\, dz.$$

Theorem 1.5.1 implies that

$$C_w P = P C_w = C_w,$$
$$(wI - A) C_w = C_w(wI - A) = P.$$

Therefore $w \notin \mathrm{Spec}(A_0)$. Hence $\mathrm{Spec}(A_0) \cap T = \emptyset$.

Now let τ be the circle with centre 0 and radius $(\|A\| + 1)$. By expanding the resolvent on powers of $1/z$ we see that

$$I = \frac{1}{2\pi i} \int_\tau R(z, A)\, dz.$$

If σ denotes the curve $(\tau - \gamma)$ then we deduce that

$$I - P = \frac{1}{2\pi i} \int_\sigma R(z, A) \, dz.$$

By following the same argument as in the first paragraph we see that if w is in S, then it is inside γ and outside σ, so $w \notin \mathrm{Spec}(A_1)$. Hence $\mathrm{Spec}(A_1)$ $\cap S = \emptyset$. $\square$

If S consists of a single point z then the restriction of A to $\mathcal{B}_0 = \mathrm{Ran}(P)$ has spectrum equal to $\{z\}$, but this does not imply that $\mathcal{B}_0$ consists entirely of eigenvectors of A. Even if $\mathcal{B}_0$ is finite-dimensional, the restriction of A to $\mathcal{B}_0$ may have a non-trivial Jordan form. The full theory of what happens under small perturbations of A is beyond the scope of this book, but the next theorem is often useful. Its proof depends upon the following lemma. The properties of orthogonal projections on a Hilbert space are studied more thoroughly in Section 5.3. We define the rank of an operator to be the possibly infinite dimension of its range.

Lemma 1.5.5 *If P and Q are two bounded projections and $\|P - Q\| < 1$ then*

$$\mathrm{rank}(P) = \mathrm{rank}(Q).$$

Proof. If $0 \neq x \in \mathrm{Ran}(P)$ then $\|Qx - x\| = \|(Q - P)x\| < \|x\|$, so $Qx \neq 0$. Therefore Q maps $\mathrm{Ran}(P)$ one-one into $\mathrm{Ran}(Q)$ and $\mathrm{rank}(P) \leq \mathrm{rank}(Q)$. The converse has a similar proof. $\square$

A more general version of the following theorem is given in Theorem 11.1.6, but even that is less general than the case treated by Rellich, in which one simply assumes that the operator depends analytically on a complex parameter z.[11]

Theorem 1.5.6 *(Rellich) Suppose that λ is an isolated eigenvalue of A and that the associated spectral projection P has rank 1. Then for any operator B and all small enough $w \in \mathbf{C}$, $(A + wB)$ has a single eigenvalue $\lambda(w)$ near to λ, and this eigenvalue depends analytically upon w.*

Proof. Let γ be a circle enclosing λ and no other point of $\mathrm{Spec}(A)$, and let P be defined as in Theorem 1.5.4. If

$$|w| < \|B\|^{-1} \min\{\|R(z, A)\|^{-1} : z \in \gamma\}$$

[11] A systematic treatment of the perturbation of eigenvalues of higher multiplicity is given in [Kato 1966A].

then $(zI - (A + wB))$ is invertible for all $z \in \gamma$ by Theorem 1.2.9. By examining the expansion (1.5) one sees that $(zI - (A + wB))^{-1}$ depends analytically upon w for every $z \in \gamma$. It follows that the projections

$$P_w := \frac{1}{2\pi i} \int_\gamma (zI - (A + wB))^{-1} dz$$

depend analytically upon w. By Lemma 1.5.5 P_w has rank 1 for all such w.

If $f \in \text{Ran}(P)$ then $f_w := P_w f$ depends analytically upon w and lies in the range of P_w for all w. Assuming $f \neq 0$ it follows that $f_w \neq 0$ for all small enough w. Therefore f_w is the eigenvector of $(A + wB)$ associated with the eigenvalue lying within γ for all small enough w. The corresponding eigenvalue satisfies

$$\langle (A + wB) f_w, \phi \rangle = \lambda_w \langle f_w, \phi \rangle$$

where ϕ is any vector in $\mathcal{B}^*$ which satisfies $\langle f, \phi \rangle = 1$. The analytic dependence of λ_w on w for all small enough w follows from this equation.   □

Example 1.5.7 The following example shows that the eigenvalues of non-self-adjoint operators may behave in counter-intuitive ways (for those brought up in self-adjoint environments). Let H be a self-adjoint $n \times n$ matrix and let $Bf := \langle f, \phi \rangle \phi$, where ϕ is a fixed vector of norm 1 in $\mathbf{C}^n$. If $A_s := H + isB$ then $\text{Im}\langle A_s f, f \rangle$ is a monotone increasing function of $s \in \mathbf{R}$ for all $f \in \mathbf{C}^n$, and this implies that every eigenvalue of A_s has a positive imaginary part for all $s > 0$. If $\{\lambda_{r,s}\}_{r=1}^n$ are the eigenvalues of A_s then

$$\sum_{r=1}^n \lambda_{r,s} = \text{tr}(A_s) = \text{tr}(H) + is$$

for all s. All these facts (wrongly) suggest that the imaginary part of each individual eigenvalue is a positive, monotonically increasing function of s for $s \geq 0$.

More careful theoretical arguments show that the eigenvalues of such an operator move from the real axis into the upper half plane as s increases from 0. All except one then turn around and converge back to the real axis as $s \to +\infty$. For $n = 2$ the calculations are elementary, but the case

$$A_s := \begin{pmatrix} -1 + is & is & is \\ is & is & is \\ is & is & 1 + is \end{pmatrix} \tag{1.12}$$

is more typical.[12]   □

---

[12] See Lemma 11.2.9 for further examples of a similar type.

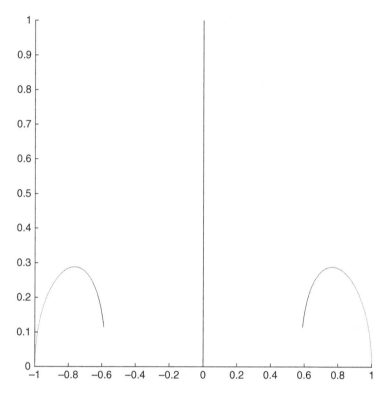

Figure 1.2: Eigenvalues of (1.12) for $0 \leq s \leq 1$

If an operator $A(z)$ has several eigenvalues $\lambda_r(z)$, all of which depend analytically on z, then generically they will only coincide in pairs, and this will happen for certain discrete values of z. One can analyze the z-dependence of two such eigenvalues by restricting attention to the two-dimensional linear span of the corresponding eigenvectors. The following example illustrates what can happen.

Example 1.5.8 If

$$A(z) := \begin{pmatrix} a(z) & b(z) \\ c(z) & d(z) \end{pmatrix}$$

where a, b, c, d are all analytic functions, then the eigenvalues of $A(z)$ are given by

$$\lambda_{\pm}(z) := (a(z) + d(z))/2 \pm \left\{ (a(z) - d(z))^2/4 + b(z)c(z) \right\}^{1/2}.$$

For most values of z the two branches are analytic functions of z, but for certain special z they coincide and one has a square root singularity. In the typical case

$$A(z) := \begin{pmatrix} 0 & z \\ 1 & 0 \end{pmatrix}$$

one has $\lambda(z) = \pm\sqrt{z}$. The two eigenvalues coincide for $z = 0$, but when this happens the matrix has a non-trivial Jordan form and the eigenvalue 0 has multiplicity 1. □

2

Function spaces

2.1 L^p spaces

The serious analysis of any operators acting in infinite-dimensional spaces has to start with the precise specification of the spaces and their norms. In this chapter we present the definitions and properties of the L^p spaces that will be used for most of the applications in the book. Although these are only a tiny fraction of the function spaces that have been used in various applications, they are by far the most important ones. Indeed a large number of books confine attention to operators acting in Hilbert space, the case $p := 2$, but this is not natural for many applications, such as those to probability theory.

Before we start this section we need to make a series of standing hypotheses of a measure-theoretic character. We recommend that the reader skims through these, and refers back to them as necessary. The conditions are satisfied in all normal contexts within measure theory.[1]

(i) We define a measure space to be a triple (X, Σ, μ) consisting of a set X, a σ-field Σ of 'measurable' subsets of X, and a non-negative countably additive measure μ on Σ. We will usually denote the measure by dx.

(ii) We will always assume that the measure μ is σ-finite in the sense that there is an increasing sequence of measurable subsets X_n with finite measures and union equal to X.

(iii) We assume that each X_n is provided with a finite partition $\mathcal{E}_n$, by which we mean a sequence of disjoint measurable subsets $\{E_1, E_2, \ldots, E_{m(n)}\}$, each of which has positive measure $|E_r| := \mu(E_r)$. The union of the subsets E_r must equal X_n.

---

[1] Lebesgue integration and measure theory date back to the beginning of the twentieth century. There are many good accounts of the subject, for example [Lieb and Loss 1997, Rudin 1966, Weir 1973.]

(iv) We assume that the partition $\mathcal{E}_{n+1}$ is finer than $\mathcal{E}_n$ for every n, in the sense that each set in $\mathcal{E}_n$ is the union of one or more sets in $\mathcal{E}_{n+1}$.

 (v) We define $\mathcal{L}_n$ to be the linear space of all functions $f := \sum_{r=1}^{m(n)} \alpha_r \chi_r$, where χ_r denotes the characteristic function of a set $E_r \in \mathcal{E}_n$. Condition (iv) is then equivalent to $\mathcal{L}_n \subseteq \mathcal{L}_{n+1}$ for all n.

(vi) We assume that the σ-field Σ is countably generated in the sense that it is generated by the totality of all sets in all partitions $\mathcal{E}_n$.

(vii) If $1 \leq p < \infty$, the expression $L^p(X, dx)$, or more briefly $L^p(X)$, denotes the space of all measurable functions $f : X \to \mathbf{C}$ such that

$$\|f\|_p := \left\{ \int_X |f(x)|^p \, dx \right\}^{1/p} < \infty,$$

two functions being identified if they are equal almost everywhere. If $f, g \in L^p(X, dx)$ and $\alpha, \beta \in \mathbf{C}$, the pointwise inequality

$$|\alpha f(x) + \beta g(x)|^p \leq 2^p |\alpha|^p |f(x)|^p + 2^p |\beta|^p |g(x)|^p$$

implies that $L^p(X, dx)$ is a vector space. We prove that $\|\cdot\|_p$ is a norm in Theorem 2.1.7. Condition (vi) is equivalent to $\bigcup_{n \geq 1} \mathcal{L}_n$ being dense in $L^p(X, dx)$ for all $1 \leq p < \infty$. It follows that $L^p(X, dx)$ is separable in the sense of containing a countable dense set.

(viii) If X is a finite or countable set, $l^p(X)$ refers to the space $L^p(X, dx)$, taking Σ to consist of all subsets of X and the measure to be the counting measure.

(ix) If $f : X \to \mathbf{C}$ is a measurable function we define its support by

$$\mathrm{supp}(f) := \{x : f(x) \neq 0\}.$$

This is only defined up to modification by a null set, i.e. a set of zero measure.

If $f, g : X \to \mathbf{C}$ are measurable functions and $fg \in L^1(X, dx)$ we will often use the notation

$$\langle f, g \rangle := \int_X f(x) g(x) \, dx.$$

Sometimes $g(x)$ should be replaced by $\overline{g(x)}$.

Example 2.1.1 The construction of Lebesgue measure on $\mathbf{R}^N$ is not elementary, but we indicate how the above conditions are satisfied in that case. We start by defining the sets X_n by

$$X_n := \{x \in \mathbf{R}^N : -n \leq x_r < n \text{ for all } 1 \leq r \leq N\}.$$

It is immediate that $|X_n| = (2n)^N$. We define the partition $\mathcal{E}_n$ of X_n to consist of all subsets E of X_n which are of the form

$$\prod_{r=1}^{N} \left[\frac{m_r - 1}{2^n}, \frac{m_r}{2^n} \right)$$

for suitable integers $m_1, \ldots, m_N$. Every such 'cube' in $\mathcal{E}_n$ is the union of 2^N disjoint cubes in $\mathcal{E}_{n+1}$. The totality of all such cubes in all $\mathcal{E}_n$ as n varies generates the Borel σ-field of $\mathbf{R}^N$. $\square$

We say that a measurable function $f : X \to \mathbf{C}$ is essentially bounded if the set $\{x : |f(x)| > c\}$ has zero measure for some c. The space $L^\infty(X)$ is defined to be the set of all measurable, essentially bounded functions on X, where we again identify two such functions if they coincide except on a null set. We define $\|f\|_\infty$ to be the smallest constant c above. The proof that $L^\infty(X)$ is a Banach space for this norm is routine.

If $1 \leq p < \infty$, the proof that $\| \cdot \|_p$ is a norm is not elementary, except in the cases $p = 1, 2$. We approach it via a series of definitions and lemmas. We say that a function $\phi : [a, b] \to [0, \infty)$ is log-convex if

$$\phi((1 - \lambda)u + \lambda v) \leq \phi(u)^{1-\lambda} \phi(v)^\lambda$$

for all $u, v \in [a, b]$ and $0 < \lambda < 1$. We first deal with a singular case. We warn the reader that when referring to the exponents in L^p spaces one often uses the notation $[p, q]$ to refer to $\{(1 - \lambda)p + \lambda q : 0 \leq \lambda \leq 1\}$ without any requirement that $p \leq q$. Similarly $[p, q)$ may refer to $\{(1 - \lambda)p + \lambda q : 0 \leq \lambda < 1\}$.

Problem 2.1.2 Prove that if ϕ is log-convex on $[a, b]$ and $\phi(c) = 0$ for some $c \in [a, b]$ then $\phi(x) = 0$ for all $x \in (a, b)$. $\square$

Problem 2.1.3 Prove that if $\phi : [a, b] \to (0, \infty)$ is C^2 then it is log-convex if and only if

$$\frac{\mathrm{d}^2}{\mathrm{d}x^2} \log(\phi(x)) \geq 0$$

for all $x \in [a, b]$. $\square$

Problem 2.1.4 Suppose that $0 < a < b < \infty$ and that $h : X \to [0, \infty)$ is measurable. If

$$\phi(s) := \int_X h(x)^s \, \mathrm{d}x$$

is finite for $s = a, b$, prove that ϕ is finite and log-convex on the interval $[a, b]$. $\square$

Problem 2.1.5 Suppose that $1 \le p < q \le \infty$ and that $f_t \in L^p(X, dx) \cap L^q(X, dx)$ for all $t \in (0, 1)$. If $\lim_{t \to 0} \|f_t\|_p = 0$ and $\sup_{0 < t < 1} \|f_t\|_q < \infty$, prove that $\lim_{t \to 0} \|f_t\|_r = 0$ for all $r \in (p, q)$. $\square$

Lemma 2.1.6 *(Hölder inequality) If $1 \le p \le \infty$ and q is the conjugate index in the sense that $1/p + 1/q = 1$, then $fg \in L^1(X, dx)$ for all $f \in L^p(X, dx)$ and $g \in L^q(X, dx)$, and*

$$|\langle f, g \rangle| \le \|f\|_p \|g\|_q. \tag{2.1}$$

Proof. The cases $p = 1$ and $p = \infty$ are elementary, so we assume that $1 < p < \infty$. Given $f \in L^p$ and $g \in L^q$ we consider the log-convex function

$$\phi(s) := \int_X |f(x)|^{sp} |g(x)|^{(1-s)q} \, dx.$$

Putting $s := 1/p$ yields $1 - s = 1/q$ and

$$\phi(1/p) \le \phi(0)^{1/q} \phi(1)^{1/p}.$$

This implies the required inequality directly. $\square$

Theorem 2.1.7 *If $1 \le p < \infty$ then the quantity $\|\cdot\|_p$ is a norm on $L^p(X, dx)$, and makes it a Banach space. If $f_r \in L^p(X, dx)$ and*

$$\sum_{r=1}^{\infty} \|f_r\|_p < \infty$$

then the partial sums $s_n := \sum_{r=1}^n f_r$ converge in L^p norm and almost everywhere to the same limit.

Proof. One can check that $\|\cdot\|_p$ satisfies all the axioms for a norm by using the identity

$$\|f\|_p = \sup \left\{ |\langle f, g \rangle| : \|g\|_q \le 1 \right\}$$

which is proved with the help of Lemma 2.1.6. The supremum is achieved for

$$g := \overline{f} |f|^{p-2} \|f\|_p^{1-p}.$$

We prove completeness and the final statement of the theorem together, using Problem 1.1.1. If f_n satisfy the stated conditions and we put

$$g_n := \sum_{r=1}^n |f_r|$$

then g_n is a monotonic increasing sequence and

$$\|g_n\|_p \le \sum_{r=1}^{\infty} \|f_r\|_p < \infty$$

for all n. By applying the monotone convergence theorem to g_n^p we conclude that g_n converges almost everywhere to a finite limit. This implies by a domination argument that s_n converges almost everywhere to a finite limit, which we call s. Given $\varepsilon > 0$, Fatou's lemma then implies that

$$\|s - s_n\|_p \le \liminf_{m \to \infty} \|s_m - s_n\|_p < \varepsilon$$

for all large enough n. Hence $s \in L^p(X, dx)$ and $\|s_n - s\|_p \to 0$ as $n \to \infty$. $\square$

Problem 2.1.8 Prove that if g_n converges in norm to g in $L^p(X, dx)$ then there exists a subsequence $g_{n(r)}$ which converges to g almost everywhere as $r \to \infty$. $\square$

Problem 2.1.9 Prove that if X contains two disjoint sets with positive measures and $L^p(X, dx)$ is isometrically isomorphic to a Hilbert space then $p = 2$. $\square$

Problem 2.1.10 Construct a sequence of continuous, non-negative functions f_n on $[0, 1]$ which converge to 0 in L^2 norm without converging pointwise anywhere in $[0, 1]$. $\square$

Problem 2.1.11 Prove that if $f \in L^p(X)$ for all large enough finite p, then $f \in L^\infty(X)$ if and only if the norms $\|f\|_p$ are uniformly bounded as $p \to \infty$. $\square$

Theorem 2.1.12 *Let* $1 \le p, q \le \infty$, $0 < \lambda < 1$ *and*

$$\frac{1}{r} := \frac{1 - \lambda}{p} + \frac{\lambda}{q}.$$

If $f \in L^p(X, dx) \cap L^q(X, dx)$ *then* $f \in L^r(X, dx)$ *and*

$$\|f\|_r \le \|f\|_p^{1-\lambda} \|f\|_q^{\lambda}. \tag{2.2}$$

Proof. If $p = \infty$ this reduces to the elementary statement that

$$\int_X |f(x)|^r \, dx \le \|f\|_\infty^{r-q} \int_X |f(x)|^q \, dx$$

provided $q < r < \infty$. A similar proof applies if $q = \infty$, so we henceforth assume that both are finite.

We may rewrite (2.2) in the form

$$\int_X |f(x)|^r \, dx \leq \left\{ \int_X |f(x)|^p \, dx \right\}^{(1-\lambda)r/p} \left\{ \int_X |f(x)|^q \, dx \right\}^{\lambda r/q}.$$

Putting $s := \lambda r/q$ this is equivalent to

$$\int_X |f(x)|^{(1-s)p+sq} \, dx \leq \left\{ \int_X |f(x)|^p \, dx \right\}^{1-s} \left\{ \int_X |f(x)|^q \, dx \right\}^{s}.$$

The proof is completed by applying Problem 2.1.4, which implies the log-convexity of the function

$$\phi(t) := \int_X |f(x)|^t \, dx. \qquad \square$$

Theorem 2.1.13 *Let $f \in L^p(X, dx)$ and $g \in L^q(X, dx)$ where $1 \leq p \leq \infty$ and $1 \leq q \leq \infty$. Also suppose that*

$$1/r := 1/p + 1/q.$$

Then $fg \in L^r(X, dx)$ and

$$\|fg\|_r \leq \|f\|_p \|g\|_q.$$

Proof. This is elementary if $p = \infty$ or $q = \infty$, so we assume that both are finite. Put $d'x := |g(x)|^q dx$ and $h := |f|^p |g|^{-q} \chi_E$, where $E := \{x : g(x) \neq 0\}$. One sees immediately that

$$\int_X 1 \, d'x = \int_X |g(x)|^q \, dx,$$

$$\int_X h(x) \, d'x = \int_E |f(x)|^p \, dx,$$

$$\int_X h(x)^s \, d'x = \int_X |f(x)g(x)|^r \, dx,$$

provided $s := q/(p+q)$. Problem 2.1.4 implies that

$$\int_X h(x)^s \, d'x \leq \left\{ \int_X 1 \, d'x \right\}^{1-s} \left\{ \int_X h(x) \, d'x \right\}^{s}.$$

This may be rewritten in the form

$$\int_X |f(x)g(x)|^r \, dx \le \left\{ \int_X |g(x)|^q \, dx \right\}^{p/(p+q)} \left\{ \int_E |f(x)|^p \, dx \right\}^{q/(p+q)},$$

which leads directly to the statement of the theorem if one uses $r = pq/(p+q)$. □

Both of the above theorems are still valid if the conditions $p, q, r \ge 1$ are relaxed to $p, q, r > 0$. The stronger assumptions are present for the following reason.

Problem 2.1.14 Prove that if X contains two disjoint sets with positive finite measures then

$$\|f\|_p := \left\{ \int_X |f(x)|^p \, dx \right\}^{1/p}$$

is not a norm if $0 < p < 1$. Prove also that

$$d(f, g) := \int_X |f(x) - g(x)|^p \, dx$$

is a metric. □

We end the section with a few results from the geometry of Banach spaces, with particular reference to $L^p(X, dx)$.

Theorem 2.1.15 *(James)*[2] *The Banach space $\mathcal{B}$ is reflexive if and only if every $\phi \in \mathcal{B}^*$ achieves its norm, i.e. there exists $x \in \mathcal{B}$ such that $\|x\| = 1$ and $\phi(x) = \|\phi\|$.*

Problem 2.1.16 Prove that the Banach spaces $l^1(\mathbf{Z})$ and $C([0, 1])$ are not reflexive. □

Theorem 2.1.17 *(Clarkson)*[3] *If $2 \le p < \infty$ and $f, g \in L^p(X, dx)$ then*

$$\|f + g\|_p^p + \|f - g\|_p^p \le 2^{p-1} \left(\|f\|_p^p + \|g\|_p^p \right). \tag{2.3}$$

---

[2] See [Diestel 1975, Chap. 1] for several proofs of James's theorem and references to even more. Another proof may be found in [Megginson 1998, Sect. 1.13]. For separable Banach spaces the proof in [Nygaard 2005] is comparatively simple.

[3] (2.3) is just one of several inequalities due to Clarkson. They are closely related to the concept of uniform convexity of Banach spaces, a subject with many ramifications, which we touch on in Problem 2.1.18.

Proof. We prove that

$$|u + v|^p + |u - v|^p \leq 2^{p-1} (|u|^p + |v|^p) \qquad (2.4)$$

for all $u, v \in \mathbf{C}$. This yields (2.3) by putting $u := f(x)$, $v := g(x)$ and integrating over X. The bound (2.4) can be rewritten in the form

$$\|Aw\|_p \leq 2^{1-1/p} \|w\|_p$$

for all $w := (u, v)' \in \mathbf{C}^2$ where

$$A := \begin{pmatrix} 1 & 1 \\ 1 & -1 \end{pmatrix}.$$

If $\|A\|_p$ denotes the norm of A regarded as an operator acting on $\mathbf{C}^2$ provided with the L^p norm, then the bound $\|A\|_p \leq 2^{1-1/p}$ is an immediate corollary of the Riesz-Thorin interpolation Theorem 2.2.14 – all the L^p spaces involved being two-dimensional. The identity $\|A\|_2 = \sqrt{2}$ is obtained by observing that the eigenvalues of the self-adjoint matrix A are $\pm\sqrt{2}$, while $\|A\|_\infty = 2$ is entirely elementary. $\qquad\square$

Problem 2.1.18 [4] Prove that $L^p(X, \mathrm{d}x)$ is uniformly convex for $2 \leq p < \infty$ in the following sense. For every $\varepsilon > 0$ there exists $\delta_\varepsilon > 0$ such that if $\|f\|_p = 1$, $\|g\|_p = 1$ and $\|f + g\|_p > 2 - \delta_\varepsilon$ then $\|f - g\|_p < \varepsilon$. $\qquad\square$

Lemma 2.1.19 *If $1 < p < \infty$, $s \in \mathbf{R}$ and $f, h \in L^p(X, \mathrm{d}x)$ then*

$$\frac{\mathrm{d}}{\mathrm{d}s} \|f + sh\|_p^p = p \operatorname{Re} \left\{ \int_X h(x) g_s(x) \, \mathrm{d}x \right\}$$

where $g_s \in L^q(X, \mathrm{d}x)$ is defined by

$$g_s(x) := \overline{f(x) + sh(x)} \, |f(x) + sh(x)|^{p-2}$$

and is a norm continuous function of s.

Proof. A direct calculation shows that

$$\frac{\mathrm{d}}{\mathrm{d}s} |f(x) + sh(x)|^p = p \operatorname{Re} \{h(x) g_s(x)\}$$

for all $x \in X$ and $s \in \mathbf{R}$. Therefore

$$|f(x) + sh(x)|^p = |f(x)|^p + p \int_{u=0}^s \operatorname{Re} \{h(x) g_u(x)\} \, \mathrm{d}u.$$

---

[4] The space $L^p(X, \mathrm{d}x)$ is also uniformly convex for all $1 < p < 2$, but the proof uses a different Clarkson inequality.

Integrating both sides with respect to x yields

$$\|f + sh\|_p^p = \|f\|_p^p + p \int_{u=0}^{s} \int_X \mathrm{Re}\,\{h(x)g_u(x)\}\,\mathrm{d}x\mathrm{d}u. \tag{2.5}$$

The interchange of the order of integration is justified by the bound

$$|h(x)g_u(x)| \leq |h(x)|\,(|f(x)| + S|h(x)|)^{p-1}$$

valid for all $x \in X$ and all $u \in [-S, S]$; the right-hand side lies in $L^1(X \times [-S, S])$. Differentiating (2.5) with respect to s yields the statement of the lemma. $\square$

Theorem 2.1.20 [5] *If $1 \leq p < \infty$ then the Banach dual space of $L^p(X, \mathrm{d}x)$ is isometrically isomorphic to $L^q(X, \mathrm{d}x)$ where $1/p + 1/q = 1$. The functional $\phi \in L^p(X, \mathrm{d}x)^*$ corresponds to the function $g \in L^q(X, \mathrm{d}x)$ according to the formula*

$$\phi(f) := \int_X f(x)g(x)\,\mathrm{d}x. \tag{2.6}$$

Proof. We start by considering the case in which $1 < p \leq 2$ and X has finite measure. Since L^2 is continuously embedded in L^p the Riesz representation theorem for Hilbert spaces implies that for any $\phi \in (L^p)^*$ there exists $g \in L^2$ such that (2.6) holds for all $f \in L^2$. If we put

$$g_n(x) := \begin{cases} g(x) & \text{if } |g(x)| \leq n, \\ 0 & \text{otherwise,} \end{cases}$$

and $f_n := \overline{g_n}|g_n|^{q-2}$ then f_n and g_n are both bounded and hence lie in all L^r spaces. Moreover

$$\|g_n\|_q^q = \int_X f_n(x)g(x)\,\mathrm{d}x = \phi(f_n) \leq \|f_n\|_p\|\phi\| = \|g_n\|_q^{q/p}\|\phi\|.$$

Therefore $\|g_n\|_q \leq \|\phi\|$ for all n. Letting $n \to \infty$ we deduce that $\|g\|_q \leq \|\phi\|$. Since $L^2(X, \mathrm{d}x)$ is dense in $L^p(X, \mathrm{d}x)$, an approximation argument implies that (2.6) holds for all $f \in L^p(X, \mathrm{d}x)$. An application of Lemma 2.1.6 finally proves that $\|g\|_q = \|\phi\|$.

If $1 < p \leq 2$ but X has infinite measure then we write X as a disjoint union of sets E_n, each of which has finite measure. By the first part of the proof, for each n there exists $g_n \in L^q(E_n, \mathrm{d}x)$ such that (2.6) holds for all

---

[5] The proof of this theorem in Dunford and Schwartz 1966, Theorems IV.8.1 and IV.8.5 is based on the Radon-Nikodym theorem. Some authors, such as [Carleson 1966, Kuttler 1997, Lieb and Loss 1997,] make use of uniform convexity. Our proof uses the former method for $1 \leq p \leq 2$ and the latter for $2 \leq p < \infty$.

$f \in L^p(X, dx)$ with support in E_n. We now concatenate the g_n to produce a function $g : X \to \mathbf{C}$. If $X_n := \bigcup_{r=1}^n E_r$ then the restriction g_n of g to X_n lies in $L^q(X_n, dx)$ and (2.6) holds for all $f \in L^p(X, dx)$ with support in X_n. Moreover $\|g_n\|_q \le \|\phi\|$ for all n. By letting $n \to \infty$ we conclude that $g \in L^q(X, dx)$ and $\|g\|_q = \|\phi\|$.

If $p = 1$ then a straightforward modification of the above argument yields the identity $(L^1)^* = L^\infty$.

The above argument establishes that if $1 < p \le 2$ every $\phi \in (L^p)^*$ is of the form $\phi(f) := \int_X f(x)g(x)\,dx$ for some $g \in L^q$. ϕ therefore achieves its norm at $f := \overline{g}g^{q-2}/\|\overline{g}g^{q-2}\|_p \in L^p$. Theorem 2.1.15 now implies that L^p is reflexive, so $(L^q)^* = L^p$ for all $2 \le q < \infty$.

Our second proof of the result proved in the last paragraph is longer but more elementary. Let $2 \le p < \infty$ and let $\phi \in (L^p)^*$ satisfy $\|\phi\| = 1$. There exists a sequence $f_n \in L^p$ such that $\|f_n\|_p = 1$ and $\phi(f_n) \to 1$ as $n \to \infty$. Since $\phi(f_m + f_n) \to 2$ as $m, n \to \infty$ we deduce that $\|f_m + f_n\|_p \to 2$. The uniform convexity of L^p implies that f_n converges to a limit $f \in L^p$ such that $\|f\|_p = 1$ and $\phi(f) = 1$; see Problem 2.1.18. If $g := \overline{f}|f|^{p-2}$ then $\|g\|_q = 1$ and the functional $\psi(h) := \int_X h(x)g(x)\,dx$ on L^p satisfies $\|\psi\| = 1$ and $\psi(f) = 1$. If we show that $\phi = \psi$, then it follows that $(L^p)^* = L^q$.

It suffices to prove that $\mathrm{Ker}(\phi) \subseteq \mathrm{Ker}(\psi)$, because both kernels have co-dimension 1. If $\phi(h) = 0$ then $\phi(f + sh) = 1$ for all $s \in \mathbf{R}$. Therefore $\|f + sh\|_p \ge 1$ for all $s \in \mathbf{R}$. The function $F(s) := \|f + sh\|_p^p$ is differentiable by Lemma 2.1.19 and takes its minimum value at $s = 0$, so $F'(0) = 0$. Lemma 2.1.19 also yields

$$\mathrm{Re}\left\{ \int_X h(x)g(x)\,dx \right\} = 0.$$

Repeating the above argument with h replaced by ih we deduce that

$$\int_X h(x)g(x)\,dx = 0,$$

so $h \in \mathrm{Ker}(\psi)$. $\square$

Problem 2.1.21 Give an elementary, ab initio proof that the dual of $l^p(X)$ is isometrically isomorphic to $l^q(X)$ for all $1 \le p < \infty$, where $1/p + 1/q = 1$. $\square$

2.2 Operators acting on L^p spaces

In this section we prove the boundedness of various operators acting between L^p spaces. We start by considering multiplication operators, whose spectrum is easy to describe. They are of central importance in our treatment of the spectral theorem for normal and self-adjoint operators in Section 5.4.

Problem 2.2.1 If m is a measurable function on X, we say that z lies in the essential range of m if $\{x : |m(x) - z| < \varepsilon\}$ has positive measure for all $\varepsilon > 0$. Equivalently z is not in the essential range if $(m(x) - z)^{-1}$ is a bounded function of x on X, possibly after alteration on a null set. Prove that if m is a bounded, measurable function on X and the multiplication operator $M : L^p(X, \mathrm{d}x) \to L^p(X, \mathrm{d}x)$ is defined by $(Mf)(x) := m(x)f(x)$, where $1 \leq p < \infty$, then $\mathrm{Spec}(M)$ equals the essential range of m. Prove also that if $m : \mathbf{R}^N \to \mathbf{C}$ is a bounded, continuous function, then $\mathrm{Spec}(M)$ is the closure of $\{m(x) : x \in \mathbf{R}^N\}$. $\quad\square$

In the theorems below we will not labour the obvious requirement that all integral kernels must be measurable. The following lemma characterizes Hilbert-Schmidt operators.

Lemma 2.2.2 *If $K \in L^2(X \times X)$ then the formula*

$$(Af)(x) := \int_X K(x, y)f(y)\,\mathrm{d}y$$

defines a bounded linear operator on $L^2(X)$ satisfying $\|A\| \leq \|K\|_2$.

Proof. If $f \in L^2(X)$ then

$$|(Af)(x)|^2 \leq \left\{ \int_X |K(x, y)f(y)|\,\mathrm{d}y \right\}^2$$

$$\leq \int_X |K(x, y)|^2\,\mathrm{d}y \int_X |f(y)|^2\,\mathrm{d}y.$$

Therefore

$$\|Af\|_2^2 \leq \|K\|_2^2 \|f\|_2^2$$

and the lemma follows. $\quad\square$

We call

$$\|A\|_2 := \left\{ \int_{X \times X} |K(x, y)|^2\,\mathrm{d}x\mathrm{d}y \right\}^{1/2}$$

the Hilbert-Schmidt or Frobenius norm of A. The notation $\| \cdot \|_{\mathrm{HS}}$ is also used.

Problem 2.2.3 Prove that the Hilbert-Schmidt norm

$$\|A\|_2 := \Big\{ \sum_{r,s=1}^{n} |A_{r,s}|^2 \Big\}^{1/2}$$

of an $n \times n$ matrix A and its operator norm

$$\|A\| := \sup\{\|Av\|/\|v\| : 0 \neq v \in \mathbf{C}^n\}$$

are related by

$$\|A\| \leq \|A\|_2 \leq n^{1/2}\|A\|. \qquad \qquad \square$$

Problem 2.2.4 Prove that if the Hilbert-Schmidt operators A_n converge in the weak operator topology to A and $\|A_n\|_2 \leq c < \infty$ for all n then A is Hilbert-Schmidt and $\|A\|_2 \leq c$. $\quad\square$

Lemma 2.2.2 is of rather limited application, because every Hilbert-Schmidt operator is compact for reasons explained in Theorem 4.2.16. Our next series of results depend upon proving boundedness in L^1 and L^∞ first, and then interpolating. As a consequence they prove that the relevant operators are bounded on *all* L^p spaces.

Theorem 2.2.5 *If $K : X \times X \to \mathbf{C}$ is measurable and*

$$c_1 := \operatorname*{ess-sup}_{y \in X} \int_X |K(x, y)|\, \mathrm{d}x < \infty$$

then the formula

$$(Af)(x) := \int_X K(x, y) f(y)\, \mathrm{d}y$$

defines a bounded linear operator on L^1 with $\|A\| = c_1$.

This is a special case of the following more general theorem.

Theorem 2.2.6 *If $\mathcal{B}$ is a separable Banach space and the measurable function $K : X \to \mathcal{B}$ satisfies*

$$c := \operatorname*{ess-sup}_{y \in X} \|K(y)\| < \infty$$

then the formula

$$Af := \int_X K(y) f(y)\, \mathrm{d}y$$

defines a bounded linear operator from L^1 to $\mathcal{B}$ with $\|A\| = c$.

Proof. If c is finite then

$$\|Af\| \leq \int_X \|K(y)f(y)\| \, dy$$

$$\leq \int_X c|f(y)| \, dy$$

$$= c\|f\|_1$$

for all $f \in L^1(X)$. Therefore $\|A\| \leq c$.

In the reverse direction if ϕ lies in the unit ball S of $\mathcal{B}^*$ then the bound

$$\left| \int_X \langle K(x), \phi \rangle f(x) \, dx \right| = |\langle Af, \phi \rangle| \leq \|A\| \, \|f\|_1$$

implies that

$$|\langle K(x), \phi \rangle| \leq \|A\|$$

for all x not in a certain null set N_ϕ. If $\{\phi_n\}_{n=1}^\infty$ is a weak* dense sequence in S and $N := \bigcup_{n=1}^\infty N_{\phi_n}$ then N is a null set and

$$|\langle K(x), \phi_n \rangle| \leq \|A\|$$

for all n and all $x \notin N$. A density argument now implies that

$$|\langle K(x), \phi \rangle| \leq \|A\|$$

for all $\phi \in S$ and all $x \notin N$. The Hahn-Banach theorem finally implies that $c \leq \|A\|$. $\square$

In spite of Theorem 2.2.5, it is not the case that every bounded operator on $L^1(X, dx)$ has an integral kernel (consider the identity operator). Nevertheless some results of this type do exist.

Theorem 2.2.7 *The formula*

$$(Af)(x) := \int_X K(x, y)f(y) \, dy$$

establishes a one-one correspondence between bounded linear operators

$$A : L^1(X, dx) \to L^\infty(X, dx)$$

and $K \in L^\infty(X \times X, d^2x)$. Moreover $\|A\| = \|K\|_\infty$.

Proof. If A has a bounded integral kernel then the final statement follows from Theorem 2.2.6 with $\mathcal{B} = L^\infty(X, dx)$, so we only have to prove that every bounded operator A between the stated spaces possesses a suitable kernel.

Let $\mathcal{E} = \{E_1, E_2, \ldots, E_n\}$ be a finite sequence of disjoint subsets of X with positive and finite measures. For each $r \in \{1, \ldots, n\}$ let χ_{E_r} and $|E_r|$ denote the characteristic function and measure respectively of E_r. Let

$$K_{\mathcal{E}}(x, y) = \sum_{r=1}^{n} A(\chi_{E_r})(x)\chi_{E_r}(y)|E_r|^{-1}.$$

It is easy to verify that

$$(Af)(x) = \int_X K_{\mathcal{E}}(x, y)f(y)\,dy$$

for all $f \in L_{\mathcal{E}} := \lim\{\chi_r : 1 \leq r \leq n\}$. Moreover $|K_{\mathcal{E}}(x, y)| \leq \|A\|$ for all $x, y \in X$.

Given two such sequences $\mathcal{E}$ and $\mathcal{F}$ we write $\mathcal{E} \leq \mathcal{F}$ if every set in $\mathcal{E}$ is the union of one or more sets in $\mathcal{F}$, or equivalently if $L_{\mathcal{E}} \subseteq L_{\mathcal{F}}$. Let $\mathcal{E}_n$ be an increasing sequence for which $\cup_{n=1}^{\infty} L_n$ is norm dense in $L^1(X)$, where henceforth the subscript n stands in for $\mathcal{E}_n$. Since K_n lie in the weak* compact ball $\{\phi \in L^{\infty}(X \times X) : \|\phi\|_{\infty} \leq \|A\|\}$, there is a subsequence which converges in the weak* topology to $K \in L^{\infty}(X \times X)$. Henceforth the letter n refers to terms in this subsequence.

If $f \in L_n$ and $g \in L^1(X)$ then

$$\langle Af, g \rangle = \int_{X \times X} K_n(x, y)f(y)g(x)\,dxdy$$

for all large enough n so

$$\langle Af, g \rangle = \int_{X \times X} K(x, y)f(y)g(x)\,dxdy.$$

Since n and g are arbitrary we deduce that

$$(Af)(x) = \int_X K(x, y)f(y)\,dy$$

for all $f \in \cup_{n=1}^{\infty} L_n$, and by density also for all $f \in L^1(X)$. $\square$

Theorem 2.2.8 *If*

$$c_{\infty} := \underset{x \in X}{\text{ess-sup}} \int_X |K(x, y)|\,dy < \infty$$

then the formula

$$(Af)(x) := \int_X K(x, y)f(y)\,dy$$

defines a bounded linear operator on L^{∞} with $\|A\| = c_{\infty}$.

Proof. Every such operator is the dual of an operator on $L^1(X, dx)$ with the kernel $K(y, x)$, to which we can apply Theorem 2.2.5. $\square$

The following theorems consider operators which act on several L^p spaces simultaneously. We need to formulate some general concepts before discussing this. We say that two Banach spaces $\mathcal{B}_1$ and $\mathcal{B}_2$ or their associated norms are compatible if $\mathcal{B} = \mathcal{B}_1 \cap \mathcal{B}_2$ is dense in each of them, and the following condition is satisfied. If $f_n \in \mathcal{B}$, $\|f_n - f\|_1 \to 0$ and $\|f_n - g\|_2 \to 0$ as $n \to \infty$ then $f = g \in \mathcal{B}$. Equivalently $\mathcal{B}$ is complete with respect to the norm

$$\|f\| := \|f\|_1 + \|f\|_2.$$

Problem 2.2.9 Prove that the spaces $L^p(X, dx)$ are compatible as p varies. $\square$

In the above context two bounded operators $A_i : \mathcal{B}_i \to \mathcal{C}_i$ are said to be consistent if $A_1 f = A_2 f$ for all $f \in \mathcal{B}_1 \cap \mathcal{B}_2$.

Problem 2.2.10 Prove that if the bounded operators $A_i : \mathcal{B}_i \to \mathcal{C}_i$ are consistent for $i = 1, 2$, then $R(z, A_1)$ and $R(z, A_2)$ are consistent for all z in the unbounded component U of

$$W := \mathbf{C} \setminus \{\mathrm{Spec}(A_1) \cup \mathrm{Spec}(A_2)\}. \qquad \square$$

Before continuing we give a word of warning about the possible p-dependence of the L^p spectrum of an operator.

Example 2.2.11 The operator $(Af)(x) := f(x + 1)$ is an invertible isometry on $L^p(\mathbf{R}, dx)$ for all $p \in [1, \infty]$. However, if we denote the 'same' operator acting on $\mathcal{B}_p := L^p(\mathbf{R}, e^{-|x|}dx)$ for $1 \le p < \infty$ by A_p, the situation changes. Direct calculations show that $\|A_p^{\pm}\| = e^{1/p}$. Defining $f_z(x) := e^{zx}$, we see that $f_z \in \mathcal{B}_p$ if $|\mathrm{Re}(z)| < 1/p$ and $A_p f_z = e^z f_z$. These facts imply that

$$\mathrm{Spec}(A_p) = \{z : e^{-1/p} \le |z| \le e^{1/p}\}. \qquad \square$$

Our next two lemmas are needed for the proof of Theorem 2.2.14.

Lemma 2.2.12 *(Three Lines Lemma) Let $S := \{z \in \mathbf{C} : 0 \le \mathrm{Re}(z) \le 1\}$. Let F be a continuous bounded function on S which is analytic in the interior of S. If*

$$c_\lambda := \sup\{|F(\lambda + iy)| : y \in \mathbf{R}\}$$

for $0 \leq \lambda \leq 1$ then

$$c_\lambda \leq c_0^{1-\lambda} c_1^\lambda.$$

Proof. Apply the maximum principle to

$$G_\varepsilon(z) := F(z) e^{\alpha z + \beta} (1 + \varepsilon z)^{-1}$$

for all $\varepsilon > 0$, where $\alpha, \beta \in \mathbf{R}$ are determined by

$$e^{-\beta} = c_0, \qquad e^{-\alpha-\beta} = c_1. \qquad \qquad \square$$

Let L^p denote the space $L^p(X, dx)$. We write $f \in \mathcal{M}$ if

$$f(x) = \sum_{r=1}^n \alpha_r \chi_{E_r}(x)$$

where $\alpha_r \in \mathbf{C}$ and $\{E_r\}_{r=1}^n$ is any (f-dependent) family of disjoint sets with finite positive measures. $\mathcal{M}$ is a linear subspace of L^p for all $1 \leq p \leq \infty$, and it is norm (resp. weak*) dense in L^p if $1 \leq p < \infty$ (resp. $p = \infty$).

Lemma 2.2.13 *Given $f \in \mathcal{M}$, $\lambda \in (0, 1)$, $p_1, p_2 \in [1, \infty)$ and $1/p := (1 - \lambda)/p_0 + \lambda/p_1$. Then there exist $\alpha, \beta \in \mathbf{R}$ such that the analytic family of functions*

$$f_z(x) := |f(x)|^{\alpha z + \beta - 1} f(x),$$

which all lie in $\mathcal{M}$, satisfy $f_\lambda = f$ and

$$\|f_{iy}\|_{p_0}^{p_0} = \|f\|_p^p = \|f_{1+iy}\|_{p_1}^{p_1}$$

for all $y \in \mathbf{R}$.

Proof. One may verify directly that the choices

$$\alpha := \frac{p}{p_1} - \frac{p}{p_0}, \qquad \beta := \frac{p}{p_0}$$

lead to the claimed result. $\square$

The following interpolation theorem is used throughout the book.

Theorem 2.2.14 *(Riesz-Thorin) Let $1 \leq p_0, p_1, q_0, q_1 \leq \infty$ and $0 < \lambda < 1$, and define p, q by*

$$\frac{1}{p} := \frac{1-\lambda}{p_0} + \frac{\lambda}{p_1},$$

$$\frac{1}{q} := \frac{1-\lambda}{q_0} + \frac{\lambda}{q_1}.$$

Let A be a linear map from $L^{p_0} \cap L^{p_1}$ to $L^{q_0} \cap L^{q_1}$. If

$$\|Af\|_{q_0} \leq c_0 \|f\|_{p_0},$$

$$\|Af\|_{q_1} \leq c_1 \|f\|_{p_1}$$

for all $f \in L^{p_0} \cap L^{p_1}$, then A can be extended to a bounded linear operator from L^p to L^q with norm at most $c_0^{1-\lambda} c_1^\lambda$.

Proof. We only treat the case in which the exponents are all finite. The other cases are similar. Let r, r_0, r_1 be the indices conjugate to q, q_0, q_1 respectively, so that

$$\frac{1}{r} = \frac{1-\lambda}{r_0} + \frac{\lambda}{r_1}.$$

Given $f \in \mathcal{M}$, let $f_z \in \mathcal{M}$ be constructed using Lemma 2.2.13. Given $g \in \mathcal{M}$, use an analogous procedure to construct $g_z \in \mathcal{M}$ which satisfy $g_\lambda = g$ and

$$\|g_{iy}\|_{r_0}^{r_0} = \|g\|_r^r = \|g_{1+iy}\|_{r_1}^{r_1}$$

for all $y \in \mathbf{R}$.

Now consider the analytic function

$$F(z) := \langle Af_z, g_{\bar{z}} \rangle.$$

Since

$$|F(iy)| \leq c_0 \|f\|_p^{p/p_0} \|g\|_r^{r/r_0},$$

$$|F(1+iy)| \leq c_1 \|f\|_p^{p/p_1} \|g\|_r^{r/r_1},$$

for all $y \in \mathbf{R}$, we deduce using the Three Lines Lemma that

$$|F(\lambda)| \leq c_0^{1-\lambda} c_1^\lambda \|f\|_p \|g\|_r.$$

Since g is arbitrary subject to $g \in \mathcal{M}$, we deduce that

$$\|Af\|_q \leq c_0^{1-\lambda} c_1^\lambda \|f\|_p$$

for all $f \in \mathcal{M}$. Density arguments now imply the same bound for all $f \in L^{p_0} \cap L^{p_1}$, and then for all $f \in L^p$. $\square$

Corollary 2.2.15 *If A is defined on $L^1(X) \cap L^\infty(X)$ by*

$$(Af)(x) := \int_X K(x, y) f(y) \, dy$$

and

$$c_1 := \text{ess-sup}_{y \in X} \int_X |K(x, y)| \, dx < \infty,$$

$$c_\infty := \text{ess-sup}_{x \in X} \int_X |K(x, y)| \, dy < \infty,$$

then A extends to a bounded operator on $L^2(X)$ satisfying

$$\|A\|_2 \leq \sqrt{c_1 c_\infty}.$$

Proof. Interpolate between the operator bounds of Theorems 2.2.5 and 2.2.8. $\square$

Problem 2.2.16 Give an elementary proof of the above corollary, using only Schwarz's lemma. $\square$

Problem 2.2.17 Let $A : L^p(0, 1) \to L^p(0, 1)$ be defined by

$$(Af)(x) := \int_0^1 K(x, y) f(y) \, dy$$

where

$$K(x, y) := \begin{cases} 1 & \text{if } x + y \leq 1, \\ 0 & \text{otherwise.} \end{cases}$$

Prove that $\|A\|_1 = \|A\|_\infty = 1$ but $\|A\|_2 = 2/\pi$. Hence interpolation does not give the sharp value of the norm of A on $L^2(0, 1)$. $\square$

Problem 2.2.18 Let A be a bounded self-adjoint operator on $L^2(X, dx)$ and suppose that $\|Af\|_\infty \leq c\|f\|_\infty$ for all $f \in L^2(X, dx) \cap L^\infty(X)$. Prove that for every $p \in [1, \infty]$, A extends from $L^2(X, dx) \cap L^p(X, dx)$ to a bounded linear operator A_p on $L^p(X, dx)$ with $\|A_p\| \leq c$. $\square$

Corollary 2.2.19 *If X is a measurable subset of $\mathbf{R}^N$ and A is defined on $L^1(X) \cap L^\infty(X)$ by*

$$(Af)(x) := \int_X K(x, y) f(y) \, d^N y,$$

where $|K(x, y)| \leq k(x - y)$ for all $x, y \in X$ and some $k \in L^1(\mathbf{R}^N)$, then A extends to a bounded operator on $L^p(X)$ satisfying $\|A\|_p \leq \|k\|_1$ for every p satisfying $1 \leq p < \infty$.

Given $a \in L^1(\mathbf{R}^N)$ and $f \in L^p(\mathbf{R}^N)$, where $1 \leq p < \infty$, the convolution

$$(a * f)(x) := \int_{\mathbf{R}^N} a(x - y) f(y) \, d^N y$$

of a and f lies in $L^p(\mathbf{R}^N)$ by Corollary 2.2.19, and the convolution transform $T_a(f) := a * f$ is a bounded linear operator on $L^p(\mathbf{R}^N)$ with $\|T_a\| \leq \|a\|_1$. One may verify directly that the Banach space $L^1(\mathbf{R}^N)$ becomes a commutative Banach algebra under the convolution multiplication.

The above definitions can all be adapted in an obvious way if $\mathbf{R}^N$ is replaced by $\mathbf{Z}^N$ or $[-\pi, \pi]^N_{\text{per}}$. Indeed they apply to any locally compact abelian group, if one integrates with respect to the translation invariant Haar measure. A convolution operator on $l^2(\mathbf{Z})$ is often called a Laurent operator, and is associated with an infinite matrix which is constant on diagonals, as in

$$
\begin{pmatrix}
\ddots & \ddots & \ddots & \ddots & & & \\
\ddots & c & d & e & & & \\
\ddots & b & c & d & e & & \\
 & a & b & c & d & e & \\
 & & a & b & c & d & e \\
 & & & a & b & c & d & \ddots \\
 & & & & a & b & c & \ddots \\
 & & & & & \ddots & \ddots & \ddots
\end{pmatrix}.
$$

In this book we will be concerned with the mathematical applications of convolutions, but the convolution transform is also of major importance in applied science, for example image processing. If we neglect the important contribution of noise, the perceived image g is often taken to be of the form $g := a * f$, where f is the true image and a is a known function representing the degradation of the image by the camera optics. Reconstructing the true image from the perceived one amounts to inverting the convolution transform. Unfortunately this is an ill-posed inverse problem – by Theorem 3.1.19 the inverse operator is always unbounded if it exists. In order to approximate the inverse it is normal to adopt a variational approach, in which the quantity to be minimized incorporates some expectations about the nature of the image. Several such deconvolution algorithms exist.

If one takes the finite resolution of the instrument into account by replacing $\mathbf{R}^N$ by a finite set, the operator T_a becomes a matrix. If this is one-one then it is invertible, but the norm of the inverse will normally be very large, so the problem is still ill-posed in a numerical sense.

2.3 Approximation and regularization

One frequently needs to approximate functions in $L^p(\mathbf{R}^N)$ and similar spaces by a sequence of more regular functions. The following methods of doing so are often used in combination.

Lemma 2.3.1 *If* $1 \le p < \infty$, $f \in L^p(\mathbf{R}^N)$ *and*

$$f_n(x) := \begin{cases} 0 & \text{if } |x| \ge n \text{ or if } |f(x)| \ge n, \\ f(x) & \text{otherwise,} \end{cases}$$

then $\lim_{n \to \infty} \|f_n - f\|_p = 0$, *and each* f_n *is bounded with bounded support.*

We write $f \in C_c^\infty(U)$ when f is smooth (i.e. infinitely differentiable) and has compact support contained in the region $U \subseteq \mathbf{R}^N$. The existence of many such functions starts with the observation that

$$\psi(s) := \begin{cases} e^{-1/s} & \text{if } s > 0, \\ 0 & \text{if } s \le 0 \end{cases} \tag{2.7}$$

is a smooth function on $\mathbf{R}$. The function

$$\xi(s) := \frac{\psi(1-s)}{\psi(1-s) + \psi(s)}$$

is also smooth. It equals 0 if $s \ge 1$ and 1 if $s \le 0$.

Finally

$$\phi(x) := \xi(|x| - 1) \tag{2.8}$$

is a smooth function on $\mathbf{R}^N$. It equals 1 if $|x| \le 1$ and 0 if $|x| \ge 2$. For all other $x \in \mathbf{R}^N$ its value lies in $(0, 1)$.

Problem 2.3.2 Prove that if f, g are continuous functions on $\mathbf{R}$, f is differentiable on $\mathbf{R} \backslash \{0\}$ and $f'(x) = g(x)$ for all $x \ne 0$, then f is also differentiable at 0 with $f'(0) = g(0)$. Use this to deduce that the function ψ defined in (2.7) is smooth. $\square$

Problem 2.3.3 Prove that if $1 \le p < \infty$, $f \in L^p(\mathbf{R}^N)$ and

$$f_n(x) := f(x)\phi(x/n),$$

then $\lim_{n \to \infty} \|f_n - f\|_p = 0$. $\square$

Theorem 2.3.4 *Let*

$$\phi_n(x) := c^{-1} n^N \phi(nx)$$

where ϕ is defined by (2.8) and $c := \int_{\mathbf{R}^N} \phi(y) d^N y$. If $f \in L^p(\mathbf{R}^N)$ and $f_n := \phi_n * f$, then f_n is a smooth function and

$$\lim_{n \to \infty} \|f_n - f\|_p = 0. \tag{2.9}$$

If S is the support of f then the support of f_n is contained in $S + B(0, 2/n)$.

Proof. The smoothness of f_n follows by differentiating the convolution formula

$$(f * \phi_n)(x) = \int_{\mathbf{R}^N} f(y) \phi_n(x - y) \, d^N y$$

under the integral sign repeatedly. This is justified by standard methods. If f is a continuous function of compact support then f_n converges uniformly to f and the supports of f_n are uniformly bounded. This implies (2.9) for such functions. Its general validity follows by density arguments, making use of the bound

$$\|f_n\|_p \leq \|\phi_n\|_1 \|f\|_p = \|f\|_p. \qquad \square$$

Corollary 2.3.5 $C_c^\infty(U)$ *is norm dense in* $L^p(U)$ *for any region* $U \subseteq \mathbf{R}^N$ *and any* $1 \leq p < \infty$.

Proof. The first step is to approximate $f \in L^p(U)$ by f_n where

$$f_n(x) := \begin{cases} f(x) \text{ if } \operatorname{dist}(x, \mathbf{R}^N \backslash U) \geq 1/n \text{ and } |x| \leq n, \\ 0 \quad \text{otherwise.} \end{cases}$$

We may then apply the convolution procedure of Theorem 2.3.4 to approximate f_n. $\square$

Corollary 2.3.6 *If* $f \in L^{p_1}(\mathbf{R}^N) \cap L^{p_2}(\mathbf{R}^N)$ *where* $1 \leq p_1 < p_2 < \infty$, *then there exists a sequence* $f_n \in C_c^\infty(\mathbf{R}^N)$ *such that*

$$\lim_{n \to \infty} \|f_n - f\|_{p_i} = 0$$

simultaneously for $i = 1, 2$.

Proof. Given n we first apply the procedure of Lemma 2.3.1 to construct $g_n \in L_c^\infty$ such that $\|g_n - f\|_{p_i} < 1/(2n)$ for $i = 1, 2$. We then use the procedure of Theorem 2.3.4 to construct $f_n \in C_c^\infty$ such that $\|f_n - g_n\|_{p_i} < 1/(2n)$ for $i = 1, 2$. $\square$

Problem 2.3.7 In Theorem 2.3.4 replace the stated choice of ϕ by

$$\phi(x) = e^{-|x|^2}.$$

Prove that every statement of the theorem remains true except the final one. What can you say about the decay of $f_n(x)$ and its derivatives as $|x| \to \infty$ if f has compact support?   □

One can also develop approximation procedures in the periodic context. In one dimension we say that k is a trigonometric polynomial of order n on $[-\pi, \pi]$ if

$$k(\theta) := \sum_{r=-n}^{n} c_r e^{ir\theta}$$

$$= a_0 + \sum_{r=1}^{n} \{a_r \cos(r\theta) + b_r \sin(r\theta)\}$$

for suitable complex coefficients.

Lemma 2.3.8 *There exist trigonometric polynomials k_n of order n such that $k_n(\theta) = k_n(-\theta) \geq 0$ for all θ, $\int_{-\pi}^{\pi} k_n(\theta)\, d\theta = 1$ for all n, and*

$$\lim_{n \to \infty} \int_{|\theta| > \delta} k_n(\theta)\, d\theta = 0$$

for all $\delta \in (0, \pi]$.

Proof. We put

$$k_n(\theta) := c_n^{-1} (1 + \cos(\theta))^n \tag{2.10}$$

where

$$c_n := \int_{-\pi}^{\pi} (1 + \cos(\theta))^n\, d\theta \tag{2.11}$$

$$= 2 \int_0^{\pi} (1 + \cos(\theta))^n\, d\theta$$

$$\geq 2 \int_0^{\pi} (1 + \cos(\theta))^n \sin(\theta)\, d\theta$$

$$= 2 \int_{-1}^{1} (1 + x)^n\, dx$$

$$= \frac{2^{n+2}}{n+1}.$$

The fact that k_n is a trigonometric polynomial of order n follows by rewriting (2.10) in terms of complex exponentials and expanding. Finally, given $\delta > 0$, we have

$$\int_{|\theta|>\delta} k_n(\theta)\,d\theta \leq (2\pi - 2\delta)k_n(\delta)$$

$$\leq 2\pi \frac{n+1}{2^{n+2}}(1+\cos(\delta))^n$$

$$\leq 2(n+1)\cos^{2n}(\delta/2),$$

which converges to zero as $n \to \infty$ for all δ in the stated range. $\square$

Problem 2.3.9 Find an explicit expression for the constant c_n in (2.11) by expanding $(1+\cos(\theta))^n$ as a linear combination of $e^{ir\theta}$ for $-n \leq r \leq n$. Use Stirling's formula to find the asymptotic form of c_n as $n \to \infty$. $\square$

Theorem 2.3.10 *If f is a continuous periodic function on $[-\pi, \pi]$ then $f_n := k_n * f$ are trigonometric polynomials and converge uniformly to f as $n \to \infty$.*

Proof. If we write k_n in the form

$$k_n(\theta) := \sum_{r=-n}^{n} c_r e^{ir\theta},$$

then the formula

$$f_n(\theta) := (k_n * f)(\theta) = \int_{-\pi}^{\pi} \sum_{r=-n}^{n} c_r e^{ir(\theta-\phi)} f(\phi)\,d\phi$$

establishes that f_n is a trigonometric polynomial of order at most n. The uniform convergence of f_n to f uses the alternative expression

$$f_n(\theta) = \int_{-\pi}^{\pi} f(\theta-\phi)k_n(\phi)\,d\phi.$$

One estimates the difference

$$f_n(\theta) - f(\theta) = \int_{-\pi}^{\pi} \{f(\theta-\phi) - f(\theta)\}k_n(\phi)\,d\phi$$

using the uniform continuity of f and Lemma 2.3.8. $\square$

Corollary 2.3.11 *The functions*

$$e_r(\theta) := \frac{e^{ir\theta}}{\sqrt{2\pi}},$$

where $r \in \mathbf{Z}$, form a complete orthonormal set in $L^2(-\pi, \pi)$. Hence

$$f = \lim_{n \to \infty} \sum_{r=-n}^{n} \langle f, e_r \rangle e_r \text{ and } \|f\|_2^2 = \sum_{r=-\infty}^{\infty} |\langle f, e_r \rangle|^2$$

for every $f \in L^2(-\pi, \pi)$.

Proof. A direct calculation verifies that they form an orthonormal set. Completeness is equivalent to their linear span being dense in $L^2(-\pi, \pi)$. This follows by combining Theorem 2.3.10 with the density of the continuous periodic functions in $L^2(-\pi, \pi)$, a fact that depends upon the manner of construction of the Lebesgue integral. □

Example 2.3.12 The Fourier series of a continuous periodic function on $[-\pi, \pi]$ converges uniformly to f under weak regularity assumptions, given in Theorem 3.3.10. If f has a jump discontinuity this cannot be true, and in fact the behaviour of the Fourier series near the discontinuity exhibits what is called the Gibbs phenomenon. It has recently been shown[6] that if f has a single discontinuity, and it is at $\pm\pi$, then there is no Gibbs phenomenon and one obtains substantially better convergence properties if one expands f in terms of the modified orthonormal basis

$$\phi_n(x) = \begin{cases} (2\pi)^{-1/2} & \text{if } n = 0, \\ \pi^{-1/2}\cos(nx) & \text{if } n \leq -1, \\ \pi^{-1/2}\sin((n-1/2)x) & \text{if } n \geq 1. \end{cases}$$

In particular if $f \in C^2[-\pi, \pi]$ then a direct calculation using integration by parts shows that the 'Fourier' coefficients of f with respect to this basis are of order n^{-2} as $n \to \pm\infty$. Therefore the modified Fourier series of f converges uniformly, even though no boundary conditions have been imposed on f at $\pm\pi$.

Many of the classical formulae for Fourier expansions have analogues in this context. It is worth noting that this modified Fourier basis is the set of all eigenvectors of the operator $(Hf)(x) := -f''(x)$ acting in $L^2(-\pi, \pi)$ subject to Neumann boundary conditions at $\pm\pi$. □

Problem 2.3.13 Prove the Riemann-Lebesgue lemma, which states that

$$\lim_{|n| \to \infty} \int_{-\pi}^{\pi} f(x)e^{-inx}\,dx = 0$$

for all $f \in L^1(-\pi, \pi)$. □

---

[6] See [Iserles and Nørsett 2006] for the details.

Problem 2.3.14 Use Theorem 2.3.10 and an approximation argument to prove that if $f \in L^1(-\pi, \pi)$ then $k_n * f \to f$ in L^1 norm as $n \to \infty$. Deduce that if $f \in L^1(-\pi, \pi)$ and

$$\int_{-\pi}^{\pi} f(\theta) e^{in\theta} \, d\theta = 0$$

for all $n \in \mathbf{Z}$, then $f = 0$ almost everywhere. □

Problem 2.3.15 Prove that if f and its first m derivatives are continuous and periodic on $[-\pi, \pi]$ then the rth derivative $f_n^{(r)}$ of f_n converges uniformly to $f^{(r)}$ as $n \to \infty$ for all $1 \le r \le m$. □

The following is the simplest and earliest of a family of related theorems. We give a constructive proof that can be extended to higher dimensions.

Theorem 2.3.16 (*Weierstrass*) *For every continuous function f on the interval $[-a, a]$ there exists a sequence of polynomials p_n that converge uniformly to f on $[-a, a]$.*

Proof. We extend f continuously to $\mathbf{R}$ by putting

$$\tilde{f}(x) := \begin{cases} f(x) & \text{if } |x| \le a, \\ f(ax/|x|)(2a - |x|)/a & \text{if } a < |x| < 2a, \\ 0 & \text{if } |x| \ge 2a. \end{cases}$$

From this point onwards we omit the tilde. Given $s > 0$ we define

$$f_s(x) := \int_{\mathbf{R}} s^{-1} k(y/s) f(x - y) \, dy$$

where

$$k(x) := \pi^{-1/2} e^{-x^2}.$$

We have

$$|f_s(x) - f(x)| = |\int_{\mathbf{R}} s^{-1} k(y/s)\{f(x - y) - f(x)\} \, dy|$$

$$= |\int_{\mathbf{R}} k(u)\{f(x - su) - f(x)\} \, du|$$

$$\le \int_{\mathbf{R}} k(u) \delta(su) \, du,$$

where

$$\delta(u) := \sup\{|f(x - u) - f(x)| : x \in \mathbf{R}\}$$

converges to 0 as $|u| \to 0$ by the uniform continuity of f. The dominated convergence theorem now implies that f_s converges uniformly to f as $s \to 0$.

We next fix $s > 0$ and rewrite $f_s(x)$ in the form

$$f_s(x) = \int_{|y| \leq 2a} s^{-1} k((x-y)/s) f(y) \, dy$$

$$= \pi^{-1/2} \int_{|y| \leq 2a} s^{-1} e^{-(x-y)^2/s^2} f(y) \, dy \qquad (2.12)$$

and compare it with the polynomial

$$p_{n,s}(x) := \pi^{-1/2} \int_{|y| \leq 2a} s^{-1} \sum_{r=0}^{n} \frac{(-1)^r (x-y)^{2r}}{s^{2r} r!} f(y) \, dy. \qquad (2.13)$$

The difference between (2.12) and (2.13) is estimated by using

$$\left| e^{-t} - \sum_{r=0}^{n} \frac{(-t)^r}{r!} \right| \leq \frac{t^{n+1}}{(n+1)!},$$

valid for all $t \geq 0$ and $n \geq 0$. This may be proved by using Taylor's theorem. We deduce that $p_{n,s}$ converges uniformly to f_s on $\{x : |x| \leq a\}$ as $n \to \infty$. $\square$

We finally write down the general Stone-Weierstrass theorem.[7] We leave the reader to work out how this theorem contains the previous one as a special case.

Theorem 2.3.17 *(Stone-Weierstrass) Let $\mathcal{A}$ be a subalgebra of the algebra $C_{\mathbf{R}}(K)$ of all continuous, real-valued functions on the compact Hausdorff space K. Suppose that $\mathcal{A}$ contains the constants and separates points in the sense that for every pair $x \neq y \in K$ there exists $f \in \mathcal{A}$ such that $f(x) \neq f(y)$. Then $\mathcal{A}$ is norm dense in $C_{\mathbf{R}}(K)$.*

2.4 Absolutely convergent Fourier series

In this section we prove theorems of Wiener and Bernstein about the absolute convergence of certain Fourier series. A continuous periodic function

$$f(\theta) := \sum_{n \in \mathbf{Z}} f_n e^{-in\theta}$$

[7] For the proof see [Bollobas 1999], [Dunford and Schwartz 1966, Theorem IV.6.18] or [Rudin 1973, Theorem 5.7].

is said to lie in the Wiener space $\mathcal{A}$ if

$$\|f\| := \sum_{n \in \mathbf{Z}} |f_n| < \infty.$$

Clearly $\|f\|_\infty \le \|f\|$ for all $f \in \mathcal{A}$.

Given $f, g \in \mathcal{A}$, the identity

$$f(\theta)g(\theta) = \left\{ \sum_{n \in \mathbf{Z}} f_n e^{-in\theta} \right\} \left\{ \sum_{m \in \mathbf{Z}} g_m e^{-im\theta} \right\}$$

implies

$$(fg)_n = \sum_{m \in \mathbf{Z}} f_m g_{n-m}$$

and hence

$$\|fg\| \le \|f\| \, \|g\|.$$

In the application of the lemma below to Wiener's Theorem 2.4.2 one puts $c = k = 1$ in (2.14). We have written it in the more general form because this is needed for the generalization of Wiener's theorem to higher dimensions and for applications. The estimate of $\|f^{-1}\|$ obtained in the proof depends on the size of c_g and σ_g. The size of these for a given f depends upon how easy it is to approximate f by a suitable function $g \in \mathcal{D}$.[8]

Lemma 2.4.1 *(Newman)*[9] *Let X be a compact Hausdorff space and let $\mathcal{B}$ be a subalgebra of $C(X)$ that contains the constants. Suppose that $\mathcal{B}$ is a Banach algebra with respect to a norm $\|\cdot\|$, and that $\mathcal{D}$ is a dense subset of $\mathcal{B}$. Suppose that whenever $g \in \mathcal{D}$ satisfies $|g(x)| \ge \sigma_g$ for some $\sigma_g > 0$ and all $x \in X$, it follows that g is invertible in $\mathcal{B}$ and*

$$\|g^{-n}\| \le c_g n^k c^n \sigma_g^{-n} \tag{2.14}$$

for all positive integers n; we assume that k and c do not depend on g or n and that c_g and σ_g do not depend on n. Then every $f \in \mathcal{B}$ which is invertible in $C(X)$ is also invertible in $\mathcal{B}$.

---

[8] According to [El-Fallah et al. 1999] it is not possible to obtain an upper bound on $\|f^{-1}\|$ in terms of $\|f\|$ and $\min\{|f(\theta)|\}$ alone.

[9] The standard proof of Wiener's theorem uses Gel'fand's representation theorem for commutative Banach algebras, and is not constructive; see [Rudin 1973]. We adapt the beautiful proof of Newman, which provides explicit information about the constants involved, as well as being particularly elementary, [Newman 1975]. Some further applications of our version are given in [Davies 2006].

Proof. If $f \in \mathcal{B}$ and $|z| > |||f|||$ then $(z - f)$ is invertible in $\mathcal{B}$ and therefore also invertible in $C(X)$. Hence

$$\{f(x) : x \in X\} \subseteq \{z : |z| \le |||f|||\}.$$

We deduce that $\|f\|_\infty \le |||f|||$ for all $f \in \mathcal{B}$.

If $f \in \mathcal{B}$ and $|f(x)| \ge \sigma > 0$ for all $x \in X$, let $g \in \mathcal{D}$ satisfy $|||g - f||| < \delta\sigma$ where $\delta := \{2(1+c)\}^{-1}$. This implies that $|g(x)| \ge (1-\delta)\sigma > 0$ for all $x \in X$. Therefore g is invertible in $\mathcal{B}$ and (2.14) implies that

$$|||g^{-n}||| \le c_g n^k c^n (1-\delta)^{-n} \sigma^{-n}$$

for all positive integers n. The inverse of f is given by the formula

$$f^{-1} = \sum_{n=0}^{\infty} (g-f)^n g^{-n-1}$$

in the sense of pointwise convergence on X. The following estimate shows that the series is norm convergent in $\mathcal{B}$, and hence also uniformly convergent in $C(X)$.

$$|||f^{-1}||| \le \sum_{n=0}^{\infty} (\delta\sigma)^n c_g (n+1)^k c^{n+1} (1-\delta)^{-n-1} \sigma^{-n-1}$$

$$= \frac{c_g}{\delta\sigma} \sum_{n=0}^{\infty} (n+1)^k \left(\frac{\delta c}{1-\delta} \right)^{n+1}$$

$$= \frac{c_g}{\delta\sigma} \sum_{n=0}^{\infty} (n+1)^k \left(\frac{c}{1+2c} \right)^{n+1}$$

$$\le \frac{c_g}{\delta\sigma} \sum_{n=0}^{\infty} (n+1)^k 2^{-n-1}. \qquad \square$$

We now apply the abstract theorem above to the Wiener space $\mathcal{A}$.

Theorem 2.4.2 (Wiener) *If $f \in \mathcal{A}$ and $f(\theta)$ is non-zero for all $\theta \in [-\pi, \pi]$ then $1/f \in \mathcal{A}$.*

Proof. We choose the dense subset $\mathcal{D}$ of $\mathcal{A}$ in Lemma 2.4.1 to be $C^1_{\text{per}}[-\pi, \pi]$; the set of all trigonometric polynomials would do equally well. The bounds

$$\sum_{n \in \mathbf{Z}} |f_n|^2 = \frac{1}{2\pi} \int_{-\pi}^{\pi} |f(\theta)|^2 \, d\theta \le \|f\|_\infty^2,$$

$$\sum_{n \in \mathbf{Z}} n^2 |f_n|^2 = \frac{1}{2\pi} \int_{-\pi}^{\pi} |f'(\theta)|^2 \, d\theta \le \|f'\|_\infty^2,$$

imply that

$$\||f\|| = |f_0| + \sum_{n \neq 0} |n|^{-1} |n f_n|$$

$$\leq |f_0| + \sqrt{\sum_{n \neq 0} n^{-2}} \sqrt{\sum_{n \neq 0} n^2 |f_n|^2}$$

$$\leq \|f\|_{\infty} + 2\|f'\|_{\infty}$$

for all $f \in C^1_{\text{per}}[-\pi, \pi]$. Therefore $C^1_{\text{per}}[-\pi, \pi] \subseteq \mathcal{A}$.

Our only task is to verify (2.14). Assuming that $g \in \mathcal{D}$ satisfies the conditions of the lemma, we have

$$\||g^{-n}\|| \leq \|g^{-n}\|_{\infty} + 2\|(g^{-n})'\|_{\infty}$$

$$= \|g^{-n}\|_{\infty} + 2n\|g'g^{-n-1}\|_{\infty}$$

$$\leq \sigma_g^{-n} + 2n\|g'\|_{\infty}\sigma_g^{-n-1}$$

$$\leq (1 + 2\|g'\|_{\infty}/\sigma_g) n\sigma_g^{-n},$$

which is of the required form. □

The above results can be interpreted as follows. The Banach space $\mathcal{B} := l^1(\mathbf{Z})$ is a commutative Banach algebra with identity if one assigns it the convolution multiplication. The element $\delta_0 \in l^1(\mathbf{Z})$ is the multiplicative identity. Its Gel'fand representation is the algebra homomorphism $\hat{\ }: \mathcal{B} \to C_{\text{per}}[-\pi, \pi]$ defined by

$$\hat{f}(\theta) := \sum_{n \in \mathbf{Z}} f_n e^{-in\theta}.$$

Theorem 2.4.2 then states that $f \in \mathcal{B}$ has a multiplicative inverse if and only if $\hat{f}$ is invertible in $C_{\text{per}}[-\pi, \pi]$. Moreover it provides a construction for f^{-1}.

The following is yet another perspective on the problem.

Theorem 2.4.3 *The operator A on $l^1(\mathbf{Z})$ is translation invariant if and only if it is of the form $Af := a * f$ where $a \in l^1(\mathbf{Z})$. The algebra of all translation invariant operators may be identified with $l^1(\mathbf{Z})$ regarded as a commutative Banach algebra with the convolution product. Moreover*

$$\text{Spec}(A) = \{\hat{a}(\theta) : -\pi \leq \theta \leq \pi\}. \tag{2.15}$$

Proof. We say that an operator $A : l^1(\mathbf{Z}) \to l^1(\mathbf{Z})$ is translation invariant if $AT = TA$, where $(Tf)(n) := f(n+1)$ for all $f \in l^1(\mathbf{Z})$. It is immediate that the set $\mathcal{T}$ of all translation invariant operators is a Banach algebra with identity, the norm being the operator norm; we will see that it is commutative, but this is not so obvious.

If $AT = TA$ then $AT^n = T^nA$ for all $n \in \mathbf{Z}$. This is equivalent to the validity of

$$A\delta_n = (A\delta_0) * \delta_n$$

for all $f \in l^1(\mathbf{Z})$ and $n \in \mathbf{Z}$, where $\delta_n \in l^1(\mathbf{Z})$ is given by

$$(\delta_n)_m := \begin{cases} 1 \text{ if } m = n, \\ 0 \text{ otherwise.} \end{cases}$$

Every $f \in l^1(\mathbf{Z})$ has the norm convergent expansion $f := \sum_{n \in \mathbf{Z}} f_n \delta_n$. Putting $a := A\delta_0$ we obtain

$$\begin{aligned} Af &= \sum_{n \in \mathbf{Z}} f_n (A\delta_n) \\ &= \sum_{n \in \mathbf{Z}} f_n (A\delta_0) * \delta_n \\ &= \sum_{n \in \mathbf{Z}} f_n a * \delta_n \\ &= a * f. \end{aligned}$$

It follows directly from this representation that $\mathcal{T}$ is isomorphic as a Banach algebra to $l^1(\mathbf{Z})$, and hence that it is commutative.

If $Af := a * f$ and a has the multiplicative inverse b within $l^1(\mathbf{Z})$ then

$$b * (Af) = b * a * f = f = a * b * f = A(b * f)$$

for all $f \in l^1(\mathbf{Z})$. Therefore A is an invertible operator with $A^{-1}f = b * f$ for all $f \in l^1(\mathbf{Z})$. Conversely if A is translation invariant and invertible as an operator on $l^1(\mathbf{Z})$ then the inverse must be translation invariant, so there exists $b \in l^1(\mathbf{Z})$ such that $A^{-1}f = b * f$ for all $f \in l^1(\mathbf{Z})$. It follows that b is the multiplicative inverse of a.

The above results imply that $zI - A$ is an invertible operator if and only if $z\delta_0 - a$ is an invertible element of $l^1(\mathbf{Z})$. Therefore the spectrum of A as an operator equals the spectrum of a as an element of the Banach algebra $l^1(\mathbf{Z})$. This equals the RHS of (2.15) by Wiener's Theorem 2.4.2. $\square$

The last theorem may be generalized in various directions. We start with its extension to $l^p(\mathbf{Z})$. A half-line analogue for Toeplitz operators is proved in Theorem 4.4.1. See also Theorem 3.1.19, where the following theorem is extended from $\mathbf{Z}$ to $\mathbf{R}^N$, for $p = 2$. The spectra of two convolution operators on $l^2(\mathbf{Z})$ are shown in figures on pages 127 and 267.

Theorem 2.4.4 *Let $A : l^p(\mathbf{Z}) \to l^p(\mathbf{Z})$ be the bounded operator $Af := a * f$, where $a \in l^1(\mathbf{Z})$ and $1 \le p \le \infty$. Then*

$$\mathrm{Spec}(A) = \{\hat{a}(\theta) : -\pi \le \theta \le \pi\}. \tag{2.16}$$

Proof. Let S denote the RHS of (2.16) and suppose that $z \notin S$. Then Wiener's Theorem 2.4.2 implies the existence of $b \in l^1(\mathbf{Z})$ such that $b * (a - z\delta_0) = \delta_0$. Putting $Bf := b * f$ we deduce that $B(A - zI) = (A - zI)B = I$ as operators on $l^p(\mathbf{Z})$. Therefore $z \notin \mathrm{Spec}(A)$.

Conversely suppose that $z = \hat{a}(\theta)$ for some $\theta \in [-\pi, \pi]$. We prove that $z \in \mathrm{Spec}(A)$ by constructing a sequence $f_n \in l^p(\mathbf{Z})$ such that $\|f_n\|_p \to 1$ and $\|Af_n - zf_n\|_p \to 0$ as $n \to \infty$. Let $g : \mathbf{R} \to \mathbf{R}^+$ be a continuous function with support in $(-1, 1)$ and satisfying $\|g\|_{L^p} = 1$. Then define $f_n \in l^p(\mathbf{Z})$ by

$$f_n(m) := n^{-1/p} g(m/n) e^{im\theta}.$$

Straightforward calculations show that $\|f_n\|_p \to 1$. If T_r denotes the isometry $(T_r f)(m) := f(m - r)$ acting on $l^p(\mathbf{Z})$ then

$$(T_r f_n)(m) - e^{-ir\theta} f_n(m) = e^{i(m-r)\theta} n^{-1/p} \{g((m-r)/n) - g(m/n)\}$$

for all $m, n, r \in \mathbf{Z}$. Therefore

$$\lim_{n \to \infty} \|T_r f_n - e^{-ir\theta} f_n\|_p^p = \lim_{n \to \infty} \sum_{m=-\infty}^{\infty} n^{-1} |g((m-r)/n) - g(m/n)|^p$$

$$= 0$$

for every $r \in \mathbf{Z}$. Hence

$$\lim_{n \to \infty} \|Af_n - \hat{a}(\theta)f_n\|_p = \lim_{n \to \infty} \left\| \sum_{r=-\infty}^{\infty} a(r)(T_r f_n - e^{-ir\theta} f_n) \right\|_p$$

$$\le \lim_{n \to \infty} \sum_{r=-\infty}^{\infty} |a(r)| \|T_r f_n - e^{-ir\theta} f_n\|_p$$

$$= 0. \qquad \square$$

We conclude this section with a sufficient condition for a function f to lie in the Wiener space.

Theorem 2.4.5 *(Bernstein) Let f be a continuous periodic function on $[-\pi, \pi]$ satisfying*

$$|f(u) - f(v)| \leq c|u - v|^\alpha$$

for all u, $v \in [-\pi, \pi]$, some $c > 0$ and some $\alpha \in (1/2, 1]$. Then f lies in the Wiener space $\mathcal{A}$.

Proof. If we put $f_s(x) := f(x+s)$ then the hypothesis implies that $\|f_s - f\|_2 \leq c_1|s|^\alpha$ for all $s \in \mathbf{R}$. If $a \in l^2(\mathbf{Z})$ is the sequence of Fourier coefficients of f then Corollary 2.3.11 implies that

$$\sum_{n \in \mathbf{Z}} |a_n(1 - e^{ins})|^2 \leq c_2|s|^{2\alpha}$$

for all $s \in \mathbf{R}$. Equivalently

$$\sum_{n \neq 0} |a_n|^2 \{1 - \cos(ns)\} \leq c_3 s^{2\alpha}$$

for all $s > 0$. Assuming $\delta > 0$, we define

$$b_n := \int_0^1 \{1 - \cos(ns)\} s^{2\delta - 1 - 2\alpha} \, ds$$

$$\sim n^{2\alpha - 2\delta} \int_0^\infty \{1 - \cos(t)\} t^{2\delta - 1 - 2\alpha} \, dt$$

as $n \to \infty$, the last integral being finite provided $\delta > 0$ is small enough. Therefore there exists $c_4 \in (0, \infty)$ such that

$$\sum_{n \neq 0} |a_n|^2 |n|^{2\alpha - 2\delta} \leq c_4 \sum_{n \neq 0} |a_n|^2 b_n$$

$$= c_4 \int_0^1 \sum_{n \neq 0} |a_n|^2 \{1 - \cos(ns)\} s^{2\delta - 1 - 2\alpha} \, ds$$

$$\leq c_4 \int_0^1 c_3 s^{2\delta - 1} \, ds$$

$$= c_3 c_4 / (2\delta)$$

$$< \infty.$$

The Schwarz inequality now implies that

$$\left(\sum_{n \neq 0} |a_n|\right)^2 \leq \left(\sum_{n \neq 0} |a_n|^2 |n|^{2\alpha - 2\delta}\right)\left(\sum_{n \neq 0} |n|^{2\delta - 2\alpha}\right) < \infty$$

provided $\delta > 0$ is small enough. $\quad\square$

3

Fourier transforms and bases

3.1 The Fourier transform

In this chapter we treat two topics: the theory of Fourier transforms and general bases in Banach spaces. These both generalize the classical L^2 convergence theory for Fourier series, which we regard as already understood. We will see that these topics provide key ingredients for the detailed spectral analysis of many bounded and unbounded linear operators.

One of the reasons for the importance of the Fourier transform is that it simplifies the analysis of constant coefficient differential operators, discussed below and in the next chapter. One of our main goals is to establish the following results.

$$\mathcal{F} : \mathcal{S} \to \mathcal{S} \qquad \text{one-one onto,}$$

$$\mathcal{F} : L^2(\mathbf{R}^N) \to L^2(\mathbf{R}^N) \qquad \text{one-one onto and unitary,}$$

$$\mathcal{F} : L^1(\mathbf{R}^N) \to C_0(\mathbf{R}^N) \qquad \text{one-one, but not onto,}$$

$$\mathcal{F} : L^p(\mathbf{R}^N) \to L^q(\mathbf{R}^N) \qquad \text{if } 1 \leq p \leq 2 \text{ and } 1/p + 1/q = 1,$$

$$\mathcal{F} : \mathcal{S}' \to \mathcal{S}' \qquad \text{one-one onto.}$$

If $f : \mathbf{R}^N \to \mathbf{C}$ and $\alpha := (\alpha_1, \alpha_2, \ldots, \alpha_N)$ is a multi-index of non-negative integers, then we write $D^\alpha f$ to denote the result of differentiating f α_r times with respect to x_r for every r. The order of the derivative is defined to be $|\alpha| := \alpha_1 + \cdots + \alpha_N$. If $|\alpha| = 0$ then $D^\alpha f := f$ by convention. If $x \in \mathbf{R}^N$ we define

$$x^\alpha := x_1^{\alpha_1} x_2^{\alpha_2} \ldots x_N^{\alpha_N}.$$

The Schwartz space $\mathcal{S}$ is defined to be the space of all smooth (i.e. infinitely differentiable) functions $f : \mathbf{R}^N \to \mathbf{C}$ such that for every multi-index α and every $m \geq 0$ there exists $c_{\alpha, m}$ satisfying

$$|D^\alpha f(x)| \leq c_{\alpha, m}(1 + |x|)^{-m}$$

for all $x \in \mathbf{R}^N$. An equivalent condition is that there exist constants $c_{\alpha, \beta}$ such that

$$|x^\beta D^\alpha f(x)| \leq c_{\alpha, \beta}$$

for all $x \in \mathbf{R}^N$ and all α, β. One can put a topology on $\mathcal{S}$ by means of the countable family of seminorms

$$p_{\alpha, \beta}(f) := \sup\{|x^\beta D^\alpha f(x)| : x \in \mathbf{R}^N\}.$$

It may be shown that this turns $\mathcal{S}$ into a Fréchet space, but we will not need to use this fact.

Problem 3.1.1 Prove that if p is a polynomial on $\mathbf{R}^N$ then $f(x) := p(x)\mathrm{e}^{-|x|^2}$ lies in $\mathcal{S}$.   □

Problem 3.1.2 Prove that if p is a polynomial on $\mathbf{R}^N$ and there exist $c > 0$, $R > 0$ such that $p(x) \geq c|x|$ for all $|x| \geq R$ then $f(x) = \mathrm{e}^{-p(x)}$ lies in $\mathcal{S}$.   □

We omit the proofs of our next two lemmas, which are somewhat tedious but entirely elementary exercises in the use of differentiation under the integral sign and integration by parts.

Lemma 3.1.3 *If $f, g \in \mathcal{S}$ then $fg \in \mathcal{S}$ and $f * g \in \mathcal{S}$, where*

$$(f * g)(x) := \int_{\mathbf{R}^N} f(x - y)g(y)\,\mathrm{d}^N y = (g * f)(x).$$

Lemma 3.1.4 *The Fourier transform*

$$(\mathcal{F}f)(\xi) := (2\pi)^{-N/2} \int_{\mathbf{R}^N} f(x)\mathrm{e}^{-ix\cdot\xi}\,\mathrm{d}^N x$$

maps $\mathcal{S}$ into $\mathcal{S}$. It satisfies

$$(\mathcal{F}D^\alpha f)(\xi) = i^{|\alpha|}\xi^\alpha(\mathcal{F}f)(\xi)$$

for all $f \in \mathcal{S}$ and $\xi \in \mathbf{R}^N$. Moreover

$$(\mathcal{F} Q^\alpha f)(\xi) = i^{|\alpha|}(D^\alpha \mathcal{F} f)(\xi)$$

for all $f \in \mathcal{S}$ and $\xi \in \mathbf{R}^N$, where $(Q^\alpha f)(x) := x^\alpha f(x)$.

The formula

$$(\mathcal{F} \Delta f)(\xi) = -|\xi|^2 (\mathcal{F} f)(\xi) \tag{3.1}$$

is an immediate consequence of the lemma.

Our next lemma is needed in the proof of Theorem 3.1.6.

Lemma 3.1.5 *If*

$$k_t(x) := (2\pi t)^{-N/2} \exp\{-|x|^2/2t\}$$

then

$$(\mathcal{F} k_t)(\xi) = (2\pi)^{-N/2} \exp\{-t|\xi|^2/2\}.$$

If $f \in \mathcal{S}$ then

$$\lim_{t \to 0+} (k_t * f)(x) = f(x)$$

for all $x \in \mathbf{R}^N$.

Proof. One proves the first statement by separating variables in the Fourier integral and applying the well-known result in the case $N = 1$. The second is proved by applying the dominated convergence theorem to the final integral in

$$(k_t * f)(x) - f(x) = \int_{\mathbf{R}^N} \{f(x-y) - f(x)\} k_t(y) \, d^N y$$

$$= \int_{\mathbf{R}^N} \{f(x - t^{1/2}u) - f(x)\} k_1(u) \, d^N u. \qquad \square$$

Theorem 3.1.6 *(Plancherel) The operator $\mathcal{F} : \mathcal{S} \to \mathcal{S}$ extends by completion to a unitary operator on $L^2(\mathbf{R}^N)$. Its inverse is given for all $g \in \mathcal{S}$ by*

$$(\mathcal{F}^{-1} g)(x) = (2\pi)^{-N/2} \int_{\mathbf{R}^N} g(\xi) e^{ix \cdot \xi} \, d^N \xi.$$

Proof. We first establish the inversion formula, and thereby prove that $\mathcal{F}$ maps $\mathcal{S}$ one-one onto $\mathcal{S}$. If $g := \mathcal{F}f$ and $k_t(x) := (2\pi t)^{-N/2}\exp\{-|x|^2/2t\}$ then

$$(2\pi)^{-N/2}\int_{\mathbf{R}^N} g(\xi)e^{ix\cdot\xi}\,\mathrm{d}^N\xi = \lim_{t\to 0+}(2\pi)^{-N/2}\int_{\mathbf{R}^N} g(\xi)e^{-t|\xi|^2/2}e^{ix\cdot\xi}\,\mathrm{d}^N\xi$$

$$= \lim_{t\to 0+}(2\pi)^{-N}\int_{\mathbf{R}^N\times\mathbf{R}^N} f(y)e^{-t|\xi|^2/2}e^{i(x-y)\cdot\xi}\,\mathrm{d}^Ny\,\mathrm{d}^N\xi$$

$$= \lim_{t\to 0+}(2\pi t)^{-N/2}\int_{\mathbf{R}^N} f(y)e^{-|x-y|^2/2t}\,\mathrm{d}^Ny$$

$$= \lim_{t\to 0+}\int_{\mathbf{R}^N} f(y)k_t(x-y)\,\mathrm{d}^Ny$$

$$= f(x)$$

by Lemma 3.1.5.

We next prove that $\mathcal{F}$ preserves inner products. The above calculation establishes that

$$\langle f, \mathcal{F}g\rangle = \langle \mathcal{F}^{-1}f, g\rangle$$

for all $f, g \in \mathcal{S}$ by writing out the relevant double integrals on each side. Therefore

$$\langle \mathcal{F}f, \mathcal{F}g\rangle = \langle \mathcal{F}^{-1}\mathcal{F}f, g\rangle = \langle f, g\rangle.$$

Putting $f = g$ we obtain $\|\mathcal{F}f\|_2 = \|f\|_2$. Corollary 2.3.6 implies that $\mathcal{S}$ is norm dense in $L^2(\mathbf{R}^N)$. We may therefore extend $\mathcal{F}$ in a unique way to an isometric linear operator $\overline{\mathcal{F}}: L^2(\mathbf{R}^N) \to L^2(\mathbf{R}^N)$. Continuity arguments imply that the extension is also unitary. Similar considerations apply to $\mathcal{F}^{-1}$ and imply that $\overline{\mathcal{F}}$ is surjective, and hence unitary. $\square$

The space $C_c^\infty(\mathbf{R}^N)$ of smooth functions with compact support is much less useful than $\mathcal{S}$ for Fourier analysis because it is not invariant under $\mathcal{F}$.

Problem 3.1.7 Prove that if f and $\mathcal{F}f$ both lie in $C_c^\infty(\mathbf{R}^N)$ then f is identically zero. $\square$

We also define

$$\hat{f}(\xi) := \int_{\mathbf{R}^N} f(x)e^{-ix\cdot\xi}\,\mathrm{d}^Nx = (2\pi)^{N/2}(\mathcal{F}f)(\xi). \tag{3.2}$$

One can avoid having two different normalizations for the Fourier transform by replacing $e^{-ix\cdot\xi}$ by $e^{-2\pi ix\cdot\xi}$ everywhere. Unfortunately this is not the convention used in books of tables of Fourier transforms, and we will adopt the usual definitions.

Theorem 3.1.8 *The map $f \to \hat{f}$ is a linear operator of norm 1 from $L^1(\mathbf{R}^N)$ to $C_0(\mathbf{R}^N)$. It is a homomorphism of Banach algebras, if L^1 is given the convolution multiplication and C_0 is given pointwise multiplication.*

Proof. Fundamental properties of the Lebesgue integral imply that $|\hat{f}(\xi)| \leq \|f\|_1$ for all $\xi \in \mathbf{R}^N$. If $f \geq 0$ then $\|\hat{f}\|_\infty = \hat{f}(0) = \|f\|_1$. The dominated convergence theorem implies that $\hat{f}$ is continuous. If $f \in \mathcal{S}$ then $\hat{f} \in \mathcal{S} \subseteq C_0(\mathbf{R}^N)$, so the density of $\mathcal{S}$ in $L^1(\mathbf{R}^N)$, which follows from Corollary 2.3.6, implies that $\hat{f} \in C_0(\mathbf{R}^N)$ for all $f \in L^1(\mathbf{R}^N)$. This result is called the Riemann-Lebesgue lemma.

The final statement of the theorem depends on the calculation

$$\widehat{(f * g)}(\xi) = \int_{\mathbf{R}^N} (f * g)(x) e^{-ix \cdot \xi} \, d^N x$$

$$= \int_{\mathbf{R}^N \times \mathbf{R}^N} f(x - y) g(y) e^{-ix \cdot \xi} \, d^N x \, d^N y$$

$$= \int_{\mathbf{R}^N \times \mathbf{R}^N} f(x) g(y) e^{-i(x+y) \cdot \xi} \, d^N x \, d^N y$$

$$= \hat{f}(\xi) \hat{g}(\xi). \qquad \square$$

Problem 3.1.9 Prove, by a direct computation, that the Fourier transform of the function

$$g(x) := \max\{a - b|x|, 0\}$$

is non-negative for all positive constants a, b. Deduce by taking convex combinations that the Fourier transform of the even, non-negative, continuous function $f \in L^1(\mathbf{R})$ is non-negative if f is convex on $[0, \infty)$. (These conditions imply that $f(x) \to 0$ as $|x| \to \infty$.) $\square$

Problem 3.1.10 The computation of multidimensional Fourier transforms is not easy, but the following elementary observation is sometimes useful, when combined with the use of tables of Laplace transforms.

Prove that if $a \in L^1(0, \infty)$ and

$$f(x) := \int_0^\infty a(t)(4\pi t)^{-N/2} e^{-|x|^2/4t} \, dt,$$

then $f \in L^1(\mathbf{R}^N)$ and

$$\hat{f}(\xi) = \int_0^\infty a(t) e^{-t|\xi|^2} \, dt.$$

Determine f and $\hat{f}$ explicitly in the particular case $a(t) = e^{-\lambda^2 t}$ where $\lambda > 0$ and $N = 3$. $\square$

Problem 3.1.11 Prove that if

$$\int_{\mathbf{R}^N} |f(x)|(1+|x|^2)^n \, d^N x < \infty$$

for all n, then $\hat{f}$ is a smooth function, all of whose derivatives vanish at infinity. $\square$

Theorem 3.1.12 *The map $f \to \mathscr{F}f$ on $\mathcal{S}$ extends to a bounded linear operator from $L^p(\mathbf{R}^N)$ into $L^q(\mathbf{R}^N)$ if $1 \leq p \leq 2$ and $1/p + 1/q = 1$.*

Proof. One interpolates between the cases $p = 1$ (Theorem 3.1.8) and $p = 2$ (Theorem 3.1.6). $\square$

There is a logical possibility that if f lies in two function spaces, different definitions of the Fourier transform of f are not consistent with each other. Such concerns may be resolved by showing that each is the restriction of a very abstract Fourier transform defined on an extremely large space. We define $\mathcal{S}'$ to be the algebraic dual space of $\mathcal{S}$.[1] We will refer to elements of $\mathcal{S}'$ as distributions. If $g : \mathbf{R}^N \to \mathbf{C}$ is a function which is locally integrable and polynomially bounded in the sense that

$$\int_{\mathbf{R}^N} |g(x)|(1+|x|)^{-m} \, d^N x < \infty$$

for some m, then it determines a distribution by means of the formula

$$\phi_g(f) := \int_{\mathbf{R}^N} f(x)g(x) \, d^N x \tag{3.3}$$

for all $f \in \mathcal{S}$. This class of functions contains $L^p(\mathbf{R}^N)$ for all $1 \leq p \leq \infty$. However, there are many distributions that are not associated with functions, for example

$$\delta_x^\alpha(f) := (D^\alpha f)(x)$$

and

$$\mu(f) := \int_{\mathbf{R}^N} f(x)\,\mu(dx)$$

---

[1] Ignoring the standard convention, we do not impose any continuity conditions on elements of $\mathcal{S}'$. This is good enough for proving consistency, but would prevent our using the deep theorems about tempered distributions if we needed them. See [Friedlander and Joshi 1998, Hörmander 1990] for much deeper accounts of the subject.

for any finite measure μ on $\mathbf{R}^N$. We define the Fourier transform $\mathcal{F} : \mathcal{S}' \to \mathcal{S}'$ by

$$(\mathcal{F}\phi)(f) := \phi(\mathcal{F}f) \tag{3.4}$$

for all $f \in \mathcal{S}$. Clearly $\mathcal{F}$ is a one-one linear map from $\mathcal{S}'$ onto $\mathcal{S}'$. If $\phi_n \to \phi$ as $n \to \infty$ in the sense that $\lim_{n \to \infty} \phi_n(f)) = \phi(f)$ for all $f \in \mathcal{S}$, then $\mathcal{F}\phi_n \to \mathcal{F}\phi$ as $n \to \infty$.

Theorem 3.1.13 *If $1 \le p \le 2$ and $\phi : L^p(\mathbf{R}^N) \to \mathcal{S}'$ is defined by (3.3) then the Fourier transform of $g \in L^p(\mathbf{R}^N)$ as defined in Theorem 3.1.12 is consistent with the Fourier transform of ϕ_g as defined in (3.4) in the sense that $\phi_{\mathcal{F}g} = \mathcal{F}\phi_g$.*

Proof. If $g \in L^p(\mathbf{R}^N)$ then there exists a sequence $g_n \in L_c^\infty$ such that $\|g_n - g\|_p \to 0$ as $n \to \infty$. Given any $f \in \mathcal{S}$, by using the fact that $f \in L^q(\mathbf{R}^N)$ for all $q \in [1, \infty]$, we obtain

$$\phi_{\mathcal{F}g}(f) = \int_{\mathbf{R}^N} (\mathcal{F}g)(x) f(x) \, \mathrm{d}^N x$$

$$= \lim_{n \to \infty} \int_{\mathbf{R}^N} (\mathcal{F}g_n)(x) f(x) \, \mathrm{d}^N x$$

$$= \lim_{n \to \infty} \int_{\mathbf{R}^N \times \mathbf{R}^N} g_n(\xi) e^{-ix \cdot \xi} f(x) \, \mathrm{d}^N \xi \mathrm{d}^N x$$

$$= \lim_{n \to \infty} \int_{\mathbf{R}^N} g_n(\xi) (\mathcal{F}f)(\xi) \, \mathrm{d}^N \xi$$

$$= \int_{\mathbf{R}^N} g(\xi) (\mathcal{F}f)(\xi) \, \mathrm{d}^N \xi$$

$$= (\phi_g)(\mathcal{F}f)$$

$$= (\mathcal{F}\phi_g)(f).$$

Therefore $\phi_{\mathcal{F}g} = \mathcal{F}\phi_g$. $\quad\square$

Problem 3.1.14 Calculate the Fourier transform of the function $f \in L^1(\mathbf{R})$ defined for $0 < \alpha < 1$ and $\varepsilon > 0$ by

$$f(x) := |x|^{-\alpha} e^{-\varepsilon|x|}.$$

Use the result to prove that the Fourier transform of the distribution determined by the function $f(x) := |x|^{-\alpha}$ is the distribution determined by the function $g(\xi) := c_\alpha |\xi|^{-(1-\alpha)}$ where c_α is a certain positive constant. $\quad\square$

Theorem 3.1.15 *The Fourier transform map $f \to \hat{f}$ of Theorem 3.1.8 maps $L^1(\mathbf{R}^N)$ one-one into, but not onto, $C_0(\mathbf{R}^N)$.*

Proof. By Theorem 3.1.13 the map is the restriction of the Fourier transform map $\mathcal{F} : \mathcal{S}' \to \mathcal{S}'$ (up to a constant), and this is invertible, because $\mathcal{F} : \mathcal{S} \to \mathcal{S}$ is invertible. Therefore the map of the theorem is one-one. If it were onto there would be a constant $c > 0$ such that $\|\hat{f}\|_\infty \geq c\|f\|_1$ for all $f \in L^1(\mathbf{R}^N)$, by the inverse mapping theorem. We prove that this assertion is false.

We first treat the case $N = 1$. If $\alpha > 1$ and

$$f_\alpha(x) := \frac{\Gamma(\alpha)}{2\pi}(1 - ix)^{-\alpha}$$

then $f_\alpha \in L^1(\mathbf{R})$ and

$$\hat{f}_\alpha(\xi) = \begin{cases} \xi^{\alpha-1}e^{-\xi} & \text{if } \xi > 0, \\ 0 & \text{otherwise.} \end{cases}$$

(It is easier to compute the inverse Fourier transforms.) A direct calculation shows that

$$\lim_{\alpha \to 1+} \|f_\alpha\|_1 = +\infty, \qquad \lim_{\alpha \to 1+} \|\hat{f}_\alpha\|_\infty = 1.$$

For $N > 1$ one does a similar calculation for

$$F_\alpha(x) := \prod_{r=1}^{N} f_\alpha(x_r). \qquad \square$$

Problem 3.1.16 The following is a solution of the moment problem.[2] Let $f \in L^1(\mathbf{R})$ satisfy

$$\int_{\mathbf{R}} |f(x)||e^{\alpha|x|} \, dx < \infty$$

for some $\alpha > 0$. Prove that $\hat{f}(\xi)$ may be extended to an analytic function on the strip $\{\xi : |\mathrm{Im}(\xi)| < \alpha\}$. Deduce that if

$$\int_{\mathbf{R}} f(x)x^n \, dx = 0$$

for all non-negative integers n, then $f(x) = 0$ almost everywhere. $\quad \square$

---

[2] See [Simon 1998] for a systematic account of this problem.

<u>Note</u> Theorem 3.3.11 provides a more constructive proof of a closely related result.

One may define the weak derivative of an element of $\mathcal{S}'$ without imposing any differentiability conditions as follows.

Problem 3.1.17 Prove that if $\phi \in \mathcal{S}'$ then the definition

$$(D^\alpha \phi)(f) := (-1)^{|\alpha|} \phi(D^\alpha f),$$

where $f \in \mathcal{S}$ is arbitrary, is consistent with the definition of D^α on $\mathcal{S}$ in the sense that

$$D^\alpha \phi_g = \phi_{D^\alpha g}$$

for all $g \in \mathcal{S}$. $\square$

Theorem 3.1.18 *If $a \in L^1(\mathbf{R}^N)$ then the convolution transform $T_a f := a * f$ is a bounded linear operator on $L^p(\mathbf{R}^N)$ for all $1 \leq p \leq \infty$, with norm at most $\|a\|_1$. For $p = 1$ and $p = \infty$ its norm equals $\|a\|_1$. If $1 \leq p \leq 2$ then*

$$(\mathcal{F} T_a f)(\xi) = \hat{a}(\xi)(\mathcal{F} f)(\xi) \tag{3.5}$$

almost everywhere on $\mathbf{R}^N$, where $\hat{a}$ is defined by (3.2).

Proof. The first part of the theorem follows directly from the definitions if $p = 1$ or $p = \infty$. For other p one obtains it by interpolation.

The identity (3.5) holds for $f \in L^1(\mathbf{R}^N) \cap L^p(\mathbf{R}^N)$ by Theorem 3.1.8. Its validity for general $f \in L^p(\mathbf{R}^N)$ then follows by an approximation procedure. $\square$

We mention in passing that there is a large body of theory concerning singular integral operators, including those of the form

$$(Af)(x) := \int_{\mathbf{R}^N} k(x-y)f(y) \, \mathrm{d}^N y$$

where k does <u>not</u> lie in $L^1(\mathbf{R}^N)$. An example is the Hilbert transform

$$(Hf)(x) := \int_{-\infty}^{\infty} \frac{f(y)}{x-y} \, \mathrm{d}y.$$

One of the goals in the subject is to find conditions under which such operators are bounded on $L^p(\mathbf{R}^N)$ for all $p \in (1, \infty)$. Since we are more interested in operators that are bounded for all p satisfying $1 \leq p \leq +\infty$ we will not pursue this subject.[3]

---

[3] A good introduction is [Stein 1970].

Theorem 3.1.19 *If* $a \in L^1(\mathbf{R}^N)$ *and* $T_a : L^2(\mathbf{R}^N) \to L^2(\mathbf{R}^N)$ *is defined by* $T_a f := a * f$, *then*

$$\mathrm{Spec}(T_a) = \{\hat{a}(\xi) : \xi \in \mathbf{R}^N\} \cup \{0\}$$

and

$$\|T_a\| = \max\{|\hat{a}(\xi)| : \xi \in \mathbf{R}^N\}.$$

Moreover T_a *is one-one if and only if* $\{\xi : \hat{a}(\xi) = 0\}$ *has zero Lebesgue measure. If* T_a *is one-one, the inverse operator is always unbounded.*

Proof. It follows by Theorem 3.1.18 that $B := \mathcal{F} T_a \mathcal{F}^{-1}$ is a multiplication operator. Indeed

$$(Bg)(\xi) = \hat{a}(\xi)g(\xi)$$

for all $g \in L^2(\mathbf{R}^N)$. Since T_a and B have the same norm and spectrum, the statements of the theorem follow by using Problem 2.2.1. Note that $\hat{a}(\cdot)$ is continuous and vanishes at infinity by Theorem 3.1.8, so B^{-1} must be an unbounded operator if it exists. $\quad\square$

Problem 3.1.20 If $f \in \mathcal{S}$ and $f(x) \geq 0$ almost everywhere, prove that all the eigenvalues of the real symmetric matrix

$$A_{i,j} := \frac{\partial^2 \hat{f}}{\partial \xi_i \partial \xi_j}(0)$$

are negative, unless f is identically zero. $\quad\square$

Problem 3.1.21 Prove that for every $t > 0$ the Fourier transform of the function $f_t(\xi) := e^{-t|\xi|^4}$ lies in $\mathcal{S}$ and that $\hat{f}_t(x) < 0$ on a non-empty open set. Prove also that $\|\hat{f}_t\|_1$ is independent of t. $\quad\square$

Problem 3.1.22 Prove that if $k \in L^1(\mathbf{R}^N)$ then there exists a bounded operator A on $L^2(\mathbf{R}^N)$ such that

$$\langle Af, f\rangle = \int_{\mathbf{R}^N} \int_{\mathbf{R}^N} k(x - y)|f(x) - f(y)|^2 \, \mathrm{d}^N x \mathrm{d}^N y$$

for all $f \in L^2(\mathbf{R}^N)$. Determine the spectrum of A. $\quad\square$

Example 3.1.23 One can make sense of the operators defined in Problem 3.1.22 for considerably more singular functions k. In particular the formula

$$\langle Hf, f\rangle = \int_{\mathbf{R}} \int_{\mathbf{R}} \frac{|f(x) - f(y)|^2}{|x - y|^{1+\alpha}} \, \mathrm{d}x\mathrm{d}y$$

is known to define an unbounded, non-negative, self-adjoint operator H acting in $L^2(\mathbf{R})$, provided $0 \le \alpha < 2$. The operators e^{-Ht} are bounded for all $t \ge 0$ and define an extremely important Markov semigroup called a Levy semigroup.[4] □

3.2 Sobolev spaces

The Sobolev spaces $W^{n,2}(\mathbf{R}^N)$ are often useful when studying differential operators. One can define $W^{n,2}(\mathbf{R}^N)$ for any real number n but we will only need to consider the case in which n is a non-negative integer. One can also replace 2 by $p \in (1, \infty)$ and obtain analogous theorems, usually with harder proofs.[5]

Let $\mathcal{F}$ be the Fourier transform operator on $L^2(\mathbf{R}^N)$. The Sobolev space $W^{n,2}(\mathbf{R}^N)$ is defined to be the set of all $f \in L^2(\mathbf{R}^N)$ such that $\tilde{f} := \mathcal{F}f$ satisfies

$$\|f\|_n^2 := \int_{\mathbf{R}^N} (1 + |\xi|^2)^n |\tilde{f}(\xi)|^2 \, d^N\xi < \infty. \tag{3.6}$$

Note that $W^{0,2} = L^2(\mathbf{R}^N)$. Each $W^{n,2}$ is a Hilbert space with respect to the inner product

$$\langle f, g \rangle_n := \int_{\mathbf{R}^N} (1 + |\xi|^2)^n \tilde{f}(\xi) \overline{\tilde{g}(\xi)} \, d^N\xi < \infty.$$

It is easy to prove that $f \in W^{n,2}(\mathbf{R}^N)$ if and only if the functions $\xi \to \xi^\alpha \tilde{f}(\xi)$ lie in $L^2(\mathbf{R}^N)$ for all α such that $|\alpha| \le n$. This is equivalent to the condition that $D^\alpha f \in L^2(\mathbf{R}^N)$ for all such α, where $D^\alpha f$ are weak derivatives, as defined in Problem 3.1.17.

The following diagrams provide inclusions between some of the important spaces of functions on $\mathbf{R}^N$. We assume that $n \ge 0$ and $1 \le p \le \infty$.

$$C_c^\infty \longrightarrow \mathcal{S} \longrightarrow W^{n,2} \longrightarrow L^2 \longrightarrow \mathcal{S}'$$

$$C_c^\infty \longrightarrow C_c \longrightarrow L_c^\infty \longrightarrow L^p \longrightarrow \mathcal{S}'$$

For large enough values of n the derivatives of a function in $W^{n,2}$ may be calculated in the classical manner.

---

[4] See [Bañuelos and Kulczycki 2004, Song and Vondraček 2003] for information about recent research in this field.
[5] See [Adams 1975] for a comprehensive treatment of Sobolev spaces and their embedding properties.

Theorem 3.2.1 *If $n > k + N/2$ then every $f \in W^{n,2}$ is k times continuously differentiable. Indeed $W^{n,2}(\mathbf{R}^N) \subseteq C_0(\mathbf{R}^N)$ if $n > N/2$.*

Proof. By using the Schwarz inequality we see that if $|\alpha| \leq k$ then our assumption implies that

$$\int_{\mathbf{R}^N} |\xi^\alpha \tilde{f}(\xi)| \, d^N \xi \leq c \left\{ \int_{\mathbf{R}^N} (1 + |\xi|^2)^n |\tilde{f}(\xi)|^2 \, d^N \xi \right\}^{1/2}$$

where

$$c := \left\{ \int_{\mathbf{R}^N} |\xi^\alpha|^2 (1 + |\xi|^2)^{-n} \, d^N \xi \right\}^{1/2} < \infty.$$

By taking inverse Fourier transforms we deduce that $(D^\alpha f)(x)$ is a continuous function of x vanishing as $|x| \to \infty$ for all α such that $|\alpha| \leq k$. □

Let L be the nth order differential operator given formally by

$$(Lf)(x) := \sum_{|\alpha| \leq n} a_\alpha(x)(D^\alpha f)(x)$$

where a_α are bounded measurable functions on $\mathbf{R}^N$. Then L may be defined as an operator on $\mathcal{S}$ by

$$(Lf)(x) := (2\pi)^{-N/2} \int_{\mathbf{R}^N} e^{ix\xi} \sigma(x, \xi) \tilde{f}(\xi) \, d^N \xi$$

where the symbol σ of the operator is given by

$$\sigma(x, \xi) := \sum_{|\alpha| \leq n} a_\alpha(x) i^{|\alpha|} \xi^\alpha. \tag{3.7}$$

Finding the proper domain of such an operator depends on making further hypotheses concerning its coefficients.

One of the important notions in the theory of differential operators is that of ellipticity. Subject to suitable differentiability assumptions, the symbol of a real second order elliptic operator may always be written in the form

$$\sigma(x, \xi) := \sum_{r,s=1}^{N} a_{r,s}(x)\xi_r \xi_s + \sum_{r=1}^{N} b_r(x)\xi_r + c(x)$$

where $a(x)$ is a real symmetric matrix. The symbol, or the associated operator, is said to be elliptic if the eigenvalues of $a(x)$ are all positive for all relevant x.[6]

---

[6] We refer to [Hörmander 1990, Taylor 1996] for introductions to the large literature on elliptic differential operators.

The case of constant coefficient operators is particularly simple because the symbol is then a polynomial. Such an operator has symbol

$$\sigma(\xi) = \sum_{|\alpha| \leq n} a_\alpha i^{|\alpha|} \xi^\alpha.$$

It is said to be elliptic if there exist positive constants c_0, c_1 such that the principal part of the symbol, namely

$$\sigma_n(\xi) = \sum_{|\alpha|=n} a_\alpha i^{|\alpha|} \xi^\alpha$$

satisfies

$$c_0|\xi|^n \leq |\sigma_n(\xi)| \leq c_1|\xi|^n$$

for all $\xi \in \mathbf{R}^N$. We determine the spectrum of any constant coefficient differential operator acting on $L^2(\mathbf{R}^N)$ in Theorem 8.1.1.

Example 3.2.2 The Laplace operator, or Laplacian, $H_0 := -\Delta$ has the associated symbol $\sigma(\xi) := |\xi|^2$. According to the above arguments its natural domain is $W^{2,2}(\mathbf{R}^N)$. By examining the function $f(\xi) := z - |\xi|^2$ one sees that the operator $(zI - H_0)$ maps $W^{2,2}(\mathbf{R}^N)$ one-one onto $L^2(\mathbf{R}^N)$ if and only if $z \notin [0, \infty)$. This justifies the statement that the spectrum of H_0 is $[0, \infty)$. See Theorem 8.1.1, where this is put in a more general context. $\square$

Problem 3.2.3 Use Fourier transform methods, and in particular Problem 3.1.11, to prove that if $g \in \mathcal{S}(\mathbf{R}^3)$ then the differential equation $-\Delta f = g$ has a smooth solution that vanishes at infinity together with all of its partial derivatives. Prove that $f \in L^2(\mathbf{R}^3)$ if and only if

$$\int_{\mathbf{R}^3} g(x)\,\mathrm{d}^3x = 0,$$

and that in this case $f \in W^{n,2}$ for all n. $\square$

Problem 3.2.4 Suppose that L is a constant coefficient elliptic differential operator of order $2n$ whose principal symbol satisfies

$$c_0|\xi|^{2n} \leq \sigma_{2n}(\xi) \leq c_1|\xi|^{2n}$$

for some positive constants c_0, c_1 and all $\xi \in \mathbf{R}^N$. Prove that $(L + \lambda I)$ maps $W^{2n,2}$ one-one onto $L^2(\mathbf{R}^N)$ for all large enough $\lambda > 0$, and that the inverse operator is bounded on $L^2(\mathbf{R}^N)$. $\square$

Problem 3.2.5 Prove that the first order differential operator

$$(Lf)(x, y) := \frac{\partial f}{\partial x} + i\frac{\partial f}{\partial y}$$

is elliptic in $L^2(\mathbf{R}^2)$. Can a first order constant coefficient differential operator acting in $L^2(\mathbf{R}^N)$ be elliptic if $N > 2$? □

3.3 Bases of Banach spaces

We say that a sequence $\{f_n\}_{n=1}^{\infty}$ in a Banach space $\mathcal{B}$ is complete if its linear span is dense in $\mathcal{B}$. We say that it is a basis if every $f \in \mathcal{B}$ has a unique expansion

$$f = \lim_{n \to \infty} \left(\sum_{r=1}^{n} \alpha_r f_r \right).$$

The terms Schauder basis and conditional basis are also used in this context.[7] The prototypes for our study are orthonormal bases in Hilbert space, and in particular Fourier series. Our goal will be to understand the extent to which one can adapt that theory to Banach spaces and to non-orthonormal sequences in Hilbert space. Our analysis is organized around four concepts:

complete sequence

minimal complete sequence

conditional (or Schauder) basis

unconditional basis

each of which is more special than the one before it. We explain their significance and provide a range of examples to illustrate our results.

Lemma 3.3.1 *If $\{f_n\}_{n=1}^{\infty}$ is a basis in a Banach space $\mathcal{B}$ then there exist $\phi_n \in \mathcal{B}^*$ such that the 'Fourier' coefficients α_n are given by $\alpha_n := \langle f, \phi_n \rangle$. The pair of sequences $\{f_n\}_{n=1}^{\infty}, \{\phi_n\}_{n=1}^{\infty}$ is biorthogonal in the sense that*

$$\langle f_n, \phi_m \rangle = \delta_{m,n}$$

for all m, n.

---

[7] We can do no more than mention the vast literature on bases, and refer to more serious studies of the topic in [Singer 1970, Singer 1981] and, more recently, [Carothers 2005], where a wide range of examples are presented.

Proof. The idea of the following proof is to use the fact that if $s_n := \sum_{r=1}^{n} \alpha_r f_r$ then $s_n - s_{n-1} \in \mathbf{C} f_n$.

Let K be the compact space obtained by adjoining ∞ to $\mathbf{Z}^+$ and let $\mathcal{C}$ be the space of all continuous functions $s : K \to \mathcal{B}$ such that $s_1 \in \mathbf{C}.f_1$ and $(s_n - s_{n-1}) \in \mathbf{C}.f_n$ for all $n \geq 2$. Then $\mathcal{C}$ is a Banach space under the uniform norm inherited from $C(K, \mathcal{B})$, and $T : s \to s_\infty$ is a bounded linear operator from $\mathcal{C}$ to $\mathcal{B}$. The basis property implies that T is one-one onto, and we deduce that T^{-1} is bounded by the inverse mapping theorem. The identity

$$\alpha_n f_n = \{T^{-1}f\}_n - \{T^{-1}f\}_{n-1}$$

implies that the Fourier coefficient α_n of $f \in \mathcal{B}$ depends continuously on f. The remainder of the lemma is elementary. $\square$

Problem 3.3.2 If $\{f_n\}_{n=1}^{\infty}$ is a complete set in a Banach space $\mathcal{B}$, prove that it is minimal complete, in the sense that the removal of any term makes it incomplete, if and only if there exists a sequence $\{\phi_n\}_{n=1}^{\infty}$ in $\mathcal{B}^*$ such that the pair is biorthogonal. $\square$

If $\{f_n\}_{n=1}^{\infty}$, $\{\phi_n\}_{n=1}^{\infty}$ is a biorthogonal pair in the Banach space $\mathcal{B}$ then we define operators P_n for $n = 1, 2, \ldots$ by

$$P_n f := \sum_{r=1}^{n} \langle f, \phi_r \rangle f_r. \tag{3.8}$$

Lemma 3.3.3 *The operators P_n are finite rank bounded projections. If $\{f_n\}_{n=1}^{\infty}$ is a basis then P_n are uniformly bounded in norm and converge strongly to the identity operator as $n \to \infty$. If the P_n are uniformly bounded in norm and $\{f_n\}_{n=1}^{\infty}$ is complete, then $\{f_n\}_{n=1}^{\infty}$ is a basis.*

Proof. The first statement follows directly from (3.8). The sequence $\{f_n\}_{n=1}^{\infty}$ is a basis if and only if P_n converges strongly to I, by Lemma 3.3.1. The second statement is therefore a consequence of the uniform boundedness theorem.

If f lies in the linear span $\mathcal{L}$ of $\{f_n\}_{n=1}^{\infty}$ then $P_n f = f$ for all large enough n. If $\mathcal{L}$ is dense in $\mathcal{B}$ and P_n are uniformly bounded in norm, then the strong convergence of P_n to I follows by an approximation argument. $\square$

Problem 3.3.4 Prove that if $\{f_n\}_{n=1}^{\infty}$ is a basis in the reflexive Banach space $\mathcal{B}$, then the sequence $\{\phi_n\}_{n=1}^{\infty}$ is a basis in $\mathcal{B}^*$. Prove that this need not be true if $\mathcal{B}$ is non-reflexive. $\square$

Example 3.3.5 (Schauder) Let $\{v_n\}_{n=1}^{\infty}$ be a dense sequence of distinct numbers in $[0, 1]$ such that $v_1 = 0$ and $v_2 = 1$. We construct a basis in $C[0, 1]$ as follows. We put $e_1(x) := 1$ and $e_2(x) := x$ for all $x \in [0, 1]$. For each $n \geq 3$ we put

$$u_n := \max\{v_r : r < n \text{ and} v_r < v_n\},$$

$$w_n := \min\{v_r : r < n \text{ and} v_r > v_n\}.$$

We then define $e_n \in C[0, 1]$ by

$$e_n(x) := \begin{cases} 0 & \text{if } x \leq u_n, \\ (x - u_n)/(v_n - u_n) & \text{if } u_n \leq x \leq v_n, \\ (w_n - x)/(w_n - v_n) & \text{if } v_n \leq x \leq w_n, \\ 0 & \text{if } x \geq w_n. \end{cases}$$

We also define $\phi_n \in (C[0, 1])^*$ by

$$\phi_n(f) := \begin{cases} f(0) & \text{if} n = 1, \\ f(1) - f(0) & \text{if} n = 2, \\ f(v_n) - \{f(u_n) + f(w_n)\}/2 & \text{if } n \geq 3. \end{cases}$$

Using the fact that $e_n(v_m) = 0$ if $m < n$, a direct calculation shows that $\{e_n\}_{n=1}^{\infty}$, $\{\phi_n\}_{n=1}^{\infty}$ is a biorthogonal pair in $C[0, 1]$. In order to prove that $\{e_n\}_{n=1}^{\infty}$ is a basis we examine the projection P_n. This is given explicitly by $P_n f := f_n$ where f_n is the continuous, piecewise linear function on $[0, 1]$ obtained by interpolating between the values of f at $\{v_1, v_2, \ldots, v_n\}$. The uniform convergence of f_n to f as $n \to \infty$ is proved by exploiting the uniform continuity of f. We also see that $\|P_n\| = 1$ for all n. $\square$

Complete minimal sequences of eigenvectors which are not bases turn up in many applications involving non-self-adjoint differential operators. One says that $\{f_n\}_{n=1}^{\infty}$ is an Abel-Lidskii basis in the Banach space $\mathcal{B}$ if it is a part of a biorthogonal pair $\{f_n\}_{n=1}^{\infty}$, $\{\phi_n\}_{n=1}^{\infty}$ and for all $f \in \mathcal{B}$ one has

$$f = \lim_{\varepsilon \to 0} \sum_{n=1}^{\infty} e^{-\varepsilon n} \langle f, \phi_n \rangle f_n.$$

In applications one frequently has to group the terms before summing as follows. One supposes that there is an increasing sequence $N(r)$ with $N(1) = 0$, such that

$$f = \lim_{\varepsilon \to 0} \sum_{r=1}^{\infty} e^{-\varepsilon r} \left\{ \sum_{n=N(r)+1}^{N(r+1)} \langle f, \phi_n \rangle f_n \right\}.$$

The point of the grouping is that the operators

$$B_r f := \sum_{n=N(r)+1}^{N(r+1)} \langle f, \phi_n \rangle f_n$$

may have much smaller norms than the individual terms in the finite sums would suggest.[8]

Let $\{f_n\}_{n=1}^{\infty}$, $\{\phi_n\}_{n=1}^{\infty}$ be a biorthogonal pair for the Banach space $\mathcal{B}$. The rank one projection $Q_n := P_n - P_{n-1}$, defined by $Q_n f := \langle f, \phi_n \rangle f_n$, has norm

$$\|Q_n\| = \|\phi_n\| \, \|f_n\| \geq 1.$$

We say that the biorthogonal pair has a polynomial growth bound if there exist c, α such that $\|Q_n\| \leq cn^{\alpha}$ for all n. We say that it is wild if no such bound exists. A basis has a polynomial growth bound with $\alpha = 0$.

Problem 3.3.6 Prove that the existence of a polynomial growth bound is invariant under a change from the given norm of $\mathcal{B}$ to an equivalent norm, and that the infimum of all possible values of the constant α is also invariant. $\square$

Our next lemma demonstrates the importance of biorthogonal sequences in spectral theory.

Theorem 3.3.7 *Suppose that $\{f_n\}_{n=1}^{\infty}$ is a sequence in $\mathcal{B}$ and that $\phi_n \in \mathcal{B}^*$ satisfy $\langle f_n, \phi_n \rangle = 1$ for all n. Then $\{f_n\}_{n=1}^{\infty}$, $\{\phi_n\}_{n=1}^{\infty}$ is a biorthogonal pair if and only if there exist a bounded operator A and distinct constants λ_n such that $Af_n = \lambda_n f_n$ and $A^*\phi_n = \lambda_n \phi_n$ for all n.*

Proof. If there exist A and λ_n with the stated properties then

$$\lambda_n \langle f_n, \phi_m \rangle = \langle Af_n, \phi_m \rangle = \langle f_n, A^*\phi_m \rangle = \lambda_m \langle f_n, \phi_m \rangle$$

for all m, n. If $m \neq n$ then $\lambda_m \neq \lambda_n$, so $\langle f_n, \phi_m \rangle = 0$.

Suppose, conversely, that $\{f_n\}_{n=1}^{\infty}$, $\{\phi_n\}_{n=1}^{\infty}$ is a biorthogonal pair, and put $c_n := \|f_n\| \, \|\phi_n\|$ for all n. Assuming that λ, $\lambda_n \in \mathbf{C}$ satisfy

$$s := \sum_{n=1}^{\infty} |\lambda_n - \lambda| c_n < \infty,$$

---

[8] See [Lidskii 1962]. For a range of applications of Abel-Lidskii bases see [Agronovich 1996]. Note, however, that the set of eigenvectors of the NSA harmonic oscillator is not an Abel-Lidskii basis; see Corollary 14.5.2.

we can define the operator A by

$$Ag := \lambda g + \sum_{n=1}^{\infty}(\lambda_n - \lambda)\langle g, \phi_n \rangle f_n.$$

The sum is norm convergent and $\|A\| \le |\lambda| + s$. One may verify directly that $Af_n = \lambda_n f_n$ and $A^*\phi_n = \bar{\lambda}_n \phi_n$ for all n. $\square$

Problem 3.3.8 Prove that in Theorem 3.3.7 one has

$$\mathrm{Spec}(A) = \{\lambda\} \cup \{\lambda_n : n \ge 1\},$$

and write down an explicit formula for $(zI - A)^{-1}$ when $z \notin \mathrm{Spec}(A)$. $\square$

We have indicated the importance of expanding arbitrary vectors in terms of the eigenvectors of suitable operators. The remainder of this section is devoted to related expansion problems in Fourier analysis. Corollary 2.3.11 established the L^2 norm convergence of the standard Fourier series of an arbitrary function in $L^2(-\pi, \pi)$. The same holds if we replace L^2 by L^p where $1 < p < \infty$, but the proof is harder.[9] However, the situation is quite different for $p = 1$.

Theorem 3.3.9 *The sequence* $\{e_n\}_{n \in \mathbb{Z}}$ *in* $L^1(-\pi, \pi)$ *defined by* $e_n(x) := e^{inx}$ *does not form a basis.*

Proof. A direct calculation shows that the projection P_n is given by $P_n(f) = k_n * f$ where

$$k_n(x) := \frac{1}{2\pi} \sum_{r=-n}^{n} e^{irx} = \frac{\sin((n+1/2)x)}{2\pi \sin(x/2)}.$$

Theorem 3.1.18 implies that

$$\|P_n\| = \|k_n\|_1 = \int_0^{\pi} \left| \frac{\sin((n+1/2)x)}{\pi \sin(x/2)} \right| \, \mathrm{d}x.$$

Routine estimates show that this diverges logarithmically as $n \to \infty$. $\square$

The same formula for $\|P_n\|$, where P_n is considered as an operator on $C_{\mathrm{per}}[-\pi, \pi]$, proves that the sequence $\{e_n\}_{n \in \mathbb{Z}}$ does not form a basis in $C_{\mathrm{per}}[-\pi, \pi]$, a result due to du Bois Reymond. One obstacle to producing simple functions in $C_{\mathrm{per}}[-\pi, \pi]$ whose Fourier series do not converge pointwise is provided by the following result. The Dini condition holds for

---

[9] See [Zygmund 1968, Theorem VII.6.4] or [Grafakos 2004, Theorem 3.5.6]. The key issue is to establish the L^p boundedness of the Hilbert transform for $1 < p < \infty$.

all Hölder continuous periodic functions, but also for many other functions with much weaker moduli of continuity.

Theorem 3.3.10 *(Dini) If $f \in C_{\text{per}}[-\pi, \pi]$ and there exists a continuous increasing function α on $[0, \infty)$ such that $\alpha(0) = 0$,*

$$|f(x) - f(y)| \leq \alpha(|x - y|)$$

for all $x, y \in [-\pi, \pi]$, and

$$\int_0^\pi \frac{\alpha(x)}{x} \, dx < \infty,$$

then the Fourier series of f converges uniformly.

Proof. A direct calculation shows that the partial sums of the Fourier series are given by

$$s_n(x) := \int_{-\pi}^\pi k_n(y) f(x - y) \, dy$$

where

$$k_n(y) := \frac{1}{2\pi} \sum_{r=-n}^n e^{iry} = \frac{\sin((n + 1/2)y)}{2\pi \sin(y/2)}.$$

We deduce that

$$s_n(x) - f(x) = \int_{-\pi}^\pi k_n(y)\{f(x - y) - f(x)\} \, dy$$

$$= \int_{-\pi}^\pi g_x(y) \sin((n + 1/2)y) \, dy$$

$$= \int_{-\pi}^\pi g_x(y)\{\sin(ny)\cos(y/2) + \cos(ny)\sin(y/2)\} \, dy$$

where

$$g_x(y) := \frac{f(x - y) - f(x)}{2\pi \sin(y/2)}.$$

The conditions of the theorem imply that

$$\int_{-\pi}^\pi |g_x(y)| \, dy \leq 2 \int_0^\pi \frac{\alpha(y)}{2\pi \sin(y/2)} \, dy \leq \int_0^\pi \frac{\alpha(y)}{y} \, dy < \infty.$$

The pointwise convergence of the Fourier series is now a consequence of the Riemann-Lebesgue lemma; see Problem 2.3.13. The proof of its uniform convergence depends upon using the fact that g_x depend continuously on x in the L^1 norm. $\square$

The following relatively elementary example of a continuous periodic function whose Fourier series does not converge at the origin appears to be new.

Theorem 3.3.11 *If*

$$f(\theta) := \sum_{r=1}^{\infty} (r!)^{-1/2} \sin\{(2^{r!} + \tfrac{1}{2})|\theta|\} \tag{3.9}$$

then $f \in C_{\mathrm{per}}[-\pi, \pi]$. However, the Fourier series of f does not converge at $\theta = 0$.

Proof. Given $u, v > 0$, put

$$K(u, v) := \frac{1}{\pi} \int_0^\pi \frac{\sin(u\theta)\sin(v\theta)}{\sin(\theta/2)} \, d\theta.$$

If $0 < u \leq v$ then

$$
\begin{aligned}
|K(u, v)| &\leq \frac{1}{\pi} \int_0^\pi \frac{|\sin(u\theta)|}{\sin(\theta/2)} \, d\theta \\
&\leq \int_0^\pi \frac{|\sin(u\theta)|}{\theta} \, d\theta \\
&= \int_0^u \frac{|\sin(\pi s)|}{s} \, ds \\
&\leq \pi + \log(u).
\end{aligned}
$$

Therefore

$$|K(u, v)| \leq 5/4 \log(u) \tag{3.10}$$

for all sufficiently large $u > 1$, provided $u \leq v$.

On the other hand

$$
\begin{aligned}
K(u, u) &= \frac{1}{\pi} \int_0^\pi \frac{\sin^2(u\theta)}{\sin(\theta/2)} \, d\theta \\
&\geq \frac{2}{\pi} \int_0^\pi \frac{\sin^2(u\theta)}{\theta} \, d\theta \\
&\geq \frac{2}{\pi} \int_1^u \frac{\sin^2(\pi s)}{s} \, ds \\
&\sim \frac{\log(u)}{\pi}
\end{aligned}
$$

as $u \to \infty$. Therefore

$$K(u, u) \geq \frac{3}{10} \log(u) \tag{3.11}$$

for all sufficiently large $u > 0$.

The partial sums of the Fourier series of f at $\theta = 0$ are given by

$$s(n) := \frac{1}{2\pi} \int_{-\pi}^{\pi} f(\theta) \frac{\sin((n+\frac{1}{2})\theta)}{\sin(\theta/2)} \, d\theta.$$

Since f is even we can rewrite $t(n) := s(2^{n!})$ in the form

$$t(n) = \sum_{r=1}^{\infty} (r!)^{-1/2} K(2^{r!} + \tfrac{1}{2}, 2^{n!} + \tfrac{1}{2}).$$

We show that this sequence diverges. By using (3.11) we see that the $r = n$ term is greater than

$$\frac{3}{10}(n!)^{-1/2} \log(2^{n!} + \tfrac{1}{2}) \sim \frac{3}{10}(n!)^{1/2} \log(2) \geq \frac{(n!)^{1/2}}{5}$$

for all large enough n. Using (3.10) we see that the sum of the terms with $r > n$ is dominated by

$$\sum_{r=n+1}^{\infty} (r!)^{-1/2} \frac{5}{4} \log(2^{n!} + \tfrac{1}{2}) \sim \frac{5}{4} \log(2) \sum_{r=n+1}^{\infty} (r!)^{-1/2} n! \leq \frac{(n!)^{1/2}}{20}$$

for all large enough n. We obtain an upper bound of the sum of the terms with $r < n$ by using (3.10), noting that this estimate is only valid for large enough u. There exist M, c not depending on n such that the sum is dominated by

$$c + \sum_{r=M}^{n-1} (r!)^{-1/2} \frac{5}{4} \log(2^{r!} + \tfrac{1}{2}) \sim c + \frac{5}{4} \log(2) \sum_{r=M}^{n-1} (r!)^{1/2} \leq \frac{(n!)^{1/2}}{20}$$

for all large enough n. Since

$$\frac{1}{20} + \frac{1}{20} < \frac{1}{5}$$

we conclude that $t(n)$ diverges as $n \to \infty$. $\square$

Note The two $r!$ terms in the definition of f can be replaced by many other sequences of positive integers.

The choice of coefficients in the series (3.9) is illuminated to some extent by the following problem.

Problem 3.3.12 Prove that if

$$f(\theta) := \sum_{n=1}^{\infty} a_n \sin\{(n+\tfrac{1}{2})|\theta|\}$$

and $\sum_{n=1}^{\infty} |a_n| n^{\varepsilon} < \infty$ for some $\varepsilon > 0$ then the Fourier series of f converges uniformly to f. $\square$

We should not leave this topic without mentioning that Carleson has proved that the Fourier series of every L^2 function on $(-\pi, \pi)$ converges to it almost everywhere. In the reverse direction given any set $E \subseteq (-\pi, \pi)$ with zero measure, there exists a function $f \in C_{\text{per}}[-\pi, \pi]$ whose Fourier series diverges on E.[10]

Problem 3.3.13 Prove that the sequence of monomials f_n in $C[-1, 1]$ defined for $n \geq 0$ by $f_n(x) := x^n$ is complete, and that if S is obtained from $\{0, 1, 2, \ldots\}$ by the removal of a finite number of terms, then the closed linear span of $\{f_n\}_{n \in S}$ is equal to $C[-1, 1]$ or to $\{f \in C[-1, 1] : f(0) = 0\}$. These facts imply that $\{f_n\}_{n=0}^{\infty}$ cannot be a basis. $\square$

Problem 3.3.14 Prove that the sequence of functions $f_n(x) := x^n e^{-x^2/2}$, $n = 0, 1, 2, \ldots$, is complete in $L^2(\mathbf{R})$. This result implies that the sequence of Hermite functions, defined as what one obtains by applying the Gram-Schmidt procedure to $\{f_n\}_{n=1}^{\infty}$, form a complete orthonormal set in $L^2(\mathbf{R})$. $\square$

The above issues are relevant when one examines the convergence of Fourier series in weighted L^2 spaces. Let w be a non-negative measurable function on $(-\pi, \pi)$ such that $\int_{-\pi}^{\pi} w(\theta)^2 \, d\theta < \infty$ and let L_w^2 denote the Hilbert space $L^2((-\pi, \pi), w(\theta)^2 d\theta)$. We use the notations $\langle \cdot, \cdot \rangle_w$ and $\| \cdot \|_w$ to denote the inner product and norm in this space. One can then ask whether the standard Fourier expansion of every function $f \in L_w^2$ converges to f in the L_w^2 norm. This holds if

$$\lim_{n \to \infty} \left\| f - \sum_{r=-n}^{n} \alpha_r u_r \right\|_w = 0$$

where

$$u_r(\theta) := e^{ir\theta},$$

$$\alpha_r := (2\pi)^{-1} \langle f, u_r \rangle = \langle f, u_r^* \rangle_w,$$

$$u_r^*(\theta) := e^{ir\theta} / 2\pi w(\theta)^2.$$

Note that the set on which u_r^* is undefined is a null set with respect to the measure $w(\theta)^2 d\theta$, and that $\langle u_r, u_s^* \rangle_w = \delta_{r,s}$ for all $r, s \in \mathbf{Z}$.

If we put $g(\theta) := f(\theta)w(\theta)$ and $e_n(\theta) := w(\theta)e^{in\theta}$ then one may ask instead whether

$$\lim_{n \to \infty} \left\| g - \sum_{r=-n}^{n} \alpha_r e_r \right\| = 0$$

---

[10] See [Carleson 1966, Jørsboe and Mejlbro 1982, Kahane and Katznelson 1966].

for all $g \in L^2((-\pi, \pi), d\theta)$. This is the form in which we will solve the problem.

Theorem 3.3.15 *Let e_n be defined for all $n \in \mathbf{Z}$ by $e_n(x) := w(x)e^{inx}$ where $w \in \mathcal{H} = L^2(-\pi, \pi)$. Then $\{e_n\}_{n \in \mathbf{Z}}$ is a complete set in $\mathcal{H}$ if and only if $S := \{x : w(x) = 0\}$ is a Lebesgue null set. It is a minimal complete set if and only if $w \in \mathcal{H}$ and $w^{-1} \in \mathcal{H}$.*[11]

Proof. If S has positive measure then $\langle e_n, \chi_S \rangle = 0$ for all $n \in \mathbf{Z}$, so the sequence $\{e_n\}_{n \in \mathbf{Z}}$ is not complete. On the other hand if S has zero measure and $\langle e_n, g \rangle = 0$ for some $g \in \mathcal{H}$ and all $n \in \mathbf{Z}$ then $w\overline{g} \in L^1(-\pi, \pi)$ and

$$\int_{-\pi}^{\pi} w(x)\overline{g(x)}e^{inx} \, dx = 0$$

for all $n \in \mathbf{Z}$. It follows by Problem 2.3.14 that $w\overline{g} = 0$. Hence $g = 0$ and the sequence $\{e_n\}_{n \in \mathbf{Z}}$ is complete.

If $w^{\pm 1} \in \mathcal{H}$ then $\{e_n\}_{n \in \mathbf{Z}}$ is complete. If we put

$$e_n^*(x) := \frac{e^{inx}}{2\pi w(x)}$$

then $e_n^* \in \mathcal{H}$ and $\langle e_m, e_n^* \rangle = \delta_{m,n}$ for all $m, n \in \mathbf{Z}$. Problem 3.3.2 now implies that $\{e_n\}_{n \in \mathbf{Z}}$ is minimal complete.

Conversely suppose that $\{e_n\}_{n \in \mathbf{Z}}$ is minimal complete. Given $n \in \mathbf{Z}$ there must exist a non-zero $g \in \mathcal{H}$ such that $\langle e_m, g \rangle = 0$ for all $m \neq n$. Hence

$$\int_{-\pi}^{\pi} w(x)\overline{g(x)}e^{imx} \, dx = 0$$

for all $m \neq n$. By choosing the constant α appropriately we obtain

$$\int_{-\pi}^{\pi} \left(w(x)\overline{g(x)} - \alpha e^{-inx} \right) e^{imx} \, dx = 0$$

for all $m \in \mathbf{Z}$. It follows by Problem 2.3.14 that

$$w(x)\overline{g(x)} - \alpha e^{-inx} = 0$$

almost everywhere. Therefore $g(x) = \overline{\alpha} e^{inx}/w(x)$ and $w^{-1} \in \mathcal{H}$. $\square$

We discuss this sequence of functions further in Theorem 3.4.8 and a closely related sequence in Theorem 3.4.10.

---

[11] The precise condition for $\{e_n\}_{n \in \mathbf{Z}}$ to be a basis involves Mockenhaupt classes. See [Garcia-Cueva and De Francia 1985.]

Example 3.3.16 In a series of recent papers Maz'ya and Schmidt[12] have advocated the use of translated Gaussian functions for the numerical solution of a variety of integral and differential equations on $\mathbf{R}^n$. They base their analysis on the use of formulae such as

$$(Mg)(x) := (\pi\alpha)^{-n/2} \sum_{m\in\mathbf{Z}^n} g(mh) \exp\left(\frac{-|x-mh|^2}{\alpha h^2}\right), \qquad (3.12)$$

where the lattice spacing $h > 0$ and the cut-off parameter $\alpha > 0$ together determine the error in the approximation Mg to the function g, assumed to be sufficiently smooth.

One may write this approximation in the form

$$Mg := \sum_{m\in\mathbf{Z}^n} \langle g, \phi_m \rangle f_m$$

where $\langle g, \phi_m \rangle := g(hm)$ and

$$f_m(x) := (\pi\alpha)^{-n/2} \exp\left(\frac{-|x-mh|^2}{\alpha h^2}\right).$$

While this bears a superficial resemblance to the expansion of a function with respect to a given basis one should note that

$$\langle f_m, \phi_n \rangle = (\pi\alpha)^{-n/2} \exp\left(\frac{-|m-n|^2}{\alpha}\right)$$

so $\{f_m\}_{m=1}^\infty$ and $\{\phi_n\}_{n=1}^\infty$ do not form a biorthogonal set. This has the consequence that $M(f_m)$ is not even approximately equal to f_m. In spite of this the approximation (3.12) is highly accurate for slowly varying, smooth functions g provided h is small enough.

The moral of this example is that one cannot describe all series expansion procedures in the framework associated with bases and biorthogonal systems: sometimes these concepts are appropriate, but on other occasions they are not. □

3.4 Unconditional bases

We say that a basis $\{f_n\}_{n=1}^\infty$ in a Banach space $\mathcal{B}$ is unconditional if every permutation of the sequence is still a basis.[13] This property is very restrictive, but it has correspondingly strong consequences. *We assume this condition*

---

[12] See [Maz'ya and Schmidt to appear].
[13] This definition and most of the results in this section are due to Lorch. See [Lorch 1939].

throughout the section. As usual we let $\{\phi_n\}_{n=1}^\infty$ denote the other half of the biorthogonal pair.

Theorem 3.4.1 *There exist bounded operators P_E on $\mathcal{B}$ for every $E \subseteq \mathbf{Z}^+$ with the following properties. $E \to P_E$ is a uniformly bounded, countably additive, projection-valued measure. For every $E \subseteq \mathbf{Z}^+$ and $n \in \mathbf{Z}^+$ we have*

$$P_E f_n = \begin{cases} f_n \text{ if } n \in E, \\ 0 \text{ otherwise.} \end{cases} \tag{3.13}$$

Proof. If E is a finite subset of $\mathbf{Z}^+$ we define the projection P_E on $\mathcal{B}$ by

$$P_E f := \sum_{n \in E} \langle f, \phi_n \rangle f_n.$$

Our first task is to prove the existence of a constant c such that $\|P_E\| \leq c$ for all such E.

We use the method of contradiction. If no such constant exists we construct a sequence of finite sets E_n such that $\{1, 2, \ldots, n\} \subseteq E_n$, $E_n \subseteq E_{n+1}$ and $\|P_{E_n}\| \geq n$ for all n. We put $E_1 := \{1\}$ and construct E_{n+1} inductively from E_n. Put $F := E_n \cup \{n+1\}$ and put $k := \max\{\|P_H\| : H \subseteq F\}$. Now let G be any finite set such that $\|P_G\| \geq 2k + n + 1$. Using the formula

$$P_{G \cup F} + P_{G \cap F} = P_G + P_F$$

we see that

$$\|P_{G \cup F}\| \geq \|P_G\| - 2k \geq n + 1.$$

We may therefore put $E_{n+1} := G \cup F$ to complete the induction.

We next relabel (permute) the sequence $\{f_n\}_{n=1}^\infty$ so that $E_n = \{1, 2, \ldots, m_n\}$ where m_n is a strictly increasing sequence. The conclusion, that the projections P_{E_n} are not uniformly bounded in norm, contradicts the assumption that the permuted sequence is a basis. Hence the constant c mentioned above does exist.

Now let $I_n := \{1, 2, \ldots, n\}$ and for any set $E \subseteq \mathbf{Z}^+$, finite or not, put $P_E f := \lim_{n \to \infty} P_{E \cap I_n} f$. This limit certainly exists for all finite linear combinations of the f_n, and by a density argument which depends upon the uniform bound proved above it exists for all $f \in \mathcal{B}$.

The other statements of the theorem follow easily from (3.13) and the bounds $\|P_E\| \leq c$. $\square$

Theorem 3.4.2 *Let* $\gamma := \{\gamma_n\}_{n=1}^{\infty}$ *be a bounded, complex-valued sequence. Then there exists a bounded operator* T_γ *on* $\mathcal{B}$ *such that*

$$T_\gamma f_n = \gamma_n f_n \tag{3.14}$$

for all n. Moreover

$$\|T_\gamma\| \leq 6c\|\gamma\|_\infty. \tag{3.15}$$

Proof. We start by assuming that γ is real-valued. For every integer $m \geq 0$ we define the 'dyadic approximation' $\gamma^{(m)}$ to γ on $\mathbf{Z}^+$ by

$$\gamma_n^{(m)} := \begin{cases} r2^{-m}\|\gamma\|_\infty & \text{if } r2^{-m}\|\gamma\|_\infty \leq \gamma_n < (r+1)2^{-m}\|\gamma\|_\infty, \\ 0 & \text{otherwise,} \end{cases}$$

where r is taken to be an integer.

It follows from the definition that

$$\gamma^{(m)} - \gamma^{(m-1)} = 2^{-m}\chi_{E_m}\|\gamma\|_\infty$$

for some subset E_m of $\mathbf{Z}^+$ and also that $\gamma^{(m)}$ converge uniformly to γ in $l^\infty(\mathbf{Z}^+)$. The definition and boundedness of each $T_{\gamma^{(m)}}$ follows from the bound $\|P_E\| \leq c$ for all E of Theorem 3.4.1, since each $\gamma^{(m)}$ takes only a finite number of values. The same bound also yields

$$\|T_{\gamma^{(0)}}\| \leq 2c\|\gamma\|_\infty$$

and

$$\|T_{\gamma^{(m)}} - T_{\gamma^{(m-1)}}\| \leq 2^{-m}c\|\gamma\|_\infty.$$

It follows that $T_{\gamma^{(m)}}$ converge in norm as $m \to \infty$ to a bounded operator which we call T_γ.

The identity (3.14) follows from the method of definition, as does the bound $\|T_\gamma\| \leq 3c\|\gamma\|_\infty$. The weaker bound (3.15) for complex γ is deduced by considering its real and imaginary parts separately. □

Corollary 3.4.3 *If* $f := \sum_{n=1}^{\infty} \alpha_n f_n$ *in* $\mathcal{B}$ *and* $\{\gamma_n\}_{n=1}^{\infty}$ *is a bounded, complex-valued sequence then* $\sum_{n=1}^{\infty} \alpha_n \gamma_n f_n$ *is also norm convergent.*

Corollary 3.4.4 *The map* $\gamma \to T_\gamma$ *from* $l^\infty(\mathbf{Z}^+)$ *to* $\mathcal{L}(\mathcal{B})$ *is an algebra homomorphism. If* $\gamma^{(n)} \in l^\infty$ *are uniformly bounded and converge pointwise to* γ *as* $n \to \infty$, *then* $T_{\gamma^{(n)}}$ *converge to* T_γ *in the strong operator topology.*

Proof. This follows directly by the use of (3.14). □

If we assume that $\mathcal{B}$ is a Hilbert space, there is a complete characterization of unconditional bases, which are then also called Riesz bases.

Theorem 3.4.5 *Let $\{f_n\}_{n=1}^{\infty}$ be a basis in the Hilbert space $\mathcal{H}$, normalized by requiring that $\|f_n\| = 1$ for all n. Then the following properties are equivalent.*

(i) *$\{f_n\}_{n=1}^{\infty}$ is an unconditional basis.*
(ii) *Given a complete orthonormal set $\{e_n\}_{n=1}^{\infty}$ in $\mathcal{H}$, there exists a bounded invertible operator B on $\mathcal{H}$ such that $Bf_n = e_n$ for all n.*
(iii) *The series $\sum_{n=1}^{\infty} \alpha_n f_n$ is norm convergent if and only if $\sum_{n=1}^{\infty} |\alpha_n|^2 < \infty$.*
(iv) *There is a positive constant c such that*

$$c^{-1}\|f\|^2 \leq \sum_{n=1}^{\infty} |\langle f, f_n \rangle|^2 \leq c\|f\|^2$$

for all $f \in \mathcal{H}$.

Proof. (i)⇒(ii). Let $\{\gamma_n\}_{n=1}^{\infty}$ be a sequence of distinct numbers of modulus 1 and let T be the operator defined by the method of Theorem 3.4.2, so that $\|T^{\pm n}\| \leq c$ for all $n \in \mathbf{Z}^+$. For each positive integer n we put

$$Y_n := \frac{1}{n+1} \sum_{r=0}^{n} (T^*)^r T^r$$

noting that $c^{-2}I \leq Y_n \leq c^2 I$. The identity

$$\lim_{n \to \infty} \langle Y_n f_r, f_s \rangle = \delta_{r,s}$$

establishes that Y_n converges in the weak operator topology to a limit Y, which satisfies $c^{-2}I \leq Y \leq c^2 I$ and

$$\langle Y f_r, f_s \rangle = \delta_{r,s}. \tag{3.16}$$

Putting $B := Y^{1/2}$ (see Lemma 5.2.1), the identity (3.16) implies that $e_n := Bf_n$ is an orthonormal sequence, and the invertibility of B implies that it is complete.

(ii)⇒(iii) and (ii)⇒(i). The stated properties hold for complete orthonormal sets and are preserved under similarity transformations.

(iii)⇒(ii). The assumed property establishes that the map $\alpha \to \sum_{n=1}^{\infty} \alpha_n f_n$ is a linear isomorphism between $l^2(\mathbf{Z}^+)$ and $\mathcal{H}$. It is bounded and invertible by the closed graph theorem. The proof is completed by composing this with any unitary map from $l^2(\mathbf{Z}^+)$ onto $\mathcal{H}$.

(ii)$\Rightarrow$(iv). Given B as in (ii) we have

$$\sum_{n=1}^{\infty} |\langle f, f_n \rangle|^2 = \sum_{n=1}^{\infty} |\langle f, B^{-1} e_n \rangle|^2 = \sum_{n=1}^{\infty} |\langle (B^{-1})^* f, e_n \rangle|^2 = \|(B^{-1})^* f\|^2.$$

The second inequality of (iv) follows immediately. The first inequality is deduced by using

$$\|f\| = \|B^* (B^{-1})^* f\| \leq \|B^*\| \, \|(B^{-1})^* f\|.$$

(iv)$\Rightarrow$(ii). If we put define $C : \mathcal{H} \to \mathcal{H}$ by $Cf := \sum_{n=1}^{\infty} \langle f, f_n \rangle e_n$ then $C\phi_m = e_m$ for all m, so C has dense range. Combining this with the identity

$$\|Cf\|^2 = \sum_{m,n} \langle f, f_n \rangle \langle e_n, e_m \rangle \langle f_m, f \rangle = \sum_n |\langle f, f_n \rangle|^2$$

we deduce by (iv) that C is bounded and invertible. Putting $B := (C^{-1})^*$, a direct calculation shows that $Bf_n = e_n$ for all n. $\square$

Theorem 3.4.6 *Let* $\{f_n\}_{n=1}^{\infty}$ *be a sequence of unit vectors in the Hilbert space* $\mathcal{H}$. *Then* $\{f_n\}_{n=1}^{\infty}$ *is a Riesz basis for the closed linear span* $\mathcal{L}$ *of the sequence if*

$$|\langle f_n, f_m \rangle| \leq c_{n-m}$$

for all m, n *and*

$$s := \sum_{\{r : r \neq 0\}} c_r < 1.$$

Proof. Let $\mathcal{F}$ denote the dense linear subspace of $l^2(\mathbf{Z})$ consisting of all sequences of finite support. We define the linear operator $B : \mathcal{F} \to \mathcal{H}$ by $B\phi = \sum_{n \in \mathbf{Z}} \phi_n f_n$. Since

$$\langle B\phi, B\psi \rangle = \sum_{m,n} \phi_n \overline{\psi_m} \langle f_n, f_m \rangle$$

we deduce that $B^* B = I + C$, where C has the infinite matrix

$$C_{m,n} := \begin{cases} \langle f_n, f_m \rangle & \text{if } m \neq n, \\ 0 & \text{otherwise.} \end{cases}$$

Therefore

$$|C_{m,n}| \leq \begin{cases} c_{n-m} & \text{if } m \neq n, \\ 0 & \text{otherwise.} \end{cases}$$

It follows by Corollary 2.2.15 that $\|C\| < s$ and hence that $\|B^* B - I\| \leq s < 1$. Therefore B may be extended to an invertible bounded linear operator mapping $l^2(\mathbf{Z})$ onto $\mathcal{L}$. This implies that $\{f_n\}_{n=1}^{\infty}$ is a Riesz basis for $\mathcal{L}$ by a slight modification of Theorem 3.4.5(ii). $\square$

Problem 3.4.7 Let $\{f_n\}_{n=1}^\infty$ be a sequence of unit vectors in the Hilbert space $\mathcal{H}$. Prove that $\{f_n\}_{n=1}^\infty$ is a Riesz basis for the closed linear span $\mathcal{L}$ of the sequence if

$$\sum_{\{m,n:m\neq n\}} |\langle f_n, f_m \rangle|^2 < 1. \qquad \square$$

Theorem 3.4.8 *The sequence $f_n(x) := w(x)e^{inx}$, where $n \in \mathbf{Z}$, of Theorem 3.3.15 is an unconditional basis in $L^2(-\pi, \pi)$ if and only if w and w^{-1} are both bounded functions.*

Proof. If $w^{\pm 1}$ are both bounded, then $\{f_n\}_{n\in\mathbf{Z}}$ is an unconditional basis by Theorem 3.4.5(ii).

If $\{f_n\}_{n\in\mathbf{Z}}$ is an unconditional basis and $a > 0$ put

$$S_a := \{x : a^{-1} \leq |w(x)| \leq a\}.$$

If $f \in L^2(-\pi, \pi)$ has support in S_a then Theorem 3.4.5(iv) implies that

$$c^{-1}\|f\|_2^2 \leq \sum_{n=-\infty}^\infty |\langle f, f_n \rangle|^2 \leq c\|f\|^2$$

where $c > 0$ does not depend on a. Equivalently

$$c^{-1}\|f\|_2^2 \leq 2\pi\|fw\|_2^2 \leq c\|f\|^2.$$

Since f is arbitrary subject to the stated condition we deduce that

$$c^{-1} \leq 2\pi|w(x)|^2 \leq c$$

for almost all $x \in S_a$ and all $a > 0$. Since $a > 0$ is arbitrary we deduce that $w^{\pm 1}$ are both bounded. $\square$

The following application of Problem 2.3.14 is needed in the proof of our next theorem.

Problem 3.4.9 Let $f : (0, \pi) \to \mathbf{C}$ satisfy $\int_0^\pi |f(x)| \sin(x)\, dx < \infty$ and let

$$\int_0^\pi f(x)\sin(nx)\, dx = 0$$

for all $n \in \mathbf{N}$. Prove that $f(x) = 0$ almost everywhere in $(0, \pi)$. $\square$

Theorem 3.4.10 *(Nath)*[14] *Let* $e_n(x) := w(x)\sin(nx)$ *for* $n = 1, 2, \ldots$. *Then* $e_n \in L^2(0, \pi)$ *for all n if and only if*

$$\int_0^\pi |w(x)|^2 \sin(x)^2 \, dx < \infty.$$

Assuming this, the following hold.

(i) $\{e_n\}_{n=1}^\infty$ *is a complete set if and only if $S = \{x : w(x) = 0\}$ is a Lebesgue null set in $(0, \pi)$.*

(ii) $\{e_n\}_{n=1}^\infty$ *is a minimal complete set if and only if*

$$\int_0^\pi |w(x)|^{-2} \sin(x)^2 \, dx < \infty.$$

(iii) *If $\{e_n\}_{n=1}^\infty$ is a basis then $w^{\pm 1} \in L^2(0, \pi)$.*

(iv) $\{e_n\}_{n=1}^\infty$ *is an unconditional basis if and only if $w^{\pm 1}$ are both bounded on $(0, \pi)$.*

Proof. The first statement uses the fact that $\sin(nx)/\sin(x)$ is a bounded function of x for every n.

(i) If S has positive measure then $\langle e_n, \chi_S \rangle = 0$ for all n, so the sequence $\{e_n\}_{n=1}^\infty$ cannot be complete. If S has zero measure and $g \in L^2(-\pi, \pi)$ satisfies $\langle e_n, g \rangle = 0$ for all n then $f = w\bar{g}$ satisfies the conditions of Problem 3.4.9 so $f(x) = 0$ almost everywhere. This implies that $g(x) = 0$ almost everywhere, from which we deduce that the sequence $\{e_n\}_{n=1}^\infty$ is complete.

(ii) If the sequence $\{e_n\}_{n=1}^\infty$ is minimal complete then S has zero measure and there exists a non-zero $g \in L^2(0, \pi)$ such that $\langle e_n, g \rangle = 0$ for all $n \geq 2$. The function $f(x) := w(x)\overline{g(x)} - \alpha \sin(x)$ satisfies the conditions of Problem 3.4.9 provided $\alpha \in \mathbf{C}$ is chosen suitably, so $f(x) = 0$ almost everywhere. Therefore

$$g(x) = \frac{\alpha \sin(x)}{w(x)}$$

almost everywhere. Since g is non-zero and lies in $L^2(0, \pi)$ we conclude that

$$\int_0^\pi |w(x)|^{-2} \sin(x)^2 \, dx < \infty.$$

Conversely, if this condition holds and we put

$$e_n^*(x) := \frac{2\sin(nx)}{\pi w(x)},$$

[14] See [Nath 2001].

then $e_n^* \in L^2(0, \pi)$ for all n and $\langle e_m, e_n^* \rangle = \delta_{m,n}$. Therefore the sequence $\{e_n\}_{n=1}^\infty$ is minimal complete.

(iii) If $\{e_n\}_{n=1}^\infty$ is a basis then the projections

$$Q_n f := \langle f, e_n^* \rangle e_n \tag{3.17}$$

are uniformly bounded in norm. We have

$$\|Q_n\|^2 = \|e_n\|^2 \|e_n^*\|^2$$

$$= \frac{4}{\pi^2} \int_0^\pi |w(x)|^2 \sin(nx)^2 \,\mathrm{d}x \int_0^\pi |w(x)|^{-2} \sin(nx)^2 \,\mathrm{d}x.$$

If $w^{\pm 1} \in L^2(0, \pi)$ this converges to $\pi^{-2} \|w\|_2^2 \|w^{-1}\|_2^2$, and otherwise it diverges.

(iv) If $w^{\pm 1}$ are both bounded, then $\{e_n\}_{n \in \mathbf{Z}}$ is an unconditional basis by Theorem 3.4.5(ii).

If $\{e_n\}_{n \in \mathbf{Z}}$ is an unconditional basis and $a > 0$ put $S_a := \{x : |w(x)| \leq a\}$. If $f \in L^2(0, \pi)$ has support in S_a then Theorem 3.4.5(iv) states that

$$c^{-1} \|f\|_2^2 \leq \sum_{n=1}^\infty |\langle f, e_n \rangle|^2 \leq c \|f\|_2^2$$

where $c > 0$ does not depend upon the choice of a. Equivalently

$$c^{-1} \|f\|_2^2 \leq \tfrac{1}{2} \pi \|fw\|_2^2 \leq c \|f\|_2^2.$$

Since f is arbitrary subject to the stated condition we deduce that

$$c^{-1} \leq \tfrac{1}{2} \pi |w(x)|^2 \leq c$$

for almost all $x \in S_a$ and all $a > 0$. Since $a > 0$ is arbitrary we deduce that $w^{\pm 1}$ are both bounded. $\quad\square$

Problem 3.4.11 If $w(x) := x^{-\alpha}$ and $1/2 < |\alpha| < 3/2$, find the exact rate of divergence of $\|Q_n\|$ as $n \to \infty$, where Q_n is defined by (3.17). $\quad\square$

The theory of wavelets provides examples of Riesz bases, as well as a method of constructing the operator B of Theorem 3.4.5.

Theorem 3.4.12[15] *Let $\phi \in L^2(\mathbf{R})$ and put $\phi_n(x) := \phi(x - n)$ for all $n \in \mathbf{Z}$. If there exists a constant $c > 0$ such that*

$$\Gamma(\xi) := \left\{ \sum_{n \in \mathbf{Z}} |(\mathcal{F}\phi)(\xi + 2\pi n)|^2 \right\}^{1/2} \tag{3.18}$$

---

[15] The converse of this theorem is also true, and we refer to any book on wavelet theory for its proof, e.g. [Daubechies 1992, Jensen and Cour-Harbo 2001].

satisfies

$$c^{-1} \le \Gamma(\xi) \le c$$

for all $\xi \in \mathbf{R}$ then $\{\phi_n\}_{n \in \mathbf{Z}}$ is a Riesz basis in its closed linear span.

Proof. We start by defining the bounded invertible operator B on $L^2(\mathbf{R})$ by

$$(\mathcal{F}Bf)(\xi) := \frac{(\mathcal{F}f)(\xi)}{\sqrt{2\pi}\Gamma(\xi)}.$$

We then observe that $\psi_n := B\phi_n$ satisfy

$$\langle \psi_m, \psi_n \rangle = \int_{\mathbf{R}} \frac{(\mathcal{F}\phi_n)(\xi)\overline{(\mathcal{F}\phi_m)(\xi)}}{2\pi\Gamma(\xi)^2} d\xi$$

$$= \int_{\mathbf{R}} \frac{|(\mathcal{F}\phi)(\xi)|^2 e^{i(m-n)\xi}}{2\pi\Gamma(\xi)^2} d\xi.$$

Since Γ is periodic with period 2π the integral can be expressed as the sum of integrals over $(2\pi r, 2\pi(r+1))$, where $r \in \mathbf{Z}$. An application of (3.18) now implies that

$$\langle \psi_m, \psi_n \rangle = \frac{1}{2\pi} \int_0^{2\pi} e^{i(m-n)\xi} d\xi = \delta_{m,n}. \qquad \square$$

Problem 3.4.13 Prove directly that the set of functions

$$\phi_n(x) := \max\{1 - |x - n|, 0\},$$

where $n \in \mathbf{Z}$, form a Riesz basis in its closed linear span $\mathcal{L}$, and identify $\mathcal{L}$ explicitly. $\square$

4
Intermediate operator theory

4.1 The spectral radius

This chapter is devoted to three closely related topics – compact linear operators, Fredholm operators and the essential spectrum. Each of these is a classical subject, but the last is the least settled. There are several distinct definitions of the essential spectrum, and we only consider the one that is related to the notion of Fredholm operators. This section treats some preliminary material.

We have already shown in Theorem 1.2.11 that the spectrum $\mathrm{Spec}(A)$ of a bounded linear operator A on the Banach space $\mathcal{B}$ is a non-empty closed bounded set. We define the spectral radius of A by

$$\mathrm{Rad}(A) := \max\{|z| : z \in \mathrm{Spec}(A)\}.$$

In this section we present Gel'fand's classical formula for $\mathrm{Rad}(A)$, and then give some examples that illustrate the importance of the concept.

Lemma 4.1.1 *If A is a bounded linear operator on $\mathcal{B}$ then the limit*

$$\rho := \lim_{n \to \infty} \|A^n\|^{1/n}$$

exists and satisfies $0 \le \rho \le \|A\|$.

Proof. If we put

$$p(n) := \log(\|A^n\|)$$

then it is immediate that p is subadditive in the sense that $p(m+n) \le p(m) + p(n)$ for all non-negative integers m, n. The proof is completed by applying the following lemma. $\square$

99

Lemma 4.1.2 *If $p : \mathbf{Z}^+ \to [-\infty, \infty)$ is a subadditive sequence then*

$$-\infty \leq \inf_n \{n^{-1} p(n)\} = \lim_{n \to \infty} \{n^{-1} p(n)\} < \infty. \tag{4.1}$$

Proof. If $a > 0$ and $p(a) = -\infty$ then $p(n) = -\infty$ for all $n > a$ by the subadditivity, and the lemma is trivial, so let us assume that $p(n)$ is finite for all $n \geq 0$.

If $a^{-1} p(a) < \gamma$ and $na \leq t < (n+1)a$ for some positive integer n then

$$t^{-1} p(t) \leq t^{-1} \{np(a) + p(t - na)\}$$
$$\leq a^{-1} p(a) + t^{-1} \max\{p(s) : 0 \leq s \leq a\},$$

which is less than γ for all large enough t. This implies the stated result. $\square$

Theorem 4.1.3 *(Gel'fand) If A is bounded then*

$$\mathrm{Rad}(A) = \lim_{n \to \infty} \|A^n\|^{1/n}.$$

Proof. We proved in Theorem 1.2.11 that

$$R(z, A) = \sum_{n=0}^{\infty} z^{-n-1} A^n$$

for all $|z| > \|A\|$. We also proved that $R(z, A)$ is a norm analytic function of z. Theorem 1.2.10 implies that $\|R(z, A)\|$ diverges to infinity as $z \to \mathrm{Spec}(A)$. Theorem 1.4.12 now implies that the power series $\sum_{n=0}^{\infty} w^n A^n$ has radius of convergence $\mathrm{Rad}(A)^{-1}$. The proof is completed by applying Problem 1.4.11. $\square$

Problem 4.1.4 Use Theorem 4.1.3 to find the spectral radius of the Volterra operator

$$(Af)(x) := \int_0^x f(t)\, dt$$

acting on $L^2[0, 1]$.

An operator A with spectral radius 1 need not be power-bounded in the sense that $\|A^n\| \leq c$ for some c and all $n \in \mathbf{N}$.[1]

---

[1] There are some very deep results about this question. See Theorem 10.5.1 and the note on page 317 for some results about power-bounded operators, and [Ransford 2005, Ransford and Roginskaya 2006] for new results and references to the literature on operators that have spectral radius 1 but are not power-bounded.

The asymptotics of $\|A^n f\|_2$ as $n \to \infty$ can be more complicated than that of $\|A^n\|$. For the Volterra operator in Problem 4.1.4, Shkarin has proved that

$$\lim_{n \to \infty} (n! \, \|A^n f\|_2)^{1/n} = 1 - \inf \operatorname{supp}(f)$$

for all $f \in L^2(0, 1)$.[2] □

Example 4.1.5 The linear recurrence equation

$$x_{n+1} = \phi_0 x_n + \phi_1 x_{n-1} + \cdots + \phi_k x_{n-k} + \xi_n$$

arises in many areas of applied mathematics. It may be rewritten in the vector form

$$a_{n+1} = A a_n + b_n$$

where a_n is the column vector $(x_n, x_{n-1}, \ldots, x_{n-k})'$, A is a certain highly non-self-adjoint $(k+1) \times (k+1)$ matrix and b_n are suitable vectors. If $k := 4$, for example, one has

$$A := \begin{pmatrix} \phi_0 & \phi_1 & \phi_2 & \phi_3 & \phi_4 \\ 1 & 0 & & & \\ & 1 & 0 & & \\ & & 1 & 0 & \\ & & & 1 & 0 \end{pmatrix},$$

all the unmarked entries vanishing. The eigenvalues of A are obtained by solving the characteristic equation

$$z^{k+1} = \phi_0 z^k + \phi_1 z^{k-1} + \cdots + \phi_{k-1} z + \phi_k.$$

One readily sees that the stability condition

$$|\phi_0| + |\phi_1| + \cdots + |\phi_k| < 1$$

implies that every eigenvalue z satisfies $|z| < 1$. If one assumes this condition and uses the l^∞ norm on $\mathbf{C}^{k+1}$ then $\|A\| = 1$ by Theorem 2.2.8. Although $\operatorname{Rad}(A) < 1$, the norms $\|A^r\|$ do not start decreasing until $r > k$: by computing $\|A^r 1\|_\infty$ one sees that $\|A^r\| = 1$ for all $r \leq k$.

More generally one may consider the equation

$$a_{n+1} = A a_n + b_n$$

---

[2] The same holds if one replaces 2 by p for any $1 \leq p \leq \infty$, [Shkarin 2006]. Shkarin also proves analogous results for fractional integration operators.

where A is a bounded linear operator on the Banach space $\mathcal{B}$ and b_n is a sequence of vectors in $\mathcal{B}$. Assuming that b_n is sufficiently well-behaved and $\|A^n\| \to 0$ as $n \to \infty$, one can often determine the long time asymptotics of the solution of the equation, namely

$$a_n = A^n a_0 + \sum_{r=0}^{n-1} A^r b_{n-r}.$$

The obvious condition to impose is that $\|A\| < 1$, but a better condition is $\mathrm{Rad}(A) < 1$. By Theorem 4.1.3 this implies that there exist positive constants M and $c < 1$ such that

$$\|A^n\| \le M c^n \qquad (4.2)$$

for all $n \ge 1$. The constant M will be quite large if $\|A^n\|$ remains of order 1 for all n up to some critical time k, after which it starts to decrease at an exponential rate.

If $b_n = b$ for all n and $\mathrm{Rad}(A) < 1$ then

$$a_n = A^n a_0 + (I - A^n)(I - A)^{-1} b$$

and

$$\lim_{n \to \infty} a_n = (I - A)^{-1} b.$$

Numerically the convergence of a_n depends on the rate at which $\|A^n\|$ decreases and on the size of $\|(I - A)^{-1}\|$. $\quad\square$

Problem 4.1.6 Find the explicit solution for a_n and use it to determine the long time asymptotic behaviour of a_n, given that

$$a_n = A a_{n-1} + n b + c$$

for all $n \ge 1$, where $b, c \in \mathcal{B}$ and $\mathrm{Rad}(A) < 1$. $\quad\square$

4.2 Compact linear operators

Disentangling the historical development of the spectral theory of compact operators is particularly hard, because many of the results were originally proved early in the twentieth century for integral equations acting on particular Banach spaces of functions. In this section we describe the apparently final form that the subject has now taken.[3]

---

[3] We refer the interested reader to [Dunford and Schwartz 1966, pp. 609 ff.] for an account of the history.

Lemma 4.2.1[4] *In a complete metric space* (M, d), *the following three conditions on a set* $K \subseteq M$ *are equivalent.*

 (i) K is compact.

 (ii) Every sequence $x_n \in K$ *has a subsequence which converges, the limit also lying in K(sequential compactness).*

(iii) K is closed and totally bounded in the sense that for every $\varepsilon > 0$ *there exists a finite set* $\{x_r\}_{r=1}^{n} \subseteq K$ *such that*

$$K \subseteq \bigcup_{r=1}^{n} B(x_r, \varepsilon)$$

where $B(a, s)$ *denotes the open ball with centre a and radius s.*

Note that the closure of a totally bounded set is also totally bounded, and hence compact. An operator $A : \mathcal{B} \to \mathcal{B}$ is said to be compact if $A(B(0, 1))$ has compact closure in $\mathcal{B}$.

Theorem 4.2.2 *The set* $\mathcal{K}(\mathcal{B})$ *of all compact operators on a Banach space* $\mathcal{B}$ *forms a norm closed, two-sided ideal in the algebra* $\mathcal{L}(\mathcal{B})$ *of all bounded operators on* $\mathcal{B}$.

Proof. The proof is routine except for the statement that $\mathcal{K}(\mathcal{B})$ is norm closed. Let $\lim_{n\to\infty} \|A_n - A\| = 0$, where A_n are compact and A is bounded. Let $\varepsilon > 0$ and choose n so that $\|A_n - A\| < \varepsilon/2$. Putting $S := A(B(0, 1))$ we have

$$S \subseteq A_n(B(0, 1)) + B(0, \varepsilon/2) \subseteq K + B(0, \varepsilon/2)$$

where K is compact. There exists a finite set $a_1, .., a_m$ such that $K \subseteq \bigcup_{r=1}^{m} B(a_r, \varepsilon/2)$, and this implies that $S \subseteq \bigcup_{r=1}^{m} B(a_r, \varepsilon)$. Hence S is totally bounded, and A is compact. $\square$

Problem 4.2.3 Prove that if B is a compact operator and A_n converges strongly to A then $\lim_{n\to\infty} \|A_n B - AB\| = 0$. $\square$

It is easy to see that every finite rank operator on $\mathcal{B}$ lies in $\mathcal{K}(\mathcal{B})$. If every compact operator is a norm limit of finite rank operators, one says that $\mathcal{B}$ has the approximation property. All of the standard Banach spaces have the approximation property, but spaces without it do exist.[5]

---

[4] Proofs may be found in [Pitts 1972, Theorem 5.6.1] and [Sutherland 1975, Chap. 7].
[5] A famous counterexample was constructed by Enflo in [Enflo 1973].

Theorem 4.2.4 *If there exists a sequence of finite rank operators A_n on $\mathcal{B}$ which converges strongly to the identity operator, then an operator on $\mathcal{B}$ is compact if and only if it is a norm limit of finite rank operators.*

Proof. This follows directly from Problem 4.2.3 and Theorem 4.2.2. □

We recall from Problem 2.2.9 that two operators A_p on $L^p(X, dx)$ and A_q on $L^q(X, dx)$ are said to be consistent if $A_p f = A_q f$ for all $f \in L^p \cap L^q$.

Theorem 4.2.5 *Let (X, Σ, dx) be a measure space satisfying conditions (i)–(viii) in Section 2.1. Then $L^p(X, dx)$ has the approximation property for all $p \in [1, \infty)$. Indeed there exists a consistent sequence of finite rank projections $P_{p,n}$ with norm 1 acting in $L^p(X, dx)$ which converges strongly to I as $n \to \infty$ for each p.*

Proof. We use the notation of the conditions (i)–(viii) freely. For each n we use the partition $\mathcal{E}_n$ to define the operator $P_{p,n}$ on $L^p(X)$ by

$$P_{p,n} f := \sum_{r=1}^{m(n)} |E_r|^{-1} \langle f, \chi_{E_r} \rangle \chi_{E_r}.$$

It is immediate that $P_{p,n}$ is a projection with range equal to $\mathcal{L}_n$ and that the projections are consistent as p varies. If $1/p + 1/q = 1$ and $f \in L^p(X)$ then

$$\|P_{p,n} f\|_p^p = \sum_{r=1}^{m(n)} |E_r|^{-p} |\langle f, \chi_{E_r} \rangle|^p |E_r|$$

$$\leq \sum_{r=1}^{m(n)} |E_r|^{1-p} \|f \chi_{E_r}\|_p^p \|\chi_{E_r}\|_q^p$$

$$= \sum_{r=1}^{m(n)} \|f \chi_{E_r}\|_p^p$$

$$\leq \|f\|_p^p.$$

Therefore $\|P_{p,n}\| \leq 1$.

If f lies in the dense linear subspace $\mathcal{L} := \bigcup_{n \geq 1} \mathcal{L}_n$ of $L^p(X)$ then $P_{p,n} f = f$ for all large enough n. Therefore $\|P_{p,n} g - g\|_p \to 0$ as $n \to \infty$ for all $g \in L^p(X)$ and all $p \in [1, \infty)$. □

The proof of the following theorem is motivated by the finite element approximation method in numerical analysis.

Theorem 4.2.6 *Let* (K, d) *be a compact metric space and let* $\mathcal{B} = C(K)$. *Then* $\mathcal{B}$ *has the approximation property.*

Proof. Given $\delta > 0$ there exists a finite set of points $a_1, \ldots, a_n \in K$ such that

$$\min_{1 \leq r \leq n} d(x, a_r) < \delta$$

for all $x \in K$. Define the functions ψ_r on K by

$$\psi_r(x) := \begin{cases} \delta - d(x, a_r) & \text{if } d(x, a_r) < \delta, \\ 0 & \text{otherwise,} \end{cases}$$

and put

$$\phi_r(x) := \frac{\psi_r(x)}{\sum_{s=1}^{n} \psi_s(x)}.$$

Then ϕ_r are continuous and non-negative with $\sum_{r=1}^{n} \phi_r(x) = 1$ for all $x \in K$. Moreover $\phi_r(x) = 0$ if $d(x, a_r) \geq \delta$. One calls such a set of functions a (continuous) partition of the identity.

The formula

$$(Q_\delta f)(x) := \sum_{r=1}^{n} f(a_r) \phi_r(x)$$

defines a finite rank operator on $C(K)$. It is immediate from the definition that $\|Q_\delta\| \leq 1$ for all $\delta > 0$, and we have to show that $Q_\delta f$ converges uniformly to f as $\delta \to 0$ for every $f \in C(K)$.

Given $\varepsilon > 0$ and $f \in C(K)$ there exists $\delta > 0$ such that $d(u, v) < \delta$ implies $|f(u) - f(v)| < \varepsilon$. For any such δ, some choice of $a_1, \ldots, a_n$ and any $x \in K$ let

$$S(x) := \{r : d(x, a_r) < \delta\}.$$

Then

$$|(Q_\delta f)(x) - f(x)| = |\sum_{r=1}^{n} \phi_r(x)\{f(a_r) - f(x)\}|$$

$$\leq \sum_{r \in S(x)} \phi_r(x)|f(a_r) - f(x)|$$

$$< \varepsilon \sum_{r \in S(x)} \phi_r(x)$$

$$\leq \varepsilon.$$

In other words $Q_\delta f$ converges uniformly to f as $\delta \to 0$. $\square$

A set S of complex-valued functions on a topological space K is said to be equicontinuous if for all $a \in K$ and all $\varepsilon > 0$ there exists an open set $U_{a,\varepsilon} \subseteq K$ such that $x \in U_{a,\varepsilon}$ implies $|f(x) - f(a)| < \varepsilon$ for all $f \in S$.

Theorem 4.2.7 *(Arzela-Ascoli) Let K be a compact Hausdorff space. A set $S \subseteq C(K)$ is totally bounded in the uniform norm $\| \cdot \|_\infty$ if and only if it is bounded and equicontinuous.*

Proof. We do not indicate the dependence of various quantities on S and ε below. If $S \subseteq C(K)$ is totally bounded and $\varepsilon > 0$ then there exist $f_1, \ldots, f_M \in S$ such that

$$S \subseteq \bigcup_{m=1}^{M} B(f_m, \varepsilon/3).$$

By using the continuity of each f_m, we see that if $a \in K$ and $\varepsilon > 0$ there exists an open set U such that $x \in U$ implies $|f_m(x) - f_m(a)| < \varepsilon/3$ for all $m \in \{1, \ldots, M\}$. Given $f \in S$ there exists $m \in \{1, \ldots, M\}$ such that $\|f - f_m\|_\infty < \varepsilon/3$. If $x \in U$ then

$$|f(x) - f(a)| \le |f(x) - f_m(x)|$$
$$+ |f_m(x) - f_m(a)| + |f_m(a) - f(a)|$$
$$< \varepsilon.$$

Therefore S is an equicontinuous set. The boundedness of S is elementary.

Conversely suppose that S is bounded and equicontinuous and $\varepsilon > 0$. Given $a \in K$ there exists an open set $U_a \subseteq K$ such that $x \in U_a$ implies $|f(x) - f(a)| < \varepsilon/3$ for all $f \in S$. The sets $\{U_a\}_{a \in K}$ form an open cover of K, so there exists a finite subcover $\{U_{a_n}\}_{n=1}^{N}$. Let $\mathcal{J} : C(K) \to \mathbf{C}^N$ be the map defined by $(\mathcal{J}f)_n := f(a_n)$, and give $\mathbf{C}^N$ the l^∞ norm. The set $\mathcal{J}(S)$ is bounded and finite-dimensional, so there exists a finite set $f_1, \ldots, f_M \in S$ such that

$$\mathcal{J}(S) \subseteq \bigcup_{m=1}^{M} B(\mathcal{J}(f_m), \varepsilon/3).$$

Given $f \in S$ there exists $m \in \{1, \ldots, M\}$ such that $\|\mathcal{J}f - \mathcal{J}f_m\|_\infty < \varepsilon/3$. Given $x \in K$ there exists $n \in \{1, \ldots, N\}$ such that $x \in U_{a_n}$. Hence

$$|f(x) - f_m(x)| < |f(x) - f(a_n)|$$
$$+ |f(a_n) - f_m(a_n)| + |f_m(a_n) - f_m(x)|$$
$$< \varepsilon.$$

Since m does not depend on x we deduce that $\|f - f_m\|_\infty < \varepsilon$, and that S is totally bounded. $\square$

For a set S in $L^p(X, \mathrm{d}x)$ to be compact one needs to impose decay conditions at infinity and local oscillation restrictions, in both cases uniformly for all $f \in S$.[6] Characterizing the compact subsets of $l^p(X)$ is somewhat easier, and the following special case is often useful.

Theorem 4.2.8 *Suppose that X is a countable set and $1 \le p < \infty$. If $f : X \to [0, \infty)$ then $S := \{g : |g| \le f\}$ is a compact subset of $l^p(X)$ if and only if $f \in l^p(X)$.*

Proof. If $f \in l^p(X)$ then for any choice of $\varepsilon > 0$ there exists a finite set $E \subseteq X$ such that

$$\sum_{x \notin E} |f(x)|^p \le (\varepsilon/2)^p.$$

This implies that $\|g - g\chi_E\|_p \le \varepsilon/2$ for all $g \in S$.

The set $\chi_E S$ is a closed bounded set in a finite-dimensional space, so it is compact. Lemma 4.2.1 implies that there exists a finite sequence $x_1, \ldots, x_n \in \chi_E S$ such that $\chi_E S \subseteq \bigcup_{r=1}^n B(x_r, \varepsilon/2)$. We deduce that $S \subseteq \bigcup_{r=1}^n B(x_r, \varepsilon)$. The total boundedness of S implies that it is compact by a second application of Lemma 4.2.1.

Conversely, let E_n be an increasing sequence of finite sets with union equal to X. If $f \notin l^p(X)$ then the functions $f_n := \chi_{E_n} f$ lie in S and $\|f_n\|_p$ diverges as $n \to \infty$. Therefore S is not a bounded set and cannot be compact. $\square$

Problem 4.2.9 Suppose that X is a countable set and $1 \le p < \infty$. By modifying the proof of Theorem 4.2.8 prove that a closed bounded subset S of $l^p(X)$ is compact if and only if for every $\varepsilon > 0$ there exists a finite set $E \subseteq X$ depending only on S, p and ε such that

$$\sum_{x \notin E} |f(x)|^p \le \varepsilon^p$$

for all $f \in S$. $\square$

The following problem shows that the continuous analogue of Theorem 4.2.8 is false.

---

[6] See [Dunford and Schwartz 1966, Theorem IV.8.21].

Problem 4.2.10 Prove that the set $\{f \in L^p(0, 1) : |f| \leq 1\}$ is not a compact subset of $L^p(0, 1)$ for any choice of $p \in [1, \infty]$. $\square$

The closed convex hull $\overline{\text{Conv}}(S)$ of a subset S of a Banach space $\mathcal{B}$ is defined to be the smallest closed convex subset of $\mathcal{B}$ that contains S. It may be obtained by taking the norm closure of the set of all finite convex combinations $x := \sum_{r=1}^{n} \lambda_r x_r$ where $x_r \in S$, $\lambda_r \geq 0$ and $\sum_{r=1}^{n} \lambda_r = 1$.

Theorem 4.2.11 *(Mazur) The closed convex hull of a compact subset X in a Banach space $\mathcal{B}$ is also compact.*

Proof. Since X is a compact metric space it contains a countable dense set. This set generates a closed separable subspace $\mathcal{L}$ of $\mathcal{B}$, and the whole proof may be carried out within $\mathcal{L}$. We may therefore assume that $\mathcal{B}$ is separable without loss of generality. Let K denote the unit ball in $\mathcal{B}^*$. This is compact with respect to the weak* topology by Theorem 1.3.7 and the topology is associated with a metric d by Problem 1.3.8. The linear map $\mathcal{J} : \mathcal{B} \to C(K)$ defined by $(\mathcal{J}f)(\phi) := \phi(f)$ for all $\phi \in K$ is isometric by the Hahn-Banach theorem, so the image is a closed linear subspace L of $C(K)$.[7] The set $\mathcal{J}(X)$ is compact and hence equicontinuous by Theorem 4.2.7.

Given $\varepsilon > 0$ there exists $\delta > 0$ such that $|f(x) - f(y)| < \varepsilon$ for all $x, y \in K$ such that $d(x, y) < \delta$ and for all $f \in \mathcal{J}(X)$. A direct calculation shows that the same estimate holds for all $f \in \mathcal{J}(\text{Conv}(X))$. Therefore $\mathcal{J}(\text{Conv}(X))$ is equicontinuous and bounded. Theorem 4.2.7 now implies that $\mathcal{J}(\text{Conv}(X))$ is totally bounded in $C(K)$. Hence $\text{Conv}(X)$ is totally bounded in $\mathcal{B}$ and $\overline{\text{Conv}}(X)$ is compact. $\square$

Problem 4.2.12 Give a direct proof of Theorem 4.2.11 by using finite convex combinations of elements of X to show that $\text{Conv}(X)$ is totally bounded if X is totally bounded. $\square$

Theorem 4.2.13 *(Schauder) Let A be a bounded linear operator on the Banach space $\mathcal{B}$. Then A is compact if and only if A^* is compact.*

Proof. Suppose that A is compact and let K be the compact set $\overline{\{Ax : \|x\| \leq 1\}}$ in $\mathcal{B}$. Let $\mathcal{L}$ be the closed subspace of $C(K)$ consisting of all continuous functions f on K such that $f(\alpha x + \beta y) = \alpha f(x) + \beta f(y)$ provided $\alpha, \beta \in \mathbf{C}$, $x, y \in K$ and $\alpha x + \beta y \in K$. Let $E : \mathcal{B}^* \to \mathcal{L}$ be defined by $(E\phi)(k) := \langle k, \phi \rangle$

---

[7] The fact that every Banach space $\mathcal{B}$ is isometrically isomorphic to a closed subspace of $C(K)$ for some compact Hausdorff space K is called the Banach-Mazur theorem.

for all $\phi \in \mathcal{B}^*$ and all $k \in K$. Let $D : \mathcal{L} \to \mathcal{B}^*$ be defined by $(Df)(x) := f(Ax)$ for all $f \in \mathcal{L}$ and all $x \in \mathcal{B}$ such that $\|x\| \leq 1$. One sees immediately that $A^* = DE$. Since E is compact by Theorem 4.2.7 and D is bounded, we deduce that A^* is compact.

Conversely, if A^* is compact then A^{**} must be compact. But A is the restriction of A^{**} from $\mathcal{B}^{**}$ to $\mathcal{B}$, so it also is compact. $\square$

Theorem 4.2.14[8] *Let A_p be a consistent family of bounded operators on $L^p(X, dx)$, where*

$$\frac{1}{p} := \frac{1-\lambda}{p_0} + \frac{\lambda}{p_1},$$

$1 \leq p_0, p_1 \leq \infty$ *and* $0 \leq \lambda \leq 1$. *If A_{p_0} is compact, then A_p is compact on L^p for all $p \in [p_0, p_1)$.*

Proof. Let $B_{p,n} := P_{p,n} A_p$ where $P_{p,n}$ is the consistent sequence of finite rank projections defined in Theorem 4.2.5. Then

$$\lim_{n \to \infty} \|B_{p_0,n} - A_{p_0}\| = 0$$

by Problem 4.2.3. We also have

$$\|B_{p_1,n} - A_{p_1}\| \leq 2\|A_{p_1}\|$$

for all n. By interpolation, i.e. Theorem 2.2.14, we deduce that

$$\lim_{n \to \infty} \|B_{p,n} - A_p\| = 0$$

for all $p_0 < p < p_1$. Hence A_p is compact. $\square$

The fact that the spectrum of A_p is independent of p follows from the following more general theorem.[9]

Theorem 4.2.15 *Let $\mathcal{B}_1$ and $\mathcal{B}_2$ be compatible Banach spaces as defined on page 49. If A_1 and A_2 are consistent compact operators acting in $\mathcal{B}_1$ and $\mathcal{B}_2$ respectively then*

$$\mathrm{Spec}(A_1) = \mathrm{Spec}(A_2).$$

[8] See [Krasnosel'skii 1960] and [Persson 1964], which led to several further papers on this matter for abstract interpolation spaces.
[9] Note that the spectrum of A_p may depend on p if one only assumes that A_p is a consistent family of bounded operators. See also Example 2.2.11 and Theorem 12.6.2.

The range of the spectral projection P_r of A_r associated with any non-zero eigenvalue λ is independent of r and is contained in $\mathcal{B}_1 \cap \mathcal{B}_2$. Every eigenvector of either operator is also an eigenvector of the other operator.

Proof. The set $S := \operatorname{Spec}(A_1) \cup \operatorname{Spec}(A_2)$ is closed and its only possible limit point is 0 by Theorem 4.3.19. Let $0 \neq a \in S$ and let γ be a sufficiently small circle with centre at a. The resolvent operators $R(z, A_1)$ and $R(z, A_2)$ are consistent for all $z \notin S$ by Problem 2.2.10, so the spectral projections defined as in Theorem 1.5.4 by

$$P_r := \frac{1}{2\pi i} \int_\gamma R(z, A_r) \, dz$$

are also consistent. It follows from Theorem 4.3.19 that P_r are of finite rank. This implies that $\operatorname{Ran}(P_r) = P_r(\mathcal{B}_1 \cap \mathcal{B}_2)$. Therefore the two projections have the same range. The final assertion follows from the fact that any eigenvector lies in the range of the spectral projection. $\square$

We conclude the section with some miscellaneous results which will be used later.

The following compactness theorem for Hilbert-Schmidt operators strengthens Lemma 2.2.2. We will treat this class of operators again, at a greater level of abstraction, in Section 5.5.

Theorem 4.2.16 *If $K \in L^2(X \times X)$ then the formula*

$$(Af)(x) := \int_X K(x, y) f(y) \, dy$$

defines a compact linear operator on $L^2(X, dx)$.

Proof. We see that A depends linearly on K and recall from Lemma 2.2.2 that $\|A\| \leq \|K\|_2$ for all $K \in L^2(X \times X)$.

If ϕ_n, $n = 1, 2, \ldots$ is a complete orthonormal set in $L^2(X)$, then the set of functions $\psi_{m,n}(x, y) := \phi_m(x)\overline{\phi_n(y)}$ is a complete orthonormal set in $L^2(X \times X)$.[10] The kernel K has an expansion

$$K(x, y) = \sum_{m,n=1}^{\infty} \alpha_{m,n} \phi_m(x)\overline{\phi_n(y)}$$

which is norm convergent in $L^2(X \times X)$. The Fourier coefficients are given by

$$\alpha_{m,n} := \langle K, \psi_{m,n} \rangle = \langle A\phi_n, \phi_m \rangle,$$

---

[10] The proof of this depends upon the way in which the product measure $dx \times dy$ is constructed.

where the first inner product is in $L^2(X \times X)$ and the second is in $L^2(X)$. The corresponding operator expansion

$$Af = \sum_{m,n=1}^{\infty} \alpha_{m,n} \phi_m \langle f, \phi_n \rangle$$

is convergent in the operator norm by the first half of this proof. Since A is the limit of a norm convergent sequence of finite rank operators it is compact. $\square$

The Hilbert-Schmidt norm of such operators is defined by

$$\|A\|_2^2 := \int_{X \times X} |K(x, y)|^2 \, \mathrm{d}x \mathrm{d}y$$

$$= \sum_{m,n=1}^{\infty} |\langle A\phi_n, \phi_m \rangle|^2. \tag{4.3}$$

Theorem 4.2.17 *Let $\mathcal{H} := L^2(X, \mathrm{d}x)$ where X has finite measure. If $A : L^2(X, \mathrm{d}x) \to L^\infty(X, \mathrm{d}x)$ is a bounded linear operator then A is Hilbert-Schmidt and hence compact considered as an operator from $L^2(X, \mathrm{d}x)$ to $L^2(X, \mathrm{d}x)$.*

Proof. We first observe that if $\|Af\|_\infty \le c\|f\|_2$ for all $f \in L^2(X)$ then $\|A^*f\|_2 \le c\|f\|_1$ for all $f \in L^1(X) \cap L^2(X)$.[11]

If $\mathcal{E} := \{E_1, \dots, E_n\}$ is a finite partition of X and each set E_r has positive measure, then we define the finite rank projection $P_\mathcal{E}$ on $L^2(X)$ by

$$P_\mathcal{E} f := \sum_{r=1}^{n} |E_r|^{-1} \chi_{E_r} \langle f, \chi_{E_r} \rangle.$$

A direct calculation shows that

$$(P_\mathcal{E} Af)(x) = \int_X K_\mathcal{E}(x, y) f(y) \, \mathrm{d}y$$

for all $f \in L^2(X)$, where

$$K_\mathcal{E}(x, y) := \sum_{r=1}^{n} |E_r|^{-1} \chi_{E_r}(x) \overline{(A^* \chi_{E_r})(y)}.$$

For every $x \in X$ there exists r such that

$$K_\mathcal{E}(x, y) := |E_r|^{-1} \overline{(A^* \chi_{E_r})(y)}$$

---

[11] The operator A^* maps $(L^\infty(X))^*$ into $L^2(X)^* \sim L^2(X)$, but $L^1(X)$ is isometrically embedded in $(L^\infty(X))^*$, so we also have $A^* : L^1(X) \to L^2(X)$ after restriction.

for all $y \in X$, so

$$\|K_{\mathcal{E}}(x, \cdot)\|_2 = |E_r|^{-1} \|A^* \chi_{E_r}\|_2$$
$$\leq |E_r|^{-1} c \|\chi_{E_r}\|_1$$
$$= c.$$

Therefore $\|K_{\mathcal{E}}\|_2^2 \leq c^2 |X|$. We use the fact that this bound does not depend on $\mathcal{E}$.

By conditions (i)–(ix) of Section 2.1, there exists an increasing sequence of partitions $\mathcal{E}(n)$ such that $\cup_{n=1}^{\infty} \mathcal{L}_n$ is norm dense in $L^2(X)$. This implies that $P_{\mathcal{E}(n)}$ converges strongly to I. Therefore $P_{\mathcal{E}(n)} A$ converges strongly to A and $\|P_{\mathcal{E}(n)} A\|_2 \leq c^2 |X|$ for all n. Problem 2.2.4 now implies that A is Hilbert-Schmidt. The compactness of A follows by Theorem 4.2.16.   □

Problem 4.2.18 Use Theorem 4.2.17 to prove that if X has finite measure then any linear subspace of $L^2(X, \mathrm{d}x)$ closed with respect to both $\| \cdot \|_\infty$ and $\| \cdot \|_2$ is finite-dimensional.   □

In Theorem 4.2.17 one may not replace ∞ by any finite constant p. Indeed one major technique in constructive quantum field theory depends on this fact.[12]

Theorem 4.2.19 *There exists a measure space $(X, \Sigma, \mathrm{d}\mu)$ such that $\mu(X) = 1$, and a closed linear subspace $\mathcal{L}$ of $L^2(X, \mathrm{d}\mu)$ which is also a closed linear subspace of $L^p(X, \mathrm{d}\mu)$ for all $1 \leq p < \infty$. Indeed all L^p norms are equal on $\mathcal{L}$ up to multiplicative constants. The orthogonal projection P from $L^2(X, \mathrm{d}\mu)$ onto $\mathcal{L}$ is bounded from L^2 to L^p for all $1 \leq p < \infty$, but is not compact.*

Proof. Let $X := \mathbf{R}^\infty$ with the countable infinite product measure $\mu(\mathrm{d}x) := \prod_{n=1}^{\infty} \sigma(\mathrm{d}x_n)$, where $x := \{x_n\}_{n=1}^{\infty}$ and $\sigma(\mathrm{d}s) := (2\pi)^{-1/2} \mathrm{e}^{-s^2/2} \mathrm{d}s$. Then define the subspace $\mathcal{M}$ to consist of all functions

$$f(x) := \sum_{r=1}^{n} \alpha_r x_r$$

---

[12] We refer to the theory of hypercontractive semigroups, described in [Glimm and Jaffe 1981, Simon 1974]. The subspace $\mathcal{L}$ constructed in Theorem 4.2.19 is the one-particle subspace of Fock space.

where $\alpha \in \mathbf{C}^n$ and $|\alpha|$ denotes its Euclidean norm. By exploiting the rotational invariance of the measure μ we see that

$$\|f\|_p^p = \int_{\mathbf{R}} |\alpha|^p |s|^p \sigma(\mathrm{d}s) = c_p |\alpha|^p.$$

Therefore

$$\|f\|_p = c_p^{1/p} c_2^{-1/2} \|f\|_2$$

for all $f \in \mathcal{M}$. If $\mathcal{L}$ is defined to be the L^2 norm closure of $\mathcal{M}$ then the same equality extends to $\mathcal{L}$. The final statement of the theorem depends on the fact that the projection is not compact regarded as an operator on $L^2(X)$, because its range is infinite-dimensional. $\square$

Theorem 4.2.20 *Let X be a compact set in $\mathbf{R}^N$ and assume that the restriction of the Lebesgue measure to X has support equal to X; equivalently if $f \in C(X)$ and $\|f\|_2 = 0$ then $f(x) = 0$ for all $x \in X$. If K is a continuous function on $X \times X$ then the formula*

$$(Af)(x) := \int_X K(x, y) f(y) \, \mathrm{d}^N y$$

defines a compact linear operator on $C(X)$. The spectrum of A is the same as for the Hilbert-Schmidt operator on $L^2(X, \mathrm{d}^N x)$ defined by the same formula.

Proof. The initial condition of the theorem is needed to ensure that the spaces $C(X)$ and $L^2(X, \mathrm{d}x)$ are compatible; see page 49. The first statement is a special case of our next theorem, but we give a direct proof below. The second statement is a corollary of Theorems 4.2.15 and 4.2.16.

The kernel K must be uniformly continuous on $X \times X$, so given $\varepsilon > 0$ there exists $\delta > 0$ such that $d(x, x') < \delta$ implies

$$|K(x, y) - K(x', y)| < \varepsilon/|X|$$

for all $y \in X$. This implies that

$$|(Af)(x) - (Af)(x')| < \varepsilon$$

provided $\|f\|_\infty \leq 1$. Therefore the set $A\{f : \|f\|_\infty \leq 1\}$ is equicontinuous, and the first statement follows by Theorem 4.2.7. $\square$

Theorem 4.2.21 *If X is a compact set in $\mathbf{R}^N$ and K is a continuous function from X to a Banach space $\mathcal{B}$, then the formula*

$$Af := \int_X K(x) f(x) \, \mathrm{d}^N x$$

defines a compact linear operator from $C(X)$ to $\mathcal{B}$.

Proof. Given $\varepsilon > 0$ and $f \in C(X)$ satisfying $\|f\|_\infty \leq 1$, there exists a finite partition $\{E_1, \ldots, E_M\}$ of X and points $x_m \in E_m$ such that if $x \in E_m$ then $|f(x) - f(x_m)| < \varepsilon$ and $\|K(x) - K(x_m)\| < \varepsilon$. Therefore

$$\left\| Af - \sum_{m=1}^{M} |E_m| f(x_m) K(x_m) \right\| < |X|(1+c)\varepsilon$$

where

$$c := \max\{\|K(x)\| : x \in X\}.$$

Since $\sum_{m=1}^{M} |E_m|/|X| = 1$ we deduce that Af lies in the closed convex hull of the compact subset

$$T := \{zg : z \in \mathbb{C}, \ |z| \leq |X| \text{ and } g \in K(X)\}$$

of $\mathcal{B}$. The inclusion

$$A\{f \in C(X) : \|f\|_\infty \leq 1\} \subseteq \overline{\mathrm{Conv}}(T)$$

implies that A is a compact operator by Theorem 4.2.11. $\square$

Theorem 4.2.23 below is a consequence of Theorem 4.3.19, but there is also a simple direct proof, which makes use of the following lemma.

Lemma 4.2.22 *Let A be a compact, self-adjoint operator acting on the Hilbert space $\mathcal{H}$. Then there exists a non-zero vector $f \in \mathcal{H}$ such that either $Af = \|A\|f$ or $Af = -\|A\|f$.*

Proof. We assume that $c := \|A\|$ is non-zero. There must exist a sequence $f_n \in \mathcal{H}$ such that $\|f_n\| = 1$ for all n and $\|Af_n\| \to c$. Using the compactness of A and passing to a subsequence, we may assume that $Af_n \to g$, where $\|g\| = c$. We next observe that

$$\lim_{n \to \infty} \|Ag - c^2 f_n\|^2 = \lim_{n \to \infty} \{\|Ag\|^2 - c^2\langle Ag, f_n\rangle - c^2\langle f_n, Ag\rangle + c^4 \|f_n\|^2\}$$

$$\leq \lim_{n \to \infty} \{c^4 - c^2\langle g, Af_n\rangle - c^2\langle Af_n, g\rangle + c^4\}$$

$$= 2c^4 - 2c^2 \|g\|^2$$

$$= 0.$$

We deduce that

$$A^2 g - c^2 g = \lim_{n \to \infty} \{A(Ag - c^2 f_n) + c^2(Af_n - g)\} = 0.$$

Rewriting this in the form

$$(A + cI)(A - cI)g = 0,$$

we conclude either that $Ag - cg = 0$ or that $h := (A - cI)g \neq 0$ and $Ah + ch = 0$. $\square$

Theorem 4.2.23 *Let A be a compact self-adjoint operator acting in the separable Hilbert space $\mathcal{H}$. Then there exists a complete orthonormal set $\{e_n\}_{n=1}^{\infty}$ in $\mathcal{H}$ and real numbers λ_n which converge to 0 as $n \to \infty$ such that*

$$Ae_n = \lambda_n e_n$$

for all $n = 1, 2, \ldots$. Each non-zero eigenvalue of A has finite multiplicity.

Proof. We construct an orthonormal sequence $\{e_n\}_{n=1}^{\infty}$ in $\mathcal{H}$ such that $Ae_n = \pm c_n e_n$ and $0 \leq c_n \leq c_{n-1}$ for all n.

We start the construction by using Lemma 4.2.22 to choose e_1 such that $Ae_1 = \pm c_1 e_1$, where $c_1 := \|A\|$, and proceed inductively. Given $e_1, \ldots, e_{n-1}$, we put

$$\mathcal{L}_n := \{f \in \mathcal{H} : \langle f, e_r \rangle = 0 \text{ for all } r \leq n - 1\}.$$

It is immediate that $A(\mathcal{L}_n) \subseteq \mathcal{L}_n$. Let $c_n := \|A|_{\mathcal{L}_n}\|$ and then let e_n be a unit vector in $\mathcal{L}_n$ such that $Ae_n = \pm c_n e_n$. Since $\mathcal{L}_n \subseteq \mathcal{L}_{n-1}$, it follows that $c_n \leq c_{n-1}$. This completes the inductive step.

By combining the compactness of A with the equality

$$\|Ae_m - Ae_n\|^2 = c_m^2 + c_n^2$$

we deduce that the sequence c_n converges to zero. We now put

$$\mathcal{L}_\infty := \{f \in \mathcal{H} : \langle f, e_r \rangle = 0 \text{ for all } r\}.$$

If $\mathcal{L}_\infty = \{0\}$ then the proof of the main statement of the theorem is finished, so suppose that this is not the case. Since $\mathcal{L}_\infty \subseteq \mathcal{L}_n$ for all n we have

$$\|A|_{\mathcal{L}_\infty}\| \leq \|A|_{\mathcal{L}_n}\| = c_n$$

for all n. Hence $A|_{\mathcal{L}_\infty} = 0$. We now supplement $\{e_n\}_{n=1}^{\infty}$ by any complete orthonormal set of $\mathcal{L}_\infty$ to complete the proof.

The final statement of the theorem may be proved independently of the above. Let $\mathcal{M}$ be the eigenspace associated with a non-zero eigenvalue λ of A. If $\{u_n\}_{n=1}^{\infty}$ is an infinite complete orthonormal set in $\mathcal{M}$ then the equality

$$\|Au_m - Au_n\| = |\lambda| \|u_m - u_n\| = |\lambda|\sqrt{2}$$

implies that Au_n has no convergent subsequence, contradicting the compactness of A. $\square$

4.3 Fredholm operators

Our goal in this section is three-fold – to develop the theory of Fredholm operators, to prove the spectral theorem for compact operators, and to describe some properties of the essential spectrum of general bounded linear operators. By doing all three things together we hope to make the exposition easier to understand than it would otherwise be.

A bounded operator $A : \mathcal{B} \to \mathcal{C}$ between two Banach spaces is said to be a Fredholm operator if its kernel $\mathrm{Ker}(A)$ and cokernel $\mathrm{Coker}(A) := \mathcal{C}/\mathrm{Ran}(A)$ are both finite-dimensional. Fredholm operators are important for a variety of reasons, one being the role that their index

$$\mathrm{index}(A) := \dim(\mathrm{Ker}(A)) - \dim(\mathrm{Coker}(A))$$

plays in global analysis.[13]

Problem 4.3.1 Let $A : C^1[0, 1] \to C[0, 1]$ be defined by

$$(Af)(x) := f'(x) + a(x)f(x)$$

where $a \in C[0, 1]$. By solving $f'(x) + a(x)f(x) = g(x)$ explicitly prove that A is a Fredholm operator and find its index. $\square$

Problem 4.3.2 By evaluating it explicitly, prove that the index of a linear map $A : \mathbf{C}^m \to \mathbf{C}^n$ depends on m and n but not on A. $\square$

Lemma 4.3.3 *If A is a compact operator on $\mathcal{B}$ then $(\lambda I - A)$ is Fredholm for all $\lambda \neq 0$.*

Proof. We first prove that $\mathcal{L} := \mathrm{Ker}(\lambda I - A)$ is finite-dimensional by contradiction. If this were not the case there would exist an infinite sequence $x_n \in \mathcal{L}$ such that $\|x_n\| = 1$ and $\|x_m - x_n\| \geq 1/2$ for all distinct m and n. Since $Ax_n = \lambda x_n$ and $\lambda \neq 0$, we could conclude that Ax_n has no convergent subsequence. Problem 1.1.25 allows us to write $\mathcal{B} = \mathcal{L} + \mathcal{M}$, where $\mathcal{L} \cap \mathcal{M} = \{0\}$ and $\mathcal{M}$ is a closed linear subspace on which $(\lambda I - A)$ is one-one.

We next prove that $\mathcal{R} := \mathrm{Ran}(\lambda I - A)$ is closed. If $g_n \in \mathcal{R}$ and $\|g_n - g\| \to 0$ then there exist $f_n \in \mathcal{M}$ such that $g_n = (\lambda I - A)f_n$. If $\|f_n\|$ is not a bounded sequence then by passing to a subsequence (without change of notation)

---

[13] It is, perhaps, worth mentioning that Fredholm only studied compact integral operators. 'Fredholm' operators are often called Noether operators in the German and Russian literature, after Fritz Noether, who was responsible for discovering the importance of the index. We refer to [Gilkey 1996] for an account of the applications of 'Fredholm theory' to the study of pseudo-differential operators on manifolds.

we may assume that $\|f_n\| \to \infty$ as $n \to \infty$. Putting $h_n := f_n/\|f_n\|$ we have $\|h_n\| = 1$ and $k_n := (\lambda I - A)h_n \to 0$. The compactness of A implies that $h_n = \lambda^{-1}(Ah_n + k_n)$ has a convergent subsequence. Passing to this subsequence we have $h_n \to h$ where $\|h\| = 1$, $h \in \mathcal{M}$, and $h = \lambda^{-1}Ah$. We conclude that $h \in \mathcal{M} \cap \mathcal{L}$. The contradiction implies that $\|f_n\|$ is a bounded sequence.

Given this fact the compactness of A implies that the sequence $f_n = \lambda^{-1}(Af_n + g_n)$ has a convergent subsequence. Passing to this subsequence we obtain $f_n \to f$ as $n \to \infty$, so $f = \lambda^{-1}(Af + g)$, and $g = (\lambda I - A)f$. Therefore $\mathcal{R}$ is closed.

Since $\mathrm{Ran}(\lambda I - A)$ is closed, an application of the Hahn-Banach theorem implies that its codimension equals the dimension of $\mathrm{Ker}(\lambda I - A^*)$ in $\mathcal{B}^*$. But A^* is compact by Theorem 4.2.13, so this is finite by the first paragraph. $\quad\square$

Our next theorem provides a second characterization of Fredholm operators.

Theorem 4.3.4 *Every Fredholm operator has closed range. The bounded operator $A : \mathcal{B} \to \mathcal{C}$ is Fredholm if and only if there is a bounded operator $B : \mathcal{C} \to \mathcal{B}$ such that both $(AB - I)$ and $(BA - I)$ are compact.*

Proof. If A is Fredholm then $\mathcal{B}_1 := \mathrm{Ker}(A)$ is finite-dimensional and so has a complementary closed subspace $\mathcal{B}_0$ in $\mathcal{B}$ by Problem 1.1.25. Moreover A maps $\mathcal{B}_0$ one-one onto $\mathcal{C}_0 := \mathrm{Ran}(A)$. If $\mathcal{C}_1$ is a complementary finite-dimensional subspace of $\mathcal{C}_0$ in $\mathcal{C}$ then the operator $X : \mathcal{B}_0 \oplus \mathcal{C}_1 \to \mathcal{C}$ defined by

$$X(f \oplus v) := Af + v$$

is bounded and invertible. We deduce by the inverse mapping theorem that $\mathcal{C}_0 := X(\mathcal{B}_0)$ is closed. This completes the proof of the first statement of the theorem.

Still assuming that A is Fredholm, put $B(g \oplus v) := (A_0)^{-1}g$ for all $g \in \mathcal{C}_0$ and $v \in \mathcal{C}_1$, where $A_0 : \mathcal{B}_0 \to \mathcal{C}_0$ is the restriction of A to $\mathcal{B}_0$. Then B is a bounded operator from $\mathcal{C}$ to $\mathcal{B}$ and both of

$$K_1 := AB - I, \qquad K_2 := BA - I \tag{4.4}$$

are finite rank and hence compact.

Conversely suppose that A, B are bounded, K_1, K_2 are compact and (4.4) hold. Then

$$\mathrm{Ker}(A) \subseteq \mathrm{Ker}(I + K_2),$$

$$\mathrm{Ran}(A) \supseteq \mathrm{Ran}(I + K_1).$$

Since $(I + K_1)$ and $(I + K_2)$ are both Fredholm by Lemma 4.3.3, it follows that A must be Fredholm. $\square$

The proof of Theorem 4.3.4 provides an important structure theorem for Fredholm operators.

Theorem 4.3.5 *If A is a Fredholm operator then there exist decompositions $\mathcal{B} = \mathcal{B}_0 \oplus \mathcal{B}_1$ and $\mathcal{C} = \mathcal{C}_0 \oplus \mathcal{C}_1$ such that*

(i) *$\mathcal{B}_0$ and $\mathcal{C}_0$ are closed subspaces;*
(ii) *$\mathcal{B}_1$ and $\mathcal{C}_1$ are finite-dimensional subspaces;*
(iii) *$\mathcal{B}_1 = \mathrm{Ker}(A)$ and $\mathcal{C}_0 = \mathrm{Ran}(A)$;*
(iv) *$\mathrm{Index}(A) = \dim(\mathcal{B}_1) - \dim(\mathcal{C}_1)$;*
(v) *A has the matrix representation*

$$A = \begin{pmatrix} A_0 & 0 \\ 0 & 0 \end{pmatrix} \tag{4.5}$$

where $A_0 : \mathcal{B}_0 \to \mathcal{C}_0$ is one-one onto.

Problem 4.3.6 Let A be a Fredholm operator on the Banach space $\mathcal{B}$. Prove that if $\mathrm{Ker}(A) = \{0\}$ then $\mathrm{Ker}(A - \delta I) = \{0\}$ for all small enough δ. Prove also that if $\mathrm{Ran}(A) = \mathcal{B}$ then $\mathrm{Ran}(A - \delta I) = \mathcal{B}$ for all small enough δ. $\square$

Before stating our next theorem we make some definitions. We say that λ lies in the essential spectrum $\mathrm{EssSpec}(A)$ of a bounded operator A if $(\lambda I - A)$ is not a Fredholm operator.[14]
Since the set $\mathcal{K}(\mathcal{B})$ of all compact operators is a norm closed two-sided ideal in the Banach algebra $\mathcal{L}(\mathcal{B})$ of all bounded operators on $\mathcal{B}$, the quotient algebra $\mathcal{C} := \mathcal{L}(\mathcal{B})/\mathcal{K}(\mathcal{B})$ is a Banach algebra with respect to the quotient norm

$$\|\pi(A)\| := \inf\{\|A + K\| : K \in \mathcal{K}(\mathcal{B})\}$$

where $\pi : \mathcal{L}(\mathcal{B}) \to \mathcal{C}$ is the quotient map. The Calkin algebra $\mathcal{C}$ enables us to rewrite Theorem 4.3.4 in a particularly simple form.

Theorem 4.3.7 *The bounded operator A on $\mathcal{B}$ is Fredholm if and only if $\pi(A)$ is invertible in the Calkin algebra $\mathcal{C}$. If $A \in \mathcal{L}(\mathcal{B})$ then*

$$\mathrm{EssSpec}(A) = \mathrm{Spec}(\pi(A)).$$

---

[14] A systematic treatment of five different definitions of essential spectrum may be found in [Edmunds and Evans 1987, pp. 40 ff.]. See also [Kato 1966A]. For self-adjoint operators on a Hilbert space these notions all coincide, but in general they have different properties.

Proof. Both statements of the theorem are elementary consequences of Theorem 4.3.4. □

Corollary 4.3.8 *If $A : \mathcal{B} \to \mathcal{B}$ is a Fredholm operator and $B := A + K$ where K is compact, then B is a Fredholm operator and*

$$\mathrm{EssSpec}(A) = \mathrm{EssSpec}(B).$$

Theorem 4.3.9 *If A is a Fredholm operator on $\mathcal{B}$ then A^* is Fredholm.*

Proof. Suppose that $AB = I + K_1$ and $BA = I + K_2$ where K_1, K_2 are compact. Then $B^* A^* = I + K_1^*$ and $A^* B^* = I + K_2^*$. We deduce that A^* is Fredholm by applying Theorem 4.3.4 and Theorem 4.2.13. □

Problem 4.3.10 Prove directly from the definition that if A_1 and A_2 are both Fredholm operators then so is $A_1 A_2$.

Note: If $\mathcal{B}_1 = \mathcal{B}_2$ then this is an obvious consequence of Theorem 4.3.7, but there is an elementary direct proof. □

Theorem 4.3.11 *If $A : \mathcal{B} \to \mathcal{C}$ is a Fredholm operator then there exists $\varepsilon > 0$ such that every bounded operator X satisfying $\|X - A\| < \varepsilon$ is also Fredholm with*

$$\mathrm{index}(X) = \mathrm{index}(A).$$

Proof. We make use of the matrix representation (4.5) of Theorem 4.3.5. If

$$X = \begin{pmatrix} B & C \\ D & E \end{pmatrix}$$

and $\|X - A\| < \varepsilon$ then $\|B - A_0\| \le c\varepsilon$, so B is invertible provided $\varepsilon > 0$ is small enough.

If $f \in \mathcal{B}_0$ and $g \in \mathcal{B}_1$ then $X(f \oplus g) = 0$ if and only if

$$Bf + Cg = 0,$$

$$Df + Eg = 0.$$

This reduces to

$$(E - DB^{-1}C)g = 0,$$

where $(E - DB^{-1}C) : \mathcal{B}_1 \to \mathcal{C}_1$, both of these spaces being finite-dimensional. We deduce that

$$\dim(\mathrm{Ker}(X)) = \dim(\mathrm{Ker}(E - DB^{-1}C))$$

for all small enough $\varepsilon > 0$. By applying a similar argument to

$$X^* = \begin{pmatrix} B^* & D^* \\ C^* & E^* \end{pmatrix}$$

we obtain

$$\dim(\mathrm{Coker}(X)) = \dim(\mathrm{Coker}(E - DB^{-1}C))$$

for all small enough $\varepsilon > 0$. Problem 4.3.2 now implies that

$$\mathrm{index}(X) = \mathrm{index}(E - DB^{-1}C) = \dim(\mathcal{B}_1) - \dim(\mathcal{C}_1).$$

This formula establishes that $\mathrm{index}(X)$ does not depend on X, provided $\|X - A\|$ is small enough. $\square$

Theorem 4.3.11 establishes that the index is a homotopy invariant: if $t \to A_t$ is a norm continuous family of Fredholm operators then $\mathrm{index}(A_t)$ does not depend on t. In a Hilbert space context one can even identify the homotopy classes, by using some results which are only proved in the next chapter.

Theorem 4.3.12 *If A is a Fredholm operator on the Hilbert space $\mathcal{H}$ then there exists a norm continuous family of Fredholm operators A_t defined for $0 \leq t \leq 1$ with $A_0 = A$ and $A_1 = I$ if and only if* $\mathrm{index}(A) = 0$.

Proof. If such a norm continuous family A_t exists then Theorem 4.3.11 implies that $\mathrm{index}(A) = \mathrm{index}(I) = 0$.

Conversely suppose that A is a Fredholm operator with zero index. We construct a norm continuous homotopy connecting A and I in three steps. The three maps are linked together at the end of the process in an obvious manner.

Since $\dim(\mathcal{B}_1) = \dim(\mathcal{C}_1) < \infty$, there exists a bounded linear map B on $\mathcal{B}$ such that $Bf = 0$ for all $f \in \mathcal{B}_0$ and B maps $\mathcal{B}_1$ one-one onto $\mathcal{C}_1$. The operators $A_t := A + tB$ depend norm continuously on t and are invertible for $t \neq 0$. This reduces the proof to the case in which A is invertible.

Since A is invertible it has a polar decomposition $A = U|A|$ in which $|A|$ is self-adjoint, positive and invertible while U is unitary; see Theorem 5.2.4. The one-parameter family of operators $A_t := U|A|^{1-t}$ satisfies $A_0 = A$ and $A_1 = U$ and is a norm continuous function of t by Theorem 1.5.1. This reduces the proof to the case in which $A = U$ is unitary.

If A is unitary there exists a bounded self-adjoint operator H such that $A = e^{iH}$ by the corollary Problem 5.4.3 of the spectral theorem for normal operators. We finally define $A_t = e^{iH(1-t)}$ to obtain a norm continuous homotopy from A to I. $\square$

Problem 4.3.13 By considering appropriate left and right shift operators on $l^2(\mathbf{Z}^+)$, prove that for every choice of $n \in \mathbf{Z}$ there exists a Fredholm operator A with $\text{index}(A) = n$. $\square$

Problem 4.3.14 Let $\mathcal{B}$ be the space consisting of those continuous functions on the closed unit disc D that are analytic in the interior of D, provided with the sup norm. Let $g \in \mathcal{B}$ be non-zero on the boundary ∂D of D, and let $Z(g)$ denote the number of zeros of g in D, counting multiplicities. Prove that if A is defined by $Af := gf$ then

$$\text{index}(A) = -Z(g).$$ $\square$

Lemma 11.2.1 extends the following lemma to unbounded operators.

Lemma 4.3.15 *Let A be a bounded operator on $\mathcal{B}$ and let $z \in \mathbf{C}$. If the sequence of vectors f_n in $\mathcal{B}$ converges weakly to 0 as $n \to \infty$ and satisfies*

$$\lim_{n \to \infty} \|f_n\| = 1, \qquad \lim_{n \to \infty} \|Af_n - zf_n\| = 0,$$

then $z \in \text{EssSpec}(A)$.

Proof. Suppose that $z \notin \text{EssSpec}(A)$ and put $B := A - zI$. Then B is a Fredholm operator so $K := \text{Ker}(B)$ is finite-dimensional. There exists a finite rank projection P on $\mathcal{B}$ with range K. If we put $M := \text{Ker}(P)$ then $\mathcal{B} = K \oplus M$ and it follows from the proof of Theorem 4.3.4 that there exists a positive constant c such that

$$\|Bg\| \geq c\|g\| \tag{4.6}$$

for all $g \in M$. Since P is of finite rank and f_n converge weakly to 0, we see that $\|Pf_n\| \to 0$. Therefore $g_n := (I - P)f_n$ satisfy $\lim_{n \to \infty} \|g_n\| = 1$ and $\lim_{n \to \infty} \|Bg_n\| = 0$. These results contradict (4.6). $\square$

For self-adjoint operators the essential spectrum is easy to determine.

Problem 4.3.16 Prove that if A is a bounded self-adjoint operator on the Hilbert space $\mathcal{H}$, then $\lambda \in \text{Spec}(A)$ lies in the essential spectrum unless it is an isolated eigenvalue of finite multiplicity. $\square$

An analysis of Problem 1.2.16 shows that the analogous statement for more general operators may be false.

Lemma 4.3.17 *If* λ *is an isolated point of* $\mathrm{Spec}(A)$ *and* $\mathcal{L}$ *is the range of the associated spectral projection* P *then* $\lambda \in \mathrm{EssSpec}(A)$ *if and only if* $\mathcal{L}$ *is infinite-dimensional.*

Proof. Let $\mathcal{L}$ (resp. $\mathcal{M}$) be the range (resp. kernel) of the spectral projection P constructed in Theorem 1.5.4. Both subspaces are invariant with respect to A with $\mathrm{Spec}(A|_{\mathcal{L}}) = \{\lambda\}$ and $\mathrm{Spec}(A|_{\mathcal{M}}) = \mathrm{Spec}(A)\backslash\{\lambda\}$. By restricting to $\mathcal{L}$ it is sufficient to treat the case in which $\mathrm{Spec}(A) = \{\lambda\}$. The condition $\lambda \notin \mathrm{EssSpec}(A)$ is then equivalent to $\mathrm{EssSpec}(A) = \emptyset$.

It is elementary that if $\mathcal{B}$ is finite-dimensional then $\mathrm{EssSpec}(A) = \emptyset$. Conversely suppose that $\mathrm{EssSpec}(A) = \emptyset$. The $\mathcal{C}$-valued analytic function $f(z) := \pi((zI - A)^{-1})$ is defined for all $z \in \mathbf{C}$ and satisfies $f(z) = z^{-1}1 + O(z^{-2})$ as $|z| \to \infty$. The Banach space version of Liouville's theorem (Problem 1.4.9) implies that $1 = 0$ in $\mathcal{C}$. This implies that $I \in \mathcal{L}(\mathcal{B})$ is compact, so $\mathcal{B}$ must be finite-dimensional. $\square$

If $\mathcal{B}$ is infinite-dimensional then $\mathrm{EssSpec}(A)$ is closed by an application of Problem 1.2.12 to the Calkin algebra.

Theorem 4.3.18 *Let* A *be a bounded operator on the Banach space* $\mathcal{B}$. *Let* S *denote the essential spectrum of* A, *and let* U *be the unbounded component of* $\mathbf{C}\backslash S$. *Then* $(zI - A)$ *is a Fredholm operator of zero index for all* $z \in U$ *and* $\mathrm{Spec}(A) \cap U$ *consists of a finite or countable set of isolated eigenvalues with finite algebraic and geometric multiplicities.*

Proof. It follows from the definition of the essential spectrum that $(zI - A)$ is a Fredholm operator for all $z \in U$, and from Theorem 4.3.11 that the index is constant on U. We prove that every $z \in U \cap \mathrm{Spec}(A)$ is isolated. We assume that $z = 0$ by replacing A by $(A - zI)$, but return to the original operator A in the final paragraph. The operators A^n are Fredholm for all $n \geq 1$ by Problem 4.3.10, so the subspaces $M_n := \mathrm{Ran}(A^n)$ are all closed. These subspaces decrease as n increases, and their intersection M_∞ is also closed.

The inclusion $A(M_n) \subseteq M_{n+1}$ for all n implies that $A(M_\infty) \subseteq M_\infty$. We next prove that if B denotes the restriction of A to M_∞ then $B(M_\infty) = M_\infty$ and B is a Fredholm operator on M_∞. If $f \in M_\infty$ and $f = Ah$ then

$$\{g \in M_n : Ag = f\} = M_n \cap (h + \mathrm{Ker}(A)).$$

This is a decreasing sequence of non-empty finite-dimensional affine subspaces of $\mathcal{B}$, and so must be constant beyond some critical value of n. In other words there exists g such that $Ag = f$ and $g \in M_n$ for all n. Hence $g \in M_\infty$ and B is surjective. Since $\mathrm{Ker}(B) \subseteq \mathrm{Ker}(A)$, it follows that B is a Fredholm operator on M_∞.

We next relate the index of $(B - \delta I)$ with that of $(A - \delta I)$ for all small enough $\delta \neq 0$. The fact that B is Fredholm with $B(\mathcal{M}_\infty) = \mathcal{M}_\infty$ implies that $(B - \delta I)(\mathcal{M}_\infty) = \mathcal{M}_\infty$ for all small enough δ by Problem 4.3.6. If $\delta \neq 0$ and $f \in \mathrm{Ker}(A - \delta I)$ then $f = \delta^{-n} A^n f$ for all $n \geq 1$, so $f \in M_\infty$. This implies that $\mathrm{Ker}(A - \delta I) = \mathrm{Ker}(B - \delta I)$ for all $\delta \neq 0$. Therefore

$$\dim \mathrm{Ker}(A - \delta I) = \dim \mathrm{Ker}(B - \delta I)$$

$$= \mathrm{index}(B - \delta I)$$

for all small enough $\delta \neq 0$. But the index does not depend on δ, so $\dim \mathrm{Ker}(A - \delta I)$ does not depend on δ for all small enough $\delta \neq 0$. Applying a similar argument to A^* we see that $\dim \mathrm{Coker}(A - \delta I)$ does not depend on δ for all small enough $\delta \neq 0$.

We now return to the original operator A. A topological argument implies that $\dim \mathrm{Ker}(A - zI)$ and $\dim \mathrm{Coker}(A - zI)$ are constant on U except at a finite or countable sequence of points whose limit points must lie in $\mathrm{EssSpec}(A)$. Since $(A - zI)$ is invertible if $|z| > \|A\|$, we deduce that $(A - zI)$ is invertible for all $z \in U$, excluding the possible exceptional points. Therefore $\mathrm{index}(A - zI) = 0$ throughout U.

Since every $\lambda \in U \cap \mathrm{Spec}(A)$ is an isolated point, the associated spectral projection is of finite rank by Lemma 4.3.17. □

The spectral theorem for a compact operator A is a direct corollary of Theorem 4.3.18, the essential spectrum of A being equal to $\{0\}$ provided $\dim(\mathcal{B}) = \infty$. We omit the description of the restriction of A to its finite-dimensional spectral subspaces, since this amounts to the description of general $n \times n$ matrices by means of Jordan forms.

Theorem 4.3.19 (*Riesz*) *Let A be a compact operator acting on the infinite-dimensional Banach space $\mathcal{B}$. Then the spectrum of A contains 0 and is finite or countable. Every non-zero point λ in the spectrum is associated with a spectral projection P of finite rank which commutes with A. The restriction of A to the range of P has spectrum equal to $\{\lambda\}$. If $\mathrm{Spec}(A)$ is countable then*

$$\mathrm{Spec}(A) = \{0\} \cup \{\lambda_n : n = 1, 2, \dots\},$$

where $\lim_{n \to \infty} \lambda_n = 0$.

We finally mention the extension of the ideas of this section to unbounded operators; we will return to this in Section 11.2. If $\mathcal{D}$ is a linear subspace of the Banach space $\mathcal{B}$ and $A : \mathcal{D} \to \mathcal{B}$ is a possibly unbounded linear operator, we say that $z \notin \mathrm{Spec}(A)$ if $(zI - A)$ maps $\mathcal{D}$ one-one onto $\mathcal{B}$ and the inverse operator $R(z, A)$ is bounded. Suppose also that $\mathcal{D}$ is a Banach space with respect to another norm $\|| \cdot \||$ and that the inclusion map $I : (\mathcal{D}, \|| \cdot \||) \to (\mathcal{B}, \|| \cdot \||)$ is continuous. We assume that A is bounded as an operator from $\mathcal{D}$ to $\mathcal{B}$. We then define the essential spectrum of A to be the set of all z such that $(A - zI)$ is not a Fredholm operator from $\mathcal{D}$ to $\mathcal{B}$. As before this is a subset of $\mathrm{Spec}(A)$.

4.4 Finding the essential spectrum

In this section we find the essential spectra of a number of simple operators, as illustrations of the theory of the last section. We start by considering Toeplitz operators,[15] and then go on to examples which make use of the results obtained in that case. Many of the results obtained in this section can be extended to differential operators in several dimensions; the applications to Schrödinger operators whose potentials have different asymptotic forms in different directions are of importance in multi-body quantum mechanics.

Problem 1.2.16 provides the simplest example of a Toeplitz operator. More generally if $a \in l^1(\mathbf{Z})$ and $1 < p < \infty$ we define the bounded Toeplitz operator A on $l^p(\mathbf{N})$, where $\mathbf{N}$ denotes the set of natural numbers, by

$$(Af)(n) := \sum_{m=1}^{\infty} a(n - m)f(m) \qquad (4.7)$$

for all $n \in \mathbf{N}$. Alternatively $Af := P_+(a * f)$ for all $f \in l^p(\mathbf{N})$, where P_+ is the projection from $l^p(\mathbf{Z})$ onto $l^p(\mathbf{N})$ defined by $P_+f = \chi_{\mathbf{N}}f$. One sees that A has the semi-infinite matrix

$$A = \begin{pmatrix} a_0 & a_{-1} & a_{-2} & a_{-3} & \cdots \\ a_1 & a_0 & a_{-1} & a_{-2} & \cdots \\ a_2 & a_1 & a_0 & a_{-1} & \cdots \\ a_3 & a_2 & a_1 & a_0 & \cdots \\ \cdots & \cdots & \cdots & \cdots & \cdots \end{pmatrix}.$$

We define the symbol of the operator A by

$$\hat{a}(\theta) := \sum_{m=-\infty}^{\infty} a(m)e^{-im\theta}.$$

---

[15] We refer to [Böttcher and Silbermann 1999] for a much more comprehensive treatment.

The following theorems are associated with the names of Gohberg, Hartman, Krein and Wintner. The first is closely related to Theorem 2.4.4.

Theorem 4.4.1 *If $a \in l^1(\mathbf{Z})$ and $1 < p < \infty$ then the essential spectrum of the Toeplitz operator A on $l^p(\mathbf{N})$ defined by (4.7) satisfies*

$$\mathrm{EssSpec}(A) = \{\hat{a}(\theta) : \theta \in [0, 2\pi]\}.$$

Proof. We write A in the form

$$A = \sum_{m=-\infty}^{\infty} a(m) T_m$$

where the contractions T_m on $l^p(\mathbf{N})$ are defined for all $m \in \mathbf{Z}$ and $n \in \mathbf{N}$ by

$$(T_m f)(n) := \begin{cases} f(n-m) & \text{if } n-m \geq 1, \\ 0 & \text{otherwise.} \end{cases}$$

Given two bounded operators B, C we will write $B \sim C$ if $(B-C)$ is compact, or equivalently if $\pi(B-C) = 0$ in the Calkin algebra $\mathcal{L}(\mathcal{B})/\mathcal{K}(\mathcal{B})$. A direct computation shows that $T_m T_n \sim T_{m+n}$ for all $m, n \in \mathbf{Z}$.

Now suppose that $z \notin S$ where $S := \{\hat{a}(\theta) : \theta \in [0, 2\pi]\}$. Wiener's Theorem 2.4.2 implies the existence of $b \in l^1(\mathbf{Z})$ such that $(ze - a) * b = e$, where $*$ denotes convolution and $e := \delta_0$ is the identity element of the Banach algebra $l^1(\mathbf{Z})$. Putting

$$B := \sum_{m=-\infty}^{\infty} b(m) T_m$$

we see that

$$(zI - A)B = zB - \left(\sum_{m=-\infty}^{\infty} a(m) T_m\right)\left(\sum_{n=-\infty}^{\infty} b(n) T_n\right)$$

$$\sim zB - \sum_{m=-\infty}^{\infty} \sum_{n=-\infty}^{\infty} a(m) b(n) T_{m+n}$$

$$= zB - \sum_{m=-\infty}^{\infty} (a * b)(m) T_m$$

$$= \sum_{m=-\infty}^{\infty} (zb - a * b)(m) T_m$$

$$= \sum_{m=-\infty}^{\infty} e(m) T_m$$

$$= I.$$

Since $B(zI - A) \sim I$ by a similar argument, Theorem 4.3.4 implies that $z \notin$ EssSpec(A).

Conversely suppose that $z := \hat{a}(\theta)$ for some $\theta \in [0, 2\pi]$. We prove that $z \in$ EssSpec(A) by constructing a sequence f_n satisfying the conditions of Lemma 4.3.15. Let $g : \mathbf{R} \to \mathbf{R}^+$ be a continuous function with support in $(0, 1)$ and satisfying $\|g\|_p = 1$. Then define $f_n \in l^p(\mathbf{N})$ by

$$f_n(m) := n^{-1/p} g(m/n) e^{im\theta}.$$

It follows immediately that $\|f_n\|_p \to 1$, $f_n \to 0$ weakly and one sees as in the proof of Theorem 2.4.4 that $\lim_{n\to\infty} \|Af_n - \hat{a}(\theta)f_n\|_p = 0$. $\quad\square$

Theorem 4.4.2 *(Krein) Suppose that $a \in l^1(\mathbf{Z})$, $1 < p < \infty$ and $0 \notin \{\hat{a}(\theta) : \theta \in [0, 2\pi]\}$. Then the associated Toeplitz operator A on $l^p(\mathbf{N})$ is Fredholm and its index equals[16] the winding number of $\hat{a}$ around 0.*

Proof. Both the index and the winding number are invariant under homotopies, so we can prove the theorem by deforming the operator continuously into one for which the identity is easy to prove. We start by truncating a at a large enough distance from 0 so that neither the index nor the winding number are changed. This has the effect of ensuring that $\hat{a}$ is a smooth periodic function on $[0, 2\pi]$ which does not vanish anywhere. We write

$$\hat{a}(\theta) := r(\theta) e^{i\phi(\theta)}$$

where r is a positive smooth function which is periodic with period 2π, and ϕ is a smooth real-valued function such that $\phi(2\pi) = \phi(0) + 2\pi N$, where N is the winding number. Given $t \in [0, 1]$, we now put

$$\hat{a}_t(\theta) := r(\theta)^{1-t} e^{tNi\theta + (1-t)i\phi(\theta)}$$

noting that $\hat{a}_t$ has absolutely summable Fourier coefficients for every such t. Homotopy arguments imply that A has the same index as the Toeplitz operator B associated with the symbol

$$\hat{a}_1(\theta) := e^{Ni\theta}.$$

---

[16] Some accounts prove that the index is minus the winding number, because they adopt a different convention relating the symbol of the operator to its Fourier coefficients.

The operator B is given explicitly on $l^p(\mathbf{N})$ by

$$(Bf)(n) := \begin{cases} f(n+N) & \text{if } n+N \geq 1, \\ 0 & \text{otherwise,} \end{cases}$$

and its index is N by inspection. $\quad\square$

Example 4.4.3 Let A be the convolution operator $A(f) := a * f$ on $l^2(\mathbf{N})$, where

$$a_n := \begin{cases} 4 \text{ if } n = 1, \\ 1 \text{ if } n = 6 \text{ or } -4, \\ 0 \text{ otherwise.} \end{cases}$$

The essential spectrum of A is the curve shown in Figure 4.1. The spectrum also contains every point inside the closed curve, by Theorem 4.4.2. $\quad\square$

In the case $p := 2$ it is possible to analyze Toeplitz operators with much more general symbols in considerable detail. It is also more natural to formulate the

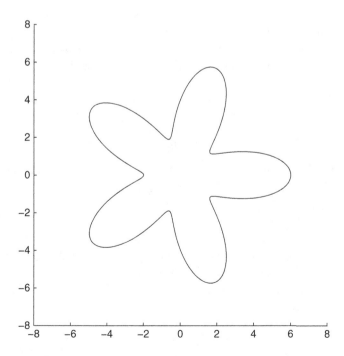

Figure 4.1: Spectrum of the operator A of Example 4.4.3

problem in a different manner. We define the Hardy space $H^2 \subseteq L^2(-\pi, \pi)$ by

$$H^2 := \overline{\text{lin}}\{e^{in\theta} : n = 0, 1, 2, \ldots\}.$$

By identifying $e^{in\theta}$ with z^n it may also be regarded as the set of boundary values of a certain space of analytic functions on $\{z : |z| < 1\}$. Given a bounded measurable function a on $(-\pi, \pi)$ one may then define the Toeplitz operator $T(a) : H^2 \to H^2$ with symbol a by

$$T(a)f := P(af)$$

where P is the orthogonal projection of $L^2(-\pi, \pi)$ onto H^2. The spectral analysis of such operators is extremely delicate, and we content ourselves with the following theorem, which already treats a larger class of symbols a than Theorem 4.4.1.

Theorem 4.4.4 *If $a, b \in C_{\text{per}}[-\pi, \pi]$ then $T(ab) - T(a)T(b)$ is a compact operator on H^2. Moreover the essential spectrum of $T(a)$ satisfies*

$$\text{EssSpec}(T(a)) = \{a(\theta) : \theta \in [0, 2\pi]\}.$$

Proof. Since $T(ab) - T(a)T(b)$ is the restriction of $Pa(I - P)bP$ to H^2, the first statement follows if $(I - P)bP$ is a compact operator on $L^2(-\pi, \pi)$. If b is a trigonometric polynomial then this operator is of finite rank, and the general statement follows by expressing b as the uniform limit of trigonometric polynomials.

The essential spectrum of $T(a)$ is equal to the spectrum of the element $\pi(T(a))$ of the Calkin algebra, by Theorem 4.3.7. By using the identity

$$\pi(T(a - z1))\pi(T(b)) = \pi(T(1)) = 1$$

where $z \notin \{a(\theta) : \theta \in [-\pi, \pi]\}$ and $b = (a - z1)^{-1}$, we deduce that $\text{EssSpec}(T(a))$ is contained in $\{a(\theta) : \theta \in [0, 2\pi]\}$. Equality is proved by constructing a suitable sequence of approximate eigenvectors, as in the proof of Theorem 4.4.1. $\square$

Although the above theorem determines the essential spectrum of such Toeplitz operators, Problem 1.2.16 shows that the spectrum may be much larger.[17] We treat the corresponding problem for differential operators on the half-line more completely in Section 11.3.

---

[17] We refer to [Böttcher and Silbermann 1999] for the full description of the spectrum and essential spectrum for piecewise continuous and more general symbols. It is remarkable that the natural generalization of Theorem 4.4.4 to piecewise continuous symbols is false.

We leave the study of Toeplitz operators at this point, and describe a miscellany of other examples for which the essential spectrum can be determined.

Example 4.4.5 The study of obstacle scattering involves operators of the following type. Let $X := \mathbf{Z}^N \backslash F$ where F is any finite set. Let A be the bounded operator on $l^2(X)$ whose matrix is given by $a_{x,y} := b_{x-y}$ for all $x, y \in X$, where b_r are the Fourier coefficients of some function $\hat{b} \in C_{\mathrm{per}}([-\pi, \pi]^N)$.

The operator A may have some eigenvalues, whose corresponding eigenfunctions decay rapidly as one moves away from F. The essential spectrum of A is easier to determine. One first replaces A by the operator $A \oplus 0$ on $l^2(\mathbf{Z}^N)$, where 0 is the zero operator in $l^2(F)$. One sees immediately that $B - (A \oplus 0)$ is of finite rank, where B is the operator of convolution by the sequence $\{b_n\}_{n \in \mathbf{Z}^N}$. We use the Fourier operator $\mathcal{F} : l^2(\mathbf{Z}^N) \to L^2([-\pi, \pi]^N)$ defined by

$$(\mathcal{F}c)(\theta) := (2\pi)^{-N/2} \sum_{n \in \mathbf{Z}^N} c_n \mathrm{e}^{-in \cdot \theta},$$

where $c \in l^2(\mathbf{Z}^N)$ and $\theta \in [-\pi, \pi]^N$. The proof that $\mathcal{F}$ is unitary uses the N-dimensional analogue of Corollary 2.3.11. Putting $\hat{B} := \mathcal{F} B \mathcal{F}^{-1}$ one sees that

$$(\hat{B}f)(\theta) = \hat{b}(\theta)f(\theta)$$

for all $f \in L^2([-\pi, \pi]^N)$. Therefore

$$\mathrm{EssSpec}(A) = \mathrm{EssSpec}(A \oplus 0)$$
$$= \mathrm{EssSpec}(B)$$
$$= \mathrm{EssSpec}(\hat{B})$$
$$= \{\hat{b}(\theta) : \theta \in [-\pi, \pi]^N\}. \qquad \square$$

We now turn to the study of operators on $l^2(\mathbf{Z})$ which have different asymptotic forms at $\pm\infty$.

Theorem 4.4.6 *Let* $A : l^2(\mathbf{Z}) \to l^2(\mathbf{Z})$ *be the bounded operator associated with an infinite matrix* $a_{m,n}$ *satisfying* $a_{m,n} = 0$ *if* $|m - n| > k$ *and*

$$\lim_{r \to \pm\infty} a_{m+r,n+r} = b_{\pm, m-n}$$

for all $m, n \in \mathbf{Z}$. *If we put*

$$b_\pm(\theta) := \sum_{r=-k}^{k} b_{\pm, r} \mathrm{e}^{-ir\theta},$$

then the essential spectrum of A is given by

$$\text{EssSpec}(A) = \{b_+(\theta) : -\pi \leq \theta \leq \pi\} \cup \{b_-(\theta) : -\pi \leq \theta \leq \pi\}. \quad (4.8)$$

Proof. [18] The boundedness of A is proved by using Corollary 2.2.15. We compare A with the operator B whose associated matrix is defined by

$$b_{m,n} := \begin{cases} b_{+,\,m-n} & \text{if } m \geq 0 \text{ and } n \geq 0, \\ b_{-,\,m-n} & \text{if } m \leq -1 \text{ and } n \leq -1, \\ 0 & \text{otherwise.} \end{cases}$$

Note that $b_{m,n} = 0$ if $|m - n| > k$. Since the matrix coefficients of the difference $C := A - B$ converge to zero as $m \to \pm\infty$ and vanish if $|m - n| > k$, it follows that $\|C_N - C\| \to 0$ as $N \to \infty$ where

$$C_{N,m,n} := \begin{cases} a_{m,n} - b_{m,n} & \text{if } |m| \leq N \text{ and } |n| \leq N, \\ 0 & \text{otherwise.} \end{cases}$$

But C_N is of finite rank, so C is compact.

We next observe that $B = B_+ \oplus B_-$ where $B_+ : l^2\{0, \infty\} \to l^2\{0, \infty\}$ and $B_- : l^2\{-\infty, -1\} \to l^2\{-\infty, -1\}$ are both Toeplitz operators. (In the case of B_- one needs to relabel the subscripts.) Corollary 4.3.8 and Theorem 4.4.1 now yield

$$\text{EssSpec}(A) = \text{EssSpec}(B)$$
$$= \text{EssSpec}(B_+) \cup \text{EssSpec}(B_-)$$
$$= \{b_+(\theta) : -\pi \leq \theta \leq \pi\} \cup \{b_-(\theta) : -\pi \leq \theta \leq \pi\}. \quad \square$$

Problem 4.4.7 Let A be the operator on $l^2(\mathbf{Z})$ associated with the infinite matrix

$$A_{r,s} := \begin{cases} 1 & \text{if } |r - s| = 1, \\ i & \text{if } r = s > 0, \\ -i & \text{if } r = s < 0, \\ 0 & \text{otherwise.} \end{cases}$$

Find the essential spectrum of A. Prove that 0 is an isolated eigenvalue of A, and that the corresponding eigenvector is concentrated around the origin. Does A have any other eigenvalues?

---

[18] An alternative proof that the LHS of (4.8) contains the RHS is indicated in Problem 14.4.5.

If one also specifies that $A_{0,0} = is$ where $-1 \leq s \leq 1$, determine how the eigenvalue found above depends on s. $\quad \square$

In the above theorem one can consider $\mathbf{Z}$ as a graph with two ends, on which A has different forms. One can also consider a graph with a larger number of ends. Mathematically one takes A to be a bounded operator on $l^2(F \cup [\{1, \ldots, k\} \times \mathbf{Z}_+])$, where k is the number of ends and F is a finite set to which they are all joined. One assumes that $A = \sum_{r=1}^{k} B_r + C$ where C is compact (or even of finite rank) and each B_r is a Toeplitz operator acting in $l^2(\{r\} \times \mathbf{Z}_+)$ with symbol $b_r \in C_{\text{per}}[-\pi, \pi]$.

Theorem 4.4.8 *Under the above assumptions the essential spectrum of A is given by*

$$\text{EssSpec}(A) = \bigcup_{r=1}^{k} \{b_r(\theta) : -\pi \leq \theta \leq \pi\}.$$

The proof is an obvious modification of the proof of Theorem 4.4.6.

One often needs to consider more complicated situations, in which each point of $\mathbf{Z}^N$ has several internal degrees of freedom attached. These are analyzed by considering matrix-valued convolution operators.

Theorem 4.4.9 *Let $\mathcal{K}$ be a finite-dimensional inner product space and let $\mathcal{H} := l^2(\mathbf{Z}^N, \mathcal{K})$ be the space of all square-summable $\mathcal{K}$-valued sequences on $\mathbf{Z}^N$. Let $A : \mathcal{H} \to \mathcal{H}$ be defined by*

$$(Af)(n) := \sum_{m \in \mathbf{Z}^N} a_{n-m} f_m$$

where $a : \mathbf{Z}^N \to \mathcal{L}(\mathcal{K})$ satisfies $\sum_{n \in \mathbf{Z}^N} \|a_n\| < \infty$. Then

$$\text{Spec}(A) = \text{EssSpec}(A) = \bigcup\{\text{Spec}(b(\theta)) : \theta \in [-\pi, \pi]^N\}$$

where

$$b(\theta) := \sum_{n \in \mathbf{Z}^N} a_n e^{-in \cdot \theta} \in \mathcal{L}(\mathcal{K}) \tag{4.9}$$

for all $\theta \in [-\pi, \pi]^N$.

Proof. We use the unitarity[19] of the Fourier operator $\mathcal{F} : l^2(\mathbf{Z}^N, \mathcal{K}) \to L^2([-\pi, \pi]^N, \mathcal{K})$ defined by

$$(\mathcal{F}c)(\theta) := (2\pi)^{-N/2} \sum_{n \in \mathbf{Z}^N} c_n e^{-in \cdot \theta},$$

where $c \in l^2(\mathbf{Z}^N, \mathcal{K})$ and $\theta \in [-\pi, \pi]^N$.

A direct calculation shows that $B := \mathcal{F}A\mathcal{F}^{-1}$ is given by

$$(Bf)(\theta) = b(\theta)f(\theta)$$

for all $f \in L^2([-\pi, \pi]^N, \mathcal{K})$, where b is defined by (4.9) and is a continuous $\mathcal{L}(\mathcal{K})$-valued function. We therefore need only establish the relevant spectral properties of B.

Identifying $\mathcal{K}$ with $\mathbf{C}^k$ for some k, we note that each $b(\theta)$ is a $k \times k$ matrix, and its spectrum is a set of k or fewer eigenvalues. These are the roots of the θ-dependent characteristic polynomial, and depend continuously on θ by Rouche's theorem. Let $S := \bigcup \{\text{Spec}(b(\theta)) : \theta \in [-\pi, \pi]^N\}$. If $z \notin S$ then the function

$$r(\theta) := (zI - b(\theta))^{-1}$$

is bounded and continuous so the corresponding multiplication operator R satisfies $(zI - B)R = R(zI - B) = I$. Therefore $z \notin \text{Spec}(B)$.

Conversely suppose that $z \in \text{Spec}(b(\theta))$ for some θ; more explicitly suppose that $b(\theta)v = zv$ for some unit vector $v \in \mathcal{K}$. Given $\varepsilon > 0$ there exists $\delta > 0$ such that $|b(\phi)v - zv| < \varepsilon$ provided $|\phi - \theta| < \delta$. Putting $f_v(x) := f(x)v$ we deduce that $\|Bf_v - zf_v\| < \varepsilon\|f_v\|$ for all $f \in L^2([-\pi, \pi]^N)$ such that $\text{supp}(f) \subseteq \{\phi : |\phi - \theta| < \delta\}$. By choosing a sequence of f whose supports decrease to $\{\theta\}$, we deduce using Lemma 4.3.15 that

$$S \subseteq \text{EssSpec}(B) \subseteq \text{Spec}(B) \subseteq S. \qquad \square$$

Problem 4.4.10 Let A be a bounded operator on $l^2(\mathbf{Z})$ with a tridiagonal matrix a, i.e. one that satisfies $a_{m,n} = 0$ if $|m - n| > 1$. Suppose also that A is periodic with period k in the sense that $a_{m+k,n+k} = a_{m,n}$ for all m, $n \in \mathbf{Z}$. Let $\sigma : \mathbf{Z} \times \{0, 1, \ldots, k-1\} \to \mathbf{Z}$ be the map $\sigma(m, j) := km + j$. Use this map to identify $l^2(\mathbf{Z})$ with $l^2(\mathbf{Z}, \mathbf{C}^k)$ and hence to rewrite A in the form considered in Theorem 4.4.9. Hence prove that the spectrum of A is the union of at most k closed curves (or intervals) in $\mathbf{C}$. $\square$

---

[19] The proof of unitarity involves choosing some orthonormal basis $\{e_1, \ldots, e_k\}$ in $\mathcal{K}$ and applying the N-dimensional analogue of Corollary 2.3.11 to each sequence $\langle c, e_r \rangle \in l^2(\mathbf{Z}^N)$.

Problem 4.4.11 Find the spectrum of the operator A on $l^2(\mathbf{Z})$ associated with the infinite, tridiagonal, period 2 matrix

$$
A := \begin{pmatrix}
\ddots & \ddots & & & & & \\
\ddots & \gamma & \alpha & & & & \\
 & \beta & -\gamma & \alpha & & & \\
 & & \beta & \gamma & \alpha & & \\
 & & & \beta & -\gamma & \alpha & \\
 & & & & \beta & \gamma & \alpha \\
 & & & & & \beta & -\gamma & \ddots \\
 & & & & & & \ddots & \ddots
\end{pmatrix}
\tag{4.10}
$$

where α, β, γ are real non-zero constants. [20] □

Figure 4.2 shows the spectrum of the $N \times N$ matrix corresponding to (4.10) for $N := 100$, $\alpha := 1$, $\beta := 5$ and $\gamma := 3.9$. We imposed periodic boundary conditions by putting $A_{N,1} := \alpha$ and $A_{1,N} := \beta$. If these entries are put equal to 0, then A is similar to a self-adjoint matrix and has real eigenvalues.

If one replaces the diagonal entries $A_{r,r} = (-1)^r \gamma$ above by the values $\pm\gamma$ chosen randomly, one would expect the eigenvalues of A to be distributed randomly in the complex plane. The fact that they are in fact regularly

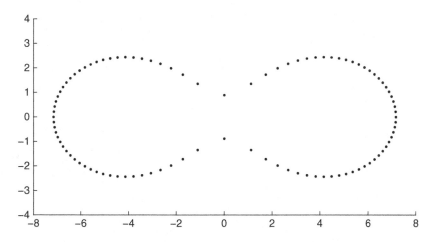

Figure 4.2: Spectrum of a period 2 matrix

[20] In spite of the explicit formula for the spectrum, this operator is not similar to a normal operator if $|\gamma| = |\alpha - \beta|$: equivalently it is not then an operator of scalar type in the sense of [Dunford and Schwartz 1971, Theorem XV.6.2]. The corresponding problem for Schrödinger operators has been fully analyzed in [Gesztesy and Tkachenko 2005].

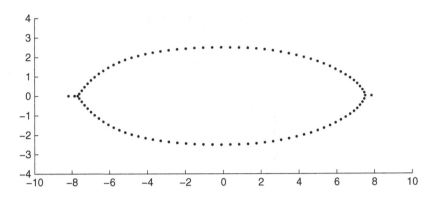

Figure 4.3: Spectrum of a matrix with random diagonal entries

distributed along certain curves was one of the real surprises in the field of random matrix theory. indexMartinez[21] Figure 4.3 shows a typical spectral diagram with the same values of N, α, β and γ as before. Once again if $A_{N,1}$ and $A_{1,N}$ are put equal to 0, then A is similar to a self-adjoint matrix and has real eigenvalues.

[21] See Goldsheid and Khoruzhenko 2000, Goldsheid and Khoruzhenko 2003 for the proof and references to the earlier literature on what is called the non-self-adjoint Anderson model. An 'exactly soluble' special case was analyzed in [Trefethen et al. 2001]. See Trefethen and Embree 2005, Chap.8for a discussion of these models in terms of pseudospectra. The spectrum of the corresponding infinite-dimensional random matrices is significantly different and was investigated in [Davies 2001A, Davies 2001B, Davies 2005B, Martinez 2005].

5

Operators on Hilbert space

5.1 Bounded operators

In this chapter we describe some of the special theorems that can be proved for operators on a Hilbert space. The best known of these is the spectral theorem for self-adjoint operators, but we also derive the basic properties of Hilbert-Schmidt and trace class operators. Some of the theorems in this chapter have clumsier versions in Banach spaces, but others have no analogues.

Lemma 5.1.1 *If A is a bounded self-adjoint operator on $\mathcal{H}$ then* $\mathrm{Spec}(A) \subset$ **R** *and*

$$\| (zI - A)^{-1} \| \leq |\mathrm{Im}(z)|^{-1} \tag{5.1}$$

*for all $z \notin$ **R**.*

Proof. Let $z := x + iy$ where x, $y \in$ **R** and $y \neq 0$. A direct calculation establishes that

$$\| (A - zI)f \|^2 = \| (A - xI)f \|^2 + y^2 \| f \|^2 = \| (A - \overline{z}I)f \|^2$$

for all $f \in \mathcal{H}$. The inequality

$$\| (A - zI)f \| \geq |y| \, \| f \| \tag{5.2}$$

establishes that $(A - zI)$ is one-one with closed range. The orthogonal complement of $\mathrm{Ran}(A - zI)$ is $\mathrm{Ker}(A - \overline{z}I)$, and this equals $\{0\}$ by a similar argument. We conclude that $z \notin \mathrm{Spec}(A)$. The bound (5.1) is equivalent to (5.2). $\square$

Lemma 5.1.2 *If the bounded self-adjoint operator B on $\mathcal{H}$ is non-negative in the sense that $\langle Bx, x \rangle \geq 0$ for all $x \in \mathcal{H}$, then*

$$\| B \| = \sup\{ \langle Bx, x \rangle : \| x \| = 1 \}.$$

It follows that if A is a bounded operator on $\mathcal{H}$ then

$$\|A\|^2 = \|A^*A\|. \tag{5.3}$$

Proof. Let γ denote the supremum of the lemma. Applying Schwarz's inequality to the semi-definite inner product

$$(x, y) := \langle Bx, y \rangle$$

we obtain

$$|\langle Bx, y \rangle|^2 \le \langle Bx, x \rangle \langle By, y \rangle \le \gamma^2$$

provided $\|x\| = \|y\| = 1$. This implies that

$$|\langle Bx, y \rangle| \le \gamma \|x\| \|y\|$$

for all $x, y \in \mathcal{H}$. Putting $y := Bx$ yields $\|B\| \le \gamma$. The reverse inequality is elementary.

If we put $\alpha := \sup\{\langle A^*Ax, x \rangle : \|x\| = 1\}$, then

$$\|Ax\|^2 = \langle Ax, Ax \rangle = \langle A^*Ax, x \rangle \le \alpha$$

provided $\|x\| = 1$. This implies that

$$\|A\|^2 \le \alpha = \|A^*A\|$$

by the first half of the lemma. Once again the reverse inequality is elementary. $\square$

Lemma 5.1.3 *If A is a bounded self-adjoint operator then*

$$\|A^m\| = \|A\|^m \tag{5.4}$$

for all positive integers m. Hence $\mathrm{Rad}(A) = \|A\|$.

Proof. Lemma 5.1.2 implies that $\|A^{2^n}\| = \|A\|^{2^n}$ for all positive integers n. If m is a positive integer and $m \le 2^n$ then

$$\|A\|^{2^n} = \|A^{2^n}\| = \|A^m A^{2^n - m}\| \le \|A^m\| \|A^{2^n - m}\| \le \|A\|^{2^n}.$$

This implies (5.4). The proof is completed by using Theorem 4.1.3. $\square$

If A and B are two bounded self-adjoint operators we will write $A \ge B$ when

$$\langle Ax, x \rangle \ge \langle Bx, x \rangle$$

for all $x \in \mathcal{H}$.

Problem 5.1.4 Prove that the set

$$\mathcal{C} := \{A \in \mathcal{L}(\mathcal{H}) : A = A^* \text{ and } A \geq 0\}$$

is a closed convex cone in $\mathcal{L}(\mathcal{H})$, and that $\mathcal{C} \cap (-\mathcal{C}) = \{0\}$. $\square$

Lemma 5.1.5 *If A is a bounded self-adjoint operator on $\mathcal{H}$ then*

$$\max\{\mathrm{Spec}(A)\} = \sup\{\langle Ax, x \rangle : \|x\| = 1\}$$

and

$$\min\{\mathrm{Spec}(A)\} = \inf\{\langle Ax, x \rangle : \|x\| = 1\}.$$

Proof. Let α (resp. β) denote the infimum (resp. supremum) of the lemma. We have $A - \alpha I \geq 0$ and

$$\sup\{\langle (A - \alpha I)x, x \rangle : \|x\| = 1\} = \beta - \alpha.$$

Therefore

$$\|A - \alpha I\| = \beta - \alpha.$$

Lemma 5.1.3 now implies that

$$\mathrm{Rad}(A - \alpha I) = \beta - \alpha.$$

Since $\mathrm{Spec}(A)$ is real it follows that $\mathrm{Spec}(A) \subseteq [2\alpha - \beta, \beta]$ and that either $2\alpha - \beta \in \mathrm{Spec}(A)$ or $\beta \in \mathrm{Spec}(A)$ (or both).

By applying a similar argument to $\beta I - A$ we obtain $\mathrm{Spec}(A) \subseteq [\alpha, 2\beta - \alpha]$ and that either $2\beta - \alpha \in \mathrm{Spec}(A)$ or $\alpha \in \mathrm{Spec}(A)$ (or both).

On combining these two statements we obtain $\mathrm{Spec}(A) \subseteq [\alpha, \beta]$, $\alpha \in \mathrm{Spec}(A)$ and $\beta \in \mathrm{Spec}(A)$. $\square$

Problem 5.1.6 Prove that a bounded operator H on the Hilbert space $\mathcal{H}$ is self-adjoint if and only if the bounded operators

$$U_t := \sum_{n=0}^{\infty} (iHt)^n / n!$$

are unitary for all $t \in \mathbf{R}$. $\square$

5.2 Polar decompositions

In this section we construct the polar decomposition of a bounded linear operator; this is the operator analogue of the formula $z := re^{i\theta}$ for complex numbers. The construction depends on the following lemma, which we prove directly, i.e. without using the spectral theorem below.

Lemma 5.2.1 (*square root lemma*) *If A is a bounded, non-negative, self-adjoint operator then there exists a bounded operator Q such that $Q = Q^* \geq 0$ and $Q^2 = A$. Moreover Q is the norm limit of a sequence of polynomials in A.*

Proof. The power series

$$\sum_{n=1}^{\infty} a_n x^n := 1 - (1 - x)^{1/2}$$

converges for all complex x such that $|x| < 1$, and the coefficients a_n are all positive. If we let $0 < x < 1$ and take the limit $x \to 1$ we deduce that $\sum_{n=1}^{\infty} a_n = 1$.

It is sufficient to prove the lemma when $A = A^* \geq 0$ and $\|A\| = 1$. Lemma 5.1.2 implies that $\|I - A\| \leq 1$, so the series of operators

$$Q := I - \sum_{n=1}^{\infty} a_n (I - A)^n$$

is norm convergent. The operator Q is clearly self-adjoint and the bound $\|Q - I\| \leq 1$ implies that $Q \geq 0$. The identity $Q^2 = A$ uses the rule for multiplying together two series term by term.   □

Problem 5.2.2 Let $A = A^* \geq 0$ and let Q be defined by the series in Lemma 5.2.1. Prove that if $B = B^* \geq 0$ and $B^2 = A$ then B commutes with Q, and then that $B = Q$.   □

Problem 5.2.3 Let A_1, A_2, B be bounded, self-adjoint operators on $\mathcal{H}$ and suppose that B is positive and invertible. Prove that $A_1 \leq A_2$ if and only if $B^{-1/2} A_1 B^{-1/2} \leq B^{-1/2} A_2 B^{-1/2}$. In particular $0 \leq A_1 \leq B$ if and only if $0 \leq B^{-1/2} A_1 B^{-1/2} \leq I$.   □

If A is a bounded linear operator on a Hilbert space $\mathcal{H}$ we define $|A|$ by

$$|A| := (A^* A)^{1/2}.$$

The formula (5.5) below is called the polar decomposition of A.

Theorem 5.2.4 *If A is a bounded operator on the Hilbert space $\mathcal{H}$ then there exists a linear operator V such that $\|V\| \leq 1$ and*

$$|A| = V^*A, \qquad\qquad A = V|A|. \qquad\qquad (5.5)$$

Moreover the compactness of any of A, $|A|$ and A^*A implies the compactness of the others. If A is invertible then $|A|$ is positive and invertible while V is unitary.

Proof. We observe that

$$\| |A|f\|^2 = \langle |A|^2 f, f \rangle = \langle A^*Af, f \rangle = \|Af\|^2$$

for all $f \in \mathcal{H}$. Hence

$$\mathrm{Ker}(A) = \mathrm{Ker}(|A|).$$

The formula $V(|A|f) := Af$ unambiguously defines an isometry mapping $\mathrm{Ran}(|A|)$ onto $\mathrm{Ran}(A)$; this can be extended to an isometry from $\overline{\mathrm{Ran}(|A|)}$ onto $\overline{\mathrm{Ran}(A)}$ and then to a contraction (actually a partial isometry as defined on page 140) on $\mathcal{H}$ by putting $Vg := 0$ for all $g \in \mathrm{Ran}(|A|)^\perp$. Note that $\mathrm{Ran}(|A|)^\perp = \mathrm{Ker}(|A|)$ is a special case of the identity $\mathrm{Ker}(B^*) = \mathrm{Ran}(B)^\perp$, valid for all bounded operators B.

The above arguments show that $A = V|A|$. The properties of V already established imply that $V^*Vg = g$ for all $g \in \mathrm{Ran}(|A|)$. Therefore

$$V^*Af = V^*V|A|f = |A|f$$

for all $f \in \mathcal{H}$, so $V^*A = |A|$.

If A is compact then A^*A is compact by Theorem 4.2.2. If $B = A^*A \geq 0$ is compact then the two identities

$$|A| = I - \sum_{n=1}^{\infty} a_n (I - B)^n,$$

$$0 = 1 - \sum_{n=1}^{\infty} a_n,$$

together yield the norm convergent expansion

$$|A| = \sum_{n=1}^{\infty} a_n \{I - (I - B)^n\}.$$

Each term in this series is compact, so $|A|$ must be compact by Theorem 4.2.2. Finally if $|A|$ is compact then A is compact by (5.5) and Theorem 4.2.2.

The last statement of the theorem follows by examining the details of the proof. $\square$

The following problem is a warning that one should not carry over results from function spaces to $\mathcal{L}(\mathcal{H})$ without proof.

Problem 5.2.5 Prove that if A, B are bounded self-adjoint operators on $\mathcal{H}$ then $\pm A \leq B$ does not imply $|A| \leq B$. (Find suitable 2×2 matrices.) Deduce that there exist self-adjoint operators S, T such that $|S + T| \not\leq |S| + |T|$. $\square$

5.3 Orthogonal projections

In Lemma 1.5.5 we established that the rank of a projection on a Banach space does not change under small enough perturbations. In this section we study projections on a Hilbert space $\mathcal{H}$ in more detail.[1] An orthogonal projection on $\mathcal{H}$ is defined to be a bounded operator P such that $P^2 = P = P^*$. We assume that the reader is familiar with the fact that every closed subspace L of a Hilbert space has an orthogonal complement $L^\perp$ such that $L \cap L^\perp = \{0\}$ and $L + L^\perp = \mathcal{H}$. The following lemma is also standard.

Lemma 5.3.1 *The map $P \to L := \mathrm{Ran}(P)$ defines a one-one correspondence between orthogonal projections P and closed subspaces L of $\mathcal{H}$. Moreover $\mathrm{Ker}(P) = L^\perp$ for all such P.*

Problem 5.3.2 Let P_1, P_2 be two orthogonal projections with ranges L_1, L_2 respectively. Prove that the following are equivalent.

(i) $L_1 \subseteq L_2$,
(ii) $P_1 \leq P_2$,
(iii) $P_1 P_2 = P_2 P_1 = P_1$. $\square$

We take the opportunity to expand on some concepts used implicitly in the last section. A partial isometry A on a Hilbert space $\mathcal{H}$ is defined to be a bounded linear operator such that $P := A^* A$ is an orthogonal projection.

[1] The ideas in this section go back to [Kato 1966A]. A complete set of unitary invariants for a pair of subspaces was obtained in [Halmos 1969], while the notion of index of a pair of subspaces was investigated in [Avron et al. 1994].

Problem 5.3.3 Prove that if $P := A^*A$ is an orthogonal projection then so is $Q := AA^*$. Moreover A maps $\text{Ran}(P)$ isometrically one-one onto $\text{Ran}(Q)$. □

An isometry A on $\mathcal{H}$ is an operator such that $A^*A = I$; equivalently it is an operator such that $\|Af\| = \|f\|$ for all $f \in \mathcal{H}$.

Problem 5.3.4 Let A be an isometry on $\mathcal{H}$ and let $\mathcal{L}_0 := \{\text{Ran}(A)\}^{\perp}$. Prove that $\mathcal{L}_n := A^n \mathcal{L}_0$ are orthogonal closed subspaces for all $n \in \mathbf{N}$, and that A maps $\mathcal{L}_n$ isometrically onto $\mathcal{L}_{n+1}$ for all such n. □

We define the distance between two closed subspaces L and M of the Hilbert space $\mathcal{H}$ by

$$d(L, M) = \|P - Q\| \tag{5.6}$$

where P, Q are the orthogonal projections with ranges L and M respectively. This defines a metric on the set of all closed subspaces.

Lemma 5.3.5 *If $d(L, M) < 1$ then*

$$\dim(L) = \dim(M), \qquad \dim(L^{\perp}) = \dim(M^{\perp}).$$

Proof. If we put

$$X := PQ + (I - P)(I - Q)$$

then

$$
\begin{aligned}
XX^* &= PQP + (I - P)(I - Q)(I - P) \\
&= PQP + I - Q - P + QP - P + PQ + P - PQP \\
&= I - Q - P + PQ + QP \\
&= I - (P - Q)^2.
\end{aligned}
$$

Since the identity

$$X^*X = I - (P - Q)^2$$

has a similar proof, we see that X is normal. If $\|P - Q\| < 1$ we deduce that X is invertible by using Problem 1.2.8 and Problem 1.2.17. Since $X(M) \subseteq L$ and $X(M^{\perp}) \subseteq L^{\perp}$, the invertibility of X implies that these are both equalities. The statements about the dimensions follow. □

Lemma 5.3.6 *If $\|P - Q\| < 1$ then there exists a canonical unitary operator U such that $U^*PU = Q$.*

Proof. The existence of such a unitary operator is equivalent to the conclusion of Lemma 5.3.5, but there are many such U, and we provide a canonical choice.

We have already proved that X is normal and invertible. Theorem 5.2.4 implies that $X = U|X|$ where $|X|$ is invertible and U is unitary. By examining the proof one sees that X, $|X|$ and U all commute. Since

$$P|X|^2 = P\{I - (P - Q)^2\} = PQP = \{I - (P - Q)^2\}P = |X|^2 P$$

we deduce by applying Lemma 5.2.1 with $A := |X|^2$ that $P|X| = |X|P$. The identity

$$|X|PU = PU|X| = PX = PQ = XQ = |X|UQ$$

finally implies that $PU = UQ$. $\quad\square$

Theorem 5.3.7 *Let $P(t)$ be a norm continuous family of orthogonal projections, where $0 \leq t \leq 1$. Then there exists a norm continuous family of unitary operators $U(t)$ such that*

$$U^*(t)P(0)U(t) = P(t)$$

for all $t \in [0, 1]$.

Proof. Since the family of projections must be uniformly continuous as a function of t, there exists a positive integer n such that $\|P(s) - P(t)\| < 1$ if $|s - t| \leq 1/n$. It is sufficient to prove the theorem separately for each interval of the form $[r/n, (r+1)/n]$ and then string the results together.

If $r/n \leq t \leq (r+1)/n$ we define $U_{r/n, t}$ by

$$U_{r/n, t} := A_{r/n, t} X_{r/n, t},$$
$$A_{r/n, t} := \{I - (P_{r/n} - P_t)^2\}^{-1/2},$$
$$X_{r/n, t} := P_{r/n} P_t + (I - P_{r/n})(I - P_t).$$

The identity

$$U^*_{r/n, t} P_{r/n} U_{r/n, t} = P_t$$

follows as in Lemma 5.3.6. The definition of $U_{r/n, t}$ implies that it is a norm continuous function of t. $\quad\square$

The following corollary follows immediately from Lemma 5.3.5 or Theorem 5.3.7, but can also be proved by using the theory of Fredholm operators.

Corollary 5.3.8 *Let* $\{L_t\}_{0 \le t \le 1}$ *be a family of closed linear subspaces which is continuous with respect to the metric (5.6). If L_0 has finite dimension n then L_t has dimension n for all $t \in [0, 1]$.*

5.4 The spectral theorem

In this section we write down the spectral theorem. The general form of the theorem was obtained independently by Stone and von Neumann between 1929 and 1932. It is undoubtedly the most important result in the subject. The theorem is used in several places in the book, but we do not give a proof, which is well documented.[2]

Recall that a bounded operator A on $\mathcal{H}$ is said to be normal if $A^*A = AA^*$, unitary if $A^*A = AA^* = I$ and self-adjoint if $A = A^*$.

Theorem 5.4.1 *Let A be a bounded normal operator acting on the separable Hilbert space $\mathcal{H}$. Then there exists a set X provided with a σ-field Σ of subsets and a σ-finite measure dx, together with a unitary map $U : \mathcal{H} \to L^2(X, dx)$ for which the following holds. The operators $M_1 := UAU^{-1}$ and $M_2 := UA^*U^{-1}$ are of the form*

$$(M_1 g)(x) = m(x)g(x), \qquad (M_2 g)(x) = \overline{m(x)}g(x)$$

where m is a bounded, complex-valued, measurable function on X. Moreover

$$(Up(A)U^{-1}g)(x) = p(m(x))g(x)$$

almost everywhere for every polynomial p and every $g \in L^2(X, dx)$. The spectrum of A equals the essential range of m.

One says informally that every normal operator is unitarily equivalent to a multiplication operator as defined in Section 2.2. The last statement of the theorem follows directly from Problem 2.2.1. It is not suggested that the above representation is unique, but its transparent character makes up for this lack to a considerable extent.

---

[2] There are many quite different statements and proofs in the literature. [Dunford and Schwartz 1963] is a standard account of the subject. [Davies 1995A, Davies 1995B, Davies 1995C] present a completely different and very explicit definition of the functional calculus that extends usefully to L^p spaces whenever certain resolvent estimates are satisfied.

The spectral theorem can be used to define a canonical functional calculus for normal operators. It is standard to write $f(N) := \mathcal{T}(f)$ where $\mathcal{T}$ is the homomorphism defined below. The uniqueness statement in the theorem is particularly important because of the large number of different constructions of the homomorphism $\mathcal{T}$.

Theorem 5.4.2 *Let N be a bounded normal operator acting in a Hilbert space $\mathcal{H}$, and let $\mathcal{A}$ denote the space of all continuous functions on the compact set $\mathrm{Spec}(N)$. We consider $\mathcal{A}$ as a commutative Banach algebra under pointwise addition and multiplication and the supremum norm. Then there exists a unique isometric algebra homomorphism $\mathcal{T}$ from $\mathcal{A}$ into $\mathcal{L}(\mathcal{H})$ such that*

$$\mathcal{T}(z) = N, \qquad \mathcal{T}(\bar{z}) = N^* \qquad \mathcal{T}(1) = I.$$

Problem 5.4.3 Prove that the normal operator A is unitary if and only if $|m(x)| = 1$ almost everywhere, and that it is self-adjoint if and only if $m(x)$ is real almost everywhere. Prove also that if A is a unitary operator on $\mathcal{H}$ then there exists a bounded self-adjoint operator H such that $A = e^{iH}$. □

The spectral theorem for bounded self-adjoint operators is a special case of that for bounded normal operators discussed above. In the unbounded case we need to make some definitions.[3] If $\mathcal{D}$ is a dense linear subspace of a Hilbert space $\mathcal{H}$, we say that $H : \mathcal{D} \to \mathcal{H}$ is a closed operator if, whenever $f_n \in \mathcal{D}$ converges in norm to f and Hf_n converges in norm to g, it follows that $f \in \mathcal{D}$ and $Hf = g$. Equivalently $\mathcal{D}$ is complete with respect to the norm

$$|||f||| := \|f\| + \|Hf\|.$$

Now suppose that $H : \mathcal{D} \to \mathcal{H}$ is symmetric in the sense that $\langle Hf, g \rangle = \langle f, Hg \rangle$ for all $f, g \in \mathcal{D}$. We say that $g \in \mathrm{Dom}(H^*)$ if $f \to \langle Hf, g \rangle$ is a bounded linear functional on $\mathcal{D}$ with respect to the standard norm on $\mathcal{H}$. The Riesz representation theorem implies that there exists a unique (g-dependent) $k \in \mathcal{H}$ such that

$$\langle Hf, g \rangle = \langle f, k \rangle$$

for all $f \in \mathcal{D}$, and we write $H^*g := k$. It is straightforward to verify that H^* is a linear operator on its domain, and that H^* is an extension of H in the

---

[3] The spectrum of an unbounded operator was defined on page 124. The following material is closely related to that on the Cayley transform for dissipative operators in Section 10.4. The duality of unbounded operators is discussed in a Banach space setting in Section 7.3.

sense that $\mathrm{Dom}(H^*) \supseteq \mathrm{Dom}(H)$ and $H^*f = Hf$ for all $f \in \mathrm{Dom}(H)$. We say that H is self-adjoint, and write $H = H^*$, if $\mathrm{Dom}(H^*) = \mathrm{Dom}(H)$.

Lemma 5.4.4 *Every symmetric operator H has a closed symmetric extension $\overline{H}$ which is minimal in the sense that every closed extension of H is also an extension of $\overline{H}$. Moreover H^* is a closed extension of $\overline{H}$.*

Proof. We start by proving that H^* is closed. If $g_n \in \mathrm{Dom}(H^*)$, $g_n \to g$ and $H^* g_n \to k$ as $n \to \infty$ then

$$\langle f, k \rangle = \lim_{n \to \infty} \langle f, H^* g_n \rangle = \lim_{n \to \infty} \langle Hf, g_n \rangle = \langle Hf, g \rangle$$

for all $f \in \mathrm{Dom}(H)$. Therefore $g \in \mathrm{Dom}(H^*)$ and $H^* g = k$.

We extend the norm $\|\|\cdot\|\|$ to $\mathrm{Dom}(H^*)$ by putting

$$\|\|f\|\| := \|f\| + \|H^* f\|.$$

Since H^* is closed $\mathrm{Dom}(H^*)$ is complete with respect to this norm. We now define $\overline{H}$ to be the restriction of H^* to the closure $\mathcal{E}$ of $\mathrm{Dom}(H)$ with respect to the norm $\|\|\cdot\|\|$. It follows immediately that $\overline{H}$ is closed and that it is the least closed extension of H.

We have finally to prove that $\overline{H}$ is symmetric. If $g \in \mathrm{Dom}(\overline{H})$ then there exist $g_n \in \mathrm{Dom}(H)$ such that $\|g_n - g\| \to 0$ and $\|Hg_n - \overline{H}g\| \to 0$ as $n \to \infty$. If also $f \in \mathrm{Dom}(H)$ then

$$\langle f, \overline{H}g \rangle = \lim_{n \to \infty} \langle f, Hg_n \rangle = \lim_{n \to \infty} \langle Hf, g_n \rangle = \langle Hf, g \rangle.$$

If $f, g \in \mathrm{Dom}(\overline{H})$ then there exist $f_n \in \mathrm{Dom}(H)$ such that $\|f_n - f\| \to 0$ and $\|Hf_n - \overline{H}f\| \to 0$ as $n \to \infty$. Therefore

$$\langle f, \overline{H}g \rangle = \lim_{n \to \infty} \langle f_n, \overline{H}g \rangle = \lim_{n \to \infty} \langle Hf_n, g \rangle = \langle \overline{H}f, g \rangle. \qquad \square$$

If H is a closed symmetric operator we define its deficiency subspaces by

$$\mathcal{L}^{\pm} := \{ f \in \mathrm{Dom}(H^*) : H^* f = \pm i f \}$$

$$= \{ f \in \mathcal{H} : \langle Hg, f \rangle = \mp i \langle g, f \rangle \text{ for all } g \in \mathrm{Dom}(H) \}.$$

The deficiency indices of H are the dimensions of the deficiency subspaces.

Theorem 5.4.5 *The closed symmetric operator H is self-adjoint if and only if its deficiency indices are both zero.*

Proof. If $H = H^*$ and $f \in \mathcal{L}^+$ then

$$i\langle f, f \rangle = \langle Hf, f \rangle - \langle f, Hf \rangle = -i\langle f, f \rangle,$$

therefore $f = 0$. The proof that $\mathcal{L}^- = \{0\}$ is similar.

Conversely suppose that $\mathcal{L}^\pm = \{0\}$. The operator $(H + iI)$ maps $\mathrm{Dom}(H)$ one-one onto a subspace M^+ and the inverse operator R is a contraction (with domain M^+) by virtue of the identity

$$\|(H + iI)f\|^2 = \|Hf\|^2 + \|f\|^2 = \|(H - iI)f\|^2.$$

Since H is closed, so is $(H + iI)$. Therefore R is a closed contraction and its domain M^+ must be a closed subspace of $\mathcal{H}$. If $g \perp M^+$ then $\langle Hf, g \rangle = \langle f, ig \rangle$ for all $f \in \mathrm{Dom}(H)$. Therefore $g \in \mathrm{Dom}(H^*)$ and $H^*g = ig$. The assumption $\mathcal{L}^+ = \{0\}$ implies that $M^+ = \mathcal{H}$.

This last identity proves that for every $g \in \mathrm{Dom}(H^*)$ there exists $f \in \mathrm{Dom}(H)$ such that $(H + iI)f = (H^* + iI)g$. Therefore $(H^* + iI)(f - g) = 0$. Since $\mathcal{L}^- = \{0\}$, we deduce that $f = g$. Therefore $\mathrm{Dom}(H^*) = \mathrm{Dom}(H)$ and $H = H^*$. $\square$

Problem 5.4.6 Let H be a symmetric operator acting in $\mathcal{H}$ and let $\{f_n\}_{n \in \mathbb{N}}$ be a complete orthonormal set in $\mathcal{H}$. Suppose also that $f_n \in \mathrm{Dom}(H)$ and $Hf_n = \lambda_n f_n$ for all $n \in \mathbb{N}$, where $\lambda_n \in \mathbb{R}$. Prove that H is essentially self-adjoint on $\mathcal{D} := \mathrm{lin}\{f_n : n \in \mathbb{N}\}$, and that $\mathrm{Spec}(\overline{H})$ is the closure of $\{\lambda_n : n \in \mathbb{N}\}$. Compare this with Problem 6.1.19. $\square$

Problem 5.4.7 Let $H := -\Delta$ act in $L^2(T)$ subject to Dirichlet boundary conditions, where T is a triangular region in $\mathbb{R}^2$. The complete list of eigenvalues is known for three choices of T, whose interior angles are 60°, 60°, 60° or 90°, 60°, 30° or 90°, 45°, 45°. We indicate how to obtain the result in the third case. It is not possible to write down the eigenvalues of the Dirichlet Laplacian explicitly for the regular hexagon or most other polygonal regions.

Let T denote the triangle

$$\{(x, y) : 0 < x < \pi, \, 0 < y < x\},$$

and let

$$\phi_{m,n}(x, y) := \sin(mx)\sin(ny) - \sin(nx)\sin(my)$$

where $1 \leq m < n \in \mathbb{N}$. The main task is to prove that $\{\phi_{m,n}\}$ is a complete orthogonal set in $L^2(T)$. Once this is done one observes that $\phi_{m,n} \in \mathrm{Dom}(H)$ and

$$H\phi_{m,n} = (m^2 + n^2)\phi_{m,n}.$$

Problem 5.4.6 then implies that

$$\mathrm{Spec}(H) = \{m^2 + n^2 : 1 \le m < n\}.$$

Note that the eigenfunction associated with the smallest eigenvalue is positive in the interior of T.[4] $\quad\square$

Example 5.4.8 Let $H := -\Delta$ act in $L^2(S_\alpha)$ subject to Dirichlet boundary conditions, where $S_\alpha \subseteq \mathbf{R}^2$ is the sector given in polar coordinates by the conditions $0 < r < 1$ and $0 < \theta < \alpha$. The eigenvalue problem may be solved by separation of variables, but the solution indicates the technical difficulties that may be associated with the definition of the domain of the operator in quite simple problems. In three dimensions these are of major concern, because of the huge range of corners that even polyhedral regions can possess.

Every eigenfunction of H is of the form $\phi(r)\sin(n\theta/\alpha)$ where $n \in \mathbf{N}$ and ϕ is a Bessel function that vanishes linearly as $r \to 1$ but like $r^{\pi/\alpha}$ as $r \to 0$. If the sector is re-entrant, i.e. $\alpha > \pi$, then the first derivatives of the eigenfunctions diverge as one approaches the origin. The definition of the precise domain of the operator is not elementary, and changes from one sector to another.[5] It is worth noting that the same analysis holds for $\alpha > 2\pi$, even though the 'sector' is no longer embeddable in $\mathbf{R}^2$. $\quad\square$

Theorem 5.4.9 *If H is a (possibly unbounded) self-adjoint operator then* $\mathrm{Spec}(H) \subseteq \mathbf{R}$. *Moreover*

$$\|(zI - H)^{-1}\| \le |\mathrm{Im}(z)|^{-1}$$

for all $z \notin \mathbf{R}$.

Proof. If $z := x + iy$ where $y \ne 0$ then the operator $K := (H - xI)/y$ is also self-adjoint. The proof of Theorem 5.4.5 implies that $(K \pm iI)$ are one-one with ranges equal to $\mathcal{H}$ and $\|(K \pm iI)^{-1}\| \le 1$. These statements are equivalent to the statement of the theorem. $\quad\square$

Theorem 5.4.10 *Let H be a (possibly unbounded) self-adjoint operator acting in the separable Hilbert space $\mathcal{H}$. Then there exists a set X provided with a σ-field of subsets and a σ-finite measure $\mathrm{d}x$, together with a unitary map*

---

[4] This is a particular case of a general fact, related to the ideas in Chapter 13 and proved in [Davies 1989].
[5] Such issues are usually resolved by using quadratic form techniques as described in [Davies 1995C], the reason being that the domain of the square root of H is $W_0^{1,2}(\Omega)$ for a wide variety of second order elliptic differential operators and regions Ω.

$U : \mathcal{H} \to L^2(X, \mathrm{d}x)$ *for which the following holds. The operator $M := UHU^{-1}$ is of the form*

$$(Mg)(x) := m(x)g(x)$$

where m is a (possibly unbounded) real-valued, measurable function on X. In particular

$$U(\mathrm{Dom}(H)) = \{g \in L^2(X, \mathrm{d}x) : mg \in L^2(X, \mathrm{d}x)\}.$$

Moreover the spectrum of H equals the essential range of m and

$$(UR(\lambda, H)U^{-1}g)(x) = (\lambda - m(x))^{-1}g(x)$$

almost everywhere, for all $\lambda \notin \mathrm{Spec}(H)$ and all $g \in L^2(X, \mathrm{d}x)$.

Problem 5.4.11 The above spectral theorem simplifies substantially in the following situation. We suppose that $\{\phi_n\}_{n=1}^{\infty}$ is an orthonormal basis in $\mathcal{H}$ and that $\{\lambda_n\}_{n=1}^{\infty}$ is any sequence of real numbers. Then we may define the self-adjoint operator H by

$$Hf := \sum_{n=1}^{\infty} \lambda_n \langle f, \phi_n \rangle \phi_n$$

with

$$\mathrm{Dom}(H) := \{f \in \mathcal{H} : \sum_{n=1}^{\infty} \lambda_n^2 |\langle f, \phi_n \rangle|^2 < \infty\}.$$

Identify the auxiliary space $L^2(X, \mathrm{d}x)$ and the unitary operator U of the general spectral theorem in this case. $\square$

Continuing with the notation of Problem 5.4.11, the spectral projections

$$P_n f := \sum_{r=1}^{n} \langle f, \phi_r \rangle \phi_r$$

converge strongly to I. The next theorem gives some information about the rate of convergence.

Theorem 5.4.12 *If $\{\lambda_n\}_{n=1}^{\infty}$ is an increasing and divergent sequence of positive real numbers and $f \in \mathrm{Dom}(H^m)$ for some m then*

$$\|P_n f - f\| \le \lambda_{n+1}^{-m} \|H^m f\|$$

for all n.

Proof. We have

$$\|P_n f - f\|^2 = \sum_{r=n+1}^{\infty} |\langle f, \phi_r \rangle|^2$$

$$\leq \lambda_{n+1}^{-2m} \sum_{r=n+1}^{\infty} \lambda_r^{2m} |\langle f, \phi_r \rangle|^2$$

$$\leq \lambda_{n+1}^{-2m} \|H^m f\|^2. \qquad \square$$

This theorem has a partial converse.

Theorem 5.4.13 *Suppose that* $\{\lambda_n\}_{n=1}^{\infty}$ *is an increasing and divergent sequence of positive real numbers and that*

$$a^2 := \sum_{r=1}^{\infty} \lambda_r^{-2k} < \infty$$

and that

$$\|P_n f - f\| \leq c \lambda_{n+1}^{-m-k}$$

for all n. Then $f \in \mathrm{Dom}(H^m)$*. Indeed*

$$\|H^m f\| \leq ca.$$

Proof. Our hypotheses imply that

$$|\langle f, \phi_n \rangle|^2 \leq c^2 \lambda_n^{-2m-2k}$$

for all *n*. Hence

$$\|H^m f\|^2 = \sum_{r=1}^{\infty} \lambda_r^{2m} |\langle f, \phi_r \rangle|^2$$

$$\leq \sum_{r=1}^{\infty} \lambda_r^{2m} c^2 \lambda_r^{-2m-2k}$$

$$\leq c^2 a^2. \qquad \square$$

Problem 5.4.14 Prove that a (possibly unbounded) self-adjoint operator H acting on a Hilbert space $\mathcal{H}$ is non-negative in the sense that $\langle Hf, f \rangle \geq 0$ for all $f \in \mathrm{Dom}(H)$ if and only if the function m in Theorem 5.4.10 satisfies $m(x) \geq 0$ for almost every $x \in X$. $\square$

The spectral theorem can be used to define a canonical functional calculus for unbounded self-adjoint operators, just as in the case of normal operators above. Once again it is standard to write $f(H) := \mathcal{T}(f)$ where $\mathcal{T}$ is the homomorphism defined below.

Theorem 5.4.15 *Let H be an unbounded self-adjoint operator acting in a Hilbert space $\mathcal{H}$, and let $\mathcal{A}$ denote the space of all continuous functions on* $\mathrm{Spec}(H)$ *which vanish at infinity. We consider $\mathcal{A}$ as a commutative Banach algebra under pointwise addition and multiplication and the supremum norm. Then there exists a unique isometric algebra homomorphism $\mathcal{T}$ from $\mathcal{A}$ into $\mathcal{L}(\mathcal{H})$ such hat*

$$\mathcal{T}(r_z) = R(z, H)$$

for all $z \notin \mathrm{Spec}(H)$, where

$$r_z(x) := (z - x)^{-1}.$$

If H is bounded then $\mathcal{T}(p) = p(H)$ for every polynomial p.

Problem 5.4.16 Use the functional calculus or the spectral theorem to prove that if A is a bounded, non-negative, self-adjoint operator on a Hilbert space $\mathcal{H}$, then there exists a bounded, non-negative, self-adjoint operator Q such that $Q^2 = A$. □

Problem 5.4.17 Prove that if A, B are bounded, self-adjoint operators on $\mathcal{H}$ and $A - B$ is compact then $f(A) - f(B)$ is compact for every continuous function f on $[-c, c]$, where $c := \max\{\|A\|, \|B\|\}$. □

We end this section by mentioning a problem whose solution is far less obvious than would be expected. If a bounded operator A is close to normal in the sense that $\|A^*A - AA^*\|$ is small then one would expect A to be close to a normal operator, and so one could apply the functional calculus to A with small errors. The most useful result in this direction is as follows.

Theorem 5.4.18 [6] *Let A be an $n \times n$ matrix satisfying*

$$\|A^*A - AA^*\| < \delta(\varepsilon, n) := \frac{\varepsilon^2}{n - 1}$$

for some $\varepsilon > 0$. Then there exists a normal matrix N such that $\|A - N\| < \varepsilon$.

---

[6] This theorem was proved by [Pearcy and Shields 1979], who also showed that there is no corresponding result for bounded operators. Later Lin proved a similar bound in which the constant $\delta(\varepsilon, n)$ does not depend on n, but unfortunately its dependence on ε is completely obscure; see [Lin 1997, Friis and Rørdam 1996].

5.5 Hilbert-Schmidt operators

In this section we prove a few of the most useful results from the large literature on classes of compact operators. In some situations below $\mathcal{H}$ denotes an abstract Hilbert space.[7] In others we assume that $\mathcal{H} = L^2(X, \mathrm{d}x)$, where X is a locally compact Hausdorff space and there is a countable basis to its topology. In this case we always assume that the Borel measure $\mathrm{d}x$ has support equal to X.

We will prove the following inclusions between different classes of operators on $\mathcal{H}$, each of which is a two-sided, self-adjoint ideal of operators in $\mathcal{L}(\mathcal{H})$.[8]

$$\text{finite rank} \longrightarrow \text{trace class} \longrightarrow \text{Hilbert-Schmidt} \longrightarrow \text{compact} \longrightarrow \text{bounded}$$

We warn the reader that we have made a tiny selection from the many classes of operators that have been found useful in various contexts. Our first few results provide an abstract version of Theorem 4.2.16; see also (4.3).

Lemma 5.5.1 *If $\{e_n\}_{n=1}^\infty$ and $\{f_n\}_{n=1}^\infty$ are two complete orthonormal sets in a Hilbert space $\mathcal{H}$ and A is a bounded operator on $\mathcal{H}$ then*

$$\sum_{n=1}^\infty \|Ae_n\|^2 = \sum_{m,n=1}^\infty |\langle Ae_n, f_m\rangle|^2 = \sum_{m=1}^\infty \|A^* f_m\|^2$$

where the two sides converge or diverge together. It follows that the values of the two outer sums do not depend upon the choice of either orthonormal set.

Proof. One simplifies the middle sum two different ways. $\qquad\square$

We say that A is Hilbert-Schmidt or that $A \in \mathcal{C}_2$ if the above series converge, and write

$$\|A\|_2^2 := \sum_{n=1}^\infty \|Ae_n\|^2.$$

The Hilbert-Schmidt norm $\|\cdot\|_2$ is also called the Frobenius norm. The notation $\|\cdot\|_{\mathrm{HS}}$ is also used.

---

[7] In all of the proofs we will assume that $\mathcal{H}$ is infinite-dimensional and separable. The finite-dimensional proofs are often simpler, but some details need modifying.

[8] See Theorem 5.6.7, Problem 5.5.3, Theorem 4.2.2 and Theorem 4.2.13.

Problem 5.5.2 Prove that $\| \cdot \|_2$ is a norm on $\mathcal{C}_2$. Prove also that $\mathcal{C}_2$ is complete for this norm. $\square$

Problem 5.5.3 Prove that $\mathcal{C}_2$ is a self-adjoint, two-sided ideal of operators in $\mathcal{L}(\mathcal{H})$. Indeed if $A \in \mathcal{C}_2$ and $B \in \mathcal{L}(\mathcal{H})$ then

$$\|A\|_2 = \|A^*\|_2,$$

$$\|AB\|_2 \le \|A\|_2 \|B\|,$$

$$\|BA\|_2 \le \|B\| \|A\|_2. \qquad \square$$

Lemma 5.5.4 *Every Hilbert-Schmidt operator A acting on a Hilbert space $\mathcal{H}$ is compact.*

Proof. Given two unit vectors $e, f \in \mathcal{H}$ we can make them the first terms of two complete orthonormal sequences. This implies

$$|\langle Ae, f \rangle|^2 \le \sum_{m,n=1}^{\infty} |\langle Ae_n, f_m \rangle|^2 = \|A\|_2^2.$$

Since e, f are arbitrary we deduce that $\|A\| \le \|A\|_2$.

For any positive integer N one may write $A := A_N + B_N$ where

$$A_N g := \sum_{m,n=1}^{N} \langle Ae_n, f_m \rangle \langle g, e_n \rangle f_m$$

for all $g \in \mathcal{H}$. Since A_N is finite rank and

$$\|B_N\|^2 \le \|B_N\|_2^2 = \sum_{m,n=1}^{\infty} |\langle Ae_n, f_m \rangle|^2 - \sum_{m,n=1}^{N} |\langle Ae_n, f_m \rangle|^2,$$

which converges to 0 as $N \to \infty$, Theorem 4.2.2 implies that A is compact. $\square$

Problem 5.5.5 If A is an $n \times n$ matrix, its Hilbert-Schmidt norm is given by

$$\|A\|_2 = \left\{ \sum_{r,s=1}^{n} |A_{r,s}|^2 \right\}^{1/2}.$$

Prove that

$$\|A\| \le \|A\|_2 \le n^{1/2} \|A\|.$$

Prove also that the second inequality becomes an equality if and only if A is a constant multiple of a unitary matrix, and that the first inequality becomes an equality if and only if A is of rank 1. □

5.6 Trace class operators

Hilbert-Schmidt operators are not the only compact operators which turn up in applications; we treated them first because the theory is the easiest to develop. One can classify compact operators A acting on a Hilbert space $\mathcal{H}$ by listing the eigenvalues s_r of $|A|$ in decreasing order, repeating each one according to its multiplicity. Note that $|A|$ is also compact by Theorem 5.2.4. The sequence $\{s_n\}_{n=1}^{\infty}$, called the singular values of A, may converge to zero at various rates, and each rate defines a corresponding class of operators. In particular one says that $A \in \mathcal{C}_p$ if $\sum_{n=1}^{\infty} s_n^p < \infty$. We will not develop the full theory of such classes, but content ourselves with a treatment of $\mathcal{C}_1$, which is the most important of the spaces after $\mathcal{C}_2$.[9]

Problem 5.6.1 Prove that the list of non-zero singular values of a compact operator A coincides with the corresponding list for A^*. □

We say that a non-negative, self-adjoint operator A is trace class if it satisfies the conditions of the following lemma.

Lemma 5.6.2 *If $A = A^* \geq 0$ and $\{e_n\}_{n=1}^{\infty}$ is a complete orthonormal set in $\mathcal{H}$ then its trace*

$$\mathrm{tr}[A] := \sum_{n=1}^{\infty} \langle Ae_n, e_n \rangle \in [0, +\infty]$$

does not depend upon the choice of $\{e_n\}_{n=1}^{\infty}$. If $\mathrm{tr}[A] < \infty$ then A is compact. If $\{\lambda_n\}_{n=1}^{\infty}$ are its eigenvalues repeated according to their multiplicities, then

$$\mathrm{tr}[A] = \sum_{n=1}^{\infty} \lambda_n.$$

Proof. Lemma 5.5.1 implies that $\mathrm{tr}[A]$ does not depend on the choice of $\{e_n\}_{n=1}^{\infty}$, because

$$\sum_{n=1}^{\infty} \langle Ae_n, e_n \rangle = \sum_{n=1}^{\infty} \|A^{1/2}e_n\|^2.$$

---

[9] See [Dunford and Schwartz 1966, Simon 2005A] for a detailed treatment of Calkin's theory of operator ideals and of the von Neumann-Schatten $\mathcal{C}_p$ classes and their many applications.

If the sum is finite then

$$\left\|A^{1/2}\right\|_2^2 = \text{tr}[A] < \infty \tag{5.7}$$

so $A^{1/2}$ is compact by Lemma 5.5.4. This implies that A is compact. The final statement of the lemma uses Theorem 4.2.23. □

Problem 5.6.3 If $\{P_n\}_{n=1}^\infty$ is an increasing sequence of orthogonal projections and $\bigcup_{n=1}^\infty \text{Ran}(P_n)$ is dense in $\mathcal{H}$, prove that

$$\text{tr}[A] = \lim_{n \to \infty} \text{tr}[P_n A P_n]$$

for every non-negative self-adjoint bounded operator A. □

Problem 5.6.4 Prove that

$$\text{tr}[AA^*] = \text{tr}[A^*A]$$

for all bounded operators A, where the two sides of this equality are finite or infinite together. □

We say that a bounded operator A on a Hilbert space $\mathcal{H}$ is trace class (or lies in $\mathcal{C}_1$) if $\text{tr}[|A|] < \infty$. We will show that $\mathcal{C}_1$ is a two-sided ideal in the algebra $\mathcal{L}(\mathcal{H})$ of all bounded operators on $\mathcal{H}$.

Lemma 5.6.5 *If A is a bounded operator on $\mathcal{H}$ then the following are equivalent.*

(i) $c_1 := \sum_{n=1}^\infty \langle |A|e_n, e_n \rangle < \infty$ *for some (or every) complete orthonormal sequence $\{e_n\}$;*

(ii) $c_2 := \inf\{\|B\|_2 \|C\|_2 : A = BC\} < \infty$;

(iii) $c_3 := \sup\{\sum_n |\langle Ae_n, f_n \rangle| : \{e_n\}, \{f_n\} \in \mathcal{O}\} < \infty$, *where $\mathcal{O}$ is the class of all (not necessarily complete) orthonormal sequences in $\mathcal{H}$.*

Moreover $c_1 = c_2 = c_3$.

Proof. We make constant reference to the properties of the polar decomposition $A = V|A|$ as described in Theorem 5.2.4.

(i)$\Rightarrow$(ii) If $c_1 < \infty$ then we may write $A = BC$ where $B := V|A|^{1/2} \in \mathcal{C}_2$, $C := |A|^{1/2} \in \mathcal{C}_2$ and V is a contraction. It follows by (5.7) and Problem 5.5.3 that $\|B\|_2 \le c_1^{1/2}$ and $\|C\|_2 \le c_1^{1/2}$. Hence $c_2 \le c_1$.

(ii)⇒(iii) If $c_2 < \infty$, $\{e_n\}$, $\{f_n\} \in \mathcal{O}$ and $A = BC$ then

$$\sum_n |\langle Ae_n, f_n \rangle| = \sum_n |\langle Ce_n, B^* f_n \rangle|$$

$$\leq \sum_n \|Ce_n\| \|B^* f_n\|$$

$$\leq \left\{ \sum_n \|Ce_n\|^2 \right\}^{1/2} \left\{ \sum_n \|B^* f_n\|^2 \right\}^{1/2}$$

$$\leq \|C\|_2 \|B^*\|_2 = \|C\|_2 \|B\|_2.$$

By taking the infimum over all decompositions $A = BC$ and then the supremum over all pairs $\{e_n\}$, $\{f_n\} \in \mathcal{O}$, we obtain $c_3 \leq c_2$.

(iii)⇒(i) If $c_3 < \infty$, we start by choosing a (possibly finite) complete orthonormal set $\{e_n\}$ for the subspace $\overline{\text{Ran}(|A|)}$. The sequence $f_n := Ve_n$ is then a complete orthonormal set for $\overline{\text{Ran}(A)}$. Also

$$\text{tr}[|A|] = \sum_n \langle |A| e_n, e_n \rangle$$

$$= \sum_n \langle V^* Ae_n, e_n \rangle$$

$$= \sum_n \langle Ae_n, Ve_n \rangle$$

$$= \sum_n \langle Ae_n, f_n \rangle$$

$$\leq c_3.$$

Therefore $c_1 \leq c_3$. □

Problem 5.6.6 Let $A : L^2(a, b) \to L^2(a, b)$ be defined by

$$(Af)(x) := \int_a^b a(x, y) f(y) \, dy.$$

Use Lemma 5.6.5(ii) and (4.3) to prove that $A \in \mathcal{C}_1$ if both $a(x, y)$ and $\frac{\partial}{\partial x} a(x, y)$ are jointly continuous on $[a, b]^2$. □

Theorem 5.6.7 *The space $\mathcal{C}_1$ is a two-sided, self-adjoint ideal of operators in $\mathcal{L}(\mathcal{H})$. Moreover $\|A\|_1 := \text{tr}[|A|]$ is a complete norm on $\mathcal{C}_1$, which satisfies*

$$\|A\|_1 = \|A^*\|_1,$$

$$\|BA\|_1 \leq \|B\| \|A^*\|_1,$$

$$\|AB\|_1 \leq \|A^*\|_1 \|B\|,$$

for all $A \in \mathcal{C}_1$ and $B \in \mathcal{L}(\mathcal{H})$.

Proof. The proof that $\mathcal{C}_1$ is closed under addition would be elementary if $|A+B| \leq |A|+|B|$, for any two operators A, B, but there is no such inequality; see Problem 5.2.5. However, it follows directly from Lemma 5.6.5(iii). The proof that $\mathcal{C}_1$ is closed under left or right multiplication by any bounded operator follows from Lemma 5.6.5(ii) and Problem 5.5.3, as does the proof that $\mathcal{C}_1$ is closed under the taking of adjoints.

The required estimates of the norms are proved by examining the above arguments in more detail. The completeness of $\mathcal{C}_1$ is proved by using Lemma 5.6.5(iii). □

Our next theorems describe methods of computing the trace of a non-negative operator given its integral kernel.

Problem 5.6.8 Suppose that the non-negative, bounded, self-adjoint operator A on $L^2(X)$ has the continuous integral kernel $a(\cdot, \cdot)$. Prove that $a(x, x) \geq 0$ for all $x \in X$ and that

$$|a(x, y)| \leq a(x, x)^{1/2} a(y, y)^{1/2}$$

for all $x, y \in X$. □

Proposition 5.6.9 (*Mercer's theorem*) *If the non-negative, bounded, self-adjoint operator A has the continuous integral kernel $a(\cdot, \cdot)$ then*

$$\mathrm{tr}[A] = \int_X a(x, x)\,\mathrm{d}x \tag{5.8}$$

where the finiteness of either side implies the finiteness of the other.

Proof. We start by considering the case in which X is compact. Let $\mathcal{E}$ be a partition of X into a finite number of disjoint Borel sets $E_1, \ldots, E_n$ with non-zero measures $|E_i|$. Let P be the orthogonal projection onto the finite-dimensional linear subspace spanned by the (orthogonal) characteristic functions χ_i of E_i. A direct calculation shows that

$$\mathrm{tr}[PAP] = \sum_{i=1}^n |E_i|^{-1} \langle A\chi_i, \chi_i \rangle$$

$$= \sum_{i=1}^n |E_i|^{-1} \int_{E_i} \int_{E_i} a(x, y)\,\mathrm{d}x\,\mathrm{d}y.$$

If P_n are the projections associated with a sequence of increasingly fine partitions $\mathcal{E}_n$ satisfying the conditions listed in Section 2.1, then $0 \leq P_n \leq P_{n+1}$ for all n and Problem 5.6.3 implies that

$$\text{tr}[A] = \lim_{n \to \infty} \text{tr}[P_n A P_n].$$

The proof of (5.8) now depends on the uniform continuity of $a(\cdot, \cdot)$.

We now treat the general case of the proposition. Let $\{K_n\}_{n=1}^{\infty}$ be an increasing sequence of compact sets with union equal to X. Let P_n denote the orthogonal projection on $L^2(X, \mathrm{d}x)$ obtained by multiplying by the characteristic function of K_n. The first part of this proof implies that

$$\text{tr}[P_n A P_n] = \int_{K_n} a(x, x) \, \mathrm{d}x.$$

The proof of the general case is completed by using Problem 5.6.3 a second time. □

Example 5.6.10 Consider the operator R acting on $L^2(0, \pi)$ according to the formula

$$(Rf)(x) := \int_0^{\pi} G(x, y) f(y) \, \mathrm{d}y$$

where the Green function G is given by

$$G(x, y) := \begin{cases} x(\pi - y)/\pi & \text{if } x \leq y, \\ (\pi - x)y/\pi & \text{if } x \geq y. \end{cases}$$

A direct calculation shows that $Rf = g$ if and only if $g(0) = g(\pi) = 0$ and $-g'' = f$. Thus R is the inverse of the non-negative, self-adjoint operator L acting in $L^2(0, \pi)$ according to the formula $Lg := -g''$, subject to the stated boundary conditions. The eigenvalues of L are $1, 4, 9, 16, \ldots$ so

$$\text{tr}(R) = \sum_{n=1}^{\infty} \frac{1}{n^2} = \frac{\pi^2}{6}.$$

This calculation is confirmed by evaluating

$$\int_0^{\pi} G(x, x) \, \mathrm{d}x.$$

The results above are extended to more general Sturm-Liouville operators in Example 11.2.8. □

If the integral kernel of A is not continuous then its values on the diagonal $x = y$ may not be well-defined and we must proceed in a more indirect manner.[10]

Theorem 5.6.11 *Let* $A = A^* \geq 0$ *be a trace class operator acting on* $L^2(\mathbf{R}^N)$ *with the kernel* $a \in L^2(\mathbf{R}^N \times \mathbf{R}^N)$. *Let*

$$a_\varepsilon(x, y) := \begin{cases} a(x, y) & \text{if } |x - y| < \varepsilon, \\ 0 & \text{otherwise,} \end{cases}$$

and let $v(\varepsilon)$ $(= v_N \varepsilon^N)$ *be the volume of any ball of radius* ε. *Then* $a_\varepsilon \in L^1(\mathbf{R}^N \times \mathbf{R}^N)$ *and*

$$\text{tr}[A] = \lim_{\varepsilon \to 0} v(\varepsilon)^{-1} \int_{\mathbf{R}^N} \int_{\mathbf{R}^N} a_\varepsilon(x, y) \, d^N x \, d^N y.$$

Proof. We first observe that the operator norm convergent spectral expansion

$$Af = \sum_{n=1}^{\infty} \lambda_n \langle f, e_n \rangle e_n$$

provided by Theorem 4.2.23 corresponds to the L^2 norm convergent expansion

$$a(x, y) = \sum_{n=1}^{\infty} \lambda_n e_n(x) \overline{e_n(y)} \tag{5.9}$$

in $L^2(\mathbf{R}^N \times \mathbf{R}^N)$.

Let K_ε be the operator with convolution kernel $k_\varepsilon(x - y)$ where

$$k_\varepsilon(x) := \begin{cases} v(\varepsilon)^{-1} & \text{if } |x| < \varepsilon, \\ 0 & \text{otherwise.} \end{cases}$$

By taking Fourier transforms we see that $\|K_\varepsilon\| \leq 1$ and that K_ε converges weakly to I as $\varepsilon \to 0$. If we put

$$t(\varepsilon) := \sum_{n=1}^{\infty} \lambda_n \langle K_\varepsilon e_n, e_n \rangle$$

then it follows that

$$\text{tr}[A] = \lim_{\varepsilon \to 0} t(\varepsilon).$$

We therefore have to prove that $a_\varepsilon \in L^1(\mathbf{R}^N \times \mathbf{R}^N)$ and that

$$t(\varepsilon) = v(\varepsilon)^{-1} \int_{\mathbf{R}^N} \int_{\mathbf{R}^N} a_\varepsilon(x, y) \, d^N x \, d^N y$$

---

[10] Compare the following with [Gohberg and Krein 1969, Theorem 10.1] .

for every $\varepsilon > 0$. We have

$$\|a_\varepsilon\|_1 = \int_{\mathbf{R}^N} \int_{\mathbf{R}^N} |a_\varepsilon(x, y)| \, \mathrm{d}^N x \mathrm{d}^N y$$

$$\leq \sum_{n=1}^\infty \lambda_n \int_{|u|<\varepsilon} \int_{x \in \mathbf{R}^N} |e_n(x-u)| |e_n(x)| \, \mathrm{d}^N x \, \mathrm{d}^N u$$

$$\leq v(\varepsilon) \sum_{n=1}^\infty \lambda_n$$

$$= v(\varepsilon) \mathrm{tr}[A].$$

Also

$$v(\varepsilon)^{-1} \int_{\mathbf{R}^N} \int_{\mathbf{R}^N} a_\varepsilon(x, y) \, \mathrm{d}^N x \mathrm{d}^N y = v(\varepsilon)^{-1} \sum_{n=1}^\infty \lambda_n \int_{|u|<\varepsilon} \int_{x \in \mathbf{R}^N} e_n(x-u)$$

$$\overline{e_n(x)} \, \mathrm{d}^N x \, \mathrm{d}^N u$$

$$= \sum_{n=1}^\infty \lambda_n \langle K_\varepsilon e_n, e_n \rangle$$

$$= t(\varepsilon). \qquad \square$$

In order to extend Theorem 5.6.11 to a more general context we need a replacement for the convolution operator K_ε used in its proof.

Lemma 5.6.12 *Let X be a locally compact, separable, metric space, $\mathrm{d}x$ a Borel measure on X with support equal to X and $\mathcal{H} = L^2(X, \mathrm{d}x)$. Let $v_{x,\varepsilon}$ denote the volume of the open ball $B(x, \varepsilon)$ with centre x and radius $\varepsilon > 0$, and let $\chi_{x,\varepsilon}$ denote the characteristic function of this ball. Suppose that there exist positive constants ε_0 and c such that*

$$v_{x,\varepsilon} \leq c v_{y,\varepsilon}$$

whenever $d(x, y) < \varepsilon < \varepsilon_0$. Put

$$k_\varepsilon(x, y) := \int_X v_{s,\varepsilon}^{-2} \chi_{s,\varepsilon}(x) \chi_{s,\varepsilon}(y) \, \mathrm{d}s.$$

Then k_ε is the integral kernel of a self-adjoint operator K_ε satisfying $0 \leq K_\varepsilon \leq cI$ and

$$\lim_{\varepsilon \to 0} \langle K_\varepsilon f, f \rangle = \|f\|^2$$

for all $f \in L^2$.

Proof. Direct calculations show that $k_\varepsilon(x, y) = k_\varepsilon(y, x) \geq 0$ and

$$0 \leq \int_X k_\varepsilon(s, y)\, ds = \int_X k_\varepsilon(x, s)\, ds \leq c$$

for all $x, y \in X$. It follows by Corollary 2.2.15 that $\|K_\varepsilon\| \leq c$. The fact that $K_\varepsilon \geq 0$ follows from

$$\langle K_\varepsilon f, f \rangle = \int_X v_{u,\varepsilon}^{-2} |\langle \chi_{u,\varepsilon}, f \rangle|^2\, du \geq 0.$$

The final statement of the lemma is proved first for $f \in C_c(X)$ and then extended to all $f \in L^2$ by a density argument. $\square$

Theorem 5.6.13 *Under the conditions of Lemma 5.6.12, if $A = A^* \geq 0$ and A is trace class with integral kernel $a(x, y)$ then*

$$\mathrm{tr}[A] = \lim_{\varepsilon \to 0} \int_X \int_X a(x, y)k_\varepsilon(x, y)\, dxdy.$$

Proof. This is a minor modification of the proof of Theorem 5.6.11, and uses the L^2 norm convergent spectral expansion (5.9) of the kernel $a(\cdot, \cdot)$. The integrand lies in $L^1(X \times X)$ for every $\varepsilon > 0$ because

$$\int_X \int_X |a(x, y)k_\varepsilon(x, y)|\, dxdy \leq \sum_{n=1}^\infty \lambda_n \int_X \int_X |e_n(x)||e_n(y)|k_\varepsilon(x, y)\, dxdy$$

$$= \sum_{n=1}^\infty \lambda_n \langle K_\varepsilon |e_n|, |e_n| \rangle$$

$$\leq c \sum_{n=1}^\infty \lambda_n. \qquad \square$$

5.7 The compactness of $f(Q)g(P)$

If f is a function on $\mathbf{R}^N$ we define the operator $f(Q)$ on $L^2(\mathbf{R}^N)$ by

$$(f(Q)\phi)(x) := f(x)\phi(x)$$

on its maximal domain, consisting of all $\phi \in L^2(\mathbf{R}^N)$ for which $f\phi$ lies in $L^2(\mathbf{R}^N)$. We define $g(P)$ by

$$g(P) := \mathcal{F}^{-1}g(Q)\mathcal{F}$$

where $\mathcal{F}$ is the Fourier transform operator; the domain of $g(P)$ is $\{f \in L^2 : g\mathcal{F}(f) \in L^2\}$. The rather strange notation is derived from quantum theory, in which P is called the momentum operator and Q is called the position operator. Operators of the form $f(Q)g(P)$ are of technical importance when proving a variety of results concerning Schrödinger operators.[11]

Theorem 5.7.1 *The operator* $A := f(Q)g(P)$ *acting on* $L^2(\mathbf{R}^N)$ *is compact if* f, g *both lie in* $L^2(\mathbf{R}^N)$, *or if both are bounded and measurable and vanish as* $|x| \to \infty$.

Proof. If $g \in L^2(\mathbf{R}^N)$ then $g(P)\phi = \check{g} * \phi$ where $\check{g}$ is the appropriately normalized inverse Fourier transform of g. If $f, g \in L^2(\mathbf{R}^N)$ then A has the Hilbert-Schmidt integral kernel

$$K(x, y) := f(x)\check{g}(x - y).$$

Therefore A is compact by Lemma 5.5.4.

Alternatively suppose that f, g are both bounded and vanish as $|x| \to \infty$. Since $f(Q)$ and $g(P)$ are bounded operators, so is A. Given $\varepsilon > 0$ suppose that $|x| > N_\varepsilon$ implies that $|f(x)| < \varepsilon$. Put

$$f_1(x) := \begin{cases} f(x) & \text{if } |x| \leq N_\varepsilon, \\ 0 & \text{otherwise,} \end{cases}$$

and $f_2 := f - f_1$. Define g_1 and g_2 similarly, so that $\|f_2\|_\infty < \varepsilon$, $\|g_2\|_\infty < \varepsilon$, while $f_1, g_1 \in L_c^\infty \subseteq L^2$. The first half of this proof shows that $f_1(Q)g_1(P)$ is compact, and we also have

$$\|f(Q)g(P) - f_1(Q)g_1(P)\| \leq \|f_1(Q)g_2(P) + f_2(Q)g_1(P) + f_2(Q)g_2(P)\|$$

$$\leq \|f\|_\infty \varepsilon + \|g\|_\infty \varepsilon + \varepsilon^2.$$

Letting $\varepsilon \to 0$ we conclude that A is compact. □

Problem 5.7.2 Assuming that f and g are sufficiently regular functions on $\mathbf{R}^N$, write down the integral kernel of the commutator $[f(Q), g(P)]$ and use the formula to prove that[12]

$$\| [f(Q), g(P)] \| \leq (2\pi)^{-N} \|\nabla f\|_\infty \int_{\mathbf{R}^N} |\xi| \, |\hat{g}(\xi)| \, d^N \xi. \qquad □$$

---

[11] The following theorem is extracted from a large literature on such operators, surveyed in [Simon 2005A].
[12] There is some evidence (in August 2006) that

$$\| [f(Q), g(P)] \| \leq \|\nabla f\|_\infty \|\nabla g\|_\infty$$

but not enough to be confident that this is true.

Theorem 5.7.3 *If* $f, g \in L^p(\mathbf{R}^N)$ *and* $2 \leq p < \infty$ *then the operator* $A :=$ $f(Q)g(P)$ *acting on* $L^2(\mathbf{R}^N)$ *is compact. Moreover*

$$\|f(Q)g(P)\| \leq c_{N,p}\|f\|_p\|g\|_p. \tag{5.10}$$

Proof. Since A is the product of two operators each of which may be unbounded, we start by proving that it is bounded.

If $\phi \in L^2$ then $\psi := \mathcal{F}(g(P)\phi)$ is given by $\psi(\xi) = g(\xi)(\mathcal{F}\phi)(\xi)$. Theorem 2.1.13 implies that it lies in L^q where $1/q := 1/2 + 1/p$, so that $1 < q < 2$. It follows from Theorem 3.1.12 that $g(P)\phi \in L^r$ where $1/r := 1/2 - 1/p$, so that $2 < r < \infty$. This finally implies that $f(Q)g(P)\phi \in L^2$, by another application of Theorem 2.1.13. If one examines the bounds in those theorems one obtains the estimate (5.10).

If $f, g \in L^p$ then there exist sequences $f_n, g_n \in L_c^\infty$ such that $\|f_n - f\|_p \to 0$ and $\|g_n - g\|_p \to 0$. (5.10) implies that $f_n(Q)g_n(P)$ converges in operator norm as $n \to \infty$. Since each operator $f_n(Q)g_n(P)$ is Hilbert-Schmidt, we deduce that the norm limit $f(Q)g(P)$ is compact. $\square$

Problem 5.7.4 Let $f \in L^p + L_0^\infty$ and $g \in L^p \cap L_0^\infty$ where $2 \leq p < \infty$. Prove that $f(Q)g(P)$ is a compact operator. (See Section 1.5 for the definition of L_0^∞.) $\square$

6

One-parameter semigroups

6.1 Basic properties of semigroups

One-parameter semigroups describe the evolution in time of many systems in applied mathematics; the problems involved range from quantum theory and the wave equation to stochastic processes. It is not our intention to describe the vast range of applications of this subject (see the preface for more information), but they demonstrate beyond doubt that the subject is an important one. In this chapter we describe what is generally regarded as the basic theory. Later chapters treat some special classes of semigroup that have proved important in a variety of applications.

One-parameter semigroups are also a useful technical device for studying unbounded linear operators. The time-dependent Schrödinger equation

$$i\frac{\partial f}{\partial t} = -\Delta f(x) + V(x)f(x)$$

is one of the few fundamental equations in physics involving $i := \sqrt{-1}$, but this very fact also makes it much harder to solve. One of the standard tricks is to replace i by -1 and study the corresponding Schrödinger semigroup. This is much better behaved analytically (it is a self-adjoint, holomorphic, contraction semigroup rather than a unitary group), and the spectral information obtained about the Schrödinger operator can then be used to analyze the original equation.

One-parameter semigroups arise as the solutions of the Cauchy problem for the differential equation

$$f'_t = Zf_t \tag{6.1}$$

where Z is a linear operator (often a differential operator) acting in a Banach space $\mathcal{B}$ and $'$ denotes the derivative with respect to time. Formally the solution of (6.1) is $f_t = T_t f_0$, where

$$T_t := e^{Zt} \tag{6.2}$$

satisfies $T_{s+t} = T_s T_t$ for all s, $t \geq 0$. However, in most applications Z is an unbounded operator, so the meaning of (6.2) is unclear. Much of this chapter is devoted to a careful treatment of problems related to the unboundedness of Z.

It might be thought that such questions are of little concern to an applied mathematician – if an evolution equation occurs in a natural context then surely it must have a solution and this solution must define a semigroup. Experience shows that adopting such a relaxed attitude to theory can lead one into serious error. If $t \in \mathbf{R}$ the expressions $e^{i\Delta t}$ define a one-parameter unitary group on $L^2(\mathbf{R}^N)$. However, the 'same' operators are not bounded on $L^p(\mathbf{R}^N)$ for any $p \neq 2$ (unless $t = 0$) by Theorem 8.3.10. The expressions $e^{\Delta t}$ define one-parameter contraction semigroups on $L^p(\mathbf{R}^N)$ for all $1 \leq p < \infty$ by Example 6.3.5; however, the L^p spectrum of Δ depends strongly on p if one replaces $\mathbf{R}^N$ by hyperbolic space or by a homogeneous tree; see Section 12.6. One must never assume without proof that properties of an unbounded operator are preserved when one considers it as acting in a different space.

Before giving the definition of a one-parameter semigroup we discuss some basic notions concerning unbounded operators acting in a Banach space $\mathcal{B}$. Such an operator is defined to be a linear map A whose domain is a linear subspace $\mathcal{L}$ of $\mathcal{B}$ (frequently a dense linear subspace) and whose range is contained in $\mathcal{B}$. In order to use the tools of analysis one needs some connection between convergence in $\mathcal{B}$ and the operator. In the absence of boundedness, we will need to assume that the operators that we study are closed, or that they can be made closed by increasing their domains.

Generalizing our earlier definition, we say that an operator A with domain $\mathrm{Dom}(A)$ in a Banach space $\mathcal{B}$ is closed if whenever $f_n \in \mathrm{Dom}(A)$ converges to a limit $f \in \mathcal{B}$ and also $\lim_{n \to \infty} A f_n = g$, it follows that $f \in \mathrm{Dom}(A)$ and $Af = g$. We say that $A \subseteq B$, or that B is an extension of A, if $\mathrm{Dom}(A) \subseteq \mathrm{Dom}(B)$ and $Af = Bf$ for all $f \in \mathrm{Dom}(A)$. We finally say that A is closable if it has a closed extension, and call its smallest closed extension $\overline{A}$ its closure.

Problem 6.1.1 Prove that A is closed if and only if its graph

$$\mathrm{Gr}(A) := \{(f, g) : f \in \mathrm{Dom}(A), g = Af\}$$

is a closed linear subspace of $\mathcal{B} \times \mathcal{B}$, which is provided with the 'product' norm

$$\|(f, g)\| := \|f\| + \|g\|.$$

Prove that this is equivalent to $\mathrm{Dom}(A)$ being complete with respect to the norm

$$\|\|f\|\| := \|f\| + \|Af\|.$$ $\square$

Problem 6.1.2 Prove that if $\lambda I - A$ is one-one from $\mathrm{Dom}(A)$ onto $\mathcal{B}$ for some $\lambda \in \mathbf{C}$, then $(\lambda I - A)^{-1}$ is bounded if and only if A is a closed operator. $\square$

Problem 6.1.3 Use the Hahn-Banach theorem to prove that an unbounded operator Z is closed if and only if it is weakly closed in the sense that if $f_n \in \mathrm{Dom}(Z)$ and

$$\text{w-}\lim_{n \to \infty} f_n = f, \qquad \text{w-}\lim_{n \to \infty} Z f_n = g$$

then $f \in \mathrm{Dom}(Z)$ and $Zf = g$. As in Section 1.3, we say that f_n converges weakly to f if

$$\lim_{n \to \infty} \langle f_n, \phi \rangle = \langle f, \phi \rangle$$

for all $\phi \in \mathcal{B}^*$. $\square$

Problem 6.1.4 Prove that if A is a closed operator on $\mathcal{B}$ and $\lambda I - A$ is one-one from $\mathrm{Dom}(A)$ onto $\mathcal{B}$ for some $\lambda \in \mathbf{C}$, then $f \to \|f\| + \|Zf\|$ and $f \to \|(\lambda I - Z)f\|$ are equivalent norms on $\mathrm{Dom}(A)$. $\square$

Problem 6.1.5 Let $m : \mathbf{R}^N \to \mathbf{C}$ be a measurable function and define the linear multiplication operator M acting in $L^2(\mathbf{R}^N)$ by $(Mf)(x) := m(x)f(x)$, where

$$\mathrm{Dom}(M) := \{f \in L^2 : mf \in L^2\}.$$

Prove that M is a closed operator. Now suppose that m is continuous, and let M_0 be the 'same' operator, but with domain $C_c^\infty(\mathbf{R}^N)$. Prove that M is the closure of M_0. Prove also that $\mathrm{Spec}(M)$, defined as on page 124, equals the essential range of m as defined in Problem 2.2.1. $\square$

Problem 6.1.6 Let A be defined in $C[0,1]$ by $(Af)(x) := f'(x) + a(x)f(x)$, where a is a continuous function. Find a domain on which A is a closed operator. (The same problem, but on $L^2(0,1)$, is much harder.) $\square$

Lemma 6.1.7 *An operator X acting in a Banach space $\mathcal{B}$ is closable if and only if $f_n \in \mathrm{Dom}(X)$, $\lim_{n \to \infty} f_n = 0$ and $\lim_{n \to \infty} X f_n = g$ together imply that $g = 0$.*

Proof. If X has a closed extension Y then the graph of Y is the closed linear subspace $\{(f, Yf) : f \in \text{Dom}(Y)\}$ of $\mathcal{B} \times \mathcal{B}$. Under the assumptions on the sequence f_n, we deduce that $(0, g) \in \mathcal{L}$. Therefore $g = Y0 = 0$.

Conversely, if X does not have a closed extension, then the closure $\mathcal{L}$ of $\text{Gr}(X)$ is not the graph of an operator. This implies that there exist f, g_1, g_2 such that $(f, g_1) \in \mathcal{L}$, $(f, g_2) \in \mathcal{L}$ and $g_1 \neq g_2$. There must exist sequences $f_n^1, f_n^2 \in \text{Dom}(X)$ such that

$$\lim_{n \to \infty} f_n^1 = f, \quad \lim_{n \to \infty} Xf_n^1 = g_1,$$

$$\lim_{n \to \infty} f_n^2 = f, \quad \lim_{n \to \infty} Xf_n^2 = g_2.$$

Putting $f_n := f_n^1 - f_n^2$ we deduce that

$$\lim_{n \to \infty} f_n = 0, \quad \lim_{n \to \infty} Xf_n = g_1 - g_2 \neq 0. \qquad \square$$

Lemma 6.1.8 *Let X and Y be operators acting in $\mathcal{B}$ and $\mathcal{B}^*$ respectively. Suppose that $\text{Dom}(X)$ is norm dense in $\mathcal{B}$, that $\text{Dom}(Y)$ is weak* dense in $\mathcal{B}^*$ and that*

$$\langle Xf, \phi \rangle = \langle f, Y\phi \rangle$$

for all $f \in \text{Dom}(X)$ and $\phi \in \text{Dom}(Y)$. Then X and Y are closable.

Proof. Suppose that $f_n \in \text{Dom}(X)$, $\lim_{n \to \infty} f_n = 0$ and $\lim_{n \to \infty} Xf_n = g$. Then

$$0 = \lim_{n \to \infty} \langle f_n, Y\phi \rangle = \lim_{n \to \infty} \langle Xf_n, \phi \rangle = \langle g, \phi \rangle$$

for all $\phi \in \text{Dom}(Y)$. Since $\text{Dom}(Y)$ is weak* dense in $\mathcal{B}^*$ we deduce that $\langle g, \phi \rangle = 0$ for all $\phi \in \mathcal{B}^*$ and then that $g = 0$ by the Hahn-Banach theorem. Lemma 6.1.7 finally establishes that X is closable. The proof that Y is closable is similar. $\square$

Example 6.1.9 Let L be a partial differential operator of the form

$$(Lf)(x) := \sum_{|\alpha| \leq n} a_\alpha(x)(D^\alpha f)(x)$$

acting in $L^p(U)$, where U is a region in $\mathbf{R}^N$ and $1 < p < \infty$. We assume that $a_\alpha(x) \in C^{|\alpha|}(\overline{U})$ for all relevant α and that $C_c^\infty(U) \subseteq \text{Dom}(L) \subseteq C^n(\overline{U})$. In order to prove that L is closable one needs to write down an operator M, acting in $L^q(U)$ where $1/p + 1/q = 1$, such that $\langle Lf, g \rangle = \langle f, Mg \rangle$ for all

$f \in \mathrm{Dom}(L)$ and $g \in \mathrm{Dom}(M)$. A suitable choice is

$$(Mg)(x) := \sum_{|\alpha| \leq n} (-1)^{|\alpha|} D^{\alpha} \{ a_{\alpha}(x) f(x) \}$$

where $\mathrm{Dom}(M) := C_c^{\infty}(U)$ is dense in $L^q(U)$. □

We now return to the main topic of the chapter: one-parameter semigroup s.[1] We define a (jointly continuous, or c_0) one-parameter semigroup on a complex Banach space $\mathcal{B}$ to be a family of bounded linear operators $T_t : \mathcal{B} \to \mathcal{B}$ parametrized by a real, non-negative parameter t and satisfying the following conditions:

(i) $T_0 = 1$;

(ii) if $0 \leq s, t < \infty$ then $T_s T_t = T_{s+t}$;

(iii) the map $t, f \to T_t f$ from $[0, \infty) \times \mathcal{B}$ to $\mathcal{B}$
is jointly continuous.

Many of the results obtained in this chapter are applicable to <u>real</u> Banach spaces, but we will only refer to this again in Chapters 11 and 12.

The following example shows that one cannot deduce strong continuity at $t = 0$ from (i), (ii) and strong continuity for all $t > 0$.

Example 6.1.10 Let $\mathcal{B} := C[0, 1]$, put $T_0 = 1$, and define

$$(T_t f)(x) := x^t f(x) - x^t \log(x) f(0)$$

for $0 < t < \infty$. After noting that $T_t f(0) = 0$ for all $t > 0$ one can easily show that T_t satisfies conditions (i) and (ii). The map $(t, f) \to T_t f$ is continuous from $(0, \infty) \times \mathcal{B}$ to $\mathcal{B}$, but

$$\lim_{t \to 0} \| T_t \| = +\infty,$$

so condition (iii) cannot hold. □

---

[1] The monograph of Hille and Phillips 1957 provides a key historical source for the following material. Many of the results in this chapter are to be found there or in one of the other texts on the subject, such as [Krein 1971, Butzer and Berens 1967, Yosida 1965, Kato 1966A, Davies 1980B.]

The parameter t is usually interpreted as time, in which case the semigroup describes the evolution of a system. The use of semigroups is appropriate if this evolution is irreversible and independent of time (autonomous). One of the obvious, and well-studied, problems about such systems is determining their long-term behaviour, that is the asymptotics of $T_t f$ as $t \to \infty$. In the context of the Schrödinger equation this is called scattering theory. On the other hand, if the evolution law is an approximation to some non-linear evolution equation, the short-term behaviour of the linear equation may well be much more important.

The (infinitesimal) generator of a one-parameter semigroup T_t is defined by

$$Zf := \lim_{t \to 0} t^{-1}(T_t f - f),$$

the domain $\mathrm{Dom}(Z)$ of Z being the set of f for which the limit exists. It is evident that $\mathrm{Dom}(Z)$ is a linear subspace of $\mathcal{B}$ and that Z is a linear operator from $\mathrm{Dom}(Z)$ into $\mathcal{B}$. Generally $\mathrm{Dom}(Z)$ is not equal to $\mathcal{B}$, but we will prove that it is always a dense linear subspace of $\mathcal{B}$.

The theory of one-parameter semigroup s can be represented by a triangle, the three vertices being the semigroup T_t, its generator Z and its resolvent operators $R_z := (zI - Z)^{-1}$. A full understanding of the subject requires one to find the conditions under which one can pass along any edge in either direction. In this chapter we avoid any mention of the resolvents, concentrating on the relationship between T_t and Z. We bring the resolvents into the picture in Section 8.1.

There are several ways of defining an invariant subspace of an unbounded operator, not all equivalent; Problem 1.2.19 gives a hint of the difficulties. If Z is the generator of a one-parameter semigroup T_t we will say that a closed subspace $\mathcal{L}$ is invariant under Z if $T_t(\mathcal{L}) \subseteq \mathcal{L}$ for all $t \geq 0$. This is a much more useful notion than the more elementary $Z(\mathcal{L} \cap \mathrm{Dom}(Z)) \subseteq \mathcal{L}$, which we will not use.

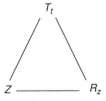

Figure 6.1: Three aspects of semigroup theory

Lemma 6.1.11 *The subspace* $\mathrm{Dom}(Z)$ *is dense in* $\mathcal{B}$, *and is invariant under* T_t *in the sense that*

$$T_t(\mathrm{Dom}(Z)) \subseteq \mathrm{Dom}(Z)$$

for all $t \geq 0$. *Moreover*

$$T_t Z f = Z T_t f$$

for all $f \in \mathrm{Dom}(Z)$ *and all* $t \geq 0$.

Proof. If $f \in \mathcal{B}$ and we define

$$f_t := \int_0^t T_x f \, \mathrm{d}x \tag{6.3}$$

then

$$
\begin{aligned}
\lim_{h \to 0} h^{-1}(T_h f_t - f_t) &= \lim_{h \to 0} \left\{ h^{-1} \int_h^{t+h} T_x f \, \mathrm{d}x - h^{-1} \int_0^t T_x f \, \mathrm{d}x \right\} \\
&= \lim_{h \to 0} \left\{ h^{-1} \int_t^{t+h} T_x f \, \mathrm{d}x - h^{-1} \int_0^h T_x f \, \mathrm{d}x \right\} \\
&= T_t f - f.
\end{aligned}
$$

Therefore $f_t \in \mathrm{Dom}(Z)$ and

$$Z(f_t) = T_t f - f. \tag{6.4}$$

Since $t^{-1} f_t \to f$ in norm as $t \to 0$, we deduce that $\mathrm{Dom}(Z)$ is dense in $\mathcal{B}$.

If $f \in \mathrm{Dom}(Z)$ and $t \geq 0$ then

$$
\begin{aligned}
\lim_{h \to 0} h^{-1}(T_h - 1)T_t f &= \lim_{h \to 0} T_t h^{-1}(T_h - 1)f \\
&= T_t Z f.
\end{aligned}
$$

Hence $T_t f \in \mathrm{Dom}(Z)$ and $T_t Z f = Z T_t f$. $\quad\square$

Lemma 6.1.12 *If* $f \in \mathrm{Dom}(Z)$ *then*

$$T_t f - f = \int_0^t T_x Z f \, \mathrm{d}x.$$

Proof. Given $f \in \mathrm{Dom}(Z)$ and $\phi \in \mathcal{B}^*$, we define the function $F : [0, \infty) \to \mathbf{C}$ by

$$F(t) := \left\langle T_t f - f - \int_0^t T_x Z f \, \mathrm{d}x, \phi \right\rangle.$$

Its right-hand derivative $D^+ F(t)$ is given by

$$D^+ F(t) = \langle Z T_t f - T_t Z f, \phi \rangle = 0.$$

Since $F(0) = 0$ and F is continuous, Lemma 1.4.4 implies that $F(t) = 0$ for all $t \in [0, \infty)$. Since $\phi \in \mathcal{B}^*$ is arbitrary, the lemma follows by applying the Hahn-Banach theorem. $\square$

Lemma 6.1.13 *If $f \in \mathrm{Dom}(Z)$ then $f_t := T_t f$ is norm continuously differentiable on $[0, \infty)$ with*

$$f_t' = Z f_t.$$

Proof. The right differentiability of $T_t f$ was proved in Lemma 6.1.11. The left derivative at points t satisfying $0 < t < \infty$ is given by

$$D^- T_t f = \lim_{h \to 0+} h^{-1} (T_t f - T_{t-h} f)$$

$$= \lim_{h \to 0+} h^{-1} \int_{t-h}^{t} T_x Z f \, dx$$

$$= T_t Z f$$

by Lemma 6.1.12. The derivative is norm continuous by virtue of the identity $f_t' = T_t(Zf)$. $\square$

Lemma 6.1.14 *The generator Z of a one-parameter semigroup T_t is a closed operator.*

Proof. Suppose that $f_n \in \mathrm{Dom}(Z)$, $\lim_{n \to \infty} f_n = f$ and $\lim_{n \to \infty} Z f_n = g$. By using Lemma 6.1.12 we obtain

$$T_t f - f = \lim_{n \to \infty} (T_t f_n - f_n)$$

$$= \lim_{n \to \infty} \int_0^t T_x Z f_n \, dx$$

$$= \int_0^t T_x g \, dx.$$

Therefore

$$\lim_{t \to 0} t^{-1} (T_t f - f) = \lim_{t \to 0} t^{-1} \int_0^t T_x g \, dx$$

$$= g,$$

so $f \in \mathrm{Dom}(Z)$ and $Zf = g$. $\square$

Lemma 6.1.15 *The space $\mathrm{Dom}(Z)$ is complete with respect to the norm*

$$|||f||| := \|f\| + \|Zf\|. \tag{6.5}$$

Moreover T_t is a one-parameter semigroup on $\mathrm{Dom}(Z)$ for this norm.

Proof. The first statement of the lemma follows by combining Problem 6.1.1 and Lemma 6.1.14. The restriction of T_t to $\mathrm{Dom}(Z)$ satisfies conditions (i) and (ii) trivially. To prove (iii) we note that if $\tilde{T}_t$ is defined on $\mathcal{B} \times \mathcal{B}$ by

$$\tilde{T}_t(f, g) := (T_t f, T_t g)$$

then $\tilde{T}_t$ satisfies (i)–(iii) and $JT_t f = \tilde{T}_t Jf$ for all $f \in \mathrm{Dom}(Z)$, where $Jf :=$ (f, Zf) is an isometry from $(\mathrm{Dom}(Z), \|| \cdot \||)$ into $\mathcal{B} \times \mathcal{B}$. $\qquad\square$

The following theorem describes the sense in which the semigroup T_t solves the Cauchy problem.

Theorem 6.1.16 *Let Z be the generator of a one-parameter semigroup T_t. If a function $f : [0, a] \to \mathrm{Dom}(Z)$ satisfies*

$$f'_t = Zf_t \tag{6.6}$$

for all $t \in [0, a]$ then

$$f_t = T_t f_0 \tag{6.7}$$

for all such t. Hence T_t is uniquely determined by Z.

Proof. Given f and $\phi \in \mathcal{B}^*$ and $t \in [0, a]$ define

$$F(s) := \langle T_s f_{t-s}, \phi \rangle$$

for all $0 \le s \le t$. By applying Lemma 6.1.11 we obtain

$$\begin{aligned}
D^+ F(s) &= \lim_{h \to 0+} \langle h^{-1}\{T_{s+h} f_{t-s-h} - T_s f_{t-s}\}, \phi \rangle \\
&= \lim_{h \to 0+} \langle T_{s+h} h^{-1}\{f_{t-s-h} - f_{t-s}\}, \phi \rangle \\
&\quad + \lim_{h \to 0+} \langle h^{-1}\{T_{s+h} - T_s\} f_{t-s}, \phi \rangle \\
&= -\langle T_s Z f_{t-s}, \phi \rangle + \langle Z T_s f_{t-s}, \phi \rangle \\
&= 0.
\end{aligned}$$

Since F is continuous on $[0, t]$, Lemma 1.4.4 implies that it is constant, so $F(t) = F(0)$. That is

$$\langle T_t f_0, \phi \rangle = \langle f_t, \phi \rangle$$

for all $\phi \in \mathcal{B}^*$. This implies (6.7).

Now suppose that T_t and S_t are two one-parameter semigroup s with the same generator Z. If $f \in \mathrm{Dom}(Z)$ then $f_t := S_t f$ satisfies the conditions of this theorem by Lemma 6.1.13, so $f_t = T_t f$. Since T_t and S_t coincide on the dense subspace $\mathrm{Dom}(Z)$ and both are bounded linear operators, they must be equal. $\square$

By Lemma 6.1.13 and Theorem 6.1.16 solving the Cauchy problem for the differential equation

$$f'_t = Zf_t$$

is equivalent to determining the semigroup T_t. We will often write

$$T_t := e^{Zt}$$

below, without suggesting that the right-hand side is more than a formal expression.

The problem of determining which operators Z are the generators of one-parameter semigroup s is highly non-trivial. It is also extremely important, since in applied mathematics one almost always starts from the Cauchy problem, that is the operator Z. There is a constant strain between theorems that are abstractly attractive and tests that can be applied to differential operators that actually arise in 'the real world'.[2]

One of the many difficulties is that the theory depends critically upon the precise choice of the domain of the operator Z, which is frequently not easy to describe explicitly. Fortunately, it is often possible to work in a slightly smaller subspace $\mathcal{D}$. One says that $\mathcal{D} \subseteq \mathrm{Dom}(Z)$ is a core for Z if for all $f \in \mathrm{Dom}(Z)$ there exists a sequence $f_n \in \mathcal{D}$ such that

$$\lim_{n \to \infty} f_n = f, \qquad \lim_{n \to \infty} Zf_n = Zf.$$

Equivalently $\mathcal{D}$ is a core for Z if it is dense in $\mathrm{Dom}(Z)$ for the norm defined in Lemma 6.1.15.

Problem 6.1.17 This extends Problem 6.1.5. Let dx be a Borel measure on the separable, locally compact Hausdorff space X. Let $m : X \to \mathbf{C}$ be a continuous function and define the multiplication operator M on the dense domain $C_c(X)$ in $L^2(X, dx)$ by $Mf(x) = m(x)f(x)$. Find necessary and sufficient conditions on the function m under which the closure of M is the generator of a one-parameter semigroup . $\square$

It is often hard to determine whether a given subspace is a core for a generator Z. The following criterion is particularly useful when the semigroup is given explicitly.

---

[2] This assumes that engineers and physicists study the real world, while mathematicians do not. Classical Platonists, of course, take exactly the opposite view of reality.

Theorem 6.1.18 *(Nelson)*[3] *If $\mathcal{D} \subseteq \mathrm{Dom}(Z)$ is dense in $\mathcal{B}$ and invariant under the semigroup T_t then $\mathcal{D}$ is a core for Z.*

Proof. We use Lemma 6.1.15 and work simultaneously with the two norms $\|\cdot\|$ and $\|\|\cdot\|\|$. Let $\overline{\mathcal{D}}$ denote the closure of $\mathcal{D}$ in $\mathrm{Dom}(Z)$ with respect to $\|\|\cdot\|\|$. If $f \in \mathrm{Dom}(Z)$ then by the density of $\mathcal{D}$ in $\mathcal{B}$ there is a sequence $f_n \in \mathcal{D}$ such that $\|f_n - f\| \to 0$. Since T_t is continuous with respect to $\|\|\cdot\|\|$ we have

$$\int_0^t T_x f_n \, dx \in \overline{\mathcal{D}}.$$

By (6.3) and (6.4)

$$\lim_{n\to\infty} \|\| \int_0^t T_x f_n \, dx - \int_0^t T_x f \, dx \|\| = \lim_{n\to\infty} \| \int_0^t T_x (f_n - f) \, dx \|$$
$$+ \lim_{n\to\infty} \| T_t f_n - f_n - T_t f + f \|$$
$$= 0$$

for every $t > 0$, so

$$\int_0^t T_x f \, dx \in \overline{\mathcal{D}}.$$

Using (6.4) again

$$\lim_{t\to 0} \|\| t^{-1} \int_0^t T_x f \, dx - f \|\| = \lim_{t\to 0} \| t^{-1} \int_0^t T_x f \, dx - f \|$$
$$+ \lim_{t\to 0} \| t^{-1} (T_t f - f) - Zf \|$$
$$= 0,$$

so $f \in \overline{\mathcal{D}}$. This proves that $\overline{\mathcal{D}} = \mathrm{Dom}(Z)$ as required. $\square$

Problem 6.1.19 Let $T_t := e^{Zt}$ be a one-parameter semigroup on $\mathcal{B}$ and let $f_n \in \mathrm{Dom}(Z)$ satisfy $Zf_n = \lambda_n f_n$ for all $n \in \mathbf{N}$, where $\lambda_n \in \mathbf{C}$. If $\mathcal{D} := \mathrm{lin}\{f_n : n \in \mathbf{N}\}$ is a dense linear subspace of $\mathcal{B}$ prove that it is a core for Z. Compare this with Problem 5.4.6. $\square$

The following example shows that it is not always easy to specify the domain of the generator explicitly.

Example 6.1.20 If V is an increasing continuous function on $\mathbf{R}$, the formula

$$(T_t f)(x) := f(x - t) e^{-V(x) + V(x-t)}$$

[3] See [Nelson 1959].

defines a one-parameter contraction semigroup on the space $C_0(\mathbf{R})$. If one differentiates formally one obtains

$$(Zf)(x) := -f'(x) - V'(x)f(x). \tag{6.8}$$

However, the function V need not be differentiable, and there may exist $f \in C_c^1(\mathbf{R})$ which do not lie in $\mathrm{Dom}(Z)$. If $f \in \mathrm{Dom}(Z)$ then $f'(x)$ must have discontinuities at the same points as $V'(x)f(x)$. If V is nowhere differentiable then $C_c^1(\mathbf{R}) \cap \mathrm{Dom}(Z) = \{0\}$ and (6.8) must be interpreted in a distributional sense. $\square$

Problem 6.1.21 Show in Example 6.1.20 that if V is bounded and has a bounded and continuous derivative, then $\mathrm{Dom}(Z)$ is the set of all continuously differentiable functions such that $f, f' \in C_0(\mathbf{R})$. $\square$

The following example uses the concept of a one-parameter group. The definition of this is almost the same as in the semigroup case, with $[0, \infty)$ replaced by $\mathbf{R}$. The lemmas and theorems already proved all extend to the group context, sometimes with simpler proofs.

Example 6.1.22 Let M be a smooth C^∞ manifold with a one-parameter group of diffeomorphisms. More precisely let there be a smooth map from $M \times \mathbf{R}$ to M such that

(i) $m \cdot 0 = m$ for all $m \in M$,
(ii) $(m \cdot s) \cdot t = m \cdot (s+t)$ for all $s, t \in \mathbf{R}$.

It is easy to show that the space $\mathcal{D} := C_c^\infty(M)$ of smooth functions of compact support is dense in $C_0(M)$. Define the one-parameter semigroup T_t on $C_0(M)$ by

$$(T_t f)(m) := f(m \cdot t).$$

It is clear that $\mathcal{D}$ is contained in $\mathrm{Dom}(Z)$ with

$$(Zf)(m) = \frac{\partial}{\partial t} f(m \cdot t)|_{t=0}.$$

Since $\mathcal{D}$ is invariant under T_t it follows by Theorem 6.1.18 that $\mathcal{D}$ is a core for Z. $\square$

The relationship between one-parameter semigroup s and groups is further clarified by the following theorem.

Theorem 6.1.23 *If* Z *and* $-Z$ *are generators of one-parameter semigroup s* S_t *and* T_t *respectively, both acting on the Banach space* $\mathcal{B}$, *then the formula*

$$U_t := \begin{cases} S_t & \text{if } t \geq 0, \\ T_{|t|} & \text{if } t < 0, \end{cases} \tag{6.9}$$

defines a one-parameter group on $\mathcal{B}$.

Proof. If U_t is defined by (6.9) then it is obvious that U_t is jointly continuous at all $t \in \mathbf{R}$, and that

$$\lim_{t \to 0} t^{-1}(U_t f - f) = Zf$$

for all $f \in \text{Dom}(Z)$. The only non-trivial fact to be proved is that

$$U_s U_t = U_{s+t}$$

when s, t have opposite signs. This follows provided

$$S_t T_t = 1 = T_t S_t \tag{6.10}$$

for all $t \geq 0$.

If $f \in \text{Dom}(Z)$ and $f_t := S_t T_t f$ then $f_t \in \text{Dom}(Z)$ for all $t \geq 0$ by Lemma 6.1.11, and

$$\begin{aligned} f_t' &= \lim_{h \to 0+} h^{-1} \{ S_{t+h} T_{t+h} f - S_t T_t f \} \\ &= \lim_{h \to 0+} S_{t+h} h^{-1} (T_{t+h} f - T_t f) \\ &\quad + \lim_{h \to 0+} h^{-1} (S_{t+h} - S_t) T_t f \\ &= S_t(-Z) T_t f + Z S_t T_t f \\ &= 0 \end{aligned}$$

by Lemma 6.1.11. Therefore $S_t T_t f = f$ for all $f \in \text{Dom}(Z)$ and all $t \geq 0$. We deduce the first part of (6.10) by using the density of $\text{Dom}(Z)$ in $\mathcal{B}$. The second part has a similar proof. $\square$

Problem 6.1.24 Let $a : \mathbf{R} \times \mathbf{R} \to \mathbf{C}$ be a continuous function. Prove that if the formula

$$(T_t f)(x) := a(x, t) f(x - t)$$

defines a one-parameter group on $C_0(\mathbf{R})$ then there exists a function $b : \mathbf{R} \to \mathbf{C}$ such that

$$a(x, t) = \frac{b(x)}{b(x - t)}$$

for all $x, t \in \mathbf{R}$. What further properties must b have? $\square$

Problem 6.1.25 Prove that if $\alpha \in \mathbf{R}$ and $1 \leq p < \infty$ then the formula

$$(T_t f)(x) := f(x - t)$$

defines a one-parameter group on $L^p(\mathbf{R}, e^{|x|^\alpha} dx)$ if and only if $0 \leq \alpha \leq 1$. Extend your analysis to the case in which $|x|^\alpha$ is replaced by a general continuous function $a : \mathbf{R} \to \mathbf{R}$.[4] $\square$

If A is a bounded operator on $\mathcal{B}$ and Z is an unbounded operator, we say that A and Z commute if A maps $\mathrm{Dom}(Z)$ into $\mathrm{Dom}(Z)$ and

$$ZAf = AZf$$

for all $f \in \mathrm{Dom}(Z)$.

Problem 6.1.26 If Z is a closed operator with $\lambda \notin \mathrm{Spec}(Z)$ and A is a bounded operator, prove that $ZA = AZ$ in the above sense if and only if $R(\lambda, Z)A = AR(\lambda, Z)$. $\square$

Theorem 6.1.27 *Let T_t be a one-parameter semigroup on $\mathcal{B}$ with generator Z. If A is a bounded operator on $\mathcal{B}$ then*

$$AT_t = T_t A \qquad (6.11)$$

for all $t \geq 0$ if and only if A and Z commute in the above sense.

Proof. The fact that (6.11) implies that A and Z commute follows directly from the definitions of Z and its domain. Conversely suppose that A and Z commute. If $f \in \mathrm{Dom}(Z)$ and $0 \leq s \leq t$ then, using the fact that $T_s(\mathrm{Dom}(Z)) \subseteq \mathrm{Dom}(Z)$ for all $s \geq 0$, we obtain

$$\frac{d}{ds}(T_{t-s} A T_s f) = T_{t-s}(-Z) A T_s f + T_{t-s} A Z T_s f = 0$$

so $s \to T_{t-s} A T_s f$ must be constant. Putting $s = 0$ and $s = t$ we obtain

$$T_t A f = A T_t f$$

for all $f \in \mathrm{Dom}(Z)$ and $t \geq 0$. The conclusion follows by a density argument. $\square$

Problem 6.1.28 Show that A and Z commute if there is a core $\mathcal{D}$ for Z such that $A\mathcal{D} \subseteq \mathrm{Dom}(Z)$ and $AZf = ZAf$ for all $f \in \mathcal{D}$. $\square$

---

[4] A general analysis of this type of phenomenon may be found in [Elst and Robinson 2006].

Problem 6.1.29 Define the one-parameter semigroup T_t on $C_0(\mathbf{R})$ by

$$T_t f(x) := e^{-x^2 t} f(x).$$

Find all bounded operators A on $C_0(\mathbf{R})$ which commute with the semigroup. □

Our final interpolation lemma for semigroups will be used several times later in the book.

Lemma 6.1.30 *Let* $1 \leq p_0 < \infty$ *and* $1 \leq p_1 \leq \infty$. *Suppose that* T_t *is a one-parameter semigroup acting on* $L^{p_0}(X, dx)$ *and satisfying*

$$\|T_t f\|_{p_i} \leq M e^{at} \|f\|_{p_i}$$

for $i = 0, 1$, *all* $t \geq 0$ *and all* $f \in L^{p_0}(X, dx) \cap L^{p_1}(X, dx)$. *Then* T_t *extends consistently to one-parameter semigroup s acting on* $L^p(X, dx)$ *for all* p *such that* $1/p := (1 - \lambda)/p_0 + \lambda/p_1$ *and* $0 < \lambda < 1$.

Proof. Let P denote the set of p satisfying the conditions of the theorem. The fact that each operator T_t extends consistently to every L^p space follows directly from Theorem 2.2.14. It is immediate that $T_{t+s} = T_t T_s$ in L^p and that $\|T_t\|_p \leq M e^{at}$ for all $p \in P$ and all $s, t \geq 0$. The remaining issue is strong continuity.

Given $g \in L^{p_0} \cap L^{p_1}$ the functions $f_t := T_t g - g$ are uniformly bounded in the L^{p_0} and L^{p_1} norms for $0 < t < 1$ and converge to 0 as $t \to 0$ in L^{p_0}. Problem 2.1.5 now implies that $\lim_{t \to 0} \|T_t g - g\|_p = 0$ for all $p \in P$. Since $L^{p_0} \cap L^{p_1}$ is dense in L^p for all $p \in P$ we conclude by the uniform boundedness of the norms that $\lim_{t \to 0} \|T_t g - g\|_p = 0$ for all $p \in P$ and all $g \in L^p$. The proof is completed by applying Theorem 6.2.1 below. □

6.2 Other continuity conditions

In this section we investigate the continuity condition (iii) in the definition of a one-parameter semigroup on page 167.

Theorem 6.2.1 *If the bounded operators* T_t *satisfy (i) and (ii) in the definition of a one-parameter semigroup , then they also satisfy (iii) if and only if*

$$\lim_{t \to 0+} T_t f = f \qquad (6.12)$$

for all $f \in \mathcal{B}$. In this case there exist constants M, a such that

$$\|T_t\| \le M e^{at} \tag{6.13}$$

for all $t \ge 0$.

Proof. If (6.12) holds and

$$c_n := \sup\{\|T_t\| : 0 \le t \le 1/n\}$$

then $c_n < \infty$ for some n. For otherwise there exist t_n with $0 \le t_n \le 1/n$ and $\|T_{t_n}\| \ge n$. This contradicts the uniform boundedness theorem when combined with (6.12). We also observe that (6.12) implies that $c_n \ge 1$ for all n.

If $c_n < \infty$ and we put $c := c_n^n$ then condition (ii) in the definition of a semigroup on page 167 implies that $\|T_t\| \le c$ for all $0 \le t \le 1$. If $[t]$ denotes the integer part of t, we deduce that

$$\|T_t\| \le c^{[t]+1} \le c^{t+1} = M e^{at}.$$

To prove condition (iii) in the definition of a semigroup we note that if $\lim_{n \to \infty} t_n = t$ and $\lim_{n \to \infty} f_n = f$ then

$$\lim_{n \to \infty} \|T_{t_n} f_n - T_t f\| \le \lim_{n \to \infty} \{\|T_{t_n}(f_n - f)\| + \|T_{t_n} f - T_t f\|\}$$

$$\le \lim_{n \to \infty} \{M e^{at_n} \|f_n - f\| + M e^{a \min(t_n, t)} \|T_{|t_n - t|} f - f\|\}$$

$$= 0. \qquad \square$$

Many arguments in the theory of one-parameter semigroup s become easier if $M = 1$, or even depend upon this. However all that one can say in general is that (6.13) implies $M \ge 1$. The following example shows that it may not be possible to put $M = 1$, and even that $\|T_t\|$ need not converge to 1 as $t \to 0$. A less artificial example is presented in Theorem 6.3.8.

The growth rate, usually denoted ω_0, of the semigroup T_t is defined as the infimum of the constants a for which (6.13) holds for some constant M_a and all $t \ge 0$. Another characterization of ω_0 is given in Theorem 10.1.6.

Problem 6.2.2 Calculate the precise value of $\|T_t\|$ in Problem 6.1.25 when $p = 1$. Prove that $\omega_0 = 0$ if $0 < \alpha < 1$ and that $M_a \to +\infty$ as $a \to 0+$. $\square$

Example 6.2.3 Let $\mathcal{B}$ be the Banach space of all continuous functions on $\mathbf{R}$ which vanish at infinity, with the norm

$$\|f\| := \|f\|_\infty + k|f(0)|,$$

where $k > 0$. This is equivalent to the usual supremum norm. We define

$$(T_t f)(x) := f(x - t)$$

for all $x \in \mathbf{R}$ and $t \in \mathbf{R}$, so that T_t is a one-parameter group on $\mathcal{B}$. Now

$$\|T_t f\| = \|f\|_\infty + k|f(-t)| \leq (k+1)\|f\|_\infty \leq (k+1)\|f\|$$

for all $f \in \mathcal{B}$. On the other hand if $f \in \mathcal{B}$ satisfies $\|f\|_\infty = 1$, $f(0) = 0$ and $f(-t) = 1$ then $\|f\| = 1$ but

$$\|T_t f\| = \|f\|_\infty + k|f(-t)| = k + 1 = (k+1)\|f\|.$$

Therefore $\|T_t\| = k + 1$ for all $t \neq 0$, while $\|T_0\| = 1$. $\quad\square$

The last two results suggest that the possibility that $M > 1$ is associated with making the 'wrong' choice of norm. However, this is not very helpful, because in applications one often does not know in advance the norm which is most appropriate for a particular semigroup. Even if one did, the norm chosen may measure some quantity of physical interest, such as energy, in which case one is not free to change it at will, even to an equivalent norm.

Problem 6.2.4 Prove that if T_t is a one-parameter semigroup on $\mathcal{B}$ then $t \to \|T_t\|$ is a lower semi-continuous function on $[0, \infty)$. Find an example of a semigroup T_t such that $\|T_t\| = 1$ for $0 \leq t < 1$ but $\|T_t\| = 0$ for $t \geq 1$. $\quad\square$

The proof of Theorem 6.2.6 below makes use of a rather advanced fact about the weak* topology.

Proposition 6.2.5 *(Krein-Šmulian theorem)*[5] *If $X : \mathcal{B}^* \to \mathbf{C}$ is linear and its restriction to the unit ball of $\mathcal{B}^*$ is continuous with respect to the weak* topology of $\mathcal{B}^*$, then there exists $f \in \mathcal{B}$ such that $X(\phi) = \langle f, \phi \rangle$ for all $\phi \in \mathcal{B}^*$.*

As far as we know, the main importance of our next theorem is to rule out a possible generalization of the notion of one-parameter semigroup.

Theorem 6.2.6 *If the bounded operators T_t on $\mathcal{B}$ satisfy (i) and (ii) in the definition of a one-parameter semigroup on page 167 then they also satisfy (iii) if and only if*

$$\underset{t \to 0+}{\text{w-}\lim}\, T_t f = f \tag{6.14}$$

for all $f \in \mathcal{B}$.

---

[5] See [Dunford and Schwartz 1966, p. 429] for the proof.

Proof. The argument in one direction is trivial so we assume that T_t satisfies (i), (ii) and (6.14). An argument similar to that of Theorem 6.2.1 establishes that there exist constants M, a such that

$$\|T_t\| \leq M e^{at} \tag{6.15}$$

for all $t \geq 0$.

If we choose a particular $f \in \mathcal{B}$, then the closed linear span of $\{T_t f : t \geq 0\}$ is invariant under T_t. It is also separable by Problem 1.3.3, being the weak and hence the norm closure of the linear subspace

$$\text{lin}\{T_t f : t \geq 0 \text{ and } t \text{ is rational}\}.$$

Since the proof that $\lim_{t \to 0} T_t f = f$ may be carried out entirely in this subspace, there is no loss of generality in assuming that $\mathcal{B}$ is separable.

If $f \in \mathcal{B}$ and $\phi \in \mathcal{B}^*$ then $\langle T_t f, \phi \rangle$ is locally bounded and right continuous as a function of t, so the integral

$$\varepsilon^{-1} \int_0^\varepsilon \langle T_t f, \phi \rangle \, dt$$

converges. It defines a bounded linear functional on $\mathcal{B}^*$ which is weak*-continuous on the unit ball of $\mathcal{B}^*$ by the dominated convergence theorem and the separability of $\mathcal{B}$; see Problem 1.3.8. The Krein-Šmulian Theorem 6.2.5 implies[6] that there exists $f_\varepsilon \in \mathcal{B}$ such that

$$\langle f_\varepsilon, \phi \rangle = \varepsilon^{-1} \int_0^\varepsilon \langle T_t f, \phi \rangle \, dt$$

for all $\phi \in \mathcal{B}^*$.

Given $h > 0$ we have

$$|\langle T_h f_\varepsilon - f_\varepsilon, \phi \rangle| = |\langle f_\varepsilon, T_h^* \phi \rangle - \langle f_\varepsilon, \phi \rangle|$$

$$= \varepsilon^{-1} \left| \int_0^\varepsilon \langle T_t f, T_h^* \phi \rangle \, dt - \int_0^\varepsilon \langle T_t f, \phi \rangle \, dt \right|$$

$$= \varepsilon^{-1} \left| \int_h^{\varepsilon+h} \langle T_t f, \phi \rangle \, dt - \int_0^\varepsilon \langle T_t f, \phi \rangle \, dt \right|$$

$$= \varepsilon^{-1} \left| \int_\varepsilon^{\varepsilon+h} \langle T_t f, \phi \rangle \, dt - \int_0^h \langle T_t f, \phi \rangle \, dt \right|$$

$$\leq \varepsilon^{-1} \|f\| \|\phi\| \left\{ \int_\varepsilon^{h+\varepsilon} M e^{at} dt + \int_0^h M e^{at} dt \right\}$$

---

[6] Note that if $\mathcal{B}$ is reflexive one does not need to appeal to the Krein-Šmulian Theorem, which is in any case trivial under this assumption.

$$\leq \varepsilon^{-1} 2hMe^{a(h+\varepsilon)} \|f\| \|\phi\|.$$

Since $\phi \in \mathcal{B}^*$ is arbitrary

$$\lim_{h \to 0} \|T_h f_\varepsilon - f_\varepsilon\| \leq \lim_{h \to 0} \varepsilon^{-1} 2hMe^{a(h+\varepsilon)} \|f\|$$
$$= 0.$$

Now let L be the set of all $g \in \mathcal{B}$ for which $\lim_{t \to 0} T_t g = g$. It is immediate that L is a linear subspace and it follows from (6.15) that L is norm closed in $\mathcal{B}$. This implies that it is weakly closed. To prove that $L = \mathcal{B}$ we have only to note that it follows directly from the definition of f_ε that it converges weakly to f as $\varepsilon \to 0$. $\square$

Problem 6.2.7 Let T_t be a one-parameter semigroup acting on the Banach space $\mathcal{B}$ and suppose that the closed linear subspace $\mathcal{L}$ is invariant under T_t for every $t \geq 0$. Prove that the family of operators $\tilde{T}_t$ induced by T_t on the quotient space $\tilde{\mathcal{B}} = \mathcal{B}/\mathcal{L}$ is also a one-parameter semigroup . $\square$

Problem 6.2.8 In the example above, if $\|T_t\| \leq Me^{at}$ for all $t \geq 0$ then the same bound holds if T_t is replaced by $T_t|_{\mathcal{L}}$ or $\tilde{T}_t$. Prove that the converse is false, and find an analogue in the reverse direction. $\square$

One may also use the *weak topology* in the definition of the generator.

Theorem 6.2.9 *Let Z be the generator of a one-parameter semigroup T_t on $\mathcal{B}$ and let $\mathcal{E}$ be a weak*-dense subset of $\mathcal{B}^*$ which is invariant under T_t^* for all $t \geq 0$. If there is a positive sequence $t_n \to 0$ such that*

$$\lim_{n \to \infty} t_n^{-1} \langle T_{t_n} f - f, \phi \rangle = \langle g, \phi \rangle$$

for some $f, g \in \mathcal{B}$ and all $\phi \in \mathcal{E}$, then $f \in \mathrm{Dom}(Z)$ and $Zf = g$.

Proof. We modify the argument of Lemma 6.1.12. If f, g and ϕ are as in the statement of the theorem and the complex-valued function F is defined by

$$F(t) := \left\langle T_t f - f - \int_0^t T_x g \, dx, \phi \right\rangle$$

then F is continuous and $F(0) = 0$. Moreover

$$\lim_{n \to \infty} t_n^{-1} \{F(t + t_n) - F(t)\} = \lim_{n \to \infty} \langle t_n^{-1}(T_{t_n}f - f), T_t^* \phi \rangle$$

$$- \lim_{n \to \infty} \left\langle t_n^{-1} \int_t^{t + t_n} T_x g \, dx, \phi \right\rangle$$

$$= \langle g, T_t^* \phi \rangle - \langle T_t g, \phi \rangle$$

$$= 0.$$

It follows by Lemma 1.4.4(i) that $F(t) = 0$ for all $t \geq 0$.

Since $\phi \in \mathcal{E}$ is arbitrary and $\mathcal{E}$ is weak*-dense in $\mathcal{B}^*$, we deduce that

$$T_t f - f = \int_0^t T_x g \, dx$$

so

$$\lim_{t \to 0} t^{-1}(T_t f - f) = \lim_{t \to 0} t^{-1} \int_0^t T_x g \, dx = g. \qquad \square$$

Problem 6.2.10 Let $u : [a, b] \to [0, 1]$ be a continuous function and define T_t on $\mathcal{B}$ by $T_0 = 1$ and

$$(T_t f)(x) := u(x)^t f(x)$$

for $t > 0$. Find the precise conditions on u under which T_t is a one-parameter semigroup   in   the   cases $\mathcal{B} := C[a, b]$, $\mathcal{B} := L^p(a, b)$ and $1 \leq p < \infty$, $\mathcal{B} := L^\infty(a, b)$. $\quad \square$

6.3 Some standard examples

We start our examples at a fairly general level, and then describe some special cases. We will be working in the function spaces $L^p(\mathbf{R}^N)$ defined for $1 \leq p < \infty$ by the finiteness of the norms

$$\|f\|_p := \left\{ \int_{\mathbf{R}^N} |f(x)|^p \, d^N x \right\}^{1/p}.$$

See Section 2.1. We also use the space $L^\infty(\mathbf{R}^N)$ of all essentially bounded functions on $\mathbf{R}^N$ with the essential supremum norm

$$\|f\|_\infty := \min\{c : \{x : |f(x)| > c\} \text{ is a Lebesgue null set}\}.$$

Let k_t be complex-valued functions on $\mathbf{R}^N$, where t is a positive real parameter. We say that the family k_t forms a convolution semigroup[7] on $\mathbf{R}^N$ if it has the following properties:

(i) $k_s * k_t = k_{s+t}$ for all $s, t > 0$, where $*$ denotes convolution.

(ii) There exists a constant c such that $\|k_t\|_1 \le c$ for all $t > 0$.

(iii) For every $r > 0$ we have

$$\lim_{t \to 0} \int_{|x| > r} |k_t(x)| \, \mathrm{d}^N x = 0.$$

(iv) For every $r > 0$ we have

$$\lim_{t \to 0} \int_{|x| < r} k_t(x) \, \mathrm{d}^N x = 1.$$

Example 6.3.1 One may prove that the Cauchy densities

$$f_t(x) := \frac{t}{\pi(t^2 + x^2)}$$

define a convolution semigroup on $\mathbf{R}$ by using the fact that

$$\int_{\mathbf{R}} f_t(x) e^{-ix\xi} \, \mathrm{d}x = e^{-t|\xi|}$$

for all $\xi \in \mathbf{R}$. Calculating the inverse Fourier transform is actually easier. One may also write

$$f_t * \phi = e^{-Ht} \phi$$

for all $\phi \in L^2(\mathbf{R})$ where $H := (-\Delta)^{1/2}$. $\square$

Theorem 6.3.2 *Under the conditions (i)–(iv) above, the formula*

$$T_t f := k_t * f$$

defines a one-parameter semigroup on $L^p(\mathbf{R}^N)$ for all $p \in [1, \infty)$. If $k_t(x) \ge 0$ for all $x \in \mathbf{R}^N$ and $t > 0$ then T_t is a positivity preserving contraction semigroup on $L^p(\mathbf{R}^N)$ for all $p \in [1, \infty)$, in the sense that $f \ge 0$ implies that $T_t f \ge 0$ for all $t \ge 0$.

---

[7] The deep classification theory for convolution semigroups on $\mathbf{R}$ is described in [Feller 1966]. See [Hunt 1956] for the extension to Lie groups.

Proof. Corollary 2.2.19 implies that $\|T_t\| \le c$ in $L^p(\mathbf{R}^N)$ for all p and all $t > 0$. The semigroup law follows immediately from condition (i). Once we prove that

$$\lim_{t \to 0} \|k_t * f - f\|_1 = 0 \qquad (6.16)$$

for all $f \in L^1(\mathbf{R}^N)$, the theorem follows by applying Lemma 6.1.30.

It is sufficient to prove (6.16) for all f in the dense subset $C_c(\mathbf{R}^N)$ of L^1. We start by proving that $k_t * f$ converges uniformly to f as $t \to 0$. Given $\varepsilon > 0$ there exists $\delta > 0$ such that $|u - v| < \delta$ implies $|f(u) - f(v)| < \varepsilon$. Given $x \in \mathbf{R}^N$ we deduce that

$$\begin{aligned}
|(k_t * f)(x) - f(x)| &\le \left| \int_{|y| \le \delta} k_t(y)\{f(x - y) - f(x)\} \, \mathrm{d}^N y \right| \\
&\quad + \left| \int_{|y| > \delta} k_t(y) f(x - y) \, \mathrm{d}^N y \right| \\
&\quad + \left| \left\{ \int_{|y| \le \delta} k_t(y) \, \mathrm{d}^N y - 1 \right\} f(x) \right| \\
&\le \varepsilon \int_{|y| \le \delta} |k_t(y)| \, \mathrm{d}^N y \\
&\quad + \|f\|_\infty \int_{|y| > \delta} |k_t(y)| \, \mathrm{d}^N y \\
&\quad + \|f\|_\infty \left| \int_{|y| \le \delta} k_t(y) \, \mathrm{d}^N y - 1 \right| \\
&< (c + 1)\varepsilon
\end{aligned}$$

for all small enough $t > 0$.

If f has support in $\{x : |x| \le R\}$ then

$$\begin{aligned}
\lim_{t \to 0} \int_{|x| \ge 2R} |(k_t * f)(x) - f(x)| \, \mathrm{d}^N x &= \lim_{t \to 0} \int_{|x| \ge 2R} |(k_t * f)(x)| \, \mathrm{d}^N x \\
&= \lim_{t \to 0} \int_{|x| \ge 2R} \int_{|y| \le R} |k_t(x - y) f(y)| \, \mathrm{d}^N y \, \mathrm{d}^N x \\
&\le \lim_{t \to 0} \int_{|u| \ge R} \int_{|y| \le R} |k_t(u) f(y)| \, \mathrm{d}^N y \, \mathrm{d}^N u \\
&= \|f\|_1 \lim_{t \to 0} \int_{|u| \ge R} |k_t(u)| \, \mathrm{d}^N u \\
&= 0.
\end{aligned}$$

Combining the above two bounds we see that

$$\|k_t * f - f\|_1 \le \|k_t * f - f\|_\infty |\{x : |x| \le 2R\}|$$
$$+ \int_{|x| \ge 2R} |(k_t * f)(x) - f(x)| \, d^N x,$$

which converges to 0 as $t \to 0$.

Problem 6.3.3 Prove that T_t acts as a one-parameter semigroup on the space $C_0(\mathbf{R}^N)$ of continuous functions on $\mathbf{R}^N$ which vanish at infinity, with the supremum norm. Prove that the corresponding statement is false for the space of all continuous bounded functions with the supremum norm. □

Problem 6.3.4 Prove that the semigroup of Theorem 6.3.2 is norm continuous for all $t > 0$. □

Example 6.3.5 The Gaussian densities

$$k_t(x) := \{4\pi t\}^{-N/2} e^{-|x|^2/4t}$$

were introduced in Lemma 3.1.5 (with a different normalization), and provide the best known example of a convolution semigroup on $\mathbf{R}^N$. Property (i) above may be proved by using the formula

$$\int_{\mathbf{R}^N} k_t(x) e^{-ix \cdot \xi} \, d^N x = e^{-|\xi|^2 t},$$

proved in Lemma 3.1.5. Properties (ii)–(iv) are verified directly. The associated semigroup T_t on $L^2(\mathbf{R}^N)$ satisfies

$$(\mathcal{F} T_t \mathcal{F}^{-1} g)(\xi) = e^{-|\xi|^2 t} g(\xi)$$

for all $g \in L^2(\mathbf{R}^N)$ and all $t > 0$. Its generator Z satisfies

$$(\mathcal{F} Z \mathcal{F}^{-1} g)(\xi) = -|\xi|^2 g(\xi)$$

and

$$\text{Dom}(Z) = \{f \in L^2 : |\xi|^2 (\mathcal{F} f)(\xi) \in L^2\} = W^{2,2}(\mathbf{R}^N).$$

We deduce by Lemma 3.1.4 that $Zf = \Delta f$ for all f in the Schwartz space $\mathcal{S}$.

It follows by Theorem 6.3.2 that T_t is a positivity preserving contraction semigroup on $L^p(\mathbf{R}^N)$ for all $1 \le p < \infty$. The rate of decay of $\|T_t f\|_p$ as $t \to \infty$ depends on the precise hypotheses about f. For example if $f \in L^1(\mathbf{R}^N)$ then

$$\|T_t f\|_2 \le \|k_t\|_2 \|f\|_1 \le (4\pi t)^{-N/4} \|f\|_1$$

for all $t > 0$. The L^p spectrum of the generator Z is determined in Example 8.4.5. □

Example 6.3.6 The Gaussian semigroup $T_t := e^{\Delta t}$ may also be constructed on $L^2(-\pi, \pi)$ subject to periodic boundary conditions. Its convolution kernel, given by

$$k_t(x) := \frac{1}{2\pi} \sum_{n \in \mathbf{Z}} e^{-n^2 t + inx},$$

is positive, smooth and periodic in x, but cannot be written in closed form. However the periodic Cauchy semigroup $T_t := e^{-Ht}$, where $H := (-\Delta)^{1/2}$ acts in $L^2(-\pi, \pi)$ subject to periodic boundary conditions, has the convolution kernel

$$k_t(x) := \frac{1}{2\pi} \sum_{n \in \mathbf{Z}} e^{-|n|t + inx}$$

$$= \frac{1}{2\pi} \frac{\sinh(t)}{\cosh(t) - \cos(x)}.$$

Both of the above are Markov semigroups on $L^1(-\pi, \pi)$ or on $C_{\mathrm{per}}[-\pi, \pi]$, in the sense of Chapter 13. □

We construct some examples with higher order generators, using our results and notation for Fourier transforms in Section 3.1.

If n is a positive integer, we define H_n to be the closure of the operator

$$H_{0,n}f := (-\Delta)^n f$$

defined on $\mathcal{S}$, so that

$$(\mathcal{F}H_n f)(\xi) = |\xi|^{2n}(\mathcal{F}f)(\xi),$$

where $\mathcal{F}$ is the unitary Fourier transform operator on $L^2(\mathbf{R}^N)$. See Lemma 3.1.4.

Lemma 6.3.7 *The closure Z of $-H_n$ is the generator of the one-parameter semigroup T_t on $L^2(\mathbf{R}^N)$ given by*

$$T_t f(x) := \int_{\mathbf{R}^N} k_t(x - y) f(y) \, d^N y = (k_t * f)(x) \tag{6.17}$$

where $k_t \in \mathcal{S}$ is defined by

$$k_t(x) := \frac{1}{(2\pi)^N} \int_{\mathbf{R}^N} e^{-|\xi|^{2n} t + ix \cdot \xi} \, d^N \xi. \tag{6.18}$$

The semigroup T_t is strongly continuous as $t \to 0+$ and norm continuous for $t > 0$. The operators T_t are contractions on $L^2(\mathbf{R}^N)$ for all $t \geq 0$.

We omit the proof, which is a routine exercise in the use of Fourier transforms.

Example 6.3.5 gives the explicit formula for $k_t(x)$ when $n = 1$. If $n > 1$ then T_t is still a contraction semigroup on $L^2(\mathbf{R}^N)$, but not on $L^1(\mathbf{R}^N)$. For the sake of simplicity we restrict to the one-dimensional case, although the same proof can be extended to higher dimensions.

Theorem 6.3.8 *The operators T_t defined by (6.17) and (6.18) form a one-parameter semigroup on $L^1(\mathbf{R})$. The norm of T_t is independent of t for $t > 0$ and is given by*

$$c = \int_{-\infty}^{\infty} |k_1(x)|\,dx.$$

This constant is greater than 1 unless $n = 1$.

Proof. We have to verify that k_t satisfy conditions (i)–(iv) for a convolution semigroup on page 183. It follows from Theorem 2.2.5 that $\|T_t\| = \|k_t\|_1$ for all $t > 0$. This is independent of t by virtue of the formula

$$k_t(x) = t^{-1/2n} k_1(t^{-1/2n}x),$$

which follows from (6.18). Since

$$\int_{-\infty}^{\infty} k_1(x)\,dx = \hat{k}_1(0) = 1$$

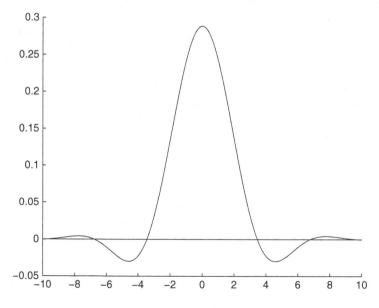

Figure 6.2: $k_1(x)$ as a function of x for $n = 2$

we must have $c \geq 1$. The identity $c = 1$ would imply that $k_1(x) \geq 0$ for all $x \in \mathbf{R}$. By Problem 3.1.20 we would conclude that

$$\frac{d^2}{d\xi^2} e^{-\xi^{2n}} \bigg|_{\xi=0} < 0.$$

This is only true for $n = 1$. Numerical calculations show that $c \sim 1.237$ if $n = 2$.

We finally extend some of the above considerations to constant coefficient differential operators of the form

$$(Af)(x) := \sum_{|\alpha| \leq n} a_\alpha (D^\alpha f)(x)$$

initially defined on the Schwartz space $\mathcal{S}$, where $a_\alpha \in \mathbf{C}$ for all multi-indices α. One has

$$(\mathcal{F}Af)(\xi) = \sigma(\xi)(\mathcal{F}f)(\xi)$$

for all $f \in \mathcal{S}$ and $\xi \in \mathbf{R}^N$, where the symbol σ of A is defined by

$$\sigma(\xi) := \sum_{|\alpha| \leq n} a_\alpha i^{|\alpha|} \xi^\alpha.$$

See Lemma 3.1.4.

Theorem 6.3.9 *Let Z denote the closure of A as an operator acting in $L^2(\mathbf{R}^N)$. Then the spectrum of Z equals the closure of $\{\sigma(\xi) : \xi \in \mathbf{R}^N\}$. Moreover Z is the generator of a one-parameter semigroup on $L^2(\mathbf{R}^N)$ if and only if there is a constant c such that $\mathrm{Re}(\sigma(\xi)) \leq c$ for all $\xi \in \mathbf{R}^N$.*

Proof. All of the statements follow directly from the fact that Z is unitarily equivalent to the multiplication operator M defined by $(Mg)(\xi) = \sigma(\xi)g(\xi)$, with the maximal domain

$$\mathrm{Dom}(M) := \{g \in L^2 : Mg \in L^2\}. \qquad \square$$

Theorem 6.3.8 provides an example in which the 'same' one-parameter semigroup has different norms when acting in $L^1(\mathbf{R})$ and $L^2(\mathbf{R})$. The following example shows more extreme behaviour of this type.

Example 6.3.10 Let T_t be the one-parameter semigroup acting in $L^p(\mathbf{R}^3)$ according to the formula $T_t f := k_t * f$, where $k_t \in \mathcal{S}$ are defined by

$$k_t(x) := \frac{1}{(2\pi)^3} \int_{\mathbf{R}^3} e^{-(|\xi|^2 - 1)^2 t + ix \cdot \xi} \, d^3\xi. \qquad (6.19)$$

The generator Z is given on $\mathcal{S}$ by $Zf := -(\Delta^2 f + 2\Delta f + f)$. It is immediate that T_t is a one-parameter contraction semigroup on $L^2(\mathbf{R}^3)$. However if $\|\cdot\|_1$ is the norm of T_t considered as an operator on $L^1(\mathbf{R}^3)$ then it may be shown[8] that there is an absolute constant $c > 0$ such that

$$\|T_t\|_1 = \|k_t\|_1 \geq ct^{1/2}$$

for all $t \geq 1$. $\quad\square$

Problem 6.3.11 Formulate and prove a vector-valued version of Theorem 6.3.9, in which a_α are all $m \times m$ matrices and the operator acts in $L^2(\mathbf{R}^N, \mathbf{C}^m)$. $\quad\square$

---

[8] See [Davies 1995D] for the proof and a description of further unexpected properties of the one-parameter semigroup s associated with higher order differential operators, and [Davies 1997] for a more general review of this subject.

7

Special classes of semigroup

7.1 Norm continuity

One-parameter semigroups arise in many different areas of applied mathematics, so it is not surprising that different types of semigroup have proved important. This chapter is devoted to a few of these, but others, for example Markov semigroups, have whole chapters to themselves.

In this section we consider norm continuous semigroups. If Z is a bounded linear operator with domain equal to $\mathcal{B}$, then it is obvious that the series

$$T_t := \sum_{n=0}^{\infty} t^n Z^n / n!$$

is norm convergent for all complex t. Restricting attention to $t \geq 0$, we obtain a one-parameter semigroup with generator Z. Clearly T_t is a norm continuous function of t. Our first result goes in the reverse direction.

Theorem 7.1.1 *A one-parameter semigroup T_t is norm continuous if and only if its generator Z is bounded.*

Proof. By taking $h > 0$ small enough that

$$\left\| h^{-1} \int_0^h T_t \, dt - 1 \right\| < 1$$

we may assume that

$$X := \int_0^h T_t \, dt$$

is invertible. We define the bounded operator Z by

$$Z := X^{-1}(T_h - 1).$$

Since

$$X(T_t - 1) = \int_t^{t+h} T_x \, dx - \int_0^h T_x \, dx$$

$$= \int_h^{t+h} T_x \, dx - \int_0^t T_x \, dx$$

$$= (T_h - 1) \int_0^t T_x \, dx$$

for all $t \geq 0$, we deduce that

$$T_t - 1 = \int_0^t Z T_x \, dx.$$

This implies that T_t is norm differentiable with

$$\frac{d}{dt} T_t = Z T_t$$

for all $t \geq 0$. The proof that $T_t = \sum_{n=0}^\infty t^n Z^n / n!$ now uses the uniqueness part of Theorem 6.1.16. $\quad\square$

While norm continuous semigroups are of limited usefulness in applications, the following class of semigroups is of considerable importance. Holomorphic semigroups are discussed in Section 8.4.

Theorem 7.1.2 *Let T_t be a one-parameter semigroup on $\mathcal{B}$ such that $T_t f \in$ Dom(Z) for all $f \in \mathcal{B}$ and all $t > 0$. Then T_t is a norm C^∞ function of t for all $t \in (0, \infty)$.*

Proof. If $0 < c < \infty$ then the operator $Z T_c$ is closed with domain $\mathcal{B}$. Therefore it is bounded. By Lemma 6.1.12

$$T_t f - f = \int_0^t T_x Z f \, dx$$

for all $f \in$ Dom(Z) and all $t > 0$. Hence

$$T_{t+c} f - T_c f = \int_0^t T_x Z T_c f \, dx \tag{7.1}$$

for all $f \in \mathcal{B}$. Using the bound $\|T_t\| \leq M e^{at}$ we deduce that

$$\|T_{t+c} - T_c\| \leq \int_0^t M e^{ax} \|Z T_c\| \, dx.$$

Therefore T_t is norm continuous at $t = c$ for all $c > 0$. We deduce from (7.1) that T_t is norm differentiable with

$$\frac{d}{dt} T_t = Z T_t$$

for all $t > 0$. If $0 < c < t$ we may rewrite this as

$$\frac{\mathrm{d}}{\mathrm{d}t}T_t = T_{t-c}ZT_c$$

and differentiate repeatedly. $\square$

Pazy has characterized the one-parameter semigroup s T_t such that $T_t x$ is differentiable for all $x \in \mathcal{B}$ and $t > a$ by means of conditions on the spectrum of the generator Z and on the norms of its resolvent operators.[1] The following is a counterexample to another possible extension of Theorem 7.1.2.

Example 7.1.3 Let T_t be the one-parameter semigroup on $C_0(\mathbf{R})$ defined by

$$(T_t f)(x) := e^{ix^2 t}(1+x^2)^{-t}f(x)$$

the generator of which is given formally by

$$(Zf)(x) := \{ix^2 - \log(1+x^2)\}f(x).$$

Although T_t is norm continuous for all $t > 0$, one only has $T_t \mathcal{B} \subseteq \mathrm{Dom}(Z^n)$ for $n \leq t < \infty$. $\square$

We turn next to the consequences of *compactness* conditions.

Theorem 7.1.4 *If T_t is a one-parameter semigroup and T_a is compact for some $a > 0$ then T_t is compact and a norm continuous function of t for all $t \geq a$.*

Proof. The compactness of T_t for $t \geq a$ follows directly from the semigroup property.

Let T_a be compact and let X be the compact closure of the image of the unit ball of $\mathcal{B}$ under T_a. Since $(t, f) \to T_t f$ is jointly continuous, given $\varepsilon > 0$, there exists $\delta > 0$ such that

$$\|T_t f - f\| < \varepsilon$$

for all $f \in X$ and all $0 \leq t < \delta$. If $a \leq b \leq t < b+\delta$ and $\|f\| \leq 1$ then

$$\|T_t f - T_b f\| = \|T_{b-a}(T_{t-b} - 1)T_a f\| \leq \|T_{b-a}\|\varepsilon.$$

Hence

$$\|T_t - T_b\| \leq \|T_{b-a}\|\varepsilon$$

and T_b is norm continuous on the right for all $b \geq a$. A similar argument proves norm continuity on the left. $\square$

Example 7.1.5 We consider a very simple *retarded differential equation*, namely

$$f'(t) = cf(t-1). \tag{7.2}$$

It is elementary that every continuous function f on $[-1,0]$ has a unique continuous extension to $[-1, \infty)$ such that (7.2) holds for all $t \geq 0$. In fact if $0 \leq t \leq 1$ then

$$f(t) = f(0) + c \int_{-1}^{t-1} f(x) \, dx.$$

For such equations the initial data at time t can be regarded as the values of $f(s)$ for $t - 1 \leq s \leq t$, and the solutions of (7.2) can be regarded as operators on the space $C[-1,0]$ of initial data: for $t \geq 0$ and $f \in C[-1,0]$ we may define $T_t f \in C[-1,0]$ by

$$(T_t f)(x) := f(x+t)$$

where f is the solution of (7.2). It is apparent that T_t is a one-parameter semigroup on $C[-1,0]$. Moreover if $0 \leq t \leq 1$

$$(T_t f)(x) = \begin{cases} f(x+t) & \text{if } -1 \leq x \leq -t, \\ f(0) + c \int_0^{x+t} f(s-1) \, ds & \text{if } -t \leq x \leq 0. \end{cases} \tag{7.3}$$

In the special case $t = 1$ we have

$$(T_1 f)(x) = f(0) + c \int_{-1}^{x} f(s) \, ds$$

for $-1 \leq x \leq 0$ and all $f \in C[-1,0]$. Theorem 4.2.7 implies that T_1 is a compact operator, while from (7.3) one may show that T_t is not compact for any $0 \leq t < 1$. $\square$

Problem 7.1.6 Show that the generator Z of the semigroup T_t of Example 7.1.5 is

$$(Zf)(x) := f'(x),$$

with domain the set of continuously differentiable functions f on $[-1,0]$ such that $f'(0) = cf(-1)$. $\square$

7.2 Trace class semigroups

This section is devoted to a very special type of one-parameter semigroup acting on a Hilbert space $\mathcal{H} := L^2(X, \mathrm{d}x)$. We assume that X is a locally compact Hausdorff space, and that there is a countable basis to its topology. We also assume that the regular Borel measure $\mathrm{d}x$ has support equal to X.

Throughout this section we assume that T_t is a self-adjoint one-parameter semigroup , or equivalently, by the spectral theorem, that $T_t = \mathrm{e}^{-Ht}$ for all $t \geq 0$, where H is a self-adjoint operator. We focus on the relationship between trace class hypotheses for such self-adjoint semigroups, as defined in Lemma 5.6.2, and properties of integral kernels K for which

$$(T_t f)(x) = \int_X K_t(x, y) f(y) \, \mathrm{d}y \tag{7.4}$$

for all $f \in L^2(X, \mathrm{d}x)$ and $t > 0$. Semigroups satisfying the conditions of the following problem are sometimes called Gibbs semigroups.[2]

Lemma 7.2.1 *If T_t is trace class for all $t > 0$ then there exists a complete orthonormal set $\{e_n\}_{n=1}^{\infty}$ in $\mathcal{H}$ and a non-decreasing sequence $\{\lambda_n\}_{n=1}^{\infty}$ of real numbers such that*

$$T_t e_n = \mathrm{e}^{-\lambda_n t} e_n, \qquad \sum_{n=1}^{\infty} \mathrm{e}^{-\lambda_n t} < \infty \tag{7.5}$$

for all n and all $t > 0$. Moreover T_t has a square integrable kernel K_t given for all such t by

$$K_t(x, y) := \sum_{n=1}^{\infty} \mathrm{e}^{-\lambda_n t} e_n(x) \overline{e_n(y)}. \tag{7.6}$$

Proof. Let $\|A\|_2$ denote the Hilbert-Schmidt norm of an operator A on $\mathcal{H}$. Since

$$\|T_t\|_2^2 = \mathrm{tr}[T_{2t}] < \infty$$

for all $t > 0$, the operators T_t are compact. The identities in (7.5) are proved by using Theorem 4.2.23. The L^2 norm convergent expansion (7.6) is a special case of the expansion used in the proof of Theorem 4.2.16. $\square$

Problem 7.2.2 Under the conditions of Lemma 7.2.1 prove that if $s, t > 0$

---

[2] See [Zagrebnov 2003] for a detailed study of such semigroups. Much of the material in this section is adapted from [Davies and Simon 1984].

then

$$\int_X K_s(x, y) K_t(y, z) \, \mathrm{d}y = K_{s+t}(x, z)$$

almost everywhere with respect to x, z. $\quad\square$

The assumption in the following theorem that all the eigenfunctions e_n are continuous and bounded holds if $T_t(L^2(X)) \subseteq L^2(X) \cap C_0(X)$. For elliptic differential operators it may be a consequence of elliptic regularity theorems.

Theorem 7.2.3 *If in addition to the hypotheses of Lemma 7.2.1 each e_n is continuous and*

$$\sum_{n=1}^{\infty} e^{-\lambda_n t} \|e_n\|_{\infty}^2 < \infty$$

for all $t > 0$, then $K_t(x, y)$ is a jointly continuous function of t, x, y for $t > 0$.

Proof. The condition of the theorem implies that the series (7.6) converges uniformly for t in any compact subinterval of $(0, \infty)$. This is enough to establish the joint continuity of K. $\quad\square$

Problem 7.2.4 If a, b, α, β are positive constants such that $\|e_n\|_{\infty} \leq a n^{\alpha}$ and $\lambda_n \geq b n^{\beta}$ for all $n \geq 1$, prove that for every $\gamma > (1 + 2\alpha)/\beta$ there exists c_{γ} such that

$$|K_t(x, y)| \leq c_{\gamma} t^{-\gamma}$$

for all $x, y \in X$ and all $t > 0$. $\quad\square$

Our remaining theorems provide partial converses to the above results.

Theorem 7.2.5[3] *If for each $t > 0$ there is a continuous integral kernel $K_t(\cdot, \cdot)$ for which (7.4) holds and T_t is of trace class for every $t > 0$, then each eigenfunction e_n is continuous and the series (7.6) is locally uniformly convergent. Moreover*

$$|e_n(x)| \leq e^{\lambda_n t/2} b_t(x) \tag{7.7}$$

for all n, x and $t > 0$, where

$$b_t(x) := K_t(x, x)^{1/2}.$$

---

[3] See [Davies and Simon 1984, Lemma 2.1].

Proof. We start with the observation that

$$\|b_t\|_2^2 = \int_X K_t(x, x) \, dx = \mathrm{tr}\left[e^{-Ht}\right] < \infty.$$

Since K_t is the kernel of a non-negative, self-adjoint operator, one has

$$|K_t(x, y)| \leq b_t(x)b_t(y)$$

for all $x, y \in X$. If S is a compact subset of X and $x \in S$ then

$$\begin{aligned}
|e_n(x)| &= \left| e^{\lambda_n t} \int_X K_t(x, y)e_n(y) \, dy \right| \\
&\leq e^{\lambda_n t} \int_X b_t(x)b_t(y)|e_n(y)| \, dy \\
&\leq e^{\lambda_n t} b_t(x)\|b_t\|_2 \\
&\leq c_t e^{\lambda_n t}
\end{aligned}$$

where

$$c_t := \sup_{x \in S} \left\{ b_t(x)\|b_t\|_2 \right\}.$$

The continuity of e_n is proved by applying the dominated convergence theorem to the formula

$$e_n(x) = e^{\lambda_n t} \int_X K_t(x, y)e_n(y) \, dy.$$

The uniform convergence of the series (7.6) for $x, y \in S$ follows from the bound

$$\sum_{n=1}^{\infty} \left| e^{-\lambda_n t} e_n(x)e_n(y) \right| \leq \sum_{n=1}^{\infty} c_{t/3}^2 e^{-\lambda_n t/3} < \infty,$$

which uses (7.8). The estimate (7.7) follows from

$$\begin{aligned}
e^{-\lambda_n t}|e_n(x)|^2 &\leq \sum_{m=1}^{\infty} e^{-\lambda_m t}|e_m(x)|^2 \\
&= K_t(x, x) \\
&= b_t(x)^2. \qquad \square
\end{aligned}$$

Corollary 7.2.6 *If X has finite measure and for each $t > 0$ there is a continuous bounded kernel $K_t(\cdot, \cdot)$ for which (7.4) holds, then the assumptions of Theorem 7.2.3 are valid.*

Proof. One repeats the argument of Theorem 7.2.5 with S replaced by X, and uses the fact that b_t is a bounded function on X. □

Corollary 7.2.7 *Under the assumptions of Theorem 7.2.5 the sets*

$$C_t := \{x : K_t(x, x) = 0\}$$

are closed, of zero measure, and independent of t.

Proof. The locally uniform convergence of the series (7.6) implies that each C_t is equal to

$$C := \{x : e_n(x) = 0 \text{ for all } n\}.$$

This set is closed because each e_n is continuous, and it has zero measure because the set $\{e_n\}_{n=1}^{\infty}$ is complete. □

Example 7.2.8 If we assume that a trace class semigroup T_t acts in $L^2(X, \mathbf{C}^n)$, one might be tempted to repeat all of the above theory with kernels $K_t(x, y)$ which takes values in the set of $n \times n$ matrices. This can be done, but it is also possible to use the scalar theory, replacing X by $X \times \{1, 2, \ldots, n\}$. □

7.3 Semigroups on dual spaces

If Z is a closed linear operator with dense domain $\mathcal{D}$ in a Banach space $\mathcal{B}$, we define its dual operator Z^* with domain $\mathcal{D}^* \subseteq \mathcal{B}^*$ as follows. $\phi \in \mathcal{D}^*$ if and only if the linear functional $f \to \langle Zf, \phi \rangle$ is norm continuous on $\mathcal{D}$; for such ϕ we define $Z^*\phi \in \mathcal{B}^*$ to be the extension of this functional to $\mathcal{B}$. Hence

$$\langle Zf, \phi \rangle = \langle f, Z^*\phi \rangle$$

for all $f \in \mathcal{D}$ and $\phi \in \mathcal{D}^*$.

Lemma 7.3.1[4] *If Z is a closed, densely defined operator, then Z^* is also closed. If $\mathcal{B}$ is reflexive then Z^* is densely defined with $Z^{**} = Z$.*

---

[4] See [Kato 1966A, p. 168].

Proof. Let $\mathcal{L}$ be the closed subspace

$$\{(f, -Zf) : f \in \mathrm{Dom}(Z)\}$$

of $\mathcal{B} \times \mathcal{B}$. The element (ψ, ϕ) of $\mathcal{B}^* \times \mathcal{B}^*$ lies in the annihilator $\mathcal{L}^\circ$ of $\mathcal{L}$ if and only if

$$\langle f, \psi \rangle = \langle Zf, \phi \rangle$$

for all $f \in \mathrm{Dom}(Z)$, or equivalently if and only if $\phi \in \mathrm{Dom}(Z^*)$ and $\psi = Z^*\phi$. Since $\mathcal{L}^\circ$ is a closed subspace, Z^* is a closed operator.

If $\mathcal{B}$ is reflexive and $f \in \mathcal{B}$ satisfies $\langle f, \phi \rangle = 0$ for all $\phi \in \mathrm{Dom}(Z^*)$ then $(0, f) \in \mathcal{L}^{\circ\circ} = \mathcal{L}$, so $Z0 = f$ and $f = 0$. This implies that $\mathrm{Dom}(Z^*)$ is dense in $\mathcal{B}^*$ by the Hahn-Banach theorem. The equality $Z^{**} = Z$ is equivalent to $\mathcal{L}^{\circ\circ} = \mathcal{L}$. $\square$

Problem 7.3.2 Give examples of closed densely defined operators Z on $C[a, b]$ and on $L^1(a, b)$ whose adjoints Z^* are not densely defined. $\square$

Theorem 7.3.3 *If Z is the generator of the one-parameter semigroup T_t on the reflexive Banach space $\mathcal{B}$, then T_t^* is a one-parameter semigroup on $\mathcal{B}^*$ and its generator is Z^*.*

Proof. If $f \in \mathcal{B}$ and $\phi \in \mathcal{B}^*$ then

$$\lim_{t \to 0} \langle f, T_t^* \phi \rangle = \lim_{t \to 0} \langle T_t f, \phi \rangle = \langle f, \phi \rangle.$$

Hence T_t^* is a one-parameter semigroup by Theorem 6.2.6. Let Z' denote the generator of T_t^*. If $f \in \mathrm{Dom}(Z)$ and $\phi \in \mathrm{Dom}(Z')$ then

$$\begin{aligned}
\langle Zf, \phi \rangle &= \lim_{t \to 0} \langle t^{-1}(T_t f - f), \phi \rangle \\
&= \lim_{t \to 0} \langle f, t^{-1}(T_t^* \phi - \phi) \rangle \\
&= \langle f, Z'\phi \rangle.
\end{aligned}$$

Therefore $\phi \in \mathrm{Dom}(Z^*)$ and $Z^*\phi = Z'\phi$. Conversely if $f \in \mathrm{Dom}(Z)$ and $\phi \in \mathrm{Dom}(Z^*)$ then

$$\begin{aligned}
\lim_{t \to 0} \langle f, t^{-1}(T_t^* \phi - \phi) \rangle &= \lim_{t \to 0} \langle t^{-1}(T_t f - f), \phi \rangle \\
&= \langle Zf, \phi \rangle \\
&= \langle f, Z^*\phi \rangle.
\end{aligned}$$

By Theorem 6.2.9 we deduce that $\phi \in \text{Dom}(Z')$ and $Z'\phi = Z^*\phi$. Hence $Z' = Z^*$. $\square$

The next example shows that the situation is less simple if $\mathcal{B}$ is not reflexive.

Example 7.3.4 If $\mathcal{B} := C_0(\mathbf{R})$ then $\mathcal{B}^*$ is the space of bounded, countably additive, complex-valued measures on $\mathbf{R}$. We define the one-parameter group T_t on $\mathcal{B}$ by

$$(T_t f)(x) := f(x - t)$$

for all $t \in \mathbf{R}$. If δ_x is the measure of mass one concentrated at x then $T_t^* \delta_x = \delta_{x-t}$ and

$$\|T_t^* \delta_x - \delta_x\| = 2$$

for all $t \neq 0$. Hence T_t^* cannot be a one-parameter semigroup as defined in Section 6.1. $\square$

If T_t is a one-parameter semigroup on $\mathcal{B}$ satisfying $\|T_t\| \leq M e^{at}$ for all $t \geq 0$ and $\mathcal{B}$ is not reflexive, the following subspaces of $\mathcal{B}^*$ are of importance.

$$\mathcal{L}_1 := \{\phi \in \mathcal{B}^* : t^{-1}(T_t^* \phi - \phi) \text{ converges in norm as } t \to 0\},$$

$$\mathcal{L}_2 := \{\phi \in \mathcal{B}^* : t^{-1}(T_t^* \phi - \phi) \text{ converges weak* as } t \to 0\},$$

$$\mathcal{L}_3 := \{\phi \in \mathcal{B}^* : T_t^* \phi \to \phi \text{ in norm as } t \to 0\}.$$

Theorem 7.3.5[5] *The subspaces are related by*

$$\mathcal{L}_1 \subseteq \mathcal{L}_2 \subseteq \mathcal{L}_3.$$

Moreover $\mathcal{L}_3$ is norm closed and $\mathcal{L}_1$ is weak dense in $\mathcal{B}^*$. The semigroup T_t^* is uniquely determined by its generator Z^* on $\mathcal{L}_1$.*

Proof. The inclusion $\mathcal{L}_1 \subseteq \mathcal{L}_2$ is trivial. If $\phi \in \mathcal{L}_2$ then the uniform boundedness theorem implies that there are positive constants c, δ such that

$$\|t^{-1}(T_t^* \phi - \phi)\| \leq c$$

for all $t \in (0, \delta)$. Hence

$$\lim_{t \to 0} \|T_t^* \phi - \phi\| \leq \lim_{t \to 0} tc = 0$$

and we conclude that $\phi \in \mathcal{L}_3$.

---

[5] See [Dynkin 1965, p. 40].

Let $F: \mathbf{R} \to [0, \infty)$ be a C^∞ function with compact support in $(0, \infty)$, satisfying

$$\int_0^\infty F(t)\, dt = 1.$$

If $\phi \in \mathcal{B}^*$ and $\varepsilon > 0$ then there exists $\phi_\varepsilon \in \mathcal{B}^*$ such that

$$\langle f, \phi_\varepsilon \rangle := \varepsilon^{-1} \int_0^\infty F(\varepsilon^{-1} t) \langle f, T_t^* \phi \rangle\, dt$$

for all $f \in \mathcal{B}$. Moreover the weak* limit of ϕ_ε as $\varepsilon \to 0$ is ϕ, so to prove that $\mathcal{L}_1$ is weak* dense in $\mathcal{B}^*$ we need only show that $\phi_\varepsilon \in \mathcal{L}_1$ for all $\varepsilon > 0$.

If we define $\psi_\varepsilon \in \mathcal{B}^*$ by

$$\langle f, \psi_\varepsilon \rangle = -\varepsilon^{-2} \int_0^\infty F'(\varepsilon^{-1} t) \langle f, T_t^* \phi \rangle\, dt$$

and define $G_{h,\varepsilon}$ by

$$G_{h,\varepsilon}(t) := \frac{F(\varepsilon^{-1}(t-h)) - F(\varepsilon^{-1} t)}{\varepsilon h} + \frac{F'(\varepsilon^{-1} t)}{\varepsilon^2}$$

then

$$|\langle f, h^{-1}(T_h^* \phi_\varepsilon - \phi_\varepsilon) - \psi_\varepsilon \rangle| = \left| \int_0^\infty G_{h,\varepsilon}(t) \langle f, T_t^* \phi \rangle\, dt \right|$$

$$\leq \int_0^\infty M e^{at} |G_{h,\varepsilon}(t)| \, \|f\| \, \|\phi\|\, dt.$$

Using the fact that F has compact support we deduce that

$$\lim_{h \to 0} \|h^{-1}(T_h^* \phi_\varepsilon - \phi_\varepsilon) - \psi_\varepsilon\|$$

$$\leq \lim_{h \to 0} M \|\phi\| \int_0^\infty |G_{h,\varepsilon}(t)| e^{at}\, dt$$

$$= \lim_{h \to 0} M \|\phi\| \int_0^\infty \left| \frac{F(\tau - \varepsilon^{-1} h) - F(\tau)}{h} + \frac{F'(\tau)}{\varepsilon} \right| e^{a\varepsilon\tau}\, d\tau$$

$$= \lim_{\delta \to 0} M \|\phi\| \int_0^\infty \left| \frac{F(\tau - \delta) - F(\tau)}{\delta} + F'(\tau) \right| \frac{e^{a\varepsilon\tau}}{\varepsilon}\, d\tau$$

$$= 0.$$

The fact that $\mathcal{L}_3$ is norm closed is elementary. The restriction S_t of T_t^* to $\mathcal{L}_3$ is jointly continuous by Theorem 6.2.1, and so is uniquely determined by its generator, the domain of which is $\mathcal{L}_1$. Finally each T_t^* is weak* continuous on $\mathcal{B}^*$, and $\mathcal{L}_3$ is weak* dense in $\mathcal{B}^*$, so T_t^* is uniquely determined by S_t. $\square$

Lemma 7.3.6 *We have $\phi \in \mathcal{L}_2$ if and only if*

$$\liminf_{t \to 0} \|t^{-1}(T_t^* \phi - \phi)\| < \infty. \tag{7.8}$$

Proof. If ϕ satisfies (7.8) then by the relative weak* compactness of bounded sets in $\mathcal{B}^*$ (Theorem 1.3.7) there exists a (generalized) sequence $t_n \to 0$ and $\psi \in \mathcal{B}^*$ such that

$$\text{w*-}\lim_{n \to \infty} t_n^{-1}(T_{t_n}^* \phi - \phi) = \psi.$$

By an argument almost identical with that of Theorem 6.2.9 we deduce that $\phi \in \mathcal{L}_2$. □

Corollary 7.3.7 *If Z is the generator of a one-parameter semigroup T_t on a reflexive Banach space $\mathcal{B}$ then*

$$\text{Dom}(Z) = \{f \in \mathcal{B} : \liminf_{t \to 0} t^{-1} \|T_t f - f\| < \infty\}.$$

Proof. By Theorem 6.2.9 the subspaces $\mathcal{L}_1$ and $\mathcal{L}_2$ coincide, so we may apply Lemma 7.3.6. □

Problem 7.3.8 Prove, in Example 7.3.4, that $\mathcal{L}_3$ is the space $L^1(\mathbf{R})$ of all finite bounded measures which are absolutely continuous with respect to Lebesgue measure. Prove also that $\mathcal{L}_1 \neq \mathcal{L}_2$ in this example. □

7.4 Differentiable and analytic vectors

In this section we apply some of the results in Section 1.5 to a one-parameter semigroup T_t acting on a Banach space $\mathcal{B}$. We say that $f \in \mathcal{B}$ is a C^∞ vector for T_t if $t \to T_t f$ is a C^∞ function on $[0, \infty)$.

Theorem 7.4.1 *The set $\mathcal{D}^\infty$ of C^∞ vectors of T_t is given by*

$$\mathcal{D}^\infty = \bigcap_{n=0}^{\infty} \text{Dom}(Z^n) \tag{7.9}$$

and is a dense subspace of $\mathcal{B}$, and a core for Z.

Proof. We start with the proof of (7.9). If $f \in \mathcal{D}^\infty$ then $f \in \mathrm{Dom}(Z)$ and

$$\frac{\mathrm{d}}{\mathrm{d}t} T_t f = T_t(Zf), \tag{7.10}$$

so Zf also lies in $\mathcal{D}^\infty$. A simple induction now establishes that $f \in \mathrm{Dom}(Z^n)$ for all n and

$$\frac{\mathrm{d}^n}{\mathrm{d}t^n} T_t f = T_t(Z^n f). \tag{7.11}$$

Conversely if $f \in \mathrm{Dom}(Z^n)$ for all n then $T_t f$ is differentiable and (7.10) holds. Once again we establish (7.11) inductively and conclude that $f \in \mathcal{D}^\infty$.

We next show that $\mathcal{D}^\infty$ is a dense linear subspace of $\mathcal{B}$. Let $F_n(t)$ be non-negative, real-valued C^∞ functions with compact support in $(0, 1/n)$ which satisfy

$$\int_0^\infty F_n(t)\,\mathrm{d}t = 1$$

for all positive integers n. Given $f \in \mathcal{B}$ we put

$$f_n := \int_0^\infty F_n(t) T_t f \,\mathrm{d}t \tag{7.12}$$

so that $\lim_{n\to\infty} f_n = f$. We also have

$$
\begin{aligned}
Zf_n &= \lim_{h\to 0+} h^{-1}(T_h f_n - f_n) \\
&= \lim_{h\to 0+} h^{-1}\left\{ \int_0^\infty F_n(t) T_{t+h} f \,\mathrm{d}t - \int_0^\infty F_n(t) T_t f \,\mathrm{d}t \right\} \\
&= \lim_{h\to 0+} \int_0^\infty h^{-1}\{F_n(t-h) - F_n(t)\} T_t f \,\mathrm{d}t \\
&= -\int_0^\infty F_n'(t) T_t f \,\mathrm{d}t.
\end{aligned}
$$

Since this is of the same general form as (7.12), we can differentiate repeatedly to conclude that $f_n \in \mathrm{Dom}(Z^m)$ for all m and n. The fact that $\mathcal{D}^\infty$ is a core then follows by applying Theorem 6.1.18. $\square$

Problem 7.4.2 Show that $\mathcal{D}^\infty$ is a complete metric space for the metric

$$\mathrm{d}(f, g) = \|f - g\| + \sum_{n=1}^\infty \frac{\|Z^n f - Z^n g\|}{2^n(1 + \|Z^n f - Z^n g\|)}$$

and that

$$\lim_{r\to\infty} \mathrm{d}(f_r, f) = 0$$

if and only if

$$\lim_{r \to \infty} Z^n f_r = Z^n f$$

for all $n \geq 0$. $\quad\square$

Combining this result with Problems 1.3.1 and 1.3.2 we see that $\mathcal{D}^\infty$ is a Fréchet space.

Problem 7.4.3 Prove that the restriction of T_t to $\mathcal{D}^\infty$ is jointly continuous with respect to the above metric topology. $\quad\square$

If T_t is a one-parameter semigroup on $\mathcal{B}$ we say that $f \in \mathcal{B}$ is an analytic vector (resp. entire vector) for T_t if the function $t \to T_t f$ can be extended to an analytic function on some neighbourhood of $[0, \infty)$ (resp. to an entire function).

Theorem 7.4.4 (*Gel'fand's theorem*)[6] *The set $\mathcal{E}$ of entire vectors of a one-parameter group T_t is a dense linear subspace of $\mathcal{B}$ and a core for the generator Z.*

Proof. Given $f \in \mathcal{B}$ we put

$$f_n := \left(\frac{n}{2\pi}\right)^{1/2} \int_{-\infty}^{\infty} e^{-nt^2/2} T_t f \, dt.$$

By (6.13) this integral is convergent and $\lim_{n \to \infty} f_n = f$, so the density of $\mathcal{E}$ follows provided we can prove that each f_n is an entire vector. To do this we note that if $s \in \mathbf{R}$ then

$$T_s f_n = \left(\frac{n}{2\pi}\right)^{1/2} \int_{-\infty}^{\infty} e^{-n(t-s)^2/2} T_t f \, dt.$$

The entire extension of this function is defined by

$$g(z) := \left(\frac{n}{2\pi}\right)^{1/2} \int_{-\infty}^{\infty} e^{-n(t-z)^2/2} T_t f \, dt$$

the convergence of this integral being a consequence of (6.13) once again. The analyticity of $g(z)$ is an exercise in differentiating under the integral sign. Since the set of entire vectors is dense and invariant under the action of T_t, it is a core for Z by Theorem 6.1.18. $\quad\square$

---

[6] See [Nelson 1959, Goodman 1971] for extensions of this theorem to Lie groups.

Problem 7.4.5 Show that if f is an entire vector for $T_t = e^{Zt}$ then

$$e^{Zt} f = \sum_{n=0}^{\infty} t^n Z^n f / n!$$

for all real t. This uses Theorem 1.4.12. $\square$

Problem 7.4.6 Use Fourier transforms to give another description of the set of all entire vectors for the one-parameter semigroup T_t acting on $L^2(\mathbf{R})$ according to the formula

$$(T_t f)(x) := (4\pi t)^{-1/2} \int_{-\infty}^{\infty} e^{-(x-y)^2/4t} f(y) \, dy.$$ $\square$

The following example shows that non-zero analytic vectors need not exist for one-parameter semigroup s.

Example 7.4.7 Let $\mathcal{B}$ be the space of continuous functions on $[0, \infty)$ which vanish at 0 and ∞, with the sup norm. Define the one-parameter semigroup T_t on $\mathcal{B}$ by

$$(T_t f)(x) := \begin{cases} f(x-t) & \text{if } 0 \leq t \leq x, \\ 0 & \text{if } t > x. \end{cases}$$

If $f \in \mathcal{B}$, $a > 0$ and $f(a) \neq 0$, we define $\phi \in \mathcal{B}^*$ by putting $\phi(g) = g(a)$ for all $g \in \mathcal{B}$. Then $t \to \langle T_t f, \phi \rangle$ is a non-zero continuous function which vanishes for $t > a$. Therefore it cannot be extended analytically to any neighbourhood of the real axis, and f is not an analytic vector. $\square$

Problem 7.4.8 Let $\mathcal{D}$ be the set of analytic vectors for the one-parameter group U_t acting on $L^p(\mathbf{R})$ according to the formula

$$(U_t f)(x) := f(x-t).$$

Prove that if $1 \leq p < \infty$ then

$$\mathcal{D} \cap L_c^p(\mathbf{R}) = \{0\}$$

where the notation L_c^p was defined on page 5. $\square$

Theorem 7.4.9 *Let T_t be a one-parameter group of isometries on $\mathcal{B}$ with generator Z, and let $\mathcal{D}$ be a dense linear subspace of $\mathcal{B}$ contained in $\mathrm{Dom}(Z)$. If $f \in \mathcal{D}$ implies that $Zf \in \mathcal{D}$ and that*

$$\sum_{n=0}^{\infty} \|Z^n f\| \alpha^n / n! < \infty$$

for some $\alpha > 0$, which may depend upon f, then $\mathcal{D}$ is a core for Z.

Proof. If $\overline{\mathcal{D}}$ denotes the closure of $\mathcal{D}$ in $\mathcal{D}^\infty$ for the metric d of Problem 7.4.2, then the hypothesis $Z\mathcal{D} \subseteq \mathcal{D}$ implies that $Z\overline{\mathcal{D}} \subseteq \overline{\mathcal{D}}$. Given $\alpha > 0$, define

$$\mathcal{D}_\alpha := \left\{ f \in \overline{\mathcal{D}} : \sum_{n=0}^\infty \|Z^n f\| \beta^n / n! < \infty \text{ for all } \beta < \alpha \right\}. \tag{7.13}$$

Then

$$\mathcal{D} \subseteq \bigcup_\alpha \mathcal{D}_\alpha \subseteq \overline{\mathcal{D}} \subseteq \text{Dom}(Z),$$

so to prove that $\mathcal{D}$ is a core it is sufficient to prove that $\cup_\alpha \mathcal{D}_\alpha$ is a core. We will actually show that if $|t| < \alpha$ then $T_t(\mathcal{D}_\alpha) \subseteq \mathcal{D}_\alpha$.

If $f \in \mathcal{D}_\alpha$ and $|t| < \alpha$ then the series

$$f(t) := \sum_{n=0}^\infty (Z^n f) t^n / n!$$

converges in the d-metric, so $f(t) \in \overline{\mathcal{D}}$. Moreover $f(t)$ is differentiable with $f'(t) = Zf(t)$, so $f(t) = T_t f$ by Theorem 6.1.16. Hence $T_t f \in \overline{\mathcal{D}}$ for all $|t| < \alpha$. Moreover

$$\|Z^n T_t f\| = \|T_t Z^n f\| = \|Z^n f\|.$$

This implies that $T_t f \in \mathcal{D}_\alpha$ for all $|t| < \alpha$. Therefore $\cup_\alpha \mathcal{D}_\alpha$ is invariant under T_t for all $t \in \mathbf{R}$, by the group law for T_t. We may then apply Theorem 6.1.18 to $\cup_\alpha \mathcal{D}_\alpha$, and conclude that $\mathcal{D}$ is a core for Z. $\square$

Problem 7.4.10 Let $T_t := e^{Zt}$ act on $L^2(\mathbf{R})$ according to the formula $(T_t f)(x) := e^{ixt} f(x)$. Prove that the set of functions of the form

$$f(x) := e^{-x^2} \sum_{r=0}^n a_r x^r,$$

where n depends upon f, is a core for Z. $\square$

7.5 Subordinated semigroups

Generalizing the ideas of Section 6.3, we define a convolution semigroup on $\mathbf{R}$ to be a family of probability measures μ_t on $\mathbf{R}$ parametrized by $t > 0$ which satisfies

(i) $\mu_s * \mu_t = \mu_{s+t}$ for all $s, t > 0$,
(ii) $\lim_{t \to 0} \mu_t(-\delta, \delta) = 1$ for all $\delta > 0$,

where the convolution $\mu * \nu$ of two measures is defined by

$$\int_{\mathbf{R}} f(x)(\mu * \nu)(\mathrm{d}x) := \int_{\mathbf{R}} \int_{\mathbf{R}} f(x+y)\mu(\mathrm{d}x)\nu(\mathrm{d}y).$$

Alternatively

$$(\mu * \nu)(E) := (\mu \times \nu)(\{(x, y) : x + y \in E\})$$

for all Borel sets E in $\mathbf{R}$. This definition is consistent with the definition of the convolution of two functions whenever the measures μ and ν have L^1 densities with respect to Lebesgue measure.

We say that μ_t is a convolution semigroup on $\mathbf{R}^+$ if $\mathrm{supp}(\mu_t) \subseteq [0, \infty)$ for all $t \geq 0$.

Convolution semigroups are closely connected with random walks on the real line. The following provides an example which is not covered by the definition in Section 6.3.

Problem 7.5.1 Prove that the Poisson distribution, i.e. the measures

$$\mu_t := \mathrm{e}^{-t} \sum_{n=0}^{\infty} \frac{t^n}{n!} \delta_n$$

define a convolution semigroup in the above sense, where $\delta_n(f) := f(n)$. $\quad\square$

Problem 7.5.2 Prove that if μ_t is a convolution semigroup and $\int_{\mathbf{R}} |x| \mu_t(\mathrm{d}x)$ is finite for any $t > 0$ then it is finite for all $t > 0$, and there is a 'drift' coefficient a such that

$$\int_{\mathbf{R}} x \mu_t(\mathrm{d}x) = at$$

for all $t > 0$. $\quad\square$

Theorem 7.5.3 *If $U_t := \mathrm{e}^{At}$ is a one-parameter group of isometries on a Banach space $\mathcal{B}$ and μ_t is a convolution semigroup on $\mathbf{R}$ then the formula*

$$T_t f := \int_{\mathbf{R}} U_s f \, \mu_t(\mathrm{d}s) \tag{7.14}$$

defines a one-parameter contraction semigroup T_t on $\mathcal{B}$. If μ_t is the Gaussian measure

$$\mu_t(\mathrm{d}x) := \{4\pi t\}^{-1/2} \mathrm{e}^{-|x|^2/4t} \, \mathrm{d}x$$

considered in Example 6.3.5, then the generator Z of T_t is given by

$$Z := A^2. \tag{7.15}$$

Proof. The bound $\|T_t\| \leq 1$ for all $t \geq 0$ follows directly from its definition. The condition (i) above implies that $t \to T_t$ is a semigroup, while (ii) implies that it is strongly continuous at $t = 0$.

We now assume that μ_t is defined as above and use the identity

$$\int_0^\infty (4\pi t)^{-1/2} e^{-s^2/4t - \lambda t} \, dt = \frac{1}{2} \lambda^{-1/2} e^{-|s|\lambda^{1/2}}$$

to compute the resolvent of Z. If $f \in \mathcal{B}$ and $\lambda > 0$ then

$$(\lambda - Z)^{-1} f = \frac{1}{2} \int_{-\infty}^\infty \lambda^{-1/2} e^{-|s|\lambda^{1/2}} e^{As} f \, ds$$

$$= \frac{1}{2} \lambda^{-1/2} (\lambda^{1/2} - A)^{-1} f + \frac{1}{2} \lambda^{-1/2} (\lambda^{1/2} + A)^{-1} f$$

$$= (\lambda - A^2)^{-1} f,$$

from which (7.15) follows. $\quad\square$

Problem 7.5.4 Prove the analogue of Theorem 7.5.3 when U_t is a one-parameter contraction semigroup on $\mathcal{B}$ and μ_t is a convolution semigroup on $\mathbf{R}^+$. Show that the Gamma distribution, i.e. the measures

$$\mu_t(dx) := \begin{cases} \Gamma(t)^{-1} x^{t-1} e^{-x} \, dx & \text{if } x > 0, \\ 0 & \text{otherwise}, \end{cases}$$

provide an example of such a convolution semigroup. $\quad\square$

Some insight into the form of the generator Z of Theorem 7.5.3 can be obtained as follows. If $\text{Re}(z) \leq 0$ and

$$\hat{\mu}_t(z) := \int_0^\infty e^{zx} \mu_t(dx)$$

then one may show that

$$\hat{\mu}_{t+s}(z) = \hat{\mu}_t(z)\hat{\mu}_s(z)$$

and

$$\lim_{t \to 0} \hat{\mu}_t(z) = 1.$$

It follows that there exists a function f defined on $\{z : \text{Re}(z) \leq 0\}$ such that

$$\hat{\mu}_t(z) = e^{tf(z)}.$$

It is customary to write $Z := f(A)$. The following provides a partial justification of this.

Problem 7.5.5 Show that if $A\phi = z\phi$ for some $\phi \in \text{Dom}(A)$ then $\text{Re}(z) = 0$ in the context of Theorem 7.5.3, and $\text{Re}(z) \leq 0$ in the context of Problem 7.5.4. In both cases prove that $\phi \in \text{Dom}(Z)$ and $Z\phi = f(z)\phi$. □

Example 7.5.6 If we define

$$f_t(x) := \frac{t}{2\sqrt{\pi x^3}} e^{-t^2/4x}$$

for all $x, t > 0$ then f_t is a probability density on $(0, \infty)$, and

$$\int_0^\infty f_t(x) e^{-zx} \, dx = e^{-z^{1/2}t}$$

for all $t > 0$ and $\text{Re}(z) \geq 0$. One may use this formula to show that

$$\mu_t(dx) := f_t(x) \, dx$$

is a convolution semigroup on $\mathbf{R}^+$. If $(-H)$ is the generator of a one-parameter contraction semigroup on $\mathcal{B}$ then the generator Z of the semigroup

$$T_t\phi = \int_0^\infty f_t(s) e^{-Hs} \phi \, ds$$

is given by $Z := -H^{1/2}$ according to the above convention.

Similar procedures may be used to define other fractional powers of H. Unfortunately the densities cannot be written down as explicitly as in the above case.[7] □

Example 7.5.7 If $H := (-\Delta)^{1/2}$ acting in $L^2(\mathbf{R}^N)$ then Example 7.5.6 implies that the Cauchy operators e^{-Ht} have the continuous integral kernels

$$K(t, x, y) := \int_{s=0}^\infty \frac{t}{2\sqrt{\pi s^3}} e^{-t^2/4s} \frac{1}{(4\pi s)^{N/2}} e^{-|x-y|^2/4s} \, ds.$$

We deduce that

$$0 < K(t, x, y) \leq ct^{-N}$$

for some $c > 0$, all $t > 0$ and all $x, y \in \mathbf{R}^N$. Indeed $K(t, x, x) = ct^{-N}$ for all $t > 0$ and $x \in \mathbf{R}^N$.

More generally if e^{-At} is a one-parameter semigroup on $L^p(X, dx)$ and

$$\|e^{-At}f\|_q \leq ct^{-\alpha/2}\|f\|_p$$

for all $f \in L^p(X, dx)$ and $t > 0$, where $1 \leq p < q \leq \infty$, then

$$\|e^{-A^{1/2}t}f\|_q \leq ct^{-\alpha}\|f\|_p$$

---

[7] See [Dunford and Schwartz 1966, p. 641] or [Yosida 1965, p. 259].

for all $f \in L^p(X, dx)$ and $t > 0.$[8]   □

All of the ideas in this section can be generalized by replacing **R** by any other locally compact group, if one has a representation of that group by isometries on some Banach space. They may also be put into a Banach algebra setting.

Problem 7.5.8 Let $f_t \in \mathcal{A}$ for all $t > 0$, where $\mathcal{A}$ is a Banach algebra. Suppose that $f_s f_t = f_{s+t}$ for all $s, t > 0$ and f_t is an approximate identity in the sense that

$$\lim_{t \to 0} \|f_t g - g\| = \lim_{t \to 0} \|g f_t - g\| = 0$$

for all $g \in \mathcal{A}$. Prove that the formulae

$$S_t g := g f_t, \qquad T_t g := f_t g$$

define two one-parameter semigroup s on $\mathcal{A}$, which commute with each other. Prove also that $t \to f_t$ is norm continuous for all $t > 0$. Deduce that both semigroups are norm continuous for $t > 0$.   □

---

[8] Bounds of the above type are closely related to Sobolev embedding theorems. See [Davies 1989] for the self-adjoint case and [Ouhabaz 2005] for the more recent non-self-adjoint theory.

8
Resolvents and generators

8.1 Elementary properties of resolvents

In the last two chapters we introduced the notion of a one-parameter semigroup T_t and defined its infinitesimal generator Z. In this chapter we complete the triangle drawn on page 168 by studying the resolvent family of Z. We use the resolvents to describe the relationship between the spectrum of Z and of the semigroup operators T_t, and also to determine which unbounded operators Z are in fact the generators of one-parameter semigroups.

Resolvent operators are particularly useful in the analysis of Sturm-Liouville operators, because in that case one can write down their integral kernels in closed form; a very simple example is written down in Example 5.6.10. In higher dimensions this is not the case, and there is the added problem that their integral kernels are singular on the diagonal. Nevertheless resolvent operators play an important theoretical role, particularly in the analysis of perturbations.

We start by studying general unbounded operators. Just as in the bounded case, the spectrum and resolvent play key roles. In some ways the resolvent operators are more fundamental, because the spectrum of an unbounded operator can be empty. We will see that the resolvent norms provide important information about many non-self-adjoint operators. This is made explicit in the study of pseudospectra in Section 9.1, but the same issue arises throughout the book.

We review some earlier definitions. Let Z be a closed linear operator with domain $\mathrm{Dom}(Z)$ and range $\mathrm{Ran}(Z)$ in a Banach space $\mathcal{B}$. A subspace $\mathcal{D}$ of $\mathrm{Dom}(Z)$ is called a core if Z is the closure of its restriction to $\mathcal{D}$. We define the resolvent set of Z to be the set of all $z \in \mathbf{C}$ such that $zI - Z$ is one-one with range equal to $\mathcal{B}$. The resolvent operator

$$R_z := (zI - Z)^{-1}$$

is bounded for such z by the closed graph theorem. The spectrum $\text{Spec}(Z)$ is by definition the complement of the resolvent set. The reader should note that we only assume that $\text{Dom}(Z)$ is dense below when this is relevant to the proof of the theorem.

We emphasize that the spectrum of an operator depends critically upon its precise domain. If one takes too small or too large a domain, the spectrum of the operator may equal $\mathbf{C}$. For a differential operator acting on functions which are defined on a region $U \subseteq \mathbf{R}^N$, boundary conditions are incorporated as conditions on the domain of the operator. Altering the boundary conditions usually changes the spectrum radically. There are only a few operators for which the spectrum is easy to determine, the following being one.

Theorem 8.1.1 *Every constant coefficient differential operator L of order n defined on the Schwartz space $S \subseteq L^2(\mathbf{R}^N)$ is closable and the domain of its closure $\overline{L}$ contains $W^{n,2}(\mathbf{R}^N)$. Moreover*

$$\text{Spec}(\overline{L}) = \overline{\{\sigma(\xi) : \xi \in \mathbf{R}^N\}}$$

where σ is the symbol of the operator, as defined in (3.7).

Proof. We first note that σ is a polynomial of degree n. Example 6.1.9 implies that L is closable. If $M := \mathcal{F}L\mathcal{F}^{-1}$ then $\text{Dom}(M) = S$ and

$$(Mg)(\xi) := \sigma(\xi)g(\xi)$$

for all $g \in S$. The closure $\overline{M}$ of M has domain

$$\{g \in L^2(\mathbf{R}^N) : \sigma g \in L^2(\mathbf{R}^N)\}.$$

This contains $W^{n,2}(\mathbf{R}^N)$ by virtue of the bound

$$|\sigma(\xi)| \leq c(1 + |\xi|^2)^{n/2}.$$

The identity

$$\overline{L} := \mathcal{F}^{-1}\overline{M}\mathcal{F}$$

implies that $\text{Spec}(\overline{L}) = \text{Spec}(\overline{M})$; the latter equals $\overline{\{\sigma(\xi) : \xi \in \mathbf{R}^N\}}$ by Problem 6.1.5. $\square$

Problem 8.1.2 Let Z be a closed operator acting in $\mathcal{B}$ and let $z \notin \text{Spec}(Z)$. Prove that the following three conditions on a subspace $\mathcal{D}$ of $\text{Dom}(Z)$ are equivalent.

(i) $\mathcal{D}$ is a core for Z;
(ii) $(zI - Z)\mathcal{D}$ is dense in $\mathcal{B}$;

(iii) $\mathcal{D}$ is dense in $\mathrm{Dom}(Z)$ for the norm

$$\|\|f\|\| := \|f\| + \|Zf\|. \qquad \qquad \square$$

Lemma 8.1.3 *The resolvent set U of a closed operator Z is open, and the resolvent operator R_z is an analytic function of z on U. If $z, w \in U$ then*

$$R_z - R_w = (w - z) R_z R_w. \qquad (8.1)$$

Proof. Let $a \in U$ and let $c := \|R_a\|^{-1}$. Then for $|z - a| < c$ the series

$$S_z := \sum_{n=0}^{\infty} (-1)^n (z - a)^n R_a^{n+1}$$

converges in norm and defines a bounded operator S_z. If $f \in \mathrm{Dom}(Z)$ then

$$S_z(zI - Z)f = \sum_{n=0}^{\infty} (-1)^n (z - a)^n R_a^{n+1}(aI - Z + zI - aI)f$$

$$= \sum_{n=0}^{\infty} (-1)^n (z - a)^n R_a^n f + (z - a) S_z f$$

$$= f - (z - a) S_z f + (z - a) S_z f$$

$$= f. \qquad (8.2)$$

On the other hand if $f \in \mathcal{B}$ and

$$g_m := \sum_{n=0}^{m} (-1)^n (z - a)^n R_a^{n+1} f$$

then $g_m \in \mathrm{Dom}(Z)$ and $\lim_{m \to \infty} g_m = S_z f$. Moreover

$$\lim_{m \to \infty} Z g_m = \lim_{m \to \infty} \sum_{n=0}^{m} (-1)^n (z - a)^n Z R_a^{n+1} f$$

$$= \lim_{m \to \infty} \sum_{n=0}^{m} (-1)^n (z - a)^n (aR_a - I) R_a^n f$$

$$= a S_z f + (z - a) S_z f - f$$

$$= z S_z f - f.$$

Since Z is closed we deduce that $S_z f$ lies in $\mathrm{Dom}(Z)$ and

$$Z S_z f = z S_z f - f$$

or equivalently

$$(zI - Z) S_z f = f.$$

Combining this with (8.2) we deduce that $\{z : |z - a| < c\} \subseteq U$ and that $S_z = R_z$ for all such z. This establishes the analyticity of R_z as a function of z and also the formula

$$R_z = \sum_{n=0}^{\infty} (-1)^n (z - a)^n R_a^{n+1} \tag{8.3}$$

for all z such that

$$|z - a| < \|R_a\|^{-1}.$$

If $f \in \mathcal{B}$ and $z, w \in U$ then

$$(zI - Z)(R_z - R_w - (w - z)R_z R_w)f$$
$$= f - (zI - wI + wI - Z)R_w f - (w - z)R_w f$$
$$= 0.$$

Since $(zI - Z)$ is one-one we deduce that

$$\{R_z - R_w - (w - z)R_z R_w\}f = 0$$

for all $f \in \mathcal{B}$. □

Corollary 8.1.4 *We have*

$$\|R_z\| \geq \text{dist}\{z, \text{Spec}(Z)\}^{-1}$$

for all $z \notin \text{Spec}(Z)$. Hence the resolvent operator cannot be analytically continued outside the resolvent set of Z.

We will see in Section 9.1 that there is no corresponding upper bound: the resolvent norm may be extremely large for z which are far from the spectrum of Z. Such phenomena are investigated under the name of pseudospectra.

   In spite of the above corollary, the 'matrix elements' $\langle R_z f, \phi \rangle$ of the resolvent operator may well be analytically continued into $\text{Spec}(Z)$ for a large class of $f \in \mathcal{B}$ and $\phi \in \mathcal{B}^*$. This is important for the theory of resonances and for quantum scattering theory. The following theorem refers to the dual of an unbounded closed operator, as defined in Section 7.3.

Theorem 8.1.5 *Let Z be a closed, densely defined operator acting in the reflexive Banach space $\mathcal{B}$. Then*

$$\text{Spec}(Z) = \text{Spec}(Z^*)$$

and

$$\{(\lambda I - Z)^{-1}\}^* = (\lambda I - Z^*)^{-1}$$

for all $\lambda \notin \text{Spec}(Z)$.

Proof. Since $(\lambda I - Z^*)$ is the dual of $(\lambda I - Z)$, it is sufficient to treat the case $\lambda = 0$. If $0 \notin \text{Spec}(Z)$ let A be the (bounded) inverse of Z. If $f \in \text{Dom}(Z)$ and $\phi \in \mathcal{B}^*$ then

$$\langle Zf, A^*\phi \rangle = \langle AZf, \phi \rangle = \langle f, \phi \rangle,$$

so $A^*\phi \in \text{Dom}(Z^*)$ and $Z^*A^*\phi = \phi$. This implies that Z^* has range equal to $\mathcal{B}^*$. If $f \in \mathcal{B}$ and $\phi \in \text{Dom}(Z^*)$ then

$$\langle f, A^*Z^*\phi \rangle = \langle Af, Z^*\phi \rangle = \langle ZAf, \phi \rangle = \langle f, \phi \rangle.$$

Therefore

$$A^*Z^*\phi = \phi$$

and Z^* has kernel $\{0\}$. We conclude that $0 \notin \text{Spec}(Z^*)$ and that A^* is the inverse of Z^*.

The converse argument uses Lemma 7.3.1. $\square$

If R_z is any family of bounded operators defined for all z in a subset U of $\mathbf{C}$ and satisfying

$$R_z - R_w = (w - z)R_z R_w \tag{8.4}$$

for all $z, w \in U$, we call R_z a pseudo-resolvent. Note that (8.4) implies that $R_z R_w = R_w R_z$ for all $z, w \in U$.

Problem 8.1.6 Show that the kernel $\text{Ker}(R_z)$ and range $\text{Ran}(R_z)$ of a pseudo-resolvent family are both independent of z. Moreover R_z is the resolvent of a closed operator Z such that $\text{Spec}(Z) \cap U = \emptyset$ if and only if $\text{Ker}(R_z) = \{0\}$. $\square$

Theorem 8.1.7 *If R_z is a pseudo-resolvent defined for all z satisfying $a < z < \infty$ and satisfying*

$$\|R_z\| \leq M(z - a)^{-1} \tag{8.5}$$

for all such z, then R_z is the resolvent of a closed, densely defined operator Z if and only if the range of R_z is dense in $\mathcal{B}$.

Proof. If Z exists and has dense domain then $\text{Ran}(R_z)$ is dense because it equals $\text{Dom}(Z)$.

Conversely if the pseudo-resolvent R_z satisfies the stated conditions and $f := R_w g$ then

$$\lim_{z \to +\infty} z R_z f = \lim_{z \to +\infty} z R_z R_w g$$
$$= \lim_{z \to +\infty} \frac{z}{w - z} (R_z g - R_w g)$$
$$= \lim_{z \to +\infty} \frac{z}{w - z} (R_z g - f)$$
$$= f.$$

Since R_w has dense range and the family of operators $z R_z$ is uniformly bounded we conclude using Problem 1.3.10 that

$$\lim_{z \to \infty} z R_z h = h$$

for all $h \in \mathcal{B}$, so $\text{Ker}(R_z) = 0$ for all z by Problem 8.1.6. We deduce, again by Problem 8.1.6, that R_z is the resolvent of an operator Z, the domain of which equals $\text{Ran}(R_z)$ and hence is dense in $\mathcal{B}$. $\square$

Problem 8.1.8 By applying Theorem 8.1.7 to R_z^* show that if R_z is a pseudo-resolvent on the reflexive Banach space $\mathcal{B}$ and satisfies (8.5) for all $a < z < \infty$, and if $\text{Ker}(R_z) = 0$, then R_z is the resolvent of a closed, densely defined operator Z. $\square$

We next describe the relationship between the spectrum of an operator Z and that of its resolvent.

Lemma 8.1.9 *If Z is a closed, unbounded operator acting in $\mathcal{B}$ and $z \notin$ Spec(Z) then*

$$\text{Spec}(R_z) = \{0\} \cup \{(z - \lambda)^{-1} : \lambda \in \text{Spec}(Z)\}.$$

Proof. Since $\text{Ran}(R_z)$ equals $\text{Dom}(Z)$ and Z is not bounded, we see that 0 lies in $\text{Spec}(R_z)$. If $w \notin \text{Spec}(Z)$ then the operator

$$S := (z - w)(zI - Z)R_w$$

is bounded and commutes with R_z. Moreover

$$\{(z - w)^{-1} I - R_z\} S = (zI - Z)R_w - (z - w)R_w$$
$$= (wI - Z)R_w$$
$$= I.$$

Therefore $(z - w)^{-1} \notin \mathrm{Spec}(R_z)$.

Conversely suppose $(z - w)^{-1} \notin \mathrm{Spec}(R_z)$, and put

$$T := \{I - (z - w)R_z\}^{-1}R_z.$$

If $f \in \mathcal{B}$ then

$$
\begin{aligned}
(wI - Z)Tf &= \{zI - Z + (w - z)I\}R_z\{I - (z - w)R_z\}^{-1}f \\
&= \{I - (z - w)R_z\}\{I - (z - w)R_z\}^{-1}f \\
&= f.
\end{aligned}
$$

On the other hand, if $f \in \mathrm{Dom}(Z)$ then

$$
\begin{aligned}
T(wI - Z)f &= \{I - (z - w)R_z\}^{-1}R_z\{zI - Z + (w - z)I\}f \\
&= \{I - (z - w)R_z\}^{-1}\{I - (z - w)R_z\}f \\
&= f.
\end{aligned}
$$

Hence $w \notin \mathrm{Spec}(Z)$. $\square$

Problem 8.1.10 Show that if R_z is compact for any $z \notin \mathrm{Spec}(Z)$ then it is compact for all such z, and that $\mathrm{Spec}(Z)$ consists of at most a countable number of eigenvalues of finite multiplicity, which diverge to infinity. $\square$

Problem 8.1.11 Show that if Z is a closed operator and $f_n \in \mathrm{Dom}(Z)$ satisfy $\|f_n\| = 1$ and

$$\lim_{n \to \infty} \|Zf_n - \lambda f_n\| = 0$$

then $\lambda \in \mathrm{Spec}(Z)$. Using Corollary 8.1.4, show conversely that if λ lies in the topological boundary of $\mathrm{Spec}(Z)$ then such a sequence f_n must exist. $\square$

We next describe a simple example in which the computation of the spectrum is far from elementary. Let $\{\mathcal{H}_n\}_{n=1}^{\infty}$ be a sequence of Hilbert spaces and put

$$\mathcal{H} := \sum_{n=1}^{\infty} \oplus \mathcal{H}_n.$$

We regard each $\mathcal{H}_n$ as a subspace of $\mathcal{H}$ in an obvious way. Let A_n be a bounded linear operator of $\mathcal{H}_n$ for each n and define the operator A acting in $\mathcal{H}$ by

$$(Af)_n := A_n f_n. \tag{8.6}$$

We do not assume that A is bounded, and choose its maximal natural domain, namely

$$\text{Dom}(A) := \{f \in \mathcal{H} : \sum_{n=1}^{\infty} \|A_n f_n\|^2 < \infty\}.$$

Our next theorem establishes that the spectrum of A need not be the limit of its truncations to the subspaces $\sum_{n=1}^{N} \oplus \mathcal{H}_n$ as $N \to \infty$. Less contrived examples of this phenomenon are described in Examples 9.3.19 and 9.3.20.

Theorem 8.1.12 *The spectrum of the operator A defined by (8.6) is given by*

$$\text{Spec}(A) = B \cup \bigcup_{n=1}^{\infty} \text{Spec}(A_n),$$

where B is the set of $z \in \mathbf{C}$ for which the sequence $n \to \|(zI_n - A_n)^{-1}\|$ is unbounded.

Proof. If $z \in \text{Spec}(A_n)$ for some n then Lemma 1.2.13 implies that either there exists a sequence of unit vectors e_r in $\mathcal{H}_n$ such that $\|A_n e_r - z e_r\| \to 0$ as $r \to \infty$, or there exists a sequence of unit vectors e_r in $\mathcal{H}_n$ such that $\|A_n^* e_r - \bar{z} e_r\| \to 0$ as $r \to \infty$. In both cases we conclude that $z \in \text{Spec}(A)$.

Now suppose that $z \in B$. There must exist a subsequence $n(r)$ and unit vectors $e_{n(r)} \in \mathcal{H}_{n(r)}$ such that

$$\lim_{r \to \infty} \|(zI_{n(r)} - A_{n(r)})^{-1} e_{n(r)}\| = +\infty.$$

Putting

$$f_{n(r)} := (zI_{n(r)} - A_{n(r)})^{-1} e_{n(r)} / \|(zI_{n(r)} - A_{n(r)})^{-1} e_{n(r)}\|$$

we see that $\|f_{n(r)}\| = 1$ and $\|A f_{n(r)} - z f_{n(r)}\| \to 0$ as $r \to \infty$. Therefore $z \in \text{Spec}(A)$.

Finally suppose that $n \to \|(zI_n - A_n)^{-1}\|$ is a bounded sequence and put

$$(Sf)_n := (zI_n - A_n)^{-1} f_n$$

for all $f \in \mathcal{H}$. Clearly S is a bounded operator. The verification that

$$(zI - A)S = S(zI - A) = I$$

is routine, and proves that $z \notin \text{Spec}(A)$. $\square$

Problem 8.1.13 In the context of (8.6), assume that $\mathcal{H}_n := \mathbf{C}^n$ and that A_n is the standard $n \times n$ Jordan matrix. Prove that although $\mathrm{Spec}(A_n) = \{0\}$ for each n, one has

$$\mathrm{Spec}(A) = \{z : |z| \leq 1\}. \qquad \square$$

8.2 Resolvents and semigroups

In this section we describe the relationship between a one-parameter semigroup T_t and the resolvent family R_z associated with its generator Z. Our first theorem provides the key formula (8.8) enabling one to pass directly from the semigroup to the resolvent operators. Most of the subsequent analysis is based on this formula or developments of it.

Theorem 8.2.1 *Let Z be the generator of a one-parameter semigroup T_t on $\mathcal{B}$ that satisfies*

$$\|T_t\| \leq M e^{at} \tag{8.7}$$

for all $t \geq 0$. Then the spectrum of Z is contained in $\{z : \mathrm{Re}(z) \leq a\}$. If $\mathrm{Re}(z) > a$ then

$$R_z f = \int_0^\infty e^{-zt} T_t f \, dt \tag{8.8}$$

for all $f \in \mathcal{B}$. Moreover

$$\|R_z\| \leq M(\mathrm{Re}(z) - a)^{-1} \tag{8.9}$$

for all such z.

Proof. In this proof we define R_z by the RHS of (8.8), which is norm convergent for all z such that $\mathrm{Re}(z) > a$ and all $f \in \mathcal{B}$, and prove that it coincides with the resolvent. If $g := R_z f$ then

$$\lim_{h \to 0} h^{-1}(T_h g - g)$$

$$= \lim_{h \to 0} \left\{ h^{-1} \int_0^\infty e^{-zt} T_{t+h} f \, dt - h^{-1} \int_0^\infty e^{-zt} T_t f \, dt \right\}$$

$$= \lim_{h \to 0} \left\{ h^{-1} \int_h^\infty e^{-z(t-h)} T_t f \, dt - h^{-1} \int_0^\infty e^{-zt} T_t f \, dt \right\}$$

$$= \lim_{h \to 0} \left\{ -h^{-1} e^{zh} \int_0^h e^{-zt} T_t f \, dt + h^{-1}(e^{zh} - 1) \int_0^\infty e^{-zt} T_t f \, dt \right\}$$

$$= -f + zg.$$

Therefore $g \in \mathrm{Dom}(Z)$ and $(zI - Z)g = f$. This establishes that $\mathrm{Ran}(R_z) \subseteq \mathrm{Dom}(Z)$ and that

$$(zI - Z)R_z = I.$$

This identity implies that $\mathrm{Ker}(R_z) = \{0\}$. If $g \in \mathrm{Dom}(Z)$ and $g' := R_z(zI - Z)g$ then $(zI - Z)(g - g') = 0$. If $f := g - g'$ is non-zero then an application of Theorem 6.1.16 implies that $T_t f = \mathrm{e}^{zt} f$ for all $t \geq 0$. Therefore $\|T_t\| \geq \mathrm{e}^{\mathrm{Re}(z)t}$ for all $t \geq 0$, which contradicts (8.7). Therefore $f = 0$, $g' = g$, and $\mathrm{Dom}(Z) \subseteq \mathrm{Ran}(R_z)$. We finally conclude that

$$R_z = (zI - Z)^{-1}.$$

The estimate (8.9) follows directly from (8.8). $\square$

Corollary 8.2.2 *If T_t is compact for all $t > 0$ then R_z is compact for all $z \notin \mathrm{Spec}(Z)$.*

Proof. First suppose that $\mathrm{Re}(z) > a$. For all $n \geq 1$ the integrand in

$$R_{n,z} := \int_{1/n}^{n} \mathrm{e}^{-zt} T_t \, \mathrm{d}t$$

is norm continuous by Theorem 7.1.4. Therefore the integral is norm convergent, and $R_{n,z}$ is compact by Theorem 4.2.2. Since $R_{n,z}$ converges in norm to R_z as $n \to \infty$, the latter operator is also compact. The compactness of R_w for all other $w \notin \mathrm{Spec}(Z)$ follows by using the resolvent formula (8.1). $\square$

Theorem 8.2.3 *Let $T_{p,t}$ be consistent one-parameter semigroup s acting on $L^p(X, \mathrm{d}x)$ for all $p \in [p_0, p_1]$, where $1 \leq p_0 < \infty$ and $1 \leq p_1 \leq \infty$. Suppose also that $T_{p_0,t}$ is compact for all $t > 0$. Then the same holds for all $p \in [p_0, p_1)$. The generators Z_p have compact resolvents for all $p \in [p_0, p_1)$ and $\mathrm{Spec}(Z_p)$ is independent of p for such p.*

Proof. The compactness of $T_{p,t}$ for $p \in [p_0, p_1)$ and $t > 0$ is proved by using Theorem 4.2.14. The compactness of the resolvents for such p uses Corollary 8.2.2. The fact that the spectrum of $R(\lambda, Z_p)$ does not depend on p for any sufficiently large λ was proved in Theorem 4.2.15. We deduce that the spectrum of Z_p does not depend on p by using Lemma 8.1.9. $\square$

The conclusion of the above theorem should not be taken for granted. Although the L^p spectrum of an operator frequently does not depend on p, there are important examples in which it does. See Theorem 12.6.2 and the comments there.

Problem 8.2.4 Show by induction that if $\mathrm{Re}(z) > a$ and $n \geq 1$ then

$$(R_z)^n f = \int_0^\infty \frac{t^{n-1}}{(n-1)!} e^{-zt} T_t f \, dt$$

for all $f \in \mathcal{B}$, and use this formula to prove that

$$\lim_{n \to \infty} \left\{ \frac{n}{s} R_{n/s} \right\}^n f = T_s f \tag{8.10}$$

for all $f \in \mathcal{B}$ and all $s > 0$. □

We mention in passing that the resolvent need not exist for one-parameter groups defined on general topological vector spaces. It is entirely possible that $\mathrm{Spec}(Z) = \mathbf{C}$, so that the resolvent set is empty.

Example 8.2.5 Let $\mathcal{B}$ be the space of all continuous functions on $\mathbf{R}$, with the topology of locally uniform convergence. If

$$(T_t f)(x) := f(x + t)$$

then T_t is a one-parameter group on $\mathcal{B}$. The generator of T_t is

$$(Zf)(x) := f'(x),$$

its domain being the space of all continuously differentiable functions on $\mathbf{R}$. If $\lambda \in \mathbf{C}$ and $f(x) := e^{\lambda x}$, then $f \in \mathrm{Dom}(Z)$ and $Zf = \lambda f$. Hence $\mathrm{Spec}(Z) = \mathbf{C}$. □

Before starting the spectral analysis of unbounded operators, we point out that it may sometimes provide little useful information.

Example 8.2.6 If $\mathcal{B} = L^2(0, c)$ and $t \geq 0$, define T_t on $\mathcal{B}$ by

$$(T_t f)(x) := \begin{cases} f(x + t) & \text{if } 0 \leq t + x < c, \\ 0 & \text{otherwise.} \end{cases}$$

Then T_t is a one-parameter semigroup with $\|T_t\| = 1$ if $0 \leq t < c$ and $T_t = 0$ if $t \geq c$. Theorem 8.2.1 is applicable for every $a \in \mathbf{R}$ provided M is chosen appropriately, so the spectrum of Z is empty. Indeed the resolvent may be written in the form

$$R_z f = \int_0^c e^{-zt} T_t f \, dt$$

for all $z \in \mathbf{C}$, and is an entire function of z. □

In spite of this example, spectral analysis is often of great interest.

If Z is the generator of a one-parameter semigroup T_t acting on $\mathcal{B}$, the relationship between the spectrum of Z and of T_t is not simple. Theorem 8.2.9 states that one cannot replace the inclusion in (8.11) below by an equality unless one imposes further conditions.

Theorem 8.2.7 *If $t \geq 0$ then*

$$\mathrm{Spec}(T_t) \supseteq \{\mathrm{e}^{\lambda t} : \lambda \in \mathrm{Spec}(Z)\}. \tag{8.11}$$

Proof. Let $\mathcal{L}(\mathcal{B})$ be the algebra of all bounded operators on $\mathcal{B}$ and let $\mathcal{A}$ be a maximal abelian subalgebra containing T_t for all $t \geq 0$ and R_z for all $z \notin \mathrm{Spec}(Z)$. Such an algebra exists by Zorn's lemma. If $X \in \mathcal{A}$ is invertible then $X^{-1} \in \mathcal{A}$ by maximality. Therefore the spectrum of X as an operator coincides with its spectrum as an element of $\mathcal{A}$. The latter equals $\{\hat{X}(m) : m \in M\}$ where M is the maximal ideal space of $\mathcal{A}$ and $\hat{X} \in C(M)$ is the Gel'fand transform of X.[1]

Let $\lambda, z \in \mathbf{C}$ be fixed numbers with $\lambda \in \mathrm{Spec}(Z)$ and $z \notin \mathrm{Spec}(Z)$. Then $(z - \lambda)^{-1} \in \mathrm{Spec}(R_z)$ so there exists $m \in M$ such that

$$\hat{R}_z(m) = (z - \lambda)^{-1} \neq 0.$$

If $\gamma(t) := \hat{T}_t(m)$ then $\gamma(0) = 1$ and

$$\gamma(s)\gamma(t) = \gamma(s + t)$$

for all $s, t \geq 0$. A simple calculation using Lemma 6.1.12 shows that $T_t R_z$ depends norm continuously on t for $0 \leq t < \infty$, so $\gamma(t)(z - \lambda)^{-1}$ also depends continuously on t for $0 \leq t < \infty$. By applying the theory of one-parameter semigroup s to the semigroup $t \to \gamma(t)$ acting on $\mathbf{C}$ we deduce that $\gamma(t) = \mathrm{e}^{\beta t}$ for some $\beta \in \mathbf{C}$. If $\mathrm{Re}(w) > a$ then

$$\int_0^\infty \mathrm{e}^{-wt} T_t R_z \, \mathrm{d}t = R_w R_z$$

as a norm convergent integral, so

$$\int_0^\infty \mathrm{e}^{\beta t}(z - \lambda)^{-1} \mathrm{e}^{-wt} \, \mathrm{d}t = (z - \lambda)^{-1} \hat{R}_w(m).$$

Therefore

$$\hat{R}_w(m) = (w - \beta)^{-1}.$$

[1] We follow the approach of [Hille and Phillips 1957, Chap. 16], which uses Gel'fand's representation of a commutative Banach algebra as an algebra of continuous functions. See [Rudin 1973] for an exposition of this subject.

On the other hand

$$\hat{R}_z(m) - \hat{R}_w(m) = (w - z)\hat{R}_z(m)\hat{R}_w(m)$$

so

$$\hat{R}_w(m) = (z - \lambda)^{-1}\{1 + (w - z)(z - \lambda)^{-1}\}^{-1}$$
$$= (w - \lambda)^{-1}.$$

This implies $\lambda = \beta$. We have now shown that

$$\hat{T}_t(m) = e^{\lambda t}$$

so $e^{\lambda t} \in \mathrm{Spec}(T_t)$. □

The following problem is used in the proof of the next theorem.

Problem 8.2.8 Let $t > 0$ and let $S := \{e^{int} : n = 1, 2, 3, \dots\}$. Then either S is a finite subgroup of $\{z : |z| = 1\}$ or it is dense in $\{z : |z| = 1\}$. In both cases there exists an increasing sequence $n(r)$ of positive integers such that $\lim_{r \to \infty} e^{in(r)t} = 1$. □

Theorem 8.2.9 *(Zabczyk)*[2] *There exists a one-parameter group $T_t := e^{Zt}$ acting on a Hilbert space $\mathcal{H}$ such that $\mathrm{Spec}(Z) \subseteq i\mathbf{R}$ and*

$$\|T_t\| = e^{|t|} \in \mathrm{Spec}(T_t) \tag{8.12}$$

for all $t \in \mathbf{R}$.

Proof. We write

$$\mathcal{H} := \sum_{n=1}^{\infty} \oplus \mathcal{H}_n$$

where $\mathcal{H}_n := \mathbf{C}^n$ and each subspace $\mathcal{H}_n$ is invariant under the group T_t to be defined. The restriction Z_n of the generator Z to $\mathcal{H}_n$ is defined to be $Z_n := J_n + in I_n$, where J_n is the standard $n \times n$ Jordan matrix and I_n is the identity operator on $\mathcal{H}_n$. The identity $\|J_n\| = 1$ implies that

$$\|e^{Z_n t}\| = \|e^{J_n t}\| \leq e^{|t|}$$

for all $t \in \mathbf{R}$ and all $n \geq 1$. By combining these groups acting on their individual spaces we obtain a one-parameter group T_t on $\mathcal{H}$ such that

$$\|T_t\| \leq e^{|t|} \tag{8.13}$$

for all $t \in \mathbf{R}$.

[2] See [Zabczyk 1975].

Our next task is to prove that $e^{|t|} \in \text{Spec}(T_t)$ for all $t > 0$; the corresponding result for $t < 0$ has a similar proof. This implies that (8.13) is actually an equality.

If $v_n \in \mathcal{H}_n$ is the unit vector $v_n := n^{-1/2}(1, 1, \ldots, 1)'$ then

$$e^{J_n t} v_n = n^{-1/2}(s_{n-1}, s_{n-2}, \ldots, s_0)'$$

where $s_m := \sum_{r=0}^{m} t^r/r!$. Therefore

$$\lim_{n \to \infty} \left\| e^{J_n t} v_n - e^t v_n \right\|^2$$

$$= \lim_{n \to \infty} n^{-1}\{(s_{n-1} - e^t)^2 + (s_{n-2} - e^t)^2 + \cdots + (s_0 - e^t)^2\}$$

$$= 0.$$

For each $t > 0$ we use Problem 8.2.8 to select an increasing sequence $n(r)$ such that $\lim_{r \to \infty} e^{in(r)t} = 1$. This implies that

$$\lim_{r \to \infty} \left\| e^{Z_{n(r)} t} v_{n(r)} - e^t v_{n(r)} \right\| = 0.$$

Therefore $e^t \in \text{Spec}(T_t)$ for all $t > 0$.

We next identify the generator Z precisely. Let $\mathcal{D}$ denote the set of all sequences $f \in \mathcal{H}$ with finite support. It is immediate that $\mathcal{D}$ is invariant under T_t and that $(Zf)_n = Z_n f_n$ for all $f \in \mathcal{D}$ and all n. Theorem 6.1.18 implies that $\mathcal{D}$ is a core for Z. The fact that Z is closed implies that

$$\text{Dom}(Z) = \{f \in \mathcal{H} : \sum_{n=1}^{\infty} \left\| Z_n f_n \right\|^2 < \infty\}.$$

It is immediate from its definition that in is an eigenvalue of Z for every positive integer n, and we will prove that every $z \notin i\mathbf{N}$ lies in the resolvent set of Z. By Theorem 8.1.12 it is sufficient to prove that $n \to \left\| (zI_n - Z_n)^{-1} \right\|$ is a bounded sequence for all such z. Since $\|J_n\| = 1$ for all n, we have

$$\limsup_{n \to \infty} \left\| (zI_n - Z_n)^{-1} \right\| \leq \limsup_{n \to \infty} (|z - in| - \|J_n\|)^{-1}$$

$$= \limsup_{n \to \infty} (|z - in| - 1)^{-1}$$

$$= 0.$$

This completes the proof. $\quad \square$

The following spectral mapping theorem is a corollary of Theorem 8.2.11 below.

Theorem 8.2.10 *Let* $\tilde{T}_t := T_t R(a, Z)$ *on* $\mathcal{B}$ *where* $a \notin \mathrm{Spec}(Z)$. *If* Z *is unbounded, one has*

$$\mathrm{Spec}(\tilde{T}_t) = \{0\} \cup \{e^{\lambda t}(a - \lambda)^{-1} : \lambda \in \mathrm{Spec}(Z)\}$$

for all $t > 0$ *and* $a > \omega_0$, *where* ω_0 *is defined as on page* **??**.

Proof. We normalize the problem by putting $Z' := Z - \gamma I$ where $\omega_0 < \gamma < a$, $a' := a - \gamma$ and $T_t' := e^{-\gamma t} T_t$, so that

$$T_t R(a, Z) = e^{\gamma t} T_t' R(a', Z').$$

The semigroup T_t' is uniformly bounded since $\omega_0' := \omega_0 - \gamma < 0$. Moreover

$$T_t' R(a', Z') = \int_0^\infty f(s) T_s \, ds$$

where

$$f(s) = \begin{cases} 0 & \text{if } 0 \le s < t, \\ e^{-a'(s-t)} & \text{if } s \ge t. \end{cases}$$

Since $f \in L^1(0, \infty)$, the stated result is implied by our next, more general, theorem. $\square$

Theorem 8.2.11[3] *Let* T_t *be a uniformly bounded one-parameter semigroup acting on* $\mathcal{B}$, *with an unbounded generator* Z. *Let* $f \in L^1(0, \infty)$ *and put*

$$X_f := \int_0^\infty f(t) T_t \, dt,$$

where the integral converges strongly in $\mathcal{L}(\mathcal{B})$. *Put*

$$\hat{f}(z) := \int_0^\infty f(t) e^{zt} \, dt$$

for all z *satisfying* $\mathrm{Re}(z) \le 0$. *Then*

$$\mathrm{Spec}(X_f) = \{0\} \cup \{\hat{f}(\lambda) : \lambda \in \mathrm{Spec}(Z)\}.$$

Proof. We follow the method of Theorem 8.2.7. Let $\mathcal{A}$ be a maximal abelian subalgebra of $\mathcal{L}(\mathcal{B})$ which contains T_t for all $t \ge 0$ and the resolvent operators R_a for all $a \notin \mathrm{Spec}(Z)$. Let M denote the maximal ideal space of $\mathcal{A}$ and let $\hat{\ }$ denote the Gel'fand transform. Then $\mathcal{A}$ is closed under the taking of inverses and strong operator limits. Hence

$$\mathrm{Spec}(D) = \{\hat{D}(m) : m \in M\}$$

---

[3] Once again we follow the ideas in [Hille and Phillips 1957, Chap. 16]. See also [Greiner and Muller 1993].

for all $D \in \mathcal{A}$.

If $a, b \notin \text{Spec}(Z)$ then the identity

$$\hat{R}_a(m) - \hat{R}_b(m) = (b-a)\hat{R}_a(m)\hat{R}_b(m) \qquad (8.14)$$

implies that the closed set

$$N := \{m \in M : \hat{R}_a(m) = 0\}$$

is independent of the choice of a. Since Z is unbounded N must be non-empty.
If $m \in M \backslash N$ then

$$\hat{R}_a(m) \in \text{Spec}(R_a) \backslash \{0\} = (a - \lambda_m)^{-1}$$

for some $\lambda_m \in \text{Spec}(Z)$. A second application of (8.14) implies that λ_m
does not depend upon a. The definition of the topology of M implies that
$\lambda : M \backslash N \to \text{Spec}(Z)$ is continuous.

Let $\mathcal{P}$ denote the set of all functions $f : [0, \infty) \to \mathbf{C}$ of the form

$$f(t) = \sum_{r=1}^{n} \alpha_r e^{-\beta_r t}$$

where $\text{Re}(\beta_r) > 0$ for all r. For such a function

$$X_f = \sum_{r=1}^{n} \alpha_r R_{\beta_r}.$$

Therefore

$$\text{Spec}(X_f) = \left\{ \hat{X}_f(m) : m \in M \right\}$$

$$= \{0\} \cup \left\{ \sum_{r=1}^{n} \alpha_r \hat{R}_{\beta_r}(m) : m \in M \backslash N \right\}$$

$$= \{0\} \cup \left\{ \sum_{r=1}^{n} \alpha_r (\beta_r - \lambda_m)^{-1} : m \in M \backslash N \right\}$$

$$= \{0\} \cup \left\{ \sum_{r=1}^{n} \alpha_r (\beta_r - \lambda)^{-1} : \lambda \in \text{Spec}(Z) \right\}$$

$$= \{0\} \cup \left\{ \int_0^{\infty} f(t) e^{\lambda t} \, dt : \lambda \in \text{Spec}(Z) \right\}$$

$$= \{0\} \cup \left\{ \hat{f}(\lambda) : \lambda \in \text{Spec}(Z) \right\}.$$

Finally let f be a general element of $L^1(0, \infty)$. There exists a sequence $f_n \in \mathcal{P}$ which converges in L^1 norm to f, and this implies that X_{f_n} converges in norm to X_f, and that $\hat{f}_n$ converges uniformly to $\hat{f}$. Hence $X_f \in \mathcal{A}$ and

$$\operatorname{Spec}(X_f) = \lim_{n \to \infty} \operatorname{Spec}(X_{f_n})$$

$$= \{0\} \cup \lim_{n \to \infty} \left\{ \hat{f}_n(\lambda) : \lambda \in \operatorname{Spec}(Z) \right\}$$

$$= \{0\} \cup \left\{ \hat{f}(\lambda) : \lambda \in \operatorname{Spec}(Z) \right\}.$$

In this final step we used the fact that $\{0\} \cup \{\hat{f}(\lambda) : \lambda \in \operatorname{Spec}(Z)\}$ is a closed set. This is because $\operatorname{Spec}(Z)$ is a closed subset of $\{z \in \mathbf{C} : \operatorname{Re}(z) \leq 0\}$, and $\hat{f}(z) \to 0$ as $|z| \to \infty$ within this set. $\quad\square$

If further conditions are imposed on the semigroup T_t, a converse to Theorem 8.2.7 can be proved.

Theorem 8.2.12 *Suppose that T_t is a one-parameter semigroup and a norm continuous function of t for $a \leq t < \infty$. Then a non-zero number c lies in $\operatorname{Spec}(T_t)$ if and only if $c = e^{\lambda t}$ for some $\lambda \in \operatorname{Spec}(Z)$.*

Proof. We continue with the notation of the proof of Theorem 8.2.7. If $0 \neq c \in \operatorname{Spec}(T_b)$ then there exists $m \in M$ such that

$$\gamma(t) := \hat{T}_t(m)$$

satisfies $\gamma(b) = c$. Moreover $\gamma(t + s) = \gamma(t)\gamma(s)$ for all $s, t \in (0, \infty)$ and $\gamma(\cdot)$ is continuous on $[a, \infty)$. This implies that $\gamma(t) = e^{\lambda t}$ for some $\lambda \in \mathbf{C}$ and all $t > 0$. Now in the equality

$$T_a(zI - Z)^{-1} = \int_0^\infty e^{-zt} T_{a+t} \, dt$$

the integral is norm convergent provided $\operatorname{Re}(z)$ is large enough. Therefore

$$e^{\lambda a} \{(zI - Z)^{-1}\}^{\hat{}}(m) = \int_0^\infty e^{-zt + (a+t)\lambda} \, dt$$

$$= e^{\lambda a}(z - \lambda)^{-1}.$$

We conclude that $(z - \lambda)^{-1} \in \operatorname{Spec}(R_z)$, so $\lambda \in \operatorname{Spec}(Z)$ by Lemma 8.1.9. $\quad\square$

If we assume that T_t is compact for some $t > 0$ even stronger conclusions can be drawn.

Theorem 8.2.13 *If T_t is a one-parameter semigroup and T_a is compact for some $a > 0$, then for all $\alpha > 0$ there exists a direct sum decomposition $\mathcal{B} = \mathcal{B}_0 \oplus \mathcal{B}_1$ with the following properties. Both $\mathcal{B}_0$ and $\mathcal{B}_1$ are invariant under T_t, $\mathcal{B}_0$ is finite-dimensional, and the restriction S_t of T_t to $\mathcal{B}_1$ satisfies*

$$\|S_t\| = o(\mathrm{e}^{-\alpha t})$$

as $t \to \infty$. The spectrum of Z consists of at most a countable, discrete set of eigenvalues, each of finite multiplicity, and if $\mathcal{B}$ is infinite-dimensional

$$\mathrm{Spec}(T_t) = \{0\} \cup \mathrm{e}^{t\,\mathrm{Spec}(Z)}.$$

If Z has an infinite number of distinct eigenvalues $\{\lambda_n\}_{n=1}^{\infty}$ then $\lim_{n\to\infty} \mathrm{Re}(\lambda_n) = -\infty$.

Proof. The last statement is an immediate consequence of Theorems 7.1.4 and 8.2.12. Since T_a is a compact operator there is a spectral decomposition $\mathcal{B} := \mathcal{B}_0 \oplus \mathcal{B}_1$ of T_a such that $\mathcal{B}_0$ is finite-dimensional and

$$\mathrm{Rad}(S_a) < \mathrm{e}^{-a\alpha}.$$

Since T_t commutes with T_a, $\mathcal{B}_0$ and $\mathcal{B}_1$ are both invariant with respect to T_t for all $t \geq 0$. By Theorem 10.1.6 below the spectral radius of S_t equals $\mathrm{e}^{-t\gamma}$ for some γ and all $t \geq 0$. Clearly $\gamma > \alpha$ and, again by Theorem 10.1.6,

$$\lim_{t\to\infty} \mathrm{e}^{\alpha t}\|S_t\| = 0.$$

This implies by Theorem 8.2.1 that the spectrum of the generator Y of S_t lies in $\{z : \mathrm{Re}(z) \leq -\alpha\}$. Since $\mathcal{B}_0$ is finite-dimensional, the spectrum of Z is the union of $\mathrm{Spec}(Y)$ with a finite set of eigenvalues of finite multiplicity. The stated properties of $\mathrm{Spec}(Z)$ now follow from the fact that $\alpha > 0$ is arbitrary. $\square$

Problem 8.2.14 Write down the simpler proof of the above theorem when T_t is assumed to be compact for all $t > 0$. $\square$

8.3 Classification of generators

When I wrote *One-Parameter Semigroups*, I referred to the first theorem below as the central result in the study of one-parameter semigroup s. Twenty-five years later, I am not so sure. The theorem gives a complete solution of the problem posed, but the criterion obtained is very difficult to apply. The reason is that one is usually given the operator Z rather than its resolvent operators, and hypotheses involving all powers of the resolvents are rarely easy to verify. In terms of their range of applications, Theorem 8.3.2 and Theorem 8.3.4 are far more useful, as well as being historically earlier.

Many of the results below have analogues for real Banach spaces, in spite of the fact that a real operator may have a complex spectrum. This is sometimes important in applications.

Theorem 8.3.1 *(Feller, Miyadera, Phillips) A closed, densely defined operator Z acting in the Banach space $\mathcal{B}$ is the generator of a one-parameter semigroup T_t satisfying*

$$\|T_t\| \leq Me^{at} \tag{8.15}$$

for all $t \geq 0$ if and only if

$$\mathrm{Spec}(Z) \subseteq \{z : \mathrm{Re}(z) \leq a\} \tag{8.16}$$

and

$$\|(\lambda I - Z)^{-m}\| \leq M(\lambda - a)^{-m} \tag{8.17}$$

for all $\lambda > a$ and all $m \geq 1$.

Proof. The proof that (8.15) implies (8.16) was given in Theorem 8.2.1. The same theorem implies that if $\lambda > a$ and $m \geq 1$ then

$$\|(\lambda I - Z)^{-m}f\| = \left\| \int_0^\infty \cdots \int_0^\infty T_{t_1 + \cdots + t_m} e^{-\lambda(t_1 + \cdots + t_m)} f \, dt_1 \ldots dt_m \right\|$$
$$\leq \int_0^\infty \cdots \int_0^\infty Me^{-(\lambda - a)(t_1 + \cdots + t_m)} \|f\| \, dt_1 \ldots dt_m$$
$$= M(\lambda - a)^{-m} \|f\|$$

for all $f \in \mathcal{B}$. This implies (8.17).

The converse is much harder since we have to construct the semigroup T_t. The idea is to approximate Z by bounded operators Z_λ and show that the semigroups

$$T_t^\lambda := e^{Z_\lambda t}$$

converge as $\lambda \to +\infty$ to a semigroup T_t whose generator is Z.

If $\lambda > a$ we define the bounded operator Z_λ by

$$Z_\lambda := \lambda Z R_\lambda.$$

We first show that

$$\lim_{\lambda \to \infty} \|Z R_\lambda f\| = 0 \tag{8.18}$$

for all $f \in \mathcal{B}$. Because

$$\|ZR_\lambda\| = \|1 - \lambda R_\lambda\|$$

$$\leq 1 + \frac{M\lambda}{\lambda - a}$$

is bounded as $\lambda \to \infty$, it is sufficient by Problem 1.3.10 to prove this for f in a dense subset of $\mathcal{B}$. If $f \in \mathrm{Dom}(Z)$ then

$$\|ZR_\lambda f\| \leq \|R_\lambda\| \|Zf\|$$

$$\leq \frac{M\|Zf\|}{\lambda - a}$$

and this converges to 0 as $\lambda \to \infty$.

We next show that

$$\lim_{\lambda \to \infty} Z_\lambda f = Zf \tag{8.19}$$

for all $f \in \mathrm{Dom}(Z)$. For any such f and any $b > a$ there exists $g \in \mathcal{B}$ such that $f = R_b g$. Hence

$$\lim_{\lambda \to \infty} \|Z_\lambda f - Zf\|$$

$$= \lim_{\lambda \to \infty} \|\lambda Z R_\lambda R_b g - Z R_b g\|$$

$$= \lim_{\lambda \to \infty} \|\lambda Z \{R_\lambda - R_b\}(b - \lambda)^{-1} g - Z R_b g\|$$

$$= \lim_{\lambda \to \infty} \left\| \left(\frac{\lambda}{\lambda - b} - 1 \right) Z R_b g - \frac{\lambda}{\lambda - b} Z R_\lambda g \right\|$$

$$\leq \lim_{\lambda \to \infty} \frac{b}{\lambda - b} \|Z R_b g\| + \lim_{\lambda \to \infty} \frac{\lambda}{\lambda - b} \|Z R_\lambda g\|$$

$$= 0$$

by (8.18).

If $T_t^\lambda := \mathrm{e}^{Z_\lambda t}$ then (8.17) implies

$$\|T_t^\lambda\| = \left\| \mathrm{e}^{\lambda\{-I + \lambda R_\lambda\}t} \right\|$$

$$\leq \mathrm{e}^{-\lambda t} \sum_{n=0}^{\infty} t^n \lambda^{2n} \|R_\lambda^n\| / n!$$

$$\leq \mathrm{e}^{-\lambda t} M \mathrm{e}^{t\lambda^2/(\lambda - a)}$$

$$= M \mathrm{e}^{t a \lambda/(\lambda - a)}$$

$$\leq M \mathrm{e}^{2at} \tag{8.20}$$

provided $\lambda \geq 2a$. Moreover

$$\limsup_{\lambda \to \infty} \|T_t^\lambda\| \leq Me^{at}. \tag{8.21}$$

We next show that if $f \in \mathcal{B}$ then $T_t^\lambda f$ converges as $\lambda \to \infty$ uniformly for t in bounded intervals. By (8.20) it is sufficient to prove this when f lies in the dense set $\text{Dom}(Z)$. For such f

$$\left\| \frac{d}{ds}\{T_{t-s}^\lambda T_s^\mu f\} \right\| = \|T_{t-s}^\lambda(-Z_\lambda + Z_\mu)T_s^\mu f\|$$

$$= \|T_{t-s}^\lambda T_s^\mu(-Z_\lambda + Z_\mu)f\|$$

$$\leq M^2 e^{2at}\|(-Z_\lambda + Z_\mu)f\|.$$

Integrating with respect to s for $0 \leq s \leq t$ we obtain

$$\|T_t^\lambda f - T_t^\mu f\| \leq tM^2 e^{2at}\|(-Z_\lambda + Z_\mu)f\|,$$

which converges to zero as $\lambda, \mu \to \infty$, uniformly for t in bounded intervals, by (8.19).

This result enables us to define the bounded operators T_t by

$$T_t f := \lim_{\lambda \to \infty} T_t^\lambda f.$$

It is an immediate consequence of (8.21) and the semigroup properties of T_t^λ that $T_0 = 1$, $\|T_t\| \leq Me^{at}$ for all $t \geq 0$ and $T_s T_t = T_{s+t}$ for all $s, t \geq 0$. The uniformity of the convergence for t in bounded intervals implies that $T_t f$ is jointly continuous in t and f, and so is a one-parameter semigroup.

Our final task is to verify that the generator B of T_t coincides with Z. We start from the equation

$$T_t^\lambda f - f = \int_0^t T_x^\lambda Z_\lambda f \, dx \tag{8.22}$$

valid for all $f \in \mathcal{B}$ by Lemma 6.1.12. If $f \in \text{Dom}(Z)$ then we let $\lambda \to \infty$ in (8.22) and use (8.19) to obtain

$$T_t f - f = \int_0^t T_x Z f \, dx.$$

Dividing by t and letting $t \to 0$ we see that $f \in \text{Dom}(B)$ and $Bf = Zf$. Hence B is an extension of Z. Since both $(xI - Z)$ and $(xI - B)$ are one-one with range equal to $\mathcal{B}$ for all $x > a$ it follows that $Z = B$. $\quad \Box$

A one-parameter contraction semigroup is defined as a one-parameter semigroup such that $\|T_t\| \leq 1$ for all $t \geq 0$.

Theorem 8.3.2 *(Hille-Yosida)*[4] *If Z is a closed, densely defined operator acting in the Banach space $\mathcal{B}$ then the following are equivalent.*

(i) $\mathrm{Spec}(Z) \cap \{z : 0 < z < \infty\} = \emptyset$ *and* $\|(\lambda I - Z)^{-1}\| \leq \lambda^{-1}$ *for all* $\lambda > 0$.
(ii) *Z is the generator of a one-parameter contraction semigroup.*
(iii) $\mathrm{Spec}(Z) \subseteq \{z : \mathrm{Re}(z) \leq 0\}$ *and*

$$\|(zI - Z)^{-1}\| \leq (\mathrm{Re}(z))^{-1} \tag{8.23}$$

for all z such that $\mathrm{Re}(z) > 0$.

Proof. (i)$\Rightarrow$(ii) follows from the case $M = 1$ and $a = 0$ of (a slight modification of) Theorem 8.3.1. (ii)$\Rightarrow$(iii) is a special case of Theorem 8.2.1, and (iii)$\Rightarrow$(i) is elementary. $\square$

Problem 8.3.3 Show that in Theorem 8.3.1 and Theorem 8.3.2 it is sufficient to assume (8.17) and (8.23) for a sequence of real λ_n such that $\lim_{n \to \infty} \lambda_n = +\infty$. $\square$

We next reformulate Theorem 8.3.2 directly in terms of the operator Z. If an operator Z acts in $\mathcal{B}$ with domain $\mathcal{D}$, we let $\mathcal{E}$ denote the set of pairs $(f, \phi) \in \mathcal{B} \times \mathcal{B}^*$ such that $f \in \mathcal{D}$, $\|f\| = 1$, $\|\phi\| = 1$ and $\langle f, \phi \rangle = 1$. Note that for each $f \in \mathcal{D}$ a suitable ϕ exists by the Hahn-Banach theorem; if $\mathcal{B}$ is a Hilbert space then ϕ is unique, but this is not true in general. We say that Z is dissipative if $\mathrm{Re}(\langle Zf, \phi \rangle) \leq 0$ for all $(f, \phi) \in \mathcal{E}$. If Z is an operator with domain $\mathcal{D}$ in a Hilbert space $\mathcal{H}$ then Z is dissipative if and only if $\mathrm{Re}(\langle Zf, f \rangle) \leq 0$ for all $f \in \mathcal{D}$.

Theorem 8.3.4 *(Lumer-Phillips)*[5] *Given an operator Z with dense domain $\mathcal{D}$ in a Banach space $\mathcal{B}$, the following are equivalent.*

(i) *Z is dissipative and the range of* $(\lambda I - Z)$ *equals $\mathcal{B}$ for all* $\lambda > 0$.
(ii) *Z is the generator of a one-parameter contraction semigroup.*

Proof. (i)$\Rightarrow$(ii) If $(f, \phi) \in \mathcal{E}$ then

$$\|(\lambda I - Z)f\| \geq |\langle (\lambda I - Z)f, \phi \rangle|$$
$$= |\lambda - \langle Zf, \phi \rangle|$$
$$\geq \lambda$$
$$= \lambda \|f\|.$$

[4] See [Hille 1948, Hille 1952, Yosida 1948].
[5] See [Lumer and Phillips 1961].

Therefore the operator $(\lambda I - Z)$ is one-one with range equal to $\mathcal{B}$, and

$$\|(\lambda I - Z)^{-1}\| \le \lambda^{-1}.$$

We may now apply Theorem 8.3.2.

(ii)$\Rightarrow$(i) If $(f, \phi) \in \mathcal{E}$ then

$$\begin{aligned}
\mathrm{Re}\langle Zf, \phi\rangle &= \mathrm{Re}\lim_{h\to 0} h^{-1}\langle T_h f - f, \phi\rangle \\
&= \mathrm{Re}\lim_{h\to 0} h^{-1}\{\langle T_h f, \phi\rangle - 1\} \\
&\le \lim_{h\to 0} h^{-1}\{\|T_h\|\,\|f\|\,\|\phi\| - 1\} \\
&\le 0.
\end{aligned}$$

The identity $\mathrm{Ran}(\lambda I - Z) = \mathcal{B}$ follows from $\lambda \notin \mathrm{Spec}(Z)$ and was proved in Theorem 8.2.1. $\quad\square$

The following modification of Theorem 8.3.4 is easier to verify because it uses a weaker notion of dissipativity and only requires one to consider a single value of λ.

Theorem 8.3.5 *(Lumer-Phillips) Let Z be a closable operator with dense domain $\mathcal{D}$ in a Banach space $\mathcal{B}$, and suppose that the range of $(\lambda I - Z)$ is dense for some $\lambda > 0$. Suppose also that for all $f \in \mathcal{D}$ there exists $\phi \in \mathcal{B}^*$ such that $\|\phi\| = 1$, $\langle f, \phi\rangle = \|f\|$ and $\mathrm{Re}(\langle Zf, \phi\rangle) \le 0$. Then Z is dissipative and the closure $\overline{Z}$ of Z is the generator of a one-parameter contraction semigroup.*

Proof. The weaker dissipativity condition still implies that

$$\|(\mu I - Z)f\| \ge \mu\|f\|$$

for all $\mu > 0$ and all $f \in \mathcal{D}$. Therefore

$$\|(\mu I - \overline{Z})f\| \ge \mu\|f\|$$

for all $\mu > 0$ and all $f \in \mathrm{Dom}(\overline{Z})$. This implies that $(\lambda I - \overline{Z})$ has range equal to $\mathcal{B}$ and that

$$\|(\lambda I - \overline{Z})^{-1}\| \le \lambda^{-1}.$$

Corollary 8.1.4 now implies that

$$\mathrm{Spec}(\overline{Z}) \cap \{z : |z - \lambda| < \lambda\} = \emptyset$$

and that $(\mu I - \overline{Z})$ has range equal to $\mathcal{B}$ for all μ such that $0 < \mu < 2\lambda$. Replacing λ by $3\lambda/2$ in the above argument, it follows by induction that

$$\mathrm{Spec}(\overline{Z}) \subseteq \{z : \mathrm{Re}(z) \le 0\}.$$

The proof is now completed by applying Theorem 8.3.2. $\quad\square$

Problem 8.3.6 If f_n, $n \geq 1$, are eigenvectors of a closable dissipative operator Z and $\mathcal{L} := \lim\{f_n : n \geq 1\}$ is dense in $\mathcal{B}$, show that the closure of Z is the generator of a one-parameter contraction semigroup. $\square$

The condition of dissipativity is also useful in relation to the Cauchy problem.

Theorem 8.3.7 *Let Z be a dissipative operator with dense domain $\mathcal{D}$ in a Banach space $\mathcal{B}$, and suppose that for all $f_0 \in \mathcal{D}$ the evolution equation*

$$f_t' = Zf_t \tag{8.24}$$

is soluble with solution $f_t \in \mathcal{D}$ for all $t > 0$. Then this solution is unique, Z is closable, and its closure is the generator of a one-parameter contraction semigroup.

Proof. Let f_t be a solution of (8.24) and let $\phi_t \in \mathcal{B}^*$ satisfy $\|\phi_t\| = 1$ and $\langle f_t, \phi_t \rangle = \|f_t\|$ for all $t \geq 0$. Then the left derivative of $\|f_t\|$ satisfies

$$D^-\|f_t\| := \lim_{h \to 0+} h^{-1}\{\|f_t\| - \|f_{t-h}\|\}$$

$$\leq \lim_{h \to 0+} h^{-1}\{\langle f_t, \phi_t \rangle - \mathrm{Re}\langle f_{t-h}, \phi_t \rangle\}$$

$$= \mathrm{Re}\langle f_t', \phi_t \rangle$$

$$= \mathrm{Re}\langle Zf_t, \phi_t \rangle$$

$$\leq 0.$$

A slight variation of Lemma 1.4.4 now implies that $\|f_t\|$ is monotonically decreasing.

If f_t and g_t are two solutions of (8.24) with $f_0 = g_0$, then their difference h_t is also a solution with $h_0 = 0$. The above argument shows that $h_t = 0$ for all $t \geq 0$, so the solution of (8.24) is unique. This implies that f_t depends linearly on f_0 and that there is a linear contraction T_t such that

$$f_t = T_t f_0 \tag{8.25}$$

for all $t \geq 0$ and $f_0 \in \mathcal{D}$. It follows routinely that T_t is a one-parameter contraction semigroup.

If B is the generator of T_t then (8.24) and (8.25) imply that B is an extension of Z. Since $\mathcal{D}$ is invariant under T_t it is a core for B by Theorem 6.1.18. In other words B is the closure of Z. $\square$

The next example shows that one may not weaken the hypothesis of Theorem 8.3.7 by assuming the solubility of the evolution equation for an interval of time which depends upon $f \in \mathcal{D}$. This is in contrast with Theorems 7.4.9.

Example 8.3.8 Let $\mathcal{D} = C_c^\infty(0, 1)$, this being dense in $L^2(0, 1)$. The operator Z defined on $\mathcal{D}$ by

$$(Zf)(x) := -f'(x)$$

is dissipative, and for all $f \in \mathcal{D}$ the evolution equation has solution $f_t(x) = f_0(x - t)$ provided $0 \le t < \varepsilon_f$. In spite of this the closure of Z is not the generator of a one-parameter semigroup . For if $\lambda \in \mathbf{C}$ the range of $(\lambda I - Z)$ is orthogonal to $\phi \in L^2(0, 1)$, where $\phi(x) := e^{\bar{\lambda}x}$, so $(\lambda I - Z)$ does not have dense range. In order to obtain a one-parameter semigroup one must enlarge $\mathcal{D}$ so that it provides appropriate information about boundary conditions at 0 and 1. $\square$

Problem 8.3.9 Modify the above example by taking $\mathcal{D}$ to be the set of all functions $f \in C^\infty[0, 1]$ such that $f(1) = cf(0)$. Write down an explicit formula for the semigroup associated with this choice of Z, and determine the values of c for which it is a one-parameter contraction semigroup. $\square$

In spite of its great theoretical value, we emphasize that the Hille-Yosida Theorem 8.3.2 is numerically fragile. An estimate which differs from that required by an unmeasurably small amount does not imply the existence of a corresponding one-parameter semigroup .

Theorem 8.3.10 *(Hörmander)*[6] *For every $\varepsilon > 0$ there exists a reflexive Banach space $\mathcal{B}$ and a closed, densely defined operator A on $\mathcal{B}$ such that*

(i) $\mathrm{Spec}(A) \subseteq i\mathbf{R}$,
(ii) $\|(\lambda I - A)^{-1}\| \le (1 + \varepsilon)/|\mathrm{Re}(\lambda)|$ *for all* $\lambda \notin i\mathbf{R}$,
(iii) A *is not the generator of a one-parameter semigroup.*

Proof. Given $1 \le p \le 2$, we define the operator A acting in $L^p(\mathbf{R})$ by

$$Af(x) := i\frac{\mathrm{d}^2 f}{\mathrm{d}x^2}.$$

---

[6] See [Hörmander 1960] for a much more general analysis of L^p multipliers. The proof here is not original.

As initial domain we choose Schwartz space $\mathcal{S}$, which is dense in $L^p(\mathbf{R})$. The closure of A, which we denote by the same symbol, has resolvents given by $R_\lambda f = g_\lambda * f$, where

$$\hat{g}_\lambda(\xi) := (\lambda - i\xi^2)^{-1}$$

for all $\lambda \notin i\mathbf{R}$. If $p = 2$ the unitarity of the Fourier transform implies that $\|R_\lambda\| \leq |\mathrm{Re}(\lambda)|^{-1}$. For $p = 1$, however, assuming for definiteness that $\mathrm{Re}(\lambda) > 0$, Theorem 2.2.5 yields

$$\|R_\lambda\| = \|g_\lambda\|_1 = \frac{1}{|\lambda|^{1/2}} \int_0^\infty \exp\left[-|x|\mathrm{Re}\{(i\lambda)^{1/2}\}\right] \mathrm{d}x.$$

Putting $\lambda := re^{i\theta}$ where $r > 0$ and $-\pi/2 < \theta < \pi/2$, we get

$$\|R_\lambda\| = \frac{1}{r\cos(\theta/2 + \pi/4)} \leq \frac{2}{|\mathrm{Re}(\lambda)|}.$$

Interpolation then implies that if $0 < \gamma < 1$ and $1/p = \gamma + (1-\gamma)/2$ then

$$\|R_\lambda\| \leq \frac{2^\gamma}{|\mathrm{Re}(\lambda)|}.$$

By taking γ close enough to 0 (or equivalently p close enough to 2) we achieve the condition (ii).

Suppose next that $1 \leq p < 2$ and that a semigroup T_t on $L^p(\mathbf{R})$ with generator A does exist. If $f \in \mathcal{S}$ and $f_t \in \mathcal{S}$ is defined for all $t \in \mathbf{R}$ by

$$\hat{f}_t(\xi) := e^{-i\xi^2 t}\hat{f}(\xi)$$

then f_t is differentiable with respect to the Schwartz space topology, and therefore with respect to the L^p norm topology, with derivative Af_t. It follows by Theorem 6.1.16 that $f_t = T_t f$. Now assume that $a > 0$ and $\hat{f}(\xi) := e^{-a\xi^2}$, so that $\hat{f}_t(\xi) = e^{-(a+it)\xi^2}$. Explicit calculations of f_t and f yield

$$\|f\|_p = (4\pi a)^{1/2p - 1/2} p^{-1/2p},$$

$$\|f_t\|_p = (4\pi)^{1/2p - 1/2} p^{-1/2p} a^{-1/2p}(a^2 + t^2)^{1/2p - 1/4}.$$

Hence

$$\|T_t\| \geq \frac{\|f_t\|_p}{\|f\|_p} = (1 + t^2/a^2)^{(2-p)/4p}.$$

But this diverges as $a \to 0$, so T_t cannot exist as a bounded operator for any $t \neq 0$. $\square$

The growth properties of one-parameter semigroup s on Hilbert space need special treatment, as we explain in Section 10.6. Theorem 8.3.1 can be used to obtain a variety of related results.

Theorem 8.3.11 *Let T_t be a one-parameter semigroup acting on a Banach space $\mathcal{B}$ and suppose that the spectrum of its generator Z is contained in $\{z : \mathrm{Re}(z) \leq 0\}$. Let $M : (0, \infty) \to [1, \infty)$ be a monotonically decreasing function. Then the bound*

$$\|T_t\| \leq \inf_{\{a:a>0\}} \{M(a)\mathrm{e}^{at}\} \tag{8.26}$$

holds for all $t \geq 0$ if and only if

$$\|(\lambda I - Z)^{-m}\| \leq \inf_{\{a:0<a<\lambda\}} \{M(a)(\lambda - a)^{-m}\} \tag{8.27}$$

for all $\lambda > 0$ and all $m \geq 1$.

Problem 8.3.12 Let T_t be a one-parameter semigroup and assume that $N \geq 1$ and $\alpha > 0$. Prove that

$$\|T_t\| \leq N(1+t)^{\alpha}$$

for all $t \geq 0$ if and only if

$$\|T_t\| \leq M(a)\mathrm{e}^{at}$$

for all $t \geq 0$, all $a > 0$ and constants $M(a)$ which you should find explicitly.
Prove that a similar result only holds for

$$\|T_t\| \leq N(1+t^{\alpha})$$

if $0 \leq \alpha \leq 1$.   □

Problem 8.3.13 Find the semigroup and resolvent bounds corresponding to the choice

$$M(a) := 1 + a^{-1}$$

in Theorem 8.3.11.   □

Problem 8.3.14 Let Z be a closed operator whose spectrum does not meet $(0, \infty)$, and suppose that

$$\|R(a, Z)\| \leq \frac{1}{a \sin(\alpha)}$$

for all $a > 0$. Prove that

$$\mathrm{Spec}(Z) \subseteq \{z : \mathrm{Arg}(z) \geq \alpha\}.$$

Give an example for which one has equality in both equations.   □

8.4 Bounded holomorphic semigroups

We consider semigroups T_z for which z takes complex values in a sector

$$S_\alpha := \{z \in \mathbf{C} : z \neq 0 \text{ and } |\mathrm{Arg}(z)| < \alpha\}.$$

We define a bounded holomorphic semigroup T_z on a Banach space $\mathcal{B}$ to be a family of bounded operators parametrized by $z \in S_\alpha$ for some $0 < \alpha \leq \pi/2$ and satisfying the following conditions.[7]

(i) $T_z T_w = T_{z+w}$ for all $z, w \in S_\alpha$.
(ii) If $\varepsilon > 0$ then $\|T_z\| \leq M_\varepsilon$ for some $M_\varepsilon < \infty$ and all $z \in S_{\alpha-\varepsilon}$.
(iii) T_z is an analytic function of z for all $z \in S_\alpha$.
(iv) If $f \in \mathcal{B}$ and $\varepsilon > 0$ then $\lim_{z \to 0} T_z f = f$ provided z remains within $S_{\alpha-\varepsilon}$.

We define the generator Z of T_z by

$$Zf := \lim_{t \to 0} t^{-1}(T_t f - f),$$

where $t > 0$ and $\mathrm{Dom}(Z)$ is the set of all $f \in \mathcal{B}$ for which the limit exists.

Our next two theorems, taken together, characterize the generators of bounded holomorphic semigroups in terms of properties of their resolvents.

Theorem 8.4.1 *If T_z is a bounded holomorphic semigroup then*

$$\mathrm{Spec}(Z) \subseteq \{w : |\mathrm{Arg}(w)| \geq \alpha + \pi/2\}. \tag{8.28}$$

For all $\varepsilon > 0$ there exists a constant $N_\varepsilon < \infty$ such that

$$\|(wI - Z)^{-1}\| \leq N_\varepsilon |w|^{-1} \tag{8.29}$$

for all $w \in S_{\alpha+\pi/2-\varepsilon}$.

Proof. If $|\theta| < \alpha - \varepsilon$, define V_t for $t > 0$ by

$$V_t = T_{e^{i\theta}t}$$

---

[7] The theory of holomorphic semigroups goes back to the earliest days of the subject, and we follow the standard approach. See [Hille and Phillips 1957, pp. 383 ff.].

and let W be the generator of V_t. If $f \in \mathrm{Dom}(W)$ and $f(z) := T_z f$ and $s > 0$ then

$$
\begin{aligned}
V_s W f &= \lim_{t \to 0+} t^{-1} V_s (V_t - 1) f \\
&= \lim_{t \to 0+} t^{-1} \left\{ f(e^{i\theta} s + e^{i\theta} t) - f(e^{i\theta} s) \right\} \\
&= e^{i\theta} f'(e^{i\theta} s) \\
&= e^{i\theta} \lim_{t \to 0+} t^{-1} \left\{ f(e^{i\theta} s + t) - f(e^{i\theta} s) \right\} \\
&= e^{i\theta} \lim_{t \to 0+} t^{-1} \left\{ T_t (V_s f) - V_s f \right\} \\
&= e^{i\theta} Z(V_s f).
\end{aligned}
$$

Therefore $V_s f \in \mathrm{Dom}(Z)$ and

$$
V_s W f = e^{i\theta} Z(V_s f)
$$

for all $s > 0$. Letting $s \to 0$ and using the fact that Z is a closed operator we conclude that $f \in \mathrm{Dom}(Z)$ and

$$
W f = e^{i\theta} Z f.
$$

Reversing the argument we find that $\mathrm{Dom}(Z) = \mathrm{Dom}(W)$ and $W = e^{i\theta} Z$.

Since $\| V_t \| \leq M_\varepsilon$ for all $t \geq 0$, it follows by Theorem 8.2.1 that

$$
\mathrm{Spec}(W) \subseteq \{ w : \mathrm{Re}(w) \leq 0 \}
$$

and

$$
\| (wI - W)^{-1} \| \leq M_\varepsilon (\mathrm{Re}(w))^{-1}
$$

for all w such that $\mathrm{Re}(w) > 0$. This implies (8.28) and (8.29). $\quad \square$

Theorem 8.4.2 *Let Z be a closed, densely defined operator acting in $\mathcal{B}$ with*

$$
\mathrm{Spec}(Z) \subseteq \{ w : |\mathrm{Arg}(w)| \geq \alpha + \pi/2 \}
$$

where $0 < \alpha \leq \pi/2$, and suppose also that for all $\varepsilon > 0$ there is a real constant N_ε such that

$$
\| (wI - Z)^{-1} \| \leq N_\varepsilon |w|^{-1}
$$

for all $w \in S_{\alpha - \varepsilon + \pi/2}$. Then Z is the generator of a bounded holomorphic semigroup on $\mathcal{B}$.

Proof. We will need to evaluate a number of integrals of the form

$$\int_\gamma g(z)\,dz$$

where g is an analytic function (often operator-valued) and $\gamma : \mathbf{R} \to S_{\alpha+\pi/2}$ is a contour such that

$$\gamma(t) := \begin{cases} te^{i\phi} & \text{for all large enough } t > 0, \\ |t|e^{-i\phi} & \text{for all large enough } t < 0. \end{cases}$$

We assume that $\phi := \alpha + \pi/2 - \varepsilon$ and that $\varepsilon > 0$ is small enough to ensure that the integral converges. Cauchy's theorem will ensure that the integral is independent of ε provided ε is small enough. The integral is evaluated by considering a closed contour γ_R and letting $R \to \infty$. The contour γ_R consists of the part of γ for which $|t| \leq R$ together with a sector of the circle with centre 0 and radius R. Sometimes this sector is the part of the circle to the right of γ and sometimes it is the part to the left, but in both cases the integral around the sector vanishes as $R \to \infty$.

Our definition of T_t is motivated by the formula

$$e^{az} = \frac{1}{2\pi i}\int_\gamma e^{zw}(w-a)^{-1}\,dw. \tag{8.30}$$

The convergence of this integral, and of those below, is ensured by the exponential factor in the integrand provided $z = re^{i\theta}$, $|\theta| \leq \alpha - 2\varepsilon$, $\alpha + \pi/2 - \varepsilon < \phi < \alpha + \pi/2$ and $|\mathrm{Arg}(a)| \geq \alpha + \pi/2$. Assuming the same conditions on z and ϕ we define the bounded operator T_z by

$$T_z f := \frac{1}{2\pi i}\int_\gamma e^{zw}(wI - Z)^{-1} f\,dw. \tag{8.31}$$

We start by proving that these operators satisfy the required bounds. By Cauchy's theorem the integral is independent of the particular contour chosen, subject to the stated constraints. Therefore

$$T_z f = \frac{1}{2\pi i}\int_\gamma \exp\{e^{i\theta}w\}(r^{-1}wI - Z)^{-1}fr^{-1}\,dw$$

and

$$\|T_z f\| \leq \frac{1}{2\pi}N_\varepsilon\|f\|\int_\gamma |w^{-1}\exp\{e^{i\theta}w\}\,dw|$$

$$\leq M_\varepsilon\|f\| \tag{8.32}$$

for some $M_\varepsilon < \infty$ and all $|\theta| \leq \alpha - 2\varepsilon$.

We next prove that T_z converge strongly to I as $z \to 0$. If $f \in \mathrm{Dom}(Z)$ and $z \in S_{\alpha-2\varepsilon}$ then

$$\int_\gamma (wI - Z)^{-1} w^{-1} Z f \, \mathrm{d}w = 0$$

by Cauchy's theorem. By combining this with the case $a = 0$ of (8.30) and (8.31) we see that if $f \in \mathrm{Dom}(Z)$ then

$$\begin{aligned}
\|T_z f - f\| &= \left\| \frac{1}{2\pi i} \int_\gamma \{ \mathrm{e}^{zw}(wI - Z)^{-1} f - \mathrm{e}^{zw} w^{-1} f \} \, \mathrm{d}w \right\| \\
&= \left\| \frac{1}{2\pi i} \int_\gamma \mathrm{e}^{zw}(wI - Z)^{-1} w^{-1} Z f \, \mathrm{d}w \right\| \\
&= \left\| \frac{1}{2\pi i} \int_\gamma (\mathrm{e}^{zw} - 1)(wI - Z)^{-1} w^{-1} Z f \, \mathrm{d}w \right\| \\
&\leq \lim_{z \to 0} \frac{1}{2\pi} \int_\gamma |\mathrm{e}^{zw} - 1| \, c \, |w^{-2} \, \mathrm{d}w| \, \|Z f\| \\
&\to 0
\end{aligned}$$

as $z \to 0$. We deduce using (8.32) that

$$\lim_{z \to 0} T_z f = f$$

for all $f \in \mathcal{B}$, provided $z \in S_{\alpha-2\varepsilon}$.

We show that T_z satisfies the semigroup law. If $z, z' \in S_{\alpha-2\varepsilon}$ and γ, γ' are two contours of the above type with γ' outside γ then

$$\begin{aligned}
T_z T_{z'} &= \left(\frac{1}{2\pi i} \right)^2 \int_\gamma \int_{\gamma'} \mathrm{e}^{zw+z'w'} (wI - Z)^{-1}(w'I - Z)^{-1} \, \mathrm{d}w \, \mathrm{d}w' \\
&= \left(\frac{1}{2\pi i} \right)^2 \int_\gamma \int_{\gamma'} \frac{\mathrm{e}^{zw+z'w'}}{w' - w} \{ (wI - Z)^{-1} - (w'I - Z)^{-1} \} \, \mathrm{d}w \, \mathrm{d}w' \\
&= \frac{1}{2\pi i} \int_\gamma \mathrm{e}^{(z+z')w} (wI - Z)^{-1} \, \mathrm{d}w \\
&= T_{z+z'}.
\end{aligned}$$

We have finally to identify the generator W of the holomorphic one-parameter semigroup T_z. If $f \in \mathrm{Dom}(Z)$ and $z \in S_{\alpha-2\varepsilon}$ then by differentiating under the integral sign we obtain

$$\begin{aligned}
\frac{\mathrm{d}}{\mathrm{d}z}(T_z f) &= \frac{1}{2\pi i} \int_\gamma \mathrm{e}^{zw} w(wI - Z)^{-1} f \, \mathrm{d}w \\
&= \frac{1}{2\pi i} \int_\gamma \mathrm{e}^{zw} \{ f + Z(wI - Z)^{-1} f \} \, \mathrm{d}w
\end{aligned}$$

$$= \frac{1}{2\pi i} \int_{\gamma} e^{zw} Z(wI - Z)^{-1} f \, dw \tag{8.33}$$

$$= T_z Z f.$$

Therefore $T_z f \in \mathrm{Dom}(W)$ and $WT_z f = T_z Z f$. Letting $z \to 0$ and using the fact that W is closed we deduce that W is an extension of Z. Since $(W - I)$ is one-one and extends $(Z - I)$, which has range equal to $\mathcal{B}$, we conclude that $W = Z$. $\square$

Problem 8.4.3 Show that condition (iv) in the definition of a bounded holomorphic semigroup on page 237 is implied by conditions (i)–(iii) together with the assumption that

$$\bigcup_{0 < t < \infty} \mathrm{Ran}(T_t)$$

is dense in $\mathcal{B}$. $\square$

Problem 8.4.4 Let T_z be a holomorphic semigroup defined for all z in the open sector $S_{\pi/2}$ and satisfying

$$\|T_z\| \leq M \tag{8.34}$$

for all such z. Show that there is a one-parameter group U_t on $\mathcal{B}$ such that

$$U_t T_z = T_{z+it}$$

for all $z \in S_{\pi/2}$ and all $t \in \mathbf{R}$. Show also that

$$U_t f = \lim_{s \downarrow 0} T_{s+it} f$$

for all $f \in \mathcal{B}$ and $t \in \mathbf{R}$. $\square$

Example 8.4.5 This is a continuation of Example 6.3.5. We define the operators T_z on $L^p(\mathbf{R}^N)$ for $\mathrm{Re}(z) > 0$ and $1 \leq p < \infty$ by $T_z f := k_z * f$ where

$$k_z(x) := (4\pi z)^{-N/2} e^{-|x|^2/4z}.$$

A direct calculation shows that

$$\|k_z\|_1 = \left(\frac{|z|}{\mathrm{Re}(z)} \right)^{N/2}.$$

Corollary 2.2.19 implies that T_z is a bounded operator on $L^p(\mathbf{R}^N)$ for all p, z in the stated ranges. One establishes that T_z is a bounded holomorphic semigroup on each L^p space by adapting the procedure followed in Example 6.3.5. The generators Z_p are consistent as p varies, and in fact $Z_p f = \Delta f$ for all $f \in \mathcal{S}$.

Theorem 8.4.1 implies that $\mathrm{Spec}(Z_p) \subseteq (-\infty, 0]$ for every p. In order to prove that $\mathrm{Spec}(Z_p) = (-\infty, 0]$ for every p it is sufficient to construct $f_n \in \mathrm{Dom}(Z_p)$ such that

$$\lim_{n \to \infty} \frac{\|Z_p f_n - \mu f_n\|_p}{\|f_n\|_p} = 0$$

for all $\mu \leq 0$; this proves that the resolvent operator cannot be bounded, if it exists. We actually construct $f_n \in \mathbf{C}_c^\infty(\mathbf{R}^N)$ such that

$$\lim_{n \to \infty} \frac{\|\Delta f_n + |\xi|^2 f_n\|_p}{\|f_n\|_p} = 0 \tag{8.35}$$

for all $\xi \in \mathbf{R}^N$. Put

$$f_n(x) := \phi(x/n) e^{ix \cdot \xi}$$

where $\phi \in C_c^\infty(\mathbf{R}^N)$ is not identically zero. A direct calculation shows that $\|f_n\|_p = c_1 n^{N/p}$ for all $n > 0$, where $c_1 > 0$. Moreover

$$(\Delta f_n)(x) + |\xi|^2 f_n(x) = 2n^{-1} i\xi \cdot (\nabla\phi)(x/n) e^{ix \cdot \xi} + n^{-2} (\Delta\phi)(x/n) e^{ix \cdot \xi}.$$

Therefore

$$\|\Delta f_n + |\xi|^2 f_n\|_p \leq c_2 n^{-1} \|(\nabla\phi)(\cdot/n)\|_p + c_3 n^{-2} \|(\Delta\phi)(\cdot/n)\|_p$$
$$= c_4 n^{-1+N/p} + c_5 n^{-2+N/p}.$$

The formula (8.35) follows.   □

Theorem 8.4.6 *Let T_z be a bounded holomorphic semigroup acting on $\mathcal{B}$ for $z \in S_\alpha$. Then*

$$T_z \mathcal{B} \subseteq \mathrm{Dom}(Z)$$

for all $z \in S_\alpha$. For every $\varepsilon > 0$ there is a constant C_ε such that

$$\|ZT_z\| \leq C_\varepsilon |z|^{-1} \tag{8.36}$$

for all $z \in S_{\alpha-\varepsilon}$.

Proof. If $\varepsilon > 0$ then there exists $c > 0$ such that for all $z \in S_{\alpha-\varepsilon}$ the circle σ with centre z and radius $c|z|$ lies inside $S_{\alpha-\varepsilon/2}$. If $f \in \mathcal{B}$ then $T_z f$ is an analytic function of z, so it lies in the domain of the generator Z. Cauchy's integral formula now implies that

$$ZT_z f = \frac{d}{dz} T_z f = \frac{1}{2\pi i} \int_\sigma \frac{T_w f}{(w-z)^2} \, dw.$$

Therefore

$$\|ZT_z f\| \leq \frac{1}{2\pi} \int_\sigma \frac{\|T_w f\|}{|w - z|^2} |dw|$$

$$\leq \frac{1}{2\pi} \int_\sigma \frac{M_{\varepsilon/2}\|f\|}{c^2 |z|^2} |dw|$$

$$= M_{\varepsilon/2} c^{-1} |z|^{-1} \|f\|,$$

which yields (8.36) with $C_\varepsilon = M_{\varepsilon/2} c^{-1}$. $\square$

The following is one of many results to the effect that the long time properties of $T_t f$ may depend upon the choice of f.

Problem 8.4.7 Let T_z be a bounded holomorphic semigroup acting on $\mathcal{B}$. Prove that

$$\|T_t f\| = O(t^{-n})$$

as $t \to \infty$ for all $f \in \mathrm{Ran}(Z^n)$. $\square$

Theorem 8.4.8 *Let T_z be a bounded holomorphic semigroup on $\mathcal{B}$ with generator Z. Then T_z is compact for all non-zero $z \in S_\alpha$ if and only if $R(w, Z)$ is compact for some (equivalently all) $w \notin \mathrm{Spec}(Z)$.*

Proof. In the forward direction we use Corollary 8.2.2. In the reverse direction we use Theorem 8.4.6 to write

$$T_t = \{(wI - Z)T_t\}R(w, Z).$$

We then observe that the product of a bounded and a compact operator is compact. $\square$

Theorem 8.4.9 *Let T_t be a one-parameter semigroup on $\mathcal{B}$ with generator Z, satisfying*

$$T_t \mathcal{B} \subseteq \mathrm{Dom}(Z)$$

for all $t > 0$. If

$$\|T_t\| \leq M, \qquad \|ZT_t\| \leq c/t$$

for some $c, M < \infty$ and all $t > 0$, then there exists $\alpha > 0$ such that T_t may be extended to a bounded holomorphic semigroup on S_α.

Proof. By Theorem 6.2.9 and its proof we see that T_t is an operator-valued C^∞ function of t for $0 < t < \infty$ with

$$\frac{d^n}{dt^n} T_t = (ZT_{t/n})^n.$$

Hence

$$\left\| \frac{d^n}{dt^n} T_t \right\| \leq n^n c^n t^{-n}.$$

An application of Stirling's formula shows that T_t can be analytically continued to the disc

$$\{z : |z - t| < e^{-1} c^{-1} t\}$$

by defining

$$T_z := \sum_{n=0}^\infty \frac{(z - t)^n}{n!} \frac{d^n}{dt^n} T_t. \tag{8.37}$$

The union of these discs is a sector S_α.

We now have to verify that T_z satisfies the conditions (i)–(iv) on page 237 for small enough $\alpha > 0$. Condition (i) holds by analytic continuation from the case when z_1 and z_2 are both real.

If $z := r e^{i\theta}$ where $|\theta| < \alpha$ then $|z - r| < \alpha r$, so

$$\|T_z\| \leq \|T_t\| + \sum_{n=1}^\infty \frac{|z - r|^n}{n!} \|ZT_{t/n}\|^n$$

$$\leq M + \sum_{n=1}^\infty \frac{c^n \alpha^n n^n}{n!}$$

$$\leq N < \infty$$

if $c\alpha e < 1$. This proves condition (ii). Condition (iii) is trivial and condition (iv) is proved by using Problem 8.4.3. □

Problem 8.4.10 Let T_t be a one-parameter semigroup on $\mathcal{B}$ with generator Z, satisfying $T_t \mathcal{B} \subseteq \mathrm{Dom}(Z)$ for all $t > 0$. If

$$\limsup_{t \to 0} t \|ZT_t\| < \infty$$

show that $e^{-\delta t} T_t$ extends analytically to a bounded holomorphic semigroup on $\mathcal{B}$ for all $\delta > 0$. □

9
Quantitative bounds on operators

9.1 Pseudospectra

The increasing availability of numerical software such as Matlab since 1990 has provided the stimulus for the investigation of quantitative aspects of operator theory. Most of the contents of this chapter date from this period. Many of the theorems have resulted from interactions between pure mathematicians, applied mathematicians and numerical analysts, and their value can only be fully appreciated with the help of numerical examples.

The notion of pseudospectra arose as a result of the realization that several 'pathological' properties of highly non-self-adjoint operators were closely related.[1] These include the existence of approximate eigenvalues far from the spectrum; the instability of the spectrum under small perturbations; the anomalous response of systems subject to a periodic driving term; the importance of resolvent norm estimates in many areas of operator theory, and in particular in semigroup theory. The connections between these can be demonstrated at a very general level.

We start by discussing the stability of solutions of the operator equation $Ax - \lambda x = b$ under perturbations of b and A. The existence of a solution and its uniqueness are guaranteed by $\lambda \notin \mathrm{Spec}(A)$, and we have $x = -(\lambda I - A)^{-1}b$. It is better known to numerical analysts than to pure mathematicians that this is not the end of the story. Suppose that $\lambda \notin \mathrm{Spec}(A)$ and that b is slightly altered, or that it is only known to a finite precision. One then has

---

[1] One can only talk about pathology if one knows what constitutes normality. Once one fully accepts that intuitions about self-adjoint matrices and operators are a very poor guide to understanding the much greater variability of non-self-adjoint problems, the pathology disappears. We follow [Trefethen and Embree 2005, Chap. 48] in not trying to give a precise meaning to the phrase 'highly non-self-adjoint', which should really be a measure of the failure of the spectral theorem and its consequences.

the perturbed equation $Ax' - \lambda x' = b + r$ where $\|r\| \leq \varepsilon$, say. We deduce immediately that

$$\|x - x'\| \leq \varepsilon \|(\lambda I - A)^{-1}\|$$

and for this to be small we need to know that $\|(\lambda I - A)^{-1}\|$ is not too big. Unfortunately for one's intuition this norm can be very large even if λ is not at all close to the spectrum of A. This phenomenon is commonplace if $\dim(\mathcal{B}) \geq 30$ and can be important for smaller matrices.[2]

One can also consider the effect of small changes in the operator A, or of only knowing A to finite precision. Suppose that $\|B\| \leq \varepsilon$ and $(A + B)x' - \lambda x' = b$. Then

$$x - x' = (\lambda I - A - B)^{-1}b - (\lambda I - A)^{-1}b$$
$$= (\lambda I - A - B)^{-1}B(\lambda I - A)^{-1}b$$
$$= (\lambda I - A)^{-1}CB(\lambda I - A)^{-1}b$$

where

$$C := (I - B(\lambda I - A)^{-1})^{-1}$$

exists provided $\delta := \varepsilon \|(\lambda I - A)^{-1}\| < 1$. Assuming this, we obtain

$$\|x - x'\| \leq \|(\lambda I - A)^{-1}\| \frac{\delta \|b\|}{1 - \delta}.$$

Once again we can only deduce that x' is close to x if $\|(\lambda I - A)^{-1}\|$ is not too big.

The size of the resolvent norm is also relevant when calculating the response of a system to a periodic driving term.

Example 9.1.1 Consider the evolution equation

$$f'(t) = Zf(t) + e^{i\omega t}a \tag{9.1}$$

in a Banach space $\mathcal{B}$, where $a \in \mathcal{B}$, ω is the frequency of the driving term and Z is the generator of a one-parameter semigroup which is stable in the sense that $\lim_{t \to +\infty} \|e^{Zt}b\| = 0$ for all $b \in \mathcal{B}$. The solution of (9.1) is

$$f(t) = e^{Zt}f(0) + (e^{i\omega t} - e^{Zt})(i\omega I - Z)^{-1}a.$$

If one ignores a transient term which decays as $t \to +\infty$ one obtains the steady state response

$$f(t) = e^{i\omega t}(i\omega I - Z)^{-1}a.$$

---

[2] See [Trefethen and Embree 2005] for a large number of examples providing support for this assertion.

We will see that $\|(i\omega I - Z)^{-1}\|$ can be very large, and that one can therefore have a very large response to the driving term, even when $i\omega$ is not close to the spectrum of Z. The moral of this example is that the stability of a driven system is not controlled by the spectrum of Z but by the size of the resolvent norms. $\quad\square$

Examples 8.2.6 and 9.1.7 show that knowing the spectrum of an operator Z provides very little guidance to the behaviour of $\|e^{Zt}\|$ for small $t > 0$. In the second case the semigroup norm is very close to 1 for all $t \in [0, 1]$, but it is extremely small for $t \geq 4$. Section 10.2 explores the relevance of pseudospectral ideas in this context. Example 10.2.1 presents a matrix Z of moderate size such that $\|e^{Zt}\|$ grows rapidly for small $t > 0$ even though it decays exponentially for large t.

Considerations such as those above motivate one to define the pseudospectra of an operator A to be the collection of sets

$$\mathrm{Spec}_\varepsilon(A) := \mathrm{Spec}(A) \cup \{z \in \mathbf{C} : \|(zI - A)^{-1}\| > \varepsilon^{-1}\},$$

parametrized by $\varepsilon > 0$. It is clear that the pseudospectra of an operator change if one replaces the given norm on $\mathcal{B}$ by an equivalent norm. However, there is often a good physical or mathematical reason to choose a particular norm, and the pseudospectra of an operator with respect to two standard norms are frequently very similar.

One can also describe the pseudospectra in terms of approximate eigenvalues.

Lemma 9.1.2 *If $\varepsilon > 0$ and $\lambda \notin \mathrm{Spec}(A)$ then $\lambda \in \mathrm{Spec}_\varepsilon(A)$ if and only if there exists $x \in \mathcal{B}$ such that*

$$\|Ax - \lambda x\| < \varepsilon \|x\|. \tag{9.2}$$

Proof. If $\|(\lambda I - A)^{-1}\| > \varepsilon^{-1}$ then there exists a non-zero vector $y \in \mathcal{B}$ such that $\|(\lambda I - A)^{-1}y\| > \varepsilon^{-1}\|y\|$. Putting $x := (\lambda I - A)^{-1}y$ this may be rewritten in the form (9.2). The converse is similar. $\quad\square$

If A is a normal operator then it follows quickly from the spectral theorem that

$$\|(A - zI)^{-1}\| = \{\mathrm{dist}(z, \mathrm{Spec}(A))\}^{-1}$$

but for operators which are far from normal the LHS can be very large even when the RHS is small; equivalently an approximate eigenvalue λ of a moderately large matrix need not be close to a true eigenvalue. In such circumstances, which are commonplace rather than exceptional, the pseudospectra contain much more information about the behaviour of a non-self-adjoint operator A than the spectrum alone. The recent monograph of Trefethen and Embree provides the first comprehensive account of the subject, and makes full use of the EigTool software developed by Wright for computing the pseudospectra of large matrices.[3] We do not attempt to compete with this monograph, but refer to Theorem 14.5.4, which describes the non-self-adjoint harmonic oscillator from this point of view and contains a diagram of the associated pseudospectra. The convection-diffusion operator provides an even simpler illustration of the importance of pseudospectral ideas and is discussed in Example 9.3.20 and Theorem 9.3.21.

Computations of pseudospectra make use of the following observations. Let A be an $n \times n$ matrix, considered as an operator acting on $\mathbf{C}^n$ provided with the Euclidean norm. Then $z \in \operatorname{Spec}_\varepsilon(A)$ if and only if the smallest singular value $\mu(z)$ of $zI - A$ is less than ε, or equivalently if and only if the smallest eigenvalue of

$$B_z := (zI - A)^*(zI - A)$$

is less than ε^2. The corresponding eigenvector f of B_z satisfies (9.10). The smallest eigenvalue of B_z may be computed using inverse power iteration or other methods; a major speedup is obtained by first reducing A to triangular form with respect to a suitable choice of orthonormal basis. One finally plots the level curves of $\mu(z)$ as a function of z within a chosen region of the complex plane.

We discuss some examples that show that $\operatorname{Spec}_\varepsilon(A)$ may be a much larger set than $\operatorname{Spec}(A)$ even for very small $\varepsilon > 0$.

Let a, b be non-orthogonal vectors of norm 1 in a Hilbert space $\mathcal{H}$ and put

$$Px := \frac{\langle x, b \rangle}{\langle a, b \rangle} a$$

for all $x \in \mathcal{H}$. It is immediate that $P^2 = P$ and $\operatorname{Spec}(P) = \{0, 1\}$. Therefore

$$(zI - P)^{-1} = (z - 1)^{-1}P + z^{-1}(I - P).$$

If $a = b$ then $P = P^*$ and $z \in \operatorname{Spec}_\varepsilon(P)$ if and only if

$$\operatorname{dist}(z, \operatorname{Spec}(P)) < \varepsilon.$$

The situation when a, b are nearly orthogonal is quite different.

Problem 9.1.3 Find an explicit formula for $\|(zI - P)^{-1}\|$ when $a \neq b$, and sketch the boundary of $\operatorname{Spec}_\varepsilon(P)$ when $\varepsilon > 0$ is small but a and b are nearly

---

[3] See [Trefethen and Embree 2005]. EigTool is available at [Wright 2002].

orthogonal. Compute the contours $\{z : \|(zI - P)^{-1}\| = \varepsilon\}$ numerically for various $\varepsilon > 0$ when $\mathcal{H} = \mathbf{C}^2$. $\square$

Example 9.1.4 We consider the standard $n \times n$ Jordan matrix J_n defined by

$$(J_n)_{r,s} := \begin{cases} 1 \text{ if } s = r+1, \\ 0 \text{ otherwise.} \end{cases}$$

The resolvent norm is most easily computed if one uses the l^1 norm on $\mathbf{C}^n$. Starting from the formula

$$((zI - J_n)^{-1})_{r,s} = \begin{cases} z^{r-s-1} & \text{if } r \leq s, \\ 0 & \text{otherwise,} \end{cases}$$

we obtain

$$\|(zI - J_n)^{-1}\|_1 = \frac{|z|^{-n} - 1}{1 - |z|}.$$

This diverges at an exponential rate as $n \to \infty$ for every z satisfying $|z| < 1$. One says that the pseudospectra fill up the unit circle at an exponential rate even though $\mathrm{Spec}(J_n) = \{0\}$ for every n. $\square$

Problem 9.1.5 Prove that if one uses the l^2 norm on $\mathbf{C}^n$ then $\|(zI - J)^{-1}\|$ once again diverges at an exponential rate as $n \to \infty$ for every z satisfying $|z| < 1$. $\square$

Problem 9.1.6 Let V denote the Volterra integral operator

$$(Vf)(x) := \int_0^x f(y)\, dy$$

acting on $L^2(0, 1)$. Prove that $\mathrm{Spec}(V) = \{0\}$ by writing down the explicit solution f of $Vf - zf = g$ for every $g \in L^2(0, 1)$ and $z \neq 0$.

If $z \neq 0$ define $f_z(x) := \mathrm{e}^{x/z}$. Calculate the size of

$$\|Vf_z - zf_z\| / \|f_z\|$$

and use this to obtain bounds on the pseudospectra $\mathrm{Spec}_\varepsilon(V)$ for every $\varepsilon > 0$. $\square$

Our definition of pseudospectra applies equally well to unbounded linear operators. In that case the difference between the spectrum and the pseudospectra can be even sharper, because the spectrum of such an operator can be empty.

Example 9.1.7 The evolution equation for the Airy operator

$$(Af)(x) := f''(x) + ixf(x) \tag{9.3}$$

acting in $L^2(\mathbf{R})$ can be solved explicitly. We take the domain of A to be the Schwartz space S. Using the Fourier transform $\mathcal{F}$ and putting $\hat{A} := \mathcal{F}A\mathcal{F}^{-1}$, a direct calculation yields

$$(\hat{A}g)(\xi) = -g'(\xi) - \xi^2 g(\xi)$$

for all $g \in S$. One may verify directly that the evolution equation $g'_t := \hat{A}g_t$ has the solution $g_t := \hat{T}_t g_0$ for all $t \geq 0$, where

$$(\hat{T}_t g)(\xi) := \exp\left(-\xi^2 t + \xi t^2 - t^3/3\right) g(\xi - t)$$
$$= \exp\left(-t(\xi - t/2)^2 - t^3/12\right) g(\xi - t) \tag{9.4}$$

for all $g \in S$. One sees immediately that $\hat{T}_t$ are bounded operators on $L^2(\mathbf{R})$ for all $t \geq 0$ and that (9.4) defines a one-parameter semigroup on $L^2(\mathbf{R})$. The formula

$$\|\hat{T}_t\| = e^{-t^3/12}$$

implies that the spectrum of the generator $\hat{A}$ is empty by Theorem 8.2.1: the constant a of that theorem can be chosen arbitrarily large and negative.

The results can now all be transferred to the original problem by putting $T_t := \mathcal{F}^{-1}\hat{T}_t\mathcal{F}$. $\square$

Problem 9.1.8 Use Theorem 6.1.18 to prove that the generator Z of T_t is the closure of A; in other words S is a core for Z. $\square$

Problem 9.1.9 Use the formula (9.4) and Fourier transforms to prove that

$$(T_t f)(x) = \int_{\mathbf{R}} K(t, x, y) f(y)\, dy$$

for all $f \in L^2(\mathbf{R})$, where

$$K(t, x, y) := (4\pi t)^{-1/2} \exp\left(-\frac{(x-y)^2}{4t} + i\frac{(x-y)t}{2} - \frac{t^3}{12} + iyt\right). \square$$

Although the Airy operator (9.3) has empty spectrum, it has a large, explicit family of approximate eigenfunctions. It is best to carry out the calculations using $\hat{A}$, but the conclusions may all be restated in terms of A. For every $z \in \mathbf{C}$, $\hat{A}g = zg$ has the solution

$$g_z(\xi) := \exp\left(-\xi^3/3 - z\xi\right).$$

This does not lie in $L^2(\mathbf{R})$, but if $\mathrm{Re}(z) = -a \ll 0$ we may replace it by the approximate eigenfunction

$$h_z(\xi) := \begin{cases} \exp\left(-\xi^3/3 - z\xi\right) & \text{if } \xi \geq 0, \\ \exp\left(-z\xi\right) & \text{if } \xi < 0. \end{cases}$$

One sees immediately that $h_z \in L^2(\mathbf{R}) \cap C^1(\mathbf{R})$ and that $|h_z(\xi)|$ takes its maximum value at $\xi = \sqrt{a}$, with

$$|h_z(\sqrt{a})| = \exp\left(2a^{3/2}/3\right).$$

Problem 9.1.10 Find an asymptotic formula for

$$\|\hat{A}h_z - zh_z\|/\|h_z\|$$

as $\mathrm{Re}(z) \to -\infty$. Also prove that $\|(zI - \hat{A})^{-1}\|$ does not depend upon the imaginary part of z. $\quad\square$

9.2 Generalized spectra and pseudospectra

The standard definition of pseudospectra starts with a closed linear operator and defines the set $\mathrm{Spec}_\varepsilon(A)$ for all $\varepsilon > 0$ by

$$\mathrm{Spec}_\varepsilon(A) := \mathrm{Spec}(A) \cup \{z \in \mathbf{C} : \|(zI - A)^{-1}\| > \varepsilon^{-1}\}.$$

In this section we take a more general point of view, which has advantages in a wide range of applications. Polynomial eigenvalue problems, for example, lie at the heart of many dynamical problems concerning engineering structures, and pose a continuing source of new challenges for engineers. Such problems also arise as a result of efforts to reduce the computational difficulty of linear eigenvalue equations, as we will see in examples below.

The standard definitions of spectrum and of pseudospectra are obtained from the theory below by putting $\Lambda = \mathbf{C}$ and $A(\lambda) = \lambda I - A$. Let Λ be a parameter space and let $\{A(\lambda)\}_{\lambda \in \Lambda}$ be a family of closed operators acting from a Banach space $\mathcal{B}_1$ to a Banach space $\mathcal{B}_2$. We define the spectrum of the family to be the set

$$\mathrm{Spec}(A(\cdot)) := \{\lambda \in \Lambda : A(\lambda) \text{ is not invertible}\}.$$

As usual invertibility means that $A(\lambda)$ maps $\mathrm{Dom}(A(\lambda))$ one-one onto $\mathcal{B}_2$ with a bounded inverse.[4] One often refers to the above as non-linear eigenvalue

---

[4] A family of operators depending non-linearly on one or more parameters is often called an operator pencil. The spectral theory of polynomial operator pencils is the main topic in [Markus 1988], which contains an English translation of the pioneering 1951 article of Keldysh on the subject. For further information see [Gohberg et al. 1983] and [Tisseur and Meerbergen 2001].

problems, meaning that they are non-linear in the eigenvalue parameter. In the case of polynomial pencils of operators one puts $\Lambda = \mathbf{C}$ and

$$A(z) := \sum_{r=0}^{n} A_r z^r.$$

This includes the linear eigenvalue problem $Af = zBf$, where A, $B : \mathcal{B}_1 \to \mathcal{B}_2$ are linear operators and z is a complex parameter. If B is invertible then the eigenvalue problem is equivalent to $B^{-1}Af = zf$, or to $B^{-1/2}AB^{-1/2}g = zg$ if B is self-adjoint, positive and invertible. If neither A nor B is invertible then the problem poses major difficulties.

Problem 9.2.1[5] Prove that if

$$A(z) := z^2 I + A_1 z + A_2 = (zI - B_1)(zI - B_2)$$

are $n \times n$ matrices then

$$\operatorname{Spec}(A(\cdot)) = \operatorname{Spec}(B_1) \cup \operatorname{Spec}(B_2). \qquad \square$$

Example 9.2.2 Quadratic pencils of operators arise naturally in the study of an abstract wave equation.[6] Given operators A, B on some space $\mathcal{B}$ one seeks solutions of an equation of the form

$$\frac{\partial^2 f}{\partial t^2} + B \frac{\partial f}{\partial t} + Af = 0$$

that are of the form $f_t := e^{kt} g$. Assuming that suitable technical conditions are satisfied this leads one directly to the non-linear eigenvalue equation

$$k^2 g + kBg + Ag = 0. \qquad \square$$

Example 9.2.3 The following method of simplifying 'matrix' eigenvalue problems is well established. Suppose that $\mathcal{B} := \mathcal{B}_0 \oplus \mathcal{B}_1$, where we assume for simplicity that $\mathcal{B}_1$ is finite-dimensional. Let us define $L : \mathcal{B} \to \mathcal{B}$ by

$$L := \begin{pmatrix} A & B \\ C & D \end{pmatrix}$$

---

[5] This method of finding the spectrum of a quadratic matrix pencil depends on producing a suitable factorization, which is often far from easy. See [Gohberg et al. 1983, Markus 1988].

[6] The review of [Tisseur and Meerbergen 2001] on quadratic eigenvalue problems starts by discussing the uncontrolled wobbling of the Millennium Bridge in London when it was opened in June 2000.

where A, B, C, D are bounded operators between the appropriate spaces. Assume that

$$S := \text{Spec}(A) = \text{EssSpec}(A) \subseteq \mathbf{R}.$$

The fact that L is a finite rank perturbation of

$$L_0 := \begin{pmatrix} A & 0 \\ 0 & 0 \end{pmatrix}$$

implies that $\text{EssSpec}(L) = S$. A simple algebraic manipulation implies that the eigenvalues of L in $\mathbf{C} \backslash S$ are precisely those z for which the matrix equation

$$C(zI - A)^{-1} Bg + Dg - zg = 0$$

has a non-zero solution $g \in \mathcal{B}_1$. In other words the linear eigenvalue problem for L can be reduced to a non-linear eigenvalue problem for the analytic family of matrices

$$M(z) := C(zI - A)^{-1} B + D - zI$$

where $z \in \mathbf{C} \backslash S$. If $\mathcal{B}_1$ is one-dimensional then

$$f(z) := C(zI - A)^{-1} B + D - z$$

is a complex-valued analytic function defined on $\mathbf{C} \backslash S$, whose zeros are the eigenvalues of L. Compare Lemma 11.2.9. $\quad \square$

Example 9.2.4 Non-linear eigenvalue problems arise when transforming differential operators into a form which is amenable to computation. We start by describing the simplest possible example, without discussing the technical issues in any detail.[7]

Consider the eigenvalue equation

$$-f''(x) + V(x)f(x) = -\mu^2 f(x) \qquad (9.5)$$

on $\mathbf{R}$, where $\text{Re}(\mu) > 0$ and V is a complex-valued, bounded function satisfying $V(x) = 0$ if $|x| > a$. If $f \in L^2(\mathbf{R})$ is a solution to this equation then there exist constants $c_\pm$ such that $f(x) = c_+ e^{-\mu x}$ if $x \geq a$ and $f(x) = c_- e^{\mu x}$ if

---

[7] This standard technique can be applied to both eigenvalue and resonance problems in more than one space dimension, and does not depend on assuming that the differential operator is self-adjoint or that $\lambda := -\mu^2$ is real. See [Aslanyan and Davies 2001] for references to the literature applying the radiation condition to waveguides.

$x \leq -a$. Therefore solving the linear eigenvalue problem in $L^2(\mathbf{R})$ is equivalent to solving the non-linear system of equations

$$-f''(x) + V(x)f(x) + \mu^2 f(x) = 0, \tag{9.6}$$

$$f'(a) + \mu f(a) = 0, \tag{9.7}$$

$$f'(-a) - \mu f(-a) = 0, \tag{9.8}$$

in $L^2(-a, a)$, subject to the 'radiation condition' $\mathrm{Re}(\mu) > 0$.

The above procedure is exact and rigorous. It should be contrasted with the more obvious procedure of imposing Dirichlet boundary conditions at $\pm n$ and finding the eigenvalues of the associated operator on $L^2(-n, n)$ for large enough values of n. Although linear, this method is much less accurate than using the radiation condition.

The above equations may also have solutions satisfying $\mathrm{Re}(\mu) < 0$. The numbers $\lambda := -\mu^2$ are not then eigenvalues, and they are called resonances. If $\mathrm{Im}(\lambda)$ is small then λ is associated with a physical state that is nearly stationary but eventually decays.

Numerically one might solve the initial value problem (9.6) subject to (9.8) by the shooting method, and then evaluate

$$F(\mu) := f'(a) + \mu f(a).$$

On letting μ vary within $\mathbf{C}$, the points at which $F(\mu)$ vanishes are eigenvalues of (9.5).

If one has reason to expect that the eigenfunction takes its maximum value near $x = 0$ then the following modification of the above method is usually more accurate. One finds solutions $f_{\pm}$ of (9.6) that satisfy the boundary conditions at $\pm a$ by the shooting method. The eigenvalues are then the values of μ for which

$$G(\mu) := f_+(0)f'_-(0) - f'_+(0)f_-(0)$$

vanishes.

If the potential V is not of compact support then one starts by determining analytically the leading asymptotics of the solutions $f_{\pm}$ of (9.5) that decay at $\pm\infty$ respectively. These functions determine the boundary conditions that should be imposed at the points $\pm a$, where a is large enough to ensure that each eigenfunction is close to its asymptotic form. One then proceeds as before.

All of the above procedures can, by their nature, only provide approximate solutions to the eigenvalue equation. Proving that these are close to true solutions can be a major problem if the differential operator is not self-adjoint. □

Problem 9.2.5 Find a procedure for computing the spectrum of a polynomial pencil of $n \times n$ matrices, and obtain an upper bound on the number of points in the spectrum. □

In applications of the following theorem, A is often a differential operator. The theorem is still valid if B is relatively bounded with respect to A with relative bound 0 in the sense of Section 11.1. Even if A and B are self-adjoint operators on a Hilbert space, the spectrum of the pencil $C(\cdot)$ is often complex, because the operator X of the theorem is non-self-adjoint.

Theorem 9.2.6 *Let A be a closed operator acting in the Banach space $\mathcal{B}$ and let B be bounded. If $z \in \mathbf{C}$ then the operator*

$$C(z) := A + Bz + z^2 I$$

is invertible if and only if $z \notin \mathrm{Spec}(X)$, where X is the closed operator on $\mathcal{B} \oplus \mathcal{B}$ with the block matrix

$$X := \begin{pmatrix} -B & I \\ -A & 0 \end{pmatrix}.$$

Proof. The proof that $C(z)$ and $X - zI$ are closed for all $z \in \mathbf{C}$ is routine. Moreover $\mathrm{Ran}(C(z)^{-1}) = \mathrm{Dom}(A)$ provided the inverse exists. The case $z = 0$ of the theorem is trivial, so we assume that $z \neq 0$. A direct algebraic calculation shows that

$$(X - zI)^{-1} = \begin{pmatrix} -zC(z)^{-1} & -C(z)^{-1} \\ AC(z)^{-1} & -(B + zI)C(z)^{-1} \end{pmatrix}$$

and that the inverse on the LHS exists iff $C(z)^{-1}$ exists. □

We now turn to the study of the pseudospectra of a family of operators.[8] For each $\varepsilon > 0$ we define the pseudospectra by

$$\mathrm{Spec}_\varepsilon(A(\cdot)) := \mathrm{Spec}(A(\cdot)) \cup S \tag{9.9}$$

where S is the set of $\lambda \in \Lambda$ for which there exists an 'approximate eigenvector' $f \in \mathrm{Dom}(A(\lambda))$ satisfying $\|A(\lambda)f\| < \varepsilon \|f\|$. Other equivalent definitions are given in Theorem 9.2.7.

---

[8] The definition of pseudospectra given in (9.9) and the lemmas below are taken from [Davies 2005D]; a slightly different definition may be found in [Tisseur and Higham 2001].

Theorem 9.2.7 *The following three conditions on an operator family* $\{A(\lambda)\}_{\lambda \in \Lambda}$ *are equivalent.*

(i) $\lambda \in \mathrm{Spec}_\varepsilon(A(\cdot))$.

(ii) There exists a bounded operator $D : \mathcal{B}_1 \to \mathcal{B}_2$ *such that* $\|D\| < \varepsilon$ *and* $A(\lambda) + D$ *is not invertible.*

(iii) Either $\lambda \in \mathrm{Spec}(A(\cdot))$ *or* $\|A(\lambda)^{-1}\| > \varepsilon^{-1}$.

Proof. (i)$\Rightarrow$(ii) If $\lambda \in \mathrm{Spec}(A(\cdot))$ we may put $D = 0$. Otherwise let $f \in \mathrm{Dom}(A(\lambda))$, $\|f\| = 1$ and $\|A(\lambda)f\| < \varepsilon$. Let $\phi \in \mathcal{B}_1^*$ satisfy $\|\phi\| = 1$ and $\phi(f) = 1$. Then define the rank one operator $D : \mathcal{B}_1 \to \mathcal{B}_2$ by

$$Dg := -\phi(g)A(\lambda)f.$$

We see immediately that $\|D\| < \varepsilon$ and $(A(\lambda) + D)f = 0$.

(ii)$\Rightarrow$(iii) We derive a contradiction from the assumption that $\lambda \notin \mathrm{Spec}(A(\cdot))$ and $\|A(\lambda)^{-1}\| \leq \varepsilon^{-1}$. Let $B : \mathcal{B}_2 \to \mathcal{B}_1$ be the bounded operator defined by the norm convergent series

$$
\begin{aligned}
B &:= \sum_{n=0}^{\infty} A(\lambda)^{-1} \left(-DA(\lambda)^{-1}\right)^n \\
&= A(\lambda)^{-1} \left(1 + DA(\lambda)^{-1}\right)^{-1}.
\end{aligned}
$$

It is immediate from these formulae that B is one-one with range equal to $\mathrm{Dom}(A(\lambda))$. We also see that

$$B\left(1 + DA(\lambda)^{-1}\right)f = A(\lambda)^{-1}f$$

for all $f \in \mathcal{B}_2$. Putting $g := A(\lambda)^{-1}f$ we conclude that

$$B(A(\lambda) + D)g = g$$

for all $g \in \mathrm{Dom}(A(\lambda))$. The proof that

$$(A(\lambda) + D)Bh = h$$

for all $h \in \mathcal{B}_2$ is similar. Hence $A(\lambda) + D$ is invertible, with inverse B.

(iii)$\Rightarrow$(i) We assume for non-triviality that $\lambda \notin \mathrm{Spec}(A(\cdot))$. There exists $g \in \mathcal{B}_2$ such that $\|A(\lambda)^{-1}g\| > \varepsilon^{-1}\|g\|$. Putting $f := A(\lambda)^{-1}g$ we see that $\|A(\lambda)f\| < \varepsilon\|f\|$. $\quad\square$

Theorem 9.2.8 *Suppose that U is an open set in the complex plane and that $A(\cdot)$ is an operator-valued family on U, which is analytic in the sense that $A(z)^{-1}$ exists and is a bounded, operator-valued, analytic function of z on U. Then*

$$\sigma(z) := \left\| A(z)^{-1} \right\|^{-1}$$

is continuous and has no local minimum in U.

Proof. It follows immediately from the assumptions and Section 1.4 that $\rho(z) := \| A(z)^{-1} \|$ is a continuous function of z. The formula

$$\rho(z) = \sup\{ \text{Re}\langle A(z)^{-1} f, \phi \rangle : f \in \mathcal{B}, \phi \in \mathcal{B}^*, \| f \| = \| \phi \| = 1 \}$$

implies that ρ is subharmonic. Such functions have no local maxima. $\square$

Example 9.2.9 If p is a (complex-valued) polynomial of degree n and $\varepsilon > 0$ then

$$\text{Spec}_\varepsilon(p(\cdot)) := \{ z \in \mathbf{C} : |p(z)| < \varepsilon \}$$

is the union of $m \leq n$ disjoint open connected sets $U_1, \ldots, U_m$, each of which contains at least one root of $p(z) = 0$. Some of these may be very small neighbourhoods of individual roots but others may be large irregular regions containing several roots. If ε is the maximum precision to which computations can be made then one can only say that there is a root of the polynomial in an open region V if $U_r \subseteq V$ for some r. The example $p(z) := z^{20} - 10^{-20}$ shows that accurate numerical evaluation of the roots of a polynomial may be infeasible if the coefficients are only given numerically.[9] $\square$

Theorem 9.2.10 *Suppose that (Λ, d) is a metric space and that*

$$A(\lambda) := B + D(\lambda)$$

for all $\lambda \in \Lambda$, where B is a closed operator and $D(\lambda)$ are bounded operators satisfying

$$\| D(\lambda) - D(\mu) \| \leq d(\lambda, \mu)$$

for all $\lambda, \mu \in \Lambda$. Then

$$\sigma(\lambda) := \begin{cases} 0 & \text{if } \lambda \in \text{Spec}(A(\cdot)), \\ \left\| A(\lambda)^{-1} \right\|^{-1} & \text{otherwise,} \end{cases}$$

---

[9] Pseudospectral questions associated with ordinary polynomials may lead to extraordinarily deep mathematics, as can be seen from [Anderson and Eiderman 2006].

satisfies

$$|\sigma(\lambda) - \sigma(\mu)| \leq d(\lambda, \mu)$$

for all $\lambda, \mu \in \Lambda$.

Proof. If $\lambda, \mu \notin \mathrm{Spec}(A(\cdot))$ then

$$A(\lambda)^{-1} - A(\mu)^{-1} = A(\lambda)^{-1} \left(D(\mu) - D(\lambda) \right) A(\mu)^{-1}.$$

This implies

$$\left| \|A(\lambda)^{-1}\| - \|A(\mu)^{-1}\| \right| \leq d(\lambda, \mu) \|A(\lambda)^{-1}\| \, \|A(\mu)^{-1}\|$$

which yields the required result immediately.

If both λ and μ lie in $\mathrm{Spec}(A(\cdot))$ then there is nothing to prove, so suppose that $\mu \in \mathrm{Spec}(A(\cdot))$ and $\lambda \notin \mathrm{Spec}(A(\cdot))$. If $\sigma(\lambda) \geq \varepsilon$ then $\|A(\lambda)^{-1}\| \leq \varepsilon^{-1}$, so $A(\lambda) + D$ is invertible for all D such that $\|D\| < \varepsilon$ by Theorem 9.2.7. Since

$$A(\mu) = A(\lambda) + (D(\mu) - D(\lambda))$$

is assumed not to be invertible, we deduce that $d(\lambda, \mu) \geq \varepsilon$. Hence $d(\lambda, \mu) < \varepsilon$ implies $\sigma(\lambda) < \varepsilon$. □

Two similar families of operators always have the same spectra, but their pseudospectra may be very different, unless the condition number

$$\kappa(T) := \|T\| \, \|T^{-1}\|$$

of the relevant operator T is fairly close to 1.

Lemma 9.2.11 *Let T be a bounded invertible operator on $\mathcal{B}$ and put*

$$\tilde{A}(\lambda) := TA(\lambda)T^{-1}.$$

Then

$$\mathrm{Spec}(A(\cdot)) = \mathrm{Spec}(\tilde{A}(\cdot))$$

and

$$\mathrm{Spec}_{\varepsilon/\kappa}(A(\cdot)) \subseteq \mathrm{Spec}_{\varepsilon}(\tilde{A}(\cdot)) \subseteq \mathrm{Spec}_{\varepsilon\kappa}(A(\cdot))$$

for all $\varepsilon > 0$ where $\kappa \geq 1$ is the condition number of T.

We omit the proof, which is elementary.

Problem 9.2.12 Let A be a bounded, invertible operator on a Banach space $\mathcal{B}$. Prove that if λ, μ are two points in $\mathrm{Spec}(A)$ then

$$\kappa(A)^{-1} \leq |\lambda/\mu| \leq \kappa(A).$$

Prove that these bounds cannot be improved if $\mathcal{B}$ is a Hilbert space and A is normal, i.e. $A A^* = A^* A$. $\square$

We now return to the study of a single operator. By definition the pseudospectra of a closed operator A acting in a Banach space $\mathcal{B}$ are the pseudospectra of the family $A(z) := zI - A$. Each subset $\mathrm{Spec}_\varepsilon(A)$ of $\mathbf{C}$ is the union of $\mathrm{Spec}(A)$ and

$$\{z \in \mathbf{C} : \|Af - zf\| < \varepsilon\|f\| \text{ for some } f \in \mathrm{Dom}(A)\}. \tag{9.10}$$

Theorem 9.2.13 *The following three conditions on a closed operator A are equivalent.*

(i) $z \in \mathrm{Spec}_\varepsilon(A)$.
(ii) *There exists a bounded operator D such that $\|D\| < \varepsilon$ and $z \in \mathrm{Spec}(A + D)$.*
(iii) *Either $z \in \mathrm{Spec}(A)$ or $\|R(z, A)\| > \varepsilon^{-1}$.*

Proof. See Theorem 9.2.7. $\square$

Example 9.2.14 By applying the above theorem to Example 9.1.4 one deduces that there must exist very small perturbations of the Jordan matrix J_n whose spectrum is far from $\{0\}$.[10] This is confirmed in Figure 9.1, which shows the eigenvalues of $J_n + \varepsilon B$ where $n := 100$, $\varepsilon := 10^{-20}$ and

$$B_{r,s} := \begin{cases} 1 & \text{if } r = n \text{ and } s = 1, \\ -2 & \text{if } r = n \text{ and } s = 2, \\ 3 & \text{if } r = n-1 \text{ and } s = 1, \\ 5 & \text{if } r = n-1 \text{ and } s = 2, \\ 0 & \text{otherwise.} \end{cases} \qquad \square$$

At a theoretical level the study of pseudospectra is equivalent to determining the behaviour of the function

$$\sigma(z) := \begin{cases} 0 & \text{if } z \in \mathrm{Spec}(A), \\ \|R(z, A)\|^{-1} & \text{otherwise.} \end{cases}$$

---

[10] The spectrum of such matrices has been investigated in [Davies and Hager 2006], where it is shown that the spectrum of J_n plus a small perturbation typically concentrates close to a certain circle, with the exception of a few eigenvalues inside it.

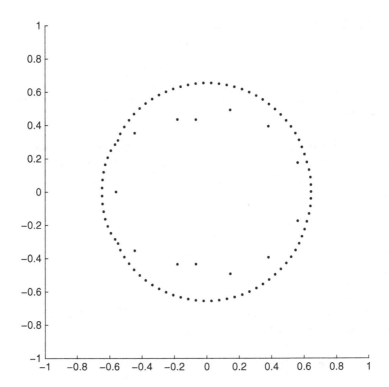

Figure 9.1: Spectrum of the matrix A of Example 9.2.14

Theorem 9.2.15 *The function σ satisfies*
$$|\sigma(z) - \sigma(w)| \leq |z - w|$$
for all $z, w \in \mathbf{C}$. It has no local minima in the complement of $\mathrm{Spec}(A)$.

Proof. This follows directly from Theorems 9.2.8 and 9.2.10. ☐

The following theorem may be used two ways. If one only knows an operator A to within an error $\delta > 0$ then its pseudospectra $\mathrm{Spec}_\varepsilon(A)$ do not have any significance for $\varepsilon < \delta$, although they are numerically stable for substantially larger ε. Conversely if one is only interested in the shape of the pseudospectra of A for $\varepsilon > \delta$, one may add any perturbation of norm significantly less than δ to A before carrying out the computation.

Theorem 9.2.16 *Let A_1, A_2 be two bounded operators on $\mathcal{B}$ satisfying $\|A_1 - A_2\| < \delta$. If we put*

$$\sigma_r(z) := \begin{cases} 0 & \text{if } z \in \mathrm{Spec}(A_r), \\ \|R(z, A_r)\|^{-1} & \text{otherwise,} \end{cases}$$

for $r = 1, 2$ *then*

$$|\sigma_1(z) - \sigma_2(z)| \leq \delta$$

for all $z \in \mathbb{C}$.

Proof. If $z \in \mathrm{Spec}(A_1)$ or $z \in \mathrm{Spec}(A_2)$ then the theorem follows directly from Theorem 9.2.13. If neither holds then we use the formula

$$R(z, A_1) - R(z, A_2) = R(z, A_1)(A_1 - A_2)R(z, A_2)$$

to obtain

$$\big|\, \|R(z, A_1)\| - \|R(z, A_2)\| \,\big| \leq \|R(z, A_1) - R(z, A_2)\|$$

$$\leq \|R(z, A_1)\| \, \|(A_1 - A_2)\| \, \|R(z, A_2)\|,$$

which is equivalent to the stated estimate. $\square$

Either Corollary 8.1.4 or Theorem 9.2.15 implies that the pseudospectra of a closed operator A satisfy

$$\mathrm{Spec}_\varepsilon(A) \supseteq \{z : \mathrm{dist}\{z, \mathrm{Spec}(A)\} < \varepsilon\}$$

for all $\varepsilon > 0$. Bounds in the reverse direction sometimes exist, but the constants involved are frequently so large that they are not useful. We start with a positive result.

Theorem 9.2.17 *Let A and S be bounded operators on the Hilbert space $\mathcal{H}$. If S is invertible and $N := SAS^{-1}$ is normal, then*

$$\mathrm{Spec}_\varepsilon(A) \subseteq \{z : \mathrm{dist}(z, \mathrm{Spec}(A)) < \varepsilon\kappa(S)\}$$

where $\kappa(S)$ is the condition number of S.

Proof. If N is normal then it has the same spectrum as A and

$$\mathrm{Spec}_\varepsilon(N) = \{z : \mathrm{dist}(z, \mathrm{Spec}(N)) < \varepsilon\}.$$

The proof is completed by applying Lemma 9.2.11. $\square$

Sometimes one wishes to restrict the possible perturbations in the definition of generalized pseudospectra on physical or mathematical grounds; for example one might only be interested in perturbations which preserve some property of the original operator family. If $\mathcal{C}$ is some class of perturbation operators with its own norm $\|\|\cdot\|\|$ then one can define the structured pseudospectra

$$\mathrm{Spec}_{\varepsilon, \mathcal{C}}(A(\cdot)) := \{\lambda \in \Lambda : \exists D \in \mathcal{C} . \|\|D\|\| < \varepsilon$$

$$\text{and } A(\lambda) + D \text{ is not invertible.}\}.$$

The choice of $\mathcal{C}$ and $\|\!|\!| \cdot |\!|\!\|$ depend heavily upon the context. One of the most obvious choices is to restrict attention to perturbations that are real (i.e. have real entries in the case of matrices). Examples demonstrate that even this change can have dramatic effects on the shape of the pseudospectral regions.[11]

The following theorem provides a procedure for computing structured pseudospectra. We suppose that A is a closed operator acting in a Banach space $\mathcal{B}$, and that $B : \mathcal{B} \to \mathcal{C}$, $C : \mathcal{D} \to \mathcal{B}$ are two given bounded operators. We also let K denote a generic bounded operator from $\mathcal{C}$ to $\mathcal{D}$. The structured pseudospectra, or spectral value sets, are defined by

$$\sigma(A, B, C, \varepsilon) := \bigcup_{\{K : \|K\| < \varepsilon\}} \mathrm{Spec}(A + CKB).$$

Theorem 9.2.18 *Under the above assumptions we have*

$$\sigma(A, B, C, \varepsilon) = \mathrm{Spec}(A) \cup \{z : \|BR(z, A)C\| > \varepsilon^{-1}\}.$$

Proof. By putting $K := 0$ one sees that

$$\mathrm{Spec}(A) \subseteq \sigma(A, B, C, \varepsilon)$$

for all $\varepsilon > 0$. If $z \notin \mathrm{Spec}(A)$ then the formula

$$(zI - A - CKB)^{-1} = R(z, A)(I - CKBR(z, A))^{-1}$$

shows that $z \notin \sigma(A, B, C, \varepsilon)$ if and only if $1 \notin \mathrm{Spec}(CKBR(z, A))$ for all K with $\|K\| < \varepsilon$. This is equivalent to $1 \notin \mathrm{Spec}(KBR(z, A)C)$ for all such K, by an adaptation of Problem 1.2.5, and hence to $\|BR(z, A)C\| \leq \varepsilon^{-1}$. $\square$

The following establishes a connection between pseudospectra and the norms of the spectral projections.

Theorem 9.2.19 *Let A be a closed operator acting in the Banach space $\mathcal{B}$. Suppose that λ is an isolated point of $\mathrm{Spec}(A)$ and that P is the corresponding spectral projection P. Let γ be a Jordan curve enclosing the point λ and no other point of $\mathrm{Spec}(A)$, and suppose that it is a pseudospectral contour in the sense that $\|R(z, A)\| = a$ for all z on γ. Then*

$$\|P\| \leq \frac{a|\gamma|}{2\pi},$$

where $|\gamma|$ is the length of γ.

---

[11] See [Trefethen and Embree 2005, Chap. 50], [Hinrichsen and Pritchard 1992] and [Hinrichsen and Pritchard 2005] for further information about structured pseudospectra and their applications to control theory.

If $Af = \lambda f$ *for all* $f \in \mathrm{Ran}(P)$ *(e.g. the Jordan form of A restricted to* $\mathrm{Ran}(P)$ *is trivial) then*

$$\|P\| = \lim_{z \to \lambda} |z - \lambda| \|R(z, A)\|.$$

Proof. The first statement is obtained by a routine estimate of the RHS of the formula

$$P = \frac{1}{2\pi i} \int_\gamma R(z, A) \, dz,$$

proved in Theorem 1.5.4.

Under the further hypothesis of the theorem we have

$$R(z, A) = P(z - \lambda)^{-1} + (I - P)R(z, A).$$

This implies

$$\lim_{z \to \lambda}(z - \lambda)R(z, A) = P \tag{9.11}$$

because

$$\lim_{z \to \lambda}(I - P)R(z, A) = R(\lambda, A')(I - P)$$

where A' is the restriction of A to $(I - P)\mathcal{B}$. Equation (9.11) implies the second statement of the theorem. $\square$

The following is a partial converse in finite dimensions. Its usefulness is limited by the fact that the constant c often increases very rapidly (e.g. exponentially) with the dimension.

Problem 9.2.20 Prove that if $\mathcal{B} = \mathbf{C}^N$ and each spectral projection P_n of A satisfy $\|P_n\| \leq c$ and $AP_n = \lambda_n P_n$ then

$$\|R(z, A)\| \leq cN\{\mathrm{dist}(z, \mathrm{Spec}(A))\}^{-1}$$

for all $z \notin \mathrm{Spec}(A)$. $\square$

Problem 9.2.21 Prove that the $n \times n$ matrix A is normal if and only if

$$\|R(z, A)\| = \{\mathrm{dist}(z, \mathrm{Spec}(A))\}^{-1}$$

for all $z \notin \mathrm{Spec}(A)$. $\square$

Computations of the norms of the spectral projections of randomly generated, non-self-adjoint matrices show that they are typically very large. The rate at which the norms increase with the dimension of the matrix depends heavily upon the class of random matrices considered, but such results cast a new light on the theorem that almost every matrix is diagonalizable. This is now a highly developed field, in which many theoretical results and numerical experiments have been published.[12]

Problem 9.2.22 Calculate the eigenvectors and then the norms of the spectral projections of the $N \times N$ matrix A defined by

$$A_{r,s} := \begin{cases} 1 & \text{if } r = s - 1, \\ 2^{-N} & \text{if } r = N \text{ and } s = 1, \\ 0 & \text{otherwise.} \end{cases} \qquad \square$$

The results in this section should not lead one to believe that the pseudospectra of an operator A control all other quantities of interest. Ransford has shown that for any $\alpha > 0$, $\beta > 0$ and $k \geq 2$ there exist $n \times n$ matrices A and B such that

$$\|R(z, A)\| = \|R(z, B)\|$$

for all $z \in \mathbf{C}$, although $\|A^k\| = \alpha$ and $\|B^k\| = \beta$. One may take $n := 2k + 3$.[13]

9.3 The numerical range

If A is a bounded operator on a Banach space $\mathcal{B}$ then Theorem 1.2.11 implies that

$$\mathrm{Spec}_\varepsilon(A) \subseteq \{z : |z| < \|A\| + \varepsilon\}.$$

Lemma 9.3.14 below provides a much stronger bound on $\mathrm{Spec}_\varepsilon(A)$ in the Hilbert space context. We define the numerical range $\mathrm{Num}(A)$ of a possibly unbounded operator A acting in a Hilbert space $\mathcal{H}$ by

$$\mathrm{Num}(A) := \{\langle Af, f \rangle : f \in \mathrm{Dom}(A) \text{ and } \|f\| = 1\}.$$

In this section and the next we introduce a variety of different sets associated with an operator. Some of the inclusions between these are summarized in the diagram on page 281.

---

[12] See [Trefethen and Embree 2005, Chaps 35–38] for a survey of current results from the pseudospectral point of view.

[13] Ransford also has counterexamples to a number of other conjectures of this type in [Ransford 2006].

Theorem 9.3.1 (*Toeplitz-Hausdorff*) *The numerical range of an operator A acting in a Hilbert space is a convex set. If A is bounded then*

$$\mathrm{Spec}(A) \subseteq \overline{\mathrm{Num}}(A) \subseteq \{z : |z| \leq \|A\|\}$$

where $\overline{\mathrm{Num}}$ *stands for the closure of the numerical range.*

Proof. Upon replacing A by $\alpha A + \beta I$ for suitable $\alpha, \beta \in \mathbf{C}$, the proof of convexity reduces to the following claim: if $\|f\| = \|g\| = 1$ and $\langle Af, f \rangle = 0$, $\langle Ag, g \rangle = 1$, then for all $\lambda \in (0, 1)$ there exists $h \in \mathrm{lin}\{f, g\}$ such that $\|h\| = 1$ and $\langle Ah, h \rangle = \lambda$.

If $\theta, s \in \mathbf{R}$ and $k_{\theta,s} := f + se^{i\theta}g$ then $k_{\theta,s} \neq 0$ and

$$\langle Ak_{\theta,s}, k_{\theta,s} \rangle = c_\theta s + s^2$$

where

$$c_\theta := e^{i\theta}\langle Ag, f \rangle + e^{-i\theta}\langle Af, g \rangle.$$

The identity $c_\pi = -c_0$ implies by the intermediate value theorem that there exists $\alpha \in [0, \pi]$ such that c_α is real. For this choice of α the real-valued function

$$F(s) := \langle Ak_{\alpha,s}, k_{\alpha,s} \rangle / \|k_{\alpha,s}\|^2$$

satisfies $F(0) = 0$ and $\lim_{s \to +\infty} F(s) = 1$. It must therefore take the value λ for some $s \in (0, \infty)$ by the intermediate value theorem. We put $h := k_{\alpha,s}/\|k_{\alpha,s}\|$ for this choice of s to complete the proof of convexity.

If $z \in \mathrm{Spec}(A)$ then Lemma 1.2.13 states that either there exists a sequence $f_n \in \mathcal{H}$ such that $\|f_n\| = 1$ and $\|Af_n - zf_n\| \to 0$, or there exists $f \in \mathcal{H}$ such that $\|f\| = 1$ and $A^*f = \bar{z}f$. In the first case it follows that $\langle Af_n, f_n \rangle \to z$ and hence $z \in \overline{\mathrm{Num}}(A)$. In the second case $\langle Af, f \rangle = z$ so $z \in \mathrm{Num}(A)$.

The final inclusion of the theorem is elementary. $\square$

The proof of the above theorem reduces to proving it for an arbitrary 2×2 matrix. In this case one can actually say much more.

Problem 9.3.2 Prove that the numerical range of a 2×2 matrix consists of the boundary and interior of a (possibly degenerate) ellipse. Prove that the numerical range of an $n \times n$ matrix is always compact and give an example of a bounded operator A whose numerical range is not closed. $\square$

If P is the orthogonal projection onto a closed subspace $\mathcal{L}$ of a Hilbert space $\mathcal{H}$ we call $PAP|_{\mathcal{L}}$ the truncation of A to $\mathcal{L}$. An enormous number of numerical computations involve truncating an operator to a large but finite-dimensional subspace, constructed using finite element or other methods, and then proving, or merely hoping, that the truncation provides useful spectral information about the original operator.[14] The next two results state that the numerical range is stable under perturbation and truncation. The spectrum is not so well behaved in either respect.[15]

Problem 9.3.3 Prove that if A and B are bounded operators on $\mathcal{H}$ and $\|A - B\| < \varepsilon$ then

$$\mathrm{Num}(A) \subseteq \{z : \mathrm{dist}(z, \mathrm{Num}(B)) < \varepsilon\}$$

and vice versa. $\square$

Theorem 9.3.4 *Let A be a bounded operator on the Hilbert space $\mathcal{H}$ and let $\mathcal{L}_n$ be an increasing sequence of closed subspaces of $\mathcal{H}$ with dense union. If A_n is the truncation of A to $\mathcal{L}_n$ then*

$$\mathrm{Spec}(A_n) \subseteq \overline{\mathrm{Num}}(A_n) \subseteq \overline{\mathrm{Num}}(A)$$

for all n, and $\overline{\mathrm{Num}}(A_n)$ is an increasing sequence of sets whose union is dense in $\overline{\mathrm{Num}}(A)$.

Proof. The first inclusion was proved in Theorem 9.3.1. The second follows from

$$\mathrm{Num}(A_n) = \{\langle Af, f \rangle : f \in \mathcal{L}_n \text{ and } \|f\| = 1\}$$
$$\subseteq \{\langle Af, f \rangle : f \in \mathcal{H} \text{ and } \|f\| = 1\}$$
$$= \mathrm{Num}(A).$$

The convergence of the sets $\overline{\mathrm{Num}}(A_n)$ to $\overline{\mathrm{Num}}(A)$ in the stated sense follows from the formula

$$\lim_{n \to \infty} \frac{\langle AP_n f, P_n f \rangle}{\|P_n f\|^2} = \frac{\langle Af, f \rangle}{\|f\|^2}$$

for all non-zero $f \in \mathcal{H}$. $\square$

---

[14] This hope is not always justified even for self-adjoint operators. If the operator has a gap in its essential spectrum the truncation regularly has a large number of spurious eigenvalues in the gap. See [Davies and Plum 2004] for methods of avoiding this pathology, called spectral pollution. We will see that the situation for non-self-adjoint operators is much worse.

[15] See [Gustafson and Rao 1997] for a further discussion of the numerical range from the point of view of both functional and numerical analysis.

Problem 9.3.5 Use the spectral theorem to prove that if A is a normal operator on a Hilbert space then $\overline{\mathrm{Num}}(A)$ is the closed convex hull of $\mathrm{Spec}(A)$. $\quad\square$

Problem 9.3.6 Use the spectral theorem to prove that if U is a unitary operator on a Hilbert space then

$$\mathrm{Spec}(U) = \overline{\mathrm{Num}}(U) \cap \{z : |z| = 1\}. \qquad\square$$

Example 9.3.7 Let A be the convolution operator $A(f) := a * f$ on $l^2(\mathbf{Z})$, where

$$a_n := \begin{cases} 2 & \text{if } n = 1, \\ 1 & \text{if } n = 5, \\ 0 & \text{otherwise.} \end{cases}$$

The spectrum of A is shown in Figure 9.2. If one truncates this operator to the space of all functions with support in $\{-n, \dots, n\}$ then one obtains a strictly upper triangular matrix A_n. Although $\mathrm{Spec}(A_n) = \{0\}$ for all n, the numerical

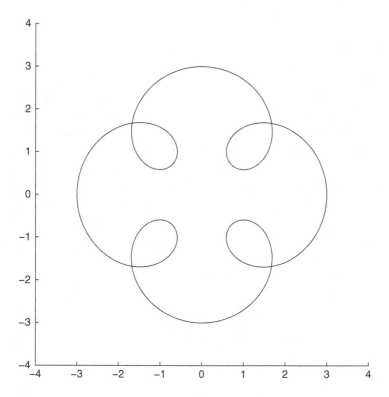

Figure 9.2: Spectrum of the operator A of Example 9.3.7

range of A_n converges to that of A by Theorem 9.3.4. This is the convex hull of the spectrum of A by Problem 9.3.5. ☐

We need a result from convexity theory, which can be extended to infinite dimensions using the Hahn-Banach theorem.

Proposition 9.3.8 *(separation theorem) If K is a compact convex set in $\mathbf{R}^n$ and $a \notin K$ then there exists a linear functional $\phi : \mathbf{R}^n \to \mathbf{R}$ such that $\phi(a) > \max\{\phi(x) : x \in K\}$.*

Problem 9.3.9 If K is a compact convex set in $\mathbf{R}^n$ and $a \notin K$ prove that there is a unique point $k \in K$ such that

$$\|a - k\| = \operatorname{dist}(a, K),$$

where $\|\cdot\|$ is the Euclidean norm. Also prove that

$$K_\varepsilon := \{x \in \mathbf{R}^n : \operatorname{dist}(x, K) < \varepsilon\}$$

is open and convex. ☐

The numerical range of operators can often be determined, or at least estimated, with the aid of the following theorem. An application of the ideas in the theorem to Schrödinger operators is described in Section 14.2. One can also determine the numerical range of certain pseudodifferential operators in the semi-classical limit; see Theorems 14.6.8 and 14.6.12.

Theorem 9.3.10 *If A is a bounded operator on $\mathcal{H}$ and $\theta \in [-\pi, \pi]$, put $\lambda_\theta := \max \operatorname{Spec}(B_\theta)$ where $B_\theta := \frac{1}{2}(e^{-i\theta} A + e^{i\theta} A^*) = B_\theta^*$. Then*

$$\overline{\operatorname{Num}}(A) = \bigcap_{\theta \in [-\pi, \pi]} H_\theta$$

where the half-space H_θ is defined by

$$H_\theta = \{z : \operatorname{Re}(e^{-i\theta} z) \le \lambda_\theta\}.$$

Proof. If $\|f\| = 1$ and $z := \langle Af, f \rangle$ then

$$\lambda_\theta \ge \langle B_\theta f, f \rangle = \tfrac{1}{2}(e^{-i\theta} z + e^{i\theta} \overline{z}) = \operatorname{Re}(e^{-i\theta} z).$$

Therefore $z \in H_\theta$ for all $\theta \in [-\pi, \pi]$.

Conversely suppose that $z \notin \overline{\mathrm{Num}}(A)$. Proposition 9.3.8 implies that there exists $\theta \in [-\pi, \pi]$ such that

$$\mathrm{Re}(e^{-i\theta}z) > \max\{\mathrm{Re}(e^{-i\theta}w) : w \in \overline{\mathrm{Num}}(A)\}$$
$$= \sup\{\mathrm{Re}(e^{-i\theta}w) : w = \langle Af, f \rangle \text{ and } \|f\| = 1\}$$
$$= \sup\{\langle B_\theta f, f \rangle : \|f\| = 1\}$$
$$= \lambda_\theta.$$

Therefore $z \notin H_\theta$. □

Note The above idea may be implemented numerically if A is a large $n \times n$ matrix. In that case λ_θ is the largest eigenvalue of B_θ, and there are efficient algorithms for computing this. If x_θ is a normalized eigenvector of B_θ associated with the eigenvalue λ_θ then $z_\theta := \langle Ax_\theta, x_\theta \rangle$ lies on the boundary of $\mathrm{Num}(A)$. Generically z_θ traces out the boundary of $\mathrm{Num}(A)$ as θ varies in $[-\pi, \pi]$, but if the boundary contains a straight line segment the situation is more complicated.

Example 9.3.11 Let J_n be the usual $n \times n$ Jordan matrix:

$$(J_n)_{r,s} := \begin{cases} 1 & \text{if } s = r+1, \\ 0 & \text{otherwise.} \end{cases}$$

If $\theta \in \mathbf{R}$ and we define $v_\theta \in \mathbf{C}^n$ by $v_{\theta,r} := n^{-1/2}e^{ir\theta}$, then a direct calculation shows that

$$\langle J_n v_\theta, v_\theta \rangle = \frac{n-1}{n}e^{i\theta}.$$

Using the convexity of $\mathrm{Num}(A)$ and the fact that $\|J_n\| = 1$ we deduce that

$$\{z : |z| \le (n-1)/n\} \subseteq \mathrm{Num}(J_n) \subseteq \{z : |z| \le 1\}.$$

This shows that $\mathrm{Num}(J_n)$ is very different from $\mathrm{Spec}(J_n) = \{0\}$. It may be explained by the extreme numerical instability of the spectrum of J_n under perturbations. □

Problem 9.3.12 Use the ideas in Theorem 9.3.10 to determine the numerical range of the Jordan matrix J_n exactly. □

Example 9.3.13 The numerical range of the Volterra operator

$$(Af)(x) := \int_0^x f(y)\,dy$$

acting on $L^2(0, 1)$ may be described in closed form by using Theorem 9.3.10. The identity

$$\mathrm{Re}\langle Af, f\rangle = \tfrac{1}{2}\left|\int_0^1 f(x)\,\mathrm{d}x\right|^2$$

implies that $\mathrm{Num}(A) \subseteq \{z : 0 \leq \mathrm{Re}(z) \leq \tfrac{1}{2}\}$. Defining λ_θ as in Theorem 9.3.10, we see that $\lambda_0 = \tfrac{1}{2}$ and $\lambda_\pi = 0$.

If $\theta \neq 0, \pi$, $\|f\| = 1$ and $B_\theta f = \lambda f$ then

$$\lambda f'(x) = \tfrac{1}{2}\mathrm{e}^{-i\theta} f(x) - \tfrac{1}{2}\mathrm{e}^{i\theta} f(x) = -i\sin(\theta) f(x).$$

Therefore $\lambda \neq 0$ and $f(x) = \mathrm{e}^{-ix\sin(\theta)/\lambda}$ for all $x \in [0, 1]$. The actual eigenvalues λ are obtained by re-substituting this into the eigenvalue equation. This yields

$$\lambda = \frac{\sin(\theta)}{2\theta + 2n\pi}$$

where $n \in \mathbf{Z}$. Therefore

$$\lambda_\theta = \frac{\sin(\theta)}{2\theta}.$$

This completes the description of $\overline{\mathrm{Num}}(A)$. $\square$

Outside the numerical range, the pseudospectra of an operator are well-behaved.

Lemma 9.3.14 *Suppose that A is closed and that $\mathbf{C}\backslash\overline{\mathrm{Num}}(A)$ is connected and contains at least one point not in $\mathrm{Spec}(A)$. Then*

$$\mathrm{Spec}(A) \subseteq \overline{\mathrm{Num}}(A).$$

Moreover

$$\mathrm{Spec}_\varepsilon(A) \subseteq \{z : \mathrm{dist}(z, \overline{\mathrm{Num}}(A)) < \varepsilon\}$$

for all $\varepsilon > 0$. Equivalently

$$\|(zI - A)^{-1}\| \leq \{\mathrm{dist}(z, \overline{\mathrm{Num}}(A))\}^{-1}$$

for all $z \in \mathbf{C}$.

Proof. We have

$$\{z : \mathrm{dist}(z, \overline{\mathrm{Num}}(A)) < \varepsilon\} = \bigcap_{H \in \mathcal{S}}\{z : \mathrm{dist}(z, H) < \varepsilon\}$$

where $\mathcal{S}$ is the set of all closed half-spaces H which contain $\overline{\mathrm{Num}}(A)$. By using this observation and replacing A by $\alpha I + \beta A$ for suitable α, $\beta \in \mathbf{C}$, one needs only to prove the following. If $\mathrm{Re}\langle Af, f \rangle \leq 0$ for all $f \in \mathrm{Dom}(A)$ and there exists $\lambda \notin \mathrm{Spec}(A)$ such that $\mathrm{Re}(\lambda) > 0$, then

$$\mathrm{Spec}(A) \subseteq \{z : \mathrm{Re}(z) \leq 0\}$$

and

$$\|(zI - A)^{-1}\| \leq \mathrm{Re}(z)^{-1}$$

for all z such that $\mathrm{Re}(z) > 0$. This is proved by combining Theorems 8.3.5 and 8.3.2. $\quad\Box$

In the next two theorems we use the numerical range to prove operator analogues of a classical theorem of Kakeya about zeros of polynomials – if $\dim(\mathcal{H}) = 1$ and the sum in (9.12) is finite then $U = \mathbf{C}$ and $\mathrm{Spec}(A(\cdot))$ is the set of zeros of the polynomial (9.12).

Theorem 9.3.15 *Suppose that A_n are bounded, self-adjoint operators on the Hilbert space $\mathcal{H}$ for all $n \geq 0$ and that $0 \leq A_{n+1} \leq A_n$ for all n. Suppose also that A_0 is invertible and that*

$$A(z) := \sum_{n=0}^{\infty} A_n z^n. \tag{9.12}$$

This series converges in norm for all z such that $|z| < 1$, and the resulting operators $A(z)$ are all invertible. If $A(z)$ can be analytically continued to a larger region U then it follows that

$$\mathrm{Spec}(A(\cdot)) \subseteq U \cap \{z : |z| \geq 1\}.$$

Proof. The norm convergence of the series if $|z| < 1$ is an immediate consequence of the bound $\|A_n\| \leq \|A_0\|$. By considering $A_0^{-1/2} A(z) A_0^{-1/2}$ and using Problem 5.2.3 we reduce to the case in which $A_0 = I$. Consider the expression

$$(1 - z)A(z) = I - \sum_{n=1}^{\infty} B_n z^n$$

where $B_n := A_{n-1} - A_n \geq 0$ and $0 \leq \sum_{n=1}^{\infty} B_n \leq I$.
  If $\|f\| = 1$ and $|z| < 1$ then

$$\mathrm{Re}\langle (1 - z)A(z)f, f \rangle = 1 - \sum_{n=1}^{\infty} \langle B_n f, f \rangle \mathrm{Re}(z^n)$$

$$\geq 1 - \sum_{n=1}^{\infty} \langle B_n f, f \rangle |z|$$

$$\geq 1 - |z|.$$

Therefore

$$\mathrm{Spec}((1-z)A(z)) \subseteq \overline{\mathrm{Num}}((1-z)A(z)) \subseteq \{w : \mathrm{Re}(w) \geq 1 - |z|\}.$$

Therefore $A(z)$ is invertible. □

Corollary 9.3.16 *Suppose that A_n are bounded, self-adjoint operators on the Hilbert space $\mathcal{H}$ for $0 \leq n \leq N$ and that $0 \leq A_{n-1} \leq A_n$ for $1 \leq n \leq N$. Suppose also that A_N is invertible and that*

$$A(z) := \sum_{n=0}^{N} A_n z^n.$$

Then

$$\mathrm{Spec}(A(\cdot)) \subseteq \{z : |z| \leq 1\}.$$

Proof. If $z \neq 0$ then

$$z^N A(z^{-1}) = \sum_{n=0}^{N} A_{N-n} z^n$$

and we may apply Theorem 9.3.15 to the RHS. □

The numerical range may be used to prove results about the zeros of orthogonal polynomials. Let μ be a probability measure which has compact support S in the complex plane, and suppose that μ is non-trivial in the sense that S is an infinite set. This condition implies that $\{z^n\}_{n=0}^{\infty}$ is a linearly independent set in $L^2(S, d\mu)$; equivalently every non-zero polynomial is also non-zero as an element of $L^2(S, d\mu)$. One may construct the sequence of orthogonal polynomials $p_n(z)$ of degree n by applying the Gram-Schmidt procedure to the monomials z^n, $n = 0, 1, 2, \ldots$, regarded as elements of the Hilbert space $L^2(S, d\mu)$. Two special cases are of particular interest: when $S \subseteq \mathbf{R}$, in which case the measure need not have compact support, but its moments must all be finite, and when $S \subseteq \{z : |z| = 1\}$.[16]

We recall some basic linear algebra. Every $n \times n$ matrix A possesses a characteristic polynomial p of degree n defined by $p(z) := \det(zI - A)$. The zeros of p coincide with the eigenvalues of A. The minimal polynomial m of A is defined to be the (unique monic) polynomial of lowest degree such that $m(A) = 0$. The minimal polynomial m is a factor of p, and if they have equal degrees then $m = p$.

---

[16] The recent 1100 page monograph of [Simon 2005B] on orthogonal polynomials demonstrates the richness of this subject.

Theorem 9.3.17 *(Fejér) All of the zeros of the polynomials p_n lie in $\overline{\text{Conv}}(S)$. In particular if $S \subseteq \mathbf{R}$ then all their zeros lie in $[a, b]$, where a (resp. b) are the maximum (resp. minimum) points of S.*

Proof. Let M be the bounded normal operator $(Mf)(z) := zf(z)$ acting on $L^2(S, d\mu)$. Then $\text{Spec}(M) = S$: if $a \notin S$ then $(aI - M)^{-1}$ is the operator of multiplication by $(a - z)^{-1}$.

Now let P_n denote the orthogonal projection onto

$$\mathcal{L}_n := \text{lin}\{1, z, z^2, \ldots, z^{n-1}\} = \text{lin}\{p_0, p_1, \ldots, p_{n-1}\}$$

and let $M_n := P_n M P_n|_{\mathcal{L}_n}$. If p is a non-zero polynomial of degree less than n then $p(M_n)1 = p$, and this is non-zero as an element of $L^2(S, d\mu)$. Therefore the minimal polynomial m_n of M_n cannot be of degree less than n. It follows that m_n equals the characteristic polynomial of M_n.

We next show that the minimal polynomial is p_n. We first observe that if a polynomial p has degree less than n then $p = p(M)1 = p(M_n)1$, and if p has degree n then $P_n p = p(M_n)1$. Taking both m_n and p_n to be monic, their difference q_n is of degree less than n. Hence

$$q_n = q(M_n)1 = p_n(M_n)1 - m_n(M_n)1 = p_n(M_n)1 = P_n p_n = 0.$$

The last equality uses the fact that $p_n \perp \mathcal{L}_n$. Therefore $m_n = p_n$.

We finally discuss the zeros of the polynomials. By the above results the zeros of p_n coincide with the eigenvalues of M_n. Since M_n is a restriction of M the set $Z(p_n)$ of zeros of p_n satisfies

$$Z(p_n) = \text{Spec}(M_n) \subseteq \overline{\text{Num}}(M_n) \subseteq \overline{\text{Num}}(M) = \overline{\text{Conv}}(S),$$

where the final equality uses Problem 9.3.5. $\square$

Example 9.3.18 The orthogonal polynomials associated with the measure $\mu := dx$ on $[-1, 1]$ are called the Legendre polynomials, and are (constant multiples of)

$$P_n(x) := \frac{1}{2^n n!} \frac{d^n}{dx^n} (x^2 - 1)^n.$$

The first few Legendre polynomials are $P_0(x) := 1$, $P_1(x) := x$, $P_2(x) := \frac{1}{2}(3x^2 - 1)$, $P_3(x) := \frac{1}{2}(5x^3 - 3x)$. The fact that these particular orthogonal polynomials are also the eigenfunctions of a Sturm-Liouville differential operator acting in $L^2(-1, 1)$, namely

$$(Lf)(x) := -\frac{d}{dx}\left\{(1 - x^2)\frac{df}{dx}\right\},$$

is highly untypical. Other examples with the same property are Hermite, Laguerre and Chebyshev polynomials. It leads to a completely different proof of Theorem 9.3.17 in these cases.[17]   □

Theorem 8.1.12 and Problem 8.1.13 show that the spectrum of an operator may change totally if one truncates it to a subspace. The remainder of the section demonstrates the same phenomenon for operators that are important in applied mathematics. We start by considering convolution operators.[18]

Example 9.3.19 Let $k : \mathbf{R} \to \mathbf{C}$ satisfy $|k(x)| \leq c e^{-\alpha|x|}$ for some positive c, α and all $x \in \mathbf{R}$. Assuming that $-\alpha < \beta < \alpha$, the function $k_\beta(x) := e^{\beta x} k(x)$ lies in $L^1(\mathbf{R})$ and is associated with a bounded convolution operator A_β defined on $L^2(\mathbf{R})$ by $A_\beta f := k_\beta * f$. It follows by Theorem 3.1.19 that

$$\mathrm{Spec}(A_\beta) = \{\hat{k}(\xi + i\beta) : \xi \in \mathbf{R}\} \cup \{0\}.$$

In particular the spectrum of k_β depends on β in a manner that is often simple to calculate.

For any positive constant n one may truncate the above operators to $L^2(-n, n)$. Denoting the truncations by $A_{\beta,n}$, we see that

$$A_{\beta,n} = S_{\beta,n} A_{0,n} S_{\beta,n}^{-1}$$

where $S_{\beta,n}$ is the operator of multiplication by $e^{\beta x}$, which is bounded and invertible on $L^2(-n, n)$. Therefore the spectrum of $A_{\beta,n}$ does not depend on β. If $k(-x) = \overline{k(x)}$ for all $x \in \mathbf{R}$ then A_0 and $A_{0,n}$ are self-adjoint operators. Therefore $\mathrm{Spec}(A_{\beta,n})$ is real for every β and n, but generically $\mathrm{Spec}(A_\beta)$ is not real except for $\beta = 0$.   □

Example 9.3.20 The following theorem about the convection-diffusion operator demonstrates that the spectrum of a non-self-adjoint differential operator sometimes provides very little information about its behaviour. It also shows that the spectrum of such an operator may not be stable under truncation. In this example the spectrum can be determined explicitly, but one would except similar behaviour for variable coefficient operators that may be harder to analyze.

---

[17] One can use the so-called oscillation properties to prove a result analogous to Theorem 9.3.17 for the eigenfunctions of a general Sturm-Liouville differential operator, even though these are generally not associated with any polynomials.

[18] The discrete analogue of the following example involves the great difference between the spectra of infinite Laurent operators and of the finite Toeplitz matrices that one obtains by truncating them. See [Schmidt and Spitzer 1960]. See also [Trefethen and Embree 2005, Chap. 7] for the detailed discussion of an example and an explanation of the connection with pseudospectra.

Let $\mathcal{D}$ denote the dense subspace of $L^2(0, a)$ consisting of all smooth functions on $[0, a]$ which vanish at 0, a. We define the operator

$$(H_a f)(x) := f''(x) + f'(x) \tag{9.13}$$

to be the closure of the operator defined initially on $\mathcal{D}$. The existence of a closure is guaranteed by Example 6.1.9. The following theorem shows that the spectrum of H_a does not converge to the spectrum of the 'same' operator H_∞ acting on $L^2(\mathbf{R})$. In fact

$$\mathrm{Spec}(H_\infty) = \mathcal{P} := \{x + iy : x = -y^2\}$$

by Theorem 8.1.1. This implies that standard methods of computing the spectrum of an operator on an infinite interval need to be reconsidered in such situations. The last statement of the theorem explains this phenomenon in terms of the pseudospectral behaviour of the operators.[19]   □

Theorem 9.3.21 *The convection-diffusion operator (9.13) satisfies*

$$\mathrm{Spec}(H_a) = \left\{ -\frac{1}{4} - \frac{\pi^2 n^2}{a^2} : n = 1, 2, 3, \ldots \right\}.$$

Its numerical range satisfies

$$\mathrm{Num}(H_a) \subseteq \mathcal{P}_1 := \{x + iy : x \le -y^2\} \tag{9.14}$$

and

$$\lim_{a \to \infty} \mathrm{Num}(H_a) = \mathcal{P}_1. \tag{9.15}$$

Given $\varepsilon > 0$, every λ in the interior $\mathcal{P}_0$ of $\mathcal{P}_1$ lies in $\mathrm{Spec}_\varepsilon(H_a)$ for all large enough a.

Proof. The operator H_a is similar to

$$(L_a f)(x) := -\frac{1}{4} f(x) + f''(x),$$

where L_a is the closure of its restriction to $\mathcal{D}$, and the similarity is given by

$$(L_a f)(x) = e^{x/2} H_a(e^{-x/2} f(x)).$$

The operator L_a is essentially self-adjoint on $\mathcal{D}$, having a completeorthonormal sequence of eigenfunctions $\phi_n(x) := (2/a)^{1/2} \sin(\pi n x/a)$, the corresponding

[19] For further information about this operator see Problem 11.3.5 and [Trefethen and Embree 2005, Chap. 12.]

eigenvalues being $\lambda_n := -1/4 - \pi^2 n^2/a^2$. This proves that the spectrum of H_a is as stated, and that it does not converge to the spectrum of H_∞.

If $f \in \mathcal{D}$ then

$$\langle H_a f, f \rangle = \int_0^a \left\{ -|f'(x)|^2 + f'(x)\overline{f(x)} \right\} dx.$$

Since $\mathcal{D} \subseteq W^{1,2}(\mathbf{R})$ this equals

$$\int_{\mathbf{R}} (-\xi^2 + i\xi)|(\mathcal{F}f)(\xi)|^2 d\xi.$$

Putting $\|f\|_2 = 1$, this implies (9.14).

We prove the last statement of the theorem by constructing approximate eigenfunctions for every point in $\mathcal{P}_0$. Given $\lambda \in \mathbf{C}$, the function

$$\phi(x) := e^{-s_1 x} - e^{-s_2 x}$$

satisfies $H_a \phi = \lambda \phi$ provided s_1, s_2 are the two solutions of $s^2 - s = \lambda$. At least one of the solutions satisfies $\text{Re}(s) > 0$. The other is purely imaginary if and only if $\lambda \in \mathcal{P}$. If $\lambda \in \mathcal{P}_0$ then both have positive imaginary parts, so ϕ decays exponentially as $x \to +\infty$. If we restrict ϕ to $[0, a]$ then it satisfies the required boundary conditions exactly at 0 but only approximately at a. We therefore define $\psi_a : [0, a] \to \mathbf{C}$ by

$$\psi_a(x) := \phi(x)\sigma(a - x)$$

where σ is a smooth function on $[0, \infty)$ that satisfies $\sigma(0) = 0$ and $\sigma(s) = 1$ if $s \geq 1$. Direct calculations establish that

$$\lim_{a \to \infty} \|\psi_a\|_2 = \|\phi\|_2 > 0$$

and that

$$\|H_a \psi_a - \lambda \psi_a\|_2 = O(e^{-\beta a})$$

as $a \to \infty$, where $\beta := \min\{\text{Re}(s_1), \text{Re}(s_2)\} > 0$.

If we put $\xi_a := \psi_a/\|\psi_a\|_2$ then $\langle H_a \xi_a, \xi_a \rangle$ converges to λ at an exponential rate as $a \to \infty$. This proves (9.15). $\square$

9.4 Higher order hulls and ranges

The topic described in this section is of very recent origin. It centres around two concepts, hulls and ranges. The discovery that they are closely related is even newer.[20] Although we start with some general theorems describing the relationship between the various concepts, we emphasize that higher order numerical ranges have already been used to determine the spectra of some physically interesting operators. This facet of the subject is treated later in the section.[21]

We start with higher order hulls. If A is a bounded operator on a Banach space $\mathcal{B}$, and p is a polynomial, then one can define

$$\text{Hull}(p, A) := \{z : |p(z)| \leq \|p(A)\|\},$$

and then

$$\text{Hull}_n(A) := \bigcap_{\deg(p) \leq n} \text{Hull}(p, A)$$

and

$$\text{Hull}_\infty(A) := \bigcap_{n \geq 1} \text{Hull}_n(A).$$

If p is a polynomial of degree n then the boundary of $\text{Hull}(p, A)$ can be determined by finding the n roots $z_r(\theta)$ of the equation

$$p(z) = e^{i\theta} \|p(A)\|$$

and then plotting them for every $r \in \{1, \ldots, n\}$ and $\theta \in [0, 2\pi]$. One can then approximate $\text{Hull}_n(A)$ by taking the intersection of $\text{Hull}(p, A)$ for a finite but representative collection of polynomials of degree n. The resulting region will be too large, so it will still contain $\text{Spec}(A)$.

Lemma 9.4.1 *We have*

$$\text{Spec}(A) \subseteq \text{Hull}_\infty(A) \subseteq \text{Hull}_n(A)$$

for all n.

Proof. It follows from the spectral mapping theorem 1.2.18 that

$$p(\text{Spec}(A)) = \text{Spec}(p(A)) \subseteq \{w : |w| \leq \|p(A)\|\}.$$

---

[20] The main theoretical result in this section is Theorem 9.4.6, identifying three a priori quite different objects. The equality $\text{Hull}_\infty(A) = \widehat{\text{Spec}}(A)$ was proved by Nevanlinna in [Nevanlinna 1993], while the equality $\text{Num}_\infty(A) = \widehat{\text{Spec}}(A)$ is due to the author, in [Davies 2005B]. It was subsequently shown in [Burke and Greenbaum 2004] that $\text{Hull}_n(A) = \text{Num}_n(A)$ for all n. From the point of view of applications, the sets $\text{Num}_n(A)$ seem to be less useful than $\text{Num}(p, A)$. See the end of this section for further discussion.

[21] There are many other generalizations of the numerical range with similar names. We refer to [Safarov 2005, Langer 2001] for discussions of some of these.

Therefore

$$\mathrm{Spec}(A) \subseteq \mathrm{Hull}(p, A)$$

for all polynomials p. The statements of the lemma follow immediately.   □

Problem 9.4.2 If $\mathcal{B}$ has finite dimension n, use the minimal polynomial of A to prove that

$$\mathrm{Spec}(A) = \mathrm{Hull}_n(A).$$ □

The definition and basic properties of the higher order numerical ranges are similar.

If A is a bounded operator on a Hilbert space $\mathcal{H}$, and p is a polynomial, then we define

$$\mathrm{Num}(p, A) := \{z : p(z) \in \overline{\mathrm{Num}}(p(A))\},$$

where $\overline{\mathrm{Num}}$ denotes the closure of the numerical range. We also put

$$\mathrm{Num}_n(A) := \bigcap_{\deg(p) \le n} \mathrm{Num}(p, A)$$

and

$$\mathrm{Num}_\infty(A) := \bigcap_{n \ge 1} \mathrm{Num}_n(A).$$

Lemma 9.4.3 *We have*

$$\mathrm{Spec}(A) \subseteq \mathrm{Num}_\infty(A) \subseteq \mathrm{Num}_n(A)$$

for all n.

Proof. It follows from Theorem 1.2.18 and Theorem 9.3.1 that

$$p(\mathrm{Spec}(A)) = \mathrm{Spec}(p(A)) \subseteq \overline{\mathrm{Num}}(p(A))$$

and this implies that

$$\mathrm{Spec}(A) \subseteq \mathrm{Num}(p, A)$$

for all polynomials p. The statements of the lemma follow immediately.   □

Lemma 9.4.4 *If A is a bounded or unbounded self-adjoint operator acting in a Hilbert space $\mathcal{H}$ then*

$$\mathrm{Spec}(A) = \mathrm{Num}_2(A).$$

Proof. We first observe that $\langle Af, f \rangle \in \mathbf{R}$ for all $f \in \mathcal{H}$, so $\mathrm{Num}_2(A) \subseteq \overline{\mathrm{Num}}(A) \subseteq \mathbf{R}$. It remains only to deal with gaps in the spectrum of A. It is sufficient to prove that if $(a, b) \cap \mathrm{Spec}(A) = \emptyset$ then $(a, b) \cap \mathrm{Num}(p, A) = \emptyset$ for some quadratic polynomial. We put $c := (a + b)/2$, $d := (b - a)/2$ and $p(z) := (z - c)^2$.

The spectrum of the self-adjoint operator $p(A)$ is contained in $[d^2, +\infty)$. It follows by Problem 9.3.5 that $\overline{\mathrm{Num}}(p(A)) \subseteq [d^2, +\infty)$. If $x \in \mathbf{R}$ then $x \in \mathrm{Num}(p, A)$ implies $(x - c)^2 \geq d^2$, so $x \notin (a, b)$. $\square$

The following theorem is rather surprising, since $\mathrm{Num}(p, A)$ may be much smaller than $\mathrm{Hull}(p, A)$ for individual polynomials p. Note, however, that the definition of higher order hulls does not make sense for unbounded operators, indicating their computational instability for matrices with very large norms.

Theorem 9.4.5 *(Burke-Greenbaum)*[22] *If A is a bounded operator acting on a Hilbert space $\mathcal{H}$ then*

$$\mathrm{Num}_n(A) = \mathrm{Hull}_n(A)$$

for all $n \in \mathbf{N}$.

Proof. We will establish that $\mathrm{Hull}_n(A) \subseteq \mathrm{Num}_n(A)$, the reverse inclusion being elementary. In the following argument p_j always refers to a polynomial of degree at most n.

If $z \notin \mathrm{Num}_n(A)$ then there exists p_1 such that $p_1(z) \notin \overline{\mathrm{Num}}(A)$. Putting $p_2(s) := p_1(s + z)$ we deduce that $p_2(0) \notin \overline{\mathrm{Num}}(p_2(A - zI))$. Therefore $0 \notin \overline{\mathrm{Num}}(p_3(A - zI))$ where $p_3(s) := p_2(s) - p_2(0)$ satisfies $p_3(0) = 0$. Using the convexity of the numerical range, one sees that if $p_4(s) := e^{i\theta} p_3(s)$ for a suitable $\theta \in \mathbf{R}$, then there exists a constant $\gamma < 0$ such that

$$\mathrm{Re}\langle Bf, f \rangle \leq \gamma \|f\|^2$$

for all $f \in \mathcal{H}$, where $B := p_4(A - zI)$. Given $\varepsilon > 0$ and $f \in \mathcal{H}$ we have

$$\begin{aligned}
\|(I + \varepsilon B)f\|^2 &= \|f\|^2 + 2\varepsilon \mathrm{Re}\langle Bf, f \rangle + \varepsilon^2 \|Bf\|^2 \\
&\leq (1 + 2\varepsilon\gamma + \varepsilon^2 \|B\|^2) \|f\|^2 \\
&= k\|f\|^2
\end{aligned}$$

---

[22] See [Burke and Greenbaum 2004]. We follow an unpublished proof of Trefethen.

where $k < 1$ if $\varepsilon > 0$ is small enough. Putting $p_5(s) := 1 + \varepsilon p_4(s)$ we deduce that $p_5(0) = 1$ and $\|p_5(A - zI)\| < 1$. Finally putting $p_6(s) := p_5(s - z)$ we obtain $p_6(z) = 1$ and $\|p_6(A)\| < 1$. Therefore $z \notin \mathrm{Hull}_n(A)$. □

The polynomial convex hull of a compact subset K of $\mathbf{C}$ is defined to be the complement of the unbounded component of $\mathbf{C} \backslash K$, i.e. K together with all open regions enclosed by this set.

Theorem 9.4.6 *(Nevanlinna)*[23] *If A is a bounded linear operator on $\mathcal{H}$ then*

$$\mathrm{Hull}_\infty(A) = \mathrm{Num}_\infty(A) = \widehat{\mathrm{Spec}}(A), \qquad (9.16)$$

where $\widehat{\mathrm{Spec}}(A)$ is the polynomial convex hull of $\mathrm{Spec}(A)$.

Proof. The first identity follows immediately from Theorem 9.4.5, so we concentrate on the second. If $a \notin \mathrm{Num}_\infty(A)$ then there exists a polynomial p such that $p(a) \notin \overline{\mathrm{Num}}(p(A))$. Since $\overline{\mathrm{Num}}(p(A))$ is closed and convex, there exists a real-linear functional $\phi : \mathbf{R}^2 \to \mathbf{R}$ such that $\phi(p(a)) < 0$ and

$$\phi(\mathrm{Spec}(p(A))) \subseteq \phi(\overline{\mathrm{Num}}(p(A))) \subseteq [0, \infty).$$

Since $\mathrm{Spec}(p(A)) = p(\mathrm{Spec}(A))$, the harmonic function $\psi : \mathbf{R}^2 \to \mathbf{R}$ defined by $\psi(z) := \phi(p(z))$ satisfies $\psi(a) < 0$ and

$$\psi(\mathrm{Spec}(A)) \subseteq [0, \infty).$$

The maximum principle for harmonic functions implies not only that $a \notin \mathrm{Spec}(A)$ but also that a does not lie in any bounded component of the complement of the spectrum. Therefore $a \notin \widehat{\mathrm{Spec}}(A)$. This implies that

$$\widehat{\mathrm{Spec}}(A) \subseteq \mathrm{Num}_\infty(A). \qquad (9.17)$$

Conversely we have to prove that if $a \notin \widehat{\mathrm{Spec}}(A)$ then there exists a polynomial p such that $p(a) \notin \overline{\mathrm{Num}}(p(A))$. We first observe that if $b \neq a$ is close enough to a and $r(z) := (b - z)^{-1}$ then

$$\max\{|r(z)| : z \in \mathrm{Spec}(A)\} < |r(a)|.$$

If we put $K := \mathrm{Spec}(A) \cup \{a\}$ then Lemma 9.4.7 below shows that there exists a sequence of polynomials which converges uniformly to r on K. Therefore there exists a polynomial q such that

$$\max\{|q(z)| : z \in \mathrm{Spec}(A)\} < |q(a)|.$$

---

[23] See [Nevanlinna 1993] and the footnote at the beginning of this section.

Since $q(\mathrm{Spec}(A)) = \mathrm{Spec}(q(A))$ this is equivalent to

$$\mathrm{Rad}(q(A)) < |q(a)|$$

where Rad denotes the spectral radius. Since

$$\mathrm{Rad}(B) = \lim_{n \to \infty} \|B^n\|^{1/n}$$

for every operator B, by Theorem 4.1.3, we deduce that if $p := q^n$ then for large enough n we have

$$\|p(A)\| < |p(a)|.$$

Since

$$\overline{\mathrm{Num}}(p(A)) \subseteq \{z : |z| \le \|p(A)\|\}$$

we finally see that $p(a)$ cannot lie in $\overline{\mathrm{Num}}(p(A))$. □

Lemma 9.4.7 *If b lies in the exterior component of a compact set K and $r(z) := (b - z)^{-1}$ then r is the uniform limit on K of a sequence of polynomials.*

Proof. Let $\gamma : [0, \infty) \to \mathbf{C} \backslash K$ be a continuous curve such that $\gamma(0) = b$ and $\gamma(s) \to \infty$ as $s \to \infty$. Consider the set S of all $s \ge 0$ such that $f_s(z) := (\gamma(s) - z)^{-1}$ is approximable uniformly by polynomials on K. This set is closed by a continuity argument and open because of the nature of the expansion of $(\gamma(s) + \varepsilon - z)^{-1}$ in powers of ε. Therefore $S = \emptyset$ or $S = [0, \infty)$. But S contains all large enough s by virtue of the uniform convergence on K of the expansion

$$\frac{1}{c - z} = \sum_{n=0}^{\infty} \frac{z^n}{c^{n+1}}$$

provided $|c| > \max\{|z| : z \in K\}$. Therefore $0 \in S$. □

We next list some of the inclusions between various sets associated with an operator A.

$$\mathrm{EssSpec}(A) \;\subseteq\; \mathrm{Spec}(A) \;\subseteq\; \widehat{\mathrm{Spec}}(A) \;=\; \mathrm{Num}_\infty(A)$$

$$\subseteq\; \mathrm{Num}_n(A) \;=\; \mathrm{Hull}_n(A) \;\subseteq\; \mathrm{Num}(A) \;\subseteq\; B(0, \|A\|)\,.$$

We conclude the section by applying the second order numerical range to obtain bounds on the spectrum of a certain type of non-self-adjoint tridiagonal operator. The results described have been applied to the non-self-adjoint

Anderson model, in which the coefficients of the operator are random variables, but the theory is also useful in other situations.[24] For example it provides simple quantitative bounds on the spectra of tridiagonal operators with periodic coefficients, which may be useful even though the exact spectrum can be computed for any particular such operator. (See Theorem 4.4.9 and the succeeding problems.) We start by formulating the results at a general level. See Examples 9.4.10 and 9.4.11 for applications involving particular operators satisfying the hypotheses. In the following theorem, one assumes that each of the operators C, S, V is easy to analyze on its own.

Theorem 9.4.8 *(Davies-Martinez) Let $A := C + iS + V$ where C, S, V are bounded, self-adjoint operators on $\mathcal{H}$. Suppose that there exist non-negative constants λ, μ such that*

$$S^2 + \lambda C^2 \leq \mu I. \tag{9.18}$$

Then

$$\operatorname{Spec}(A) \subseteq \{x + iy : y^2 \leq \tau(x)\} \tag{9.19}$$

where

$$\tau(x) := \inf_{c \in \mathbf{R}} \left\{ (x - c)^2 + \mu - \frac{\lambda \gamma_c^2}{1 + \lambda} \right\} \tag{9.20}$$

and

$$\gamma_c := \operatorname{dist}(c, \operatorname{Spec}(V)). \tag{9.21}$$

Proof. We need to prove that $x + iy \in \operatorname{Spec}(A)$ implies

$$y^2 \leq (x - c)^2 + \mu - \frac{\lambda \gamma_c^2}{1 + \lambda}$$

for every $c \in \mathbf{R}$. We reduce to the case $c = 0$ by replacing V by $V - cI$. The assumptions imply that

$$(x + iy)^2 \in \operatorname{Spec}(A^2) \subseteq \operatorname{Num}(A^2).$$

Therefore $x^2 - y^2 \geq \nu$ where ν is the bottom of the spectrum of $K := (A^2 + A^{*2})/2$. A direct computation shows that

$$K = (C + V)^2 - S^2$$
$$\geq C^2 + (CV + VC) + V^2 + \lambda C^2 - \mu I$$

---

[24] See [Davies 2005B] and [Martinez 2005] for more substantial treatments of the material here, and for references to other results about the non-self-adjoint Anderson model.

$$= \left((1+\lambda)^{1/2}C + (1+\lambda)^{-1/2}V\right)^2 + \frac{\lambda}{1+\lambda}V^2 - \mu I$$

$$\geq \frac{\lambda}{1+\lambda}V^2 - \mu I$$

$$\geq \left(\frac{\lambda\gamma_0^2}{1+\lambda} - \mu\right)I.$$

The statement of the theorem follows immediately. $\square$

One should resist the temptation to put $c := x$ in (9.20), because the minimum is often achieved for a different value of c. However, one has the following corollary.

Corollary 9.4.9 *If*

$$\text{dist}(x, \text{Spec}(V))^2 > \frac{(1+\lambda)\mu}{\lambda}$$

then

$$(x + i\mathbf{R}) \cap \text{Spec}(A) = \emptyset.$$

Proof. Put $c := x$ in Theorem 9.4.8 and use the fact that $\tau(x) < 0$. $\square$

Example 9.4.10 Let $A : l^2(\mathbf{Z}) \to l^2(\mathbf{Z})$ be defined by

$$(Af)(x) := af(x+1) + bf(x-1) + V(x)f(x)$$

where $a \neq b \in \mathbf{R}$ and $V : \mathbf{Z} \to \mathbf{R}$ is a bounded, real-valued function. The spectrum of V is the closure of $\{V(x) : x \in \mathbf{Z}\}$, and we assume that the constants γ_c defined in (9.21) are readily computable. If we put

$$(Cf)(x) := \alpha(f(x+1) + f(x-1)),$$

$$(Sf)(x) := i\beta(f(x+1) - f(x-1)),$$

where $\alpha := (a+b)/2$ and $\beta := (b-a)/2 \neq 0$, then C, S are self-adjoint and $A = C + iS + V$. Moreover (9.18) holds with $\lambda := \alpha^2/\beta^2$ and $\mu := 2\alpha^2$. $\square$

In the above example Theorem 9.4.8 yields outer bounds on the spectrum of A knowing only a, b and $\text{Spec}(V)$. The same bounds apply to all potentials whose range is contained in $\text{Spec}(V)$. This is very appropriate if the values of V are chosen randomly, because all of the relevant potentials have the same spectrum with probability one. However, the spectral bound above does not depend on the precise distribution used to choose the values $V(x)$, but only on its support, and this is not usual in the random context. Theorem 9.4.8

can equally well be used to obtain bounds on the spectrum of A when V is periodic. Whether or not the bounds are useful depends on the particular features of the operator, but there are cases in which it leads to a complete determination of the spectrum of A.[25]

Example 9.4.11 The above example was one-dimensional, but Theorem 9.4.8 can also be applied to similar operators acting on $l^2(\mathbf{Z}^n)$ for $n > 1$. This is relevant when studying the non-self-adjoint Anderson model in higher dimensions, but we describe a non-random example.[26]

We consider an operator A_M acting on $l^2(X)$ where $X := \{1, 2, \ldots, M\}^2$ and we impose periodic boundary conditions. We define A_M by

$$(A_M f)(x, y) := 2f(x+1, y) + 2f(x, y+1) + 3V(x, y)f(x, y)$$

where

$$V(x, y) := \begin{cases} 1 & \text{if } (x - M/2)^2 + (y - M/2)^2 \geq M^2/5, \\ -1 & \text{otherwise.} \end{cases}$$

The eigenvalues of A_M lie in its numerical range, and routine calculations show that this is contained in the convex hull of the union of the two circles

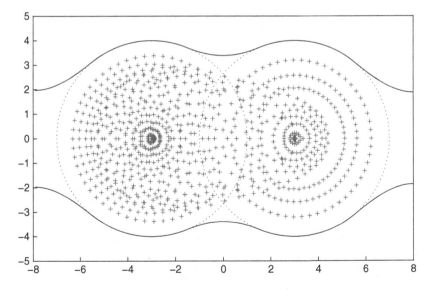

Figure 9.3: Bounds on the spectrum of A_{30} in Example 9.4.11

---

[25] See [Martinez 2005].
[26] See [Davies 2005B] for further details of this example and others of a similar type.

with centres ± 3 and radius 4. Figure 9.3 shows the eigenvalues of A_M for $M := 30$, and the boundary curves obtained by the method of Theorem 9.4.8. As $M \to \infty$ the spectrum of A_M fills up the two circles. $\square$

9.5 Von Neumann's theorem

In this section we consider a single contraction A on $\mathcal{H}$. Theorem 9.5.3 provides a classical result of von Neumann concerning the functional calculus of contractions.

Lemma 9.5.1 *If A is a contraction on $\mathcal{H}$ then*

$$B := \begin{pmatrix} A & (I - |A^*|^2)^{1/2} \\ (I - |A|^2)^{1/2} & -A^* \end{pmatrix}$$

is a unitary operator on $\mathcal{H} \oplus \mathcal{H}$.

Proof. This is elementary algebra subject to the identity

$$A(I - |A|^2)^{1/2} = (I - |A^*|^2)^{1/2} A.$$

To prove this we first note that

$$A|A|^{2n} = A(A^*A)^n = (AA^*)^n A = |A^*|^{2n} A.$$

This implies directly that

$$Ap(|A|^2) = p(|A^*|^2)A$$

for every polynomial p. This yields

$$Af(|A|) = f(|A^*|)A$$

for every continuous function f on $[0, \infty)$ by approximation. (Actually Lemma 5.2.1 suffices.) $\square$

Note One may also prove the lemma by using the polar decomposition $A = V|A|$, $A^* = |A|V^*$, $|A^*| = V|A|V^*$ and associated formulae provided V is unitary, as it is in finite dimensions. See Theorem 5.2.4.

Lemma 9.5.2 *Let A be a contraction on $\mathcal{H}$. Then there exists a norm analytic map $A(\cdot) : \mathbf{C} \to \mathcal{L}(\mathcal{H} \oplus \mathcal{H})$ such that $A(z)$ is unitary for all z such that $|z| = 1$ and*

$$A(0) = \begin{pmatrix} A & 0 \\ 0 & 0 \end{pmatrix}.$$

Proof. Put $A(z) := E(z)BE(z)$ where

$$E(z) := \begin{pmatrix} I & 0 \\ 0 & zI \end{pmatrix}, \qquad B := \begin{pmatrix} A & (I - |A^*|^2)^{1/2} \\ (I - |A|^2)^{1/2} & -A^* \end{pmatrix}.$$

All of the operators on $\mathcal{H} \oplus \mathcal{H}$ are contractions if $|z| < 1$ and unitary if $|z| = 1$. □

In the following theorem $\mathcal{A}$ denotes the space of all analytic functions that have power series expansions in z with radius of convergence greater than 1, and $\overline{\mathcal{A}}$ denotes the space of analytic functions on the open unit ball D that can be extended continuously to $\overline{D}$. Note that $\mathcal{A}$ is dense in $\overline{\mathcal{A}}$, if the latter is assigned the norm $\| \cdot \|_\infty$.

Theorem 9.5.3 *(von Neumann)*[27] *If A is a contraction on a Hilbert space $\mathcal{H}$ then $\|f(A)\| \leq \|f\|_\infty$ for all $f \in \mathcal{A}$. The map $f \to f(A)$ defined on $\mathcal{A}$ using the holomorphic functional calculus of Section 1.5 may be extended continuously to $\overline{\mathcal{A}}$.*

Proof. Given $f \in \mathcal{A}$ then, following the notation of Lemma 9.5.2, we put

$$F(z) := f(A(z)) = \sum_{n=0}^{\infty} f^{(n)}(0) A(z)^n / n!$$

for all $z \in \overline{D}$, so that

$$F(0) = \begin{pmatrix} f(A) & 0 \\ 0 & f(0)I \end{pmatrix}.$$

By applying the maximum principle to the operator-valued analytic function $F(z)$ and then the spectral theorem to the unitary operators $A(z)$ for $|z| = 1$, we deduce that

$$\|f(A)\| \leq \|F(0)\|$$
$$\leq \max\{\|F(z)\| : |z| = 1\}$$
$$\leq \max\{|f(w)| : |w| = 1\}.$$

The extension of the functional calculus from $\mathcal{A}$ to $\overline{\mathcal{A}}$ is a routine consequence of the norm bound proved above. □

---

[27] There are by now many proofs of this famous theorem, due to [von Neumann 1951]. This one was presented in [Davies and Simon 2005]. Another is given in Theorem 10.3.5. A detailed account of the functional calculus for contractions may be found in [Sz.-Nagy and Foias 1970, Sect. 3.2].

Theorem 9.5.4 *Let A be a contraction and let $f \in \overline{\mathcal{A}}$. Then the numerical range of $f(A)$ is contained in*

$$K := \overline{\mathrm{Conv}}\{f(z) : |z| = 1\}.$$

Proof. If $\phi \in \mathcal{H}$ satisfies $\|\phi\| = 1$ and $\tilde{\phi} := \phi \oplus 0 \in \mathcal{H} \oplus \mathcal{H}$, then

$$\langle f(A)\phi, \phi \rangle = \langle f(A(0))\tilde{\phi}, \tilde{\phi} \rangle = \frac{1}{2\pi} \int_{-\pi}^{\pi} \langle f(A(e^{i\theta}))\tilde{\phi}, \tilde{\phi} \rangle \mathrm{d}\theta \qquad (9.22)$$

where $A(z)$ is defined as in the proof of the last lemma. Since each operator $A(e^{i\theta})$ is unitary we see by using the spectral theorem and averaging that the right-hand side of (9.22) lies in K. $\qquad \square$

The following lemma is an immediate consequence of von Neumann's theorem, but it also has an elementary proof.

Lemma 9.5.5 *If A is a contraction on $\mathcal{H}$ and $|\lambda| < 1$ then*

$$B := (\lambda I - A)(1 - \overline{\lambda}A)^{-1}$$

is also a contraction.

Proof. The condition $\|B\| \leq 1$ is equivalent to

$$\langle Bf, Bf \rangle \leq \langle f, f \rangle$$

for all $f \in \mathcal{H}$. Putting $g = (I - \overline{\lambda}A)^{-1}f$ it is also equivalent to

$$\langle (\lambda I - A)g, (\lambda I - A)g \rangle \leq \langle (I - \overline{\lambda}A)g, (I - \overline{\lambda}A)g \rangle$$

for all $g \in \mathcal{H}$, and this is valid because

$$(1 - |\lambda|^2)\langle (I - A^*A)g, g \rangle \geq 0. \qquad \square$$

Problem 9.5.6 Construct a contraction A on $l^2(\mathbf{Z})$ such that $\|Af\| < \|f\|$ for all non-zero f, but $\mathrm{Spec}(A) \subseteq \{z : |z| = 1\}$. (Clearly such a contraction cannot have any eigenvalues.) $\qquad \square$

9.6 Peripheral point spectrum

Examples 9.1.4 and 9.2.14 imply that the $n \times n$ Jordan matrix

$$(J_n)_{r,s} := \begin{cases} 1 & \text{if } s = r + 1, \\ 0 & \text{otherwise,} \end{cases}$$

is badly behaved spectrally, in the sense that a very small perturbation may change its eigenvalues radically. Indeed any point in the interior of the unit disc is an approximate eigenvalue of J_n with error that decreases exponentially with the size of the matrix. On the other hand

$$\|(zI - J_n)^{-1}\| \leq (|z| - 1)^{-1}$$

for all z such that $|z| > 1$, by Theorem 1.2.11. Therefore z cannot be an approximate eigenvalue of J_n if $|z|$ is significantly larger than 1.

In this section we discuss approximate eigenvalues of a general $n \times n$ matrix A that are near the boundary of its numerical range. The results presented are due to Davies and Simon, and were motivated by the need to understand the distributions of the zeros of orthogonal polynomials on the unit circle.[28] The first part of the section only considers contractions, but in the second part we remove this constraint.

We start with an easy version of the type of result that we will consider in more detail below. Note that in finite dimensions an estimate of the type

$$\|(zI - A)^{-1}\| \leq c \, \{\mathrm{dist}(z, \mathrm{Spec}(A))\}^{-1} \tag{9.23}$$

implies that if $\|Af - zf\| \leq \varepsilon \|f\|$ then A has an eigenvalue λ such that $|\lambda - z| \leq c\varepsilon$. In other words if z is an approximate eigenvalue of A then it is close to a true eigenvalue. The inequality (9.23) does not make sense if $z \in \mathrm{Spec}(A)$, but we regard it as being true by convention in that case.

Theorem 9.6.1 *If A is a contraction acting on an n-dimensional inner product space $\mathcal{H}$, and $|z| = 1$, then*

$$\|(zI - A)^{-1}\| \leq \sum_{r=1}^{m} \frac{1 + |\lambda_r|}{|z - \lambda_r|}$$

$$\leq 2m \, \{\mathrm{dist}(z, \mathrm{Spec}(A))\}^{-1},$$

where $m \leq n$ is the degree of the minimal polynomial p of A, and λ_r are the eigenvalues of A, each repeated as many times as in the minimal polynomial.

Proof. On replacing $z^{-1}A$ by A, we reduce to the case $z = 1$. If A has any eigenvalues with modulus 1, we apply the argument below to sA, where $0 < s < 1$, and let $s \to 1$ at the end of the proof.

We put

$$B_r := \frac{(A - \lambda_r)}{(1 - \overline{\lambda_r} A)} \frac{(1 - \overline{\lambda_r})}{(1 - \lambda_r)}$$

---

[28] See [Davies and Simon 2005].

and use the fact that this is a contraction by Lemma 9.5.5. Moreover $B_1 B_2 \dots B_m = 0$ because its numerator is a multiple of the minimum polynomial. We have

$$(I - A)^{-1} = (I - A)^{-1}(I - B_1 B_2 \dots B_m)$$
$$= (I - A)^{-1}[(I - B_1) + B_1(1 - B_2)$$
$$+ B_1 B_2(1 - B_3) + \dots + B_1 \dots B_{m-1}(1 - B_m)].$$

Since all the operators commute we deduce that

$$\|(I - A)^{-1}\| \leq \sum_{r=1}^{m} \|(I - A)^{-1}(I - B_r)\| = \sum_{r=1}^{m} \|f_r(A)\|$$

where

$$f_r(z) := (1 - z)^{-1} \left(1 - \frac{(z - \lambda_r)}{(1 - \overline{\lambda_r} z)} \frac{(1 - \overline{\lambda_r})}{(1 - \lambda_r)} \right)$$

$$= \frac{1 - |\lambda_r|^2}{(1 - \overline{\lambda_r} z)(1 - \lambda_r)}.$$

One may estimate $\|f_r(A)\|$ by using von Neumann's Theorem 9.5.3 or by the following more elementary method, in which we omit the subscript r.

$$\|f(A)\| \leq \frac{1 - |\lambda|^2}{|1 - \lambda|}(1 + \|\overline{\lambda} A\| + \|\overline{\lambda} A\|^2 + \dots)$$

$$\leq \frac{1 - |\lambda|^2}{|1 - \lambda|}(1 + |\lambda| + |\lambda|^2 + \dots)$$

$$= \frac{1 + |\lambda|}{|1 - \lambda|}.$$

This proves the first inequality of the theorem. After the reduction to the case $z = 1$, the second inequality depends on using

$$\frac{1 + |\lambda_r|}{|1 - \lambda_r|} \leq 2 \{\text{dist}(1, \text{Spec}(A))\}^{-1}$$

for all $r \in \{1, \dots, m\}$. $\qquad \square$

The remainder of the section is devoted to obtaining improvements and generalizations of the above theorem.

Lemma 9.6.2 *Let A be an upper triangular $n \times n$ contraction with diagonal entries $\lambda_1, \ldots, \lambda_n$, and let $B := (I - A)^{-1}$. Then*

$$
|B_{i,j}| \leq \begin{cases} 0 & \text{if } i > j, \\ |1 - \lambda_i|^{-1} & \text{if } i = j, \\ \sqrt{\dfrac{(1-|\lambda_i|^2)(1-|\lambda_j|^2)}{|1-\lambda_i|^2\,|1-\lambda_j|^2}} & \text{if } i < j, \end{cases} \tag{9.24}
$$

$$
\leq \delta^{-1} T_{i,j} \tag{9.25}
$$

where

$$
T_{i,j} := \begin{cases} 0 \ \text{if } i > j, \\ 1 \ \text{if } i = j, \\ 2 \ \text{if } i < j, \end{cases}
$$

and

$$
\delta := \min_{1 \leq r \leq n} |1 - \lambda_r| = \operatorname{dist}(1, \operatorname{Spec}(A)).
$$

Proof. The first two inequalities (actually equalities) in (9.24) follow directly from the fact that A is triangular, so we concentrate on the third. If we put

$$
C := B + B^* - I
$$
$$
= (I - A)^{-1}(I - AA^*)(I - A^*)^{-1}
$$

then we see immediately that $C = C^* \geq 0$. The Schwarz inequality now implies that

$$
|C_{i,j}|^2 \leq C_{i,i} C_{j,j}
$$

for all i, j. Since $B_{i,j} = C_{i,j}$ if $i < j$, the third bound in (9.24) follows as soon as one observes that

$$
C_{i,i} = B_{i,i} + B^*_{i,i} - 1
$$
$$
= \frac{1}{1 - \lambda_i} + \frac{1}{1 - \overline{\lambda_i}} - 1
$$
$$
= \frac{1 - |\lambda_i|^2}{|1 - \lambda_i|^2}.
$$

To prove (9.25) one needs the further inequality

$$
\frac{1 - |\lambda_i|^2}{|1 - \lambda_i|^2} \leq \frac{2}{|1 - \lambda_i|} \leq \frac{2}{\delta}. \qquad \square
$$

Corollary 9.6.3 *Let A be an $n \times n$ contraction with eigenvalues $\lambda_1, \lambda_2, \ldots, \lambda_n$. If $|z| = 1$ then*

$$\|(zI - A)^{-1}\| \leq \|(zI - A)^{-1}\|_2 \leq \sqrt{2}\, n \,\{\mathrm{dist}(z, \mathrm{Spec}(A))\}^{-1} \qquad (9.26)$$

where $\|\cdot\|_2$ is the Hilbert-Schmidt norm.

Proof. We use Schur's theorem to represent A as an upper triangular matrix with respect to some orthonormal basis. The eigenvalues of A are then the diagonal entries of the matrix. By passing to $z^{-1}A$ we reduce to the case $z = 1$. The first inequality in (9.26) is elementary. The second follows from

$$\|(I - A)^{-1}\|_2 \leq \delta^{-1}\|T\|_2 \leq \delta^{-1}\sqrt{2}\, n. \qquad \square$$

Example 9.6.4 We will compute the norm of the triangular $n \times n$ matrix T exactly, but before doing this we consider the closely related Volterra operator V, defined on $L^2(0, 1)$ by

$$(Vf)(x) := \int_0^x f(y)\,dy.$$

By computing the Hilbert-Schmidt norm of V we see immediately that

$$\|V\| \leq \|V\|_2 = 1/\sqrt{2}.$$

The norm of V is equal to that of the Hankel operator H defined by $(Hf)(x) := (Vf)(1 - x)$, which has the integral kernel

$$K(x, y) := \begin{cases} 1 \text{ if } 0 \leq x + y \leq 1, \\ 0 \text{ otherwise.} \end{cases}$$

One checks directly that $Hf_n = \lambda_n f_n$ for all $n = 1, 2, \ldots$ where

$$f_n(x) := \cos((2n - 1)\pi x/2), \qquad \lambda_n := \frac{(-1)^{n-1}2}{(2n - 1)\pi}.$$

The self-adjointness of H implies that

$$\|V\| = \|H\| = \lambda_1 = 2/\pi.$$

We now exhibit the connection between V and T. Given a positive integer n, put

$$\phi_r(x) := n^{1/2}\chi_{(r-1)/n, r/n)}(x)$$

for all $1 \leq r \leq n$. This defines a finite orthonormal set and one readily checks that

$$\langle V\phi_r, \phi_s \rangle = T_{r,s}/2n.$$

If P_n is the orthogonal projection onto the linear span of $\phi_1, \ldots, \phi_n$ we deduce that

$$\|T\| = 2n\|P_n V P_n\| \leq 4n/\pi.$$

We will see that this bound is sharp in the sense that $c_n := \|T\|/n$ converges to $4/\pi$ as $n \to \infty$. $\quad\square$

Lemma 9.6.5 *The $n \times n$ triangular matrix T satisfies*

$$\|T\| = \cot(\pi/4n)$$

for all positive integers n. Hence

$$\lim_{n \to \infty} n\|T\| = 4/\pi \sim 1.2732 .$$

Proof. The norm of T is equal to that of the $n \times n$ Hankel matrix

$$H(i,j) := \begin{cases} 2 & \text{if } i+j < n, \\ 1 & \text{if } i+j = n, \\ 0 & \text{otherwise.} \end{cases}$$

Let $\theta_s := (2s-1)\pi/4n$ for $1 \leq s \leq n$, so that $0 < \theta_s < \pi/2$ for each s. Define the column vector $v_s \in \mathbf{C}^n$ by

$$(v_s)_r := \cos((2r-1)\theta_s) = (-1)^{s-1}\sin((2n+1-2r)\theta_s)$$

where $1 \leq r \leq n$. A direct computation using the trigonometric identities

$$2\{\cos(\phi) + \cos(3\phi) + \cos(5\phi) + \cdots + \cos((2t-3)\phi)\}$$
$$+ \cos((2t-1)\phi) = \cot(\phi)\sin((2t+1)\phi)$$

shows that $Hv_s = \lambda_s v_s$, where

$$\lambda_s := (-1)^{s-1}\cot(\theta_s).$$

Since H is self-adjoint we deduce that

$$\|H\| = \max\{|\lambda_s| : 1 \leq s \leq n\} = \lambda_1 = \cot(\pi/4n). \quad\square$$

We can now prove the sharp version of Theorem 9.6.1.

Theorem 9.6.6 *(Davies-Simon)*[29] *If A is a contraction acting in an n-dimensional inner product space $\mathcal{H}$ and $|z| = 1$ then*

$$\|(zI - A)^{-1}\| \leq \cot(\pi/4n)\, \{\mathrm{dist}(z, \mathrm{Spec}(A))\}^{-1}.$$

Proof. We use Schur's theorem to represent A as an upper triangular matrix with respect to some orthonormal basis. The eigenvalues of A are then the diagonal entries of the matrix. On replacing A by $z^{-1}A$ we reduce to the case $z = 1$. By using Lemmas 9.6.2 and 9.6.5 one obtains the bound

$$\|(I - A)^{-1}\| \leq \delta^{-1}\|T\| = \delta^{-1}\cot(\pi/4n). \qquad \square$$

We now describe some generalizations of the above theorem. The first is the 'correct' formulation.

Theorem 9.6.7 *Let A be an $n \times n$ matrix and let S denote the topological boundary of the numerical range of A. If z lies on or outside S then*

$$\|(zI - A)^{-1}\| \leq \cot(\pi/4n)\, \{\mathrm{dist}(z, \mathrm{Spec}(A))\}^{-1}.$$

Proof. By considering $e^{i\theta}(A - zI)$ for a suitable value of θ we reduce to the case in which $z = 0$ and $\mathrm{Re}\langle Af, f \rangle \leq 0$ for all $f \in \mathbf{C}^n$. We also use Schur's theorem to transfer to an orthonormal basis with respect to which A has an upper triangular matrix. The result is true by convention if 0 is an eigenvalue, so we assume that this is not the case. The matrix $B := -A^{-1}$ is also upper triangular with $\mathrm{Re}\langle Bf, f \rangle \geq 0$ for all $f \in \mathbf{C}^n$.

Since $B + B^* \geq 0$ the same argument as in Lemma 9.6.2 yields the bounds

$$|B_{i,j}| \leq \begin{cases} 0 & \text{if } i > j, \\ |\lambda_i|^{-1} & \text{if } i = j, \\ 2|\lambda_i\lambda_j|^{-1/2} & \text{if } i < j \end{cases}$$

$$\leq \delta^{-1}T_{i,j}$$

where $\delta = \min\{|\lambda_i| : 1 \leq i \leq n\}$. An application of Lemma 9.6.5 now completes the proof. $\square$

One may also consider situations in which $\|A^m\|$ grows subexponentially as $m \to +\infty$. This implies that $\mathrm{Spec}(A) \subseteq \{z : |z| \leq 1\}$, and suggests that one might still be able to prove bounds on the resolvent norm when $|z| = 1$. We only treat the case in which $\|A^m\|$ are uniformly bounded; a similar but

---

[29] See [Davies and Simon 2005].

weaker result holds when the norms are polynomially bounded. We do not expect that the exponents in the following theorem are sharp.

Theorem 9.6.8 *Let A be an $n \times n$ matrix and $c \geq 1$ a constant such that $\|A^m\| \leq c$ for all $m \geq 0$. If $|z| = 1$ and $\|Af - zf\| < \varepsilon\|f\|$ then A has an eigenvalue λ such that*

$$|\lambda - z| \leq 3n(c\varepsilon)^{2/3}.$$

Proof. If

$$B := \sum_{m=0}^{\infty} r^{-2m}(A^*)^m A^m$$

where $r > 1$, then

$$I \leq B \leq \frac{c^2}{1 - r^{-2}}I.$$

If we define a new inner product on $\mathbf{C}^n$ by

$$\langle f, g \rangle_1 := \langle Bf, g \rangle$$

then it follows that

$$\|f\| \leq \|f\|_1 \leq cr(r^2 - 1)^{-1/2}\|f\|$$

for all $f \in \mathbf{C}^n$. Also

$$\|Af\|_1^2 = \langle A^*BAf, f \rangle \leq r^2 \langle Bf, f \rangle = r^2\|f\|_1^2.$$

Therefore $\|A\|_1 \leq r$.

We now put $C := r^{-1}A$ so $\|C\|_1 \leq 1$. We have

$$\begin{aligned}
\|Cf - r^{-1}zf\|_1 &= r^{-1}\|Af - zf\|_1 \\
&\leq c(r^2 - 1)^{-1/2}\|Af - zf\| \\
&\leq \varepsilon c(r^2 - 1)^{-1/2}\|f\| \\
&\leq \varepsilon c(r^2 - 1)^{-1/2}\|f\|_1.
\end{aligned}$$

Therefore

$$\begin{aligned}
\|Cf - zf\|_1 &\leq \left\{ \varepsilon c(r^2 - 1)^{-1/2} + |z - r^{-1}z| \right\} \|f\|_1 \\
&\leq \left\{ \varepsilon c(2(r-1))^{-1/2} + (r-1) \right\} \|f\|_1.
\end{aligned}$$

We conclude that A has an eigenvalue λ, necessarily satisfying $|\lambda| \leq 1$, such that

$$|r^{-1}\lambda - z| \leq \frac{4n}{\pi} \left\{ \varepsilon c(2(r-1))^{-1/2} + (r-1) \right\}.$$

This implies that

$$|\lambda - z| \le (r-1) + \frac{4n}{\pi} \left\{ \varepsilon c (2(r-1))^{-1/2} + (r-1) \right\}.$$

Finally, putting $r = 1 + \frac{1}{2}(c\varepsilon)^{2/3}$, we obtain

$$|\lambda - z| \le \left(\frac{1}{2} + \frac{6n}{\pi} \right) (c\varepsilon)^{2/3}$$

$$\le 3n(c\varepsilon)^{2/3}. \qquad \square$$

10

Quantitative bounds on semigroups

10.1 Long time growth bounds

In most applications of semigroup theory one is given the generator Z explicitly, and has to infer properties of the solutions of the evolution equation $f'(t) = Zf(t)$, i.e. of the semigroup T_t. This is not an easy task, and much of the analysis depends on obtaining bounds on the resolvent norms. We devote this section to establishing a connection between the spectrum of Z and the long time asymptotics of T_t. Before starting it may be useful to summarize some of the results already obtained. These include

(i) If $\|T_t\| \leq Me^{at}$ for all $t \geq 0$ then $\mathrm{Spec}(Z) \subseteq \{z : \mathrm{Re}(z) \leq a\}$ and $\|R_z\| \leq M/(\mathrm{Re}(z) - a)$ for all z such that $\mathrm{Re}(z) > a$. (Theorem 8.2.1)

(ii) If R_z is the resolvent of a densely defined operator Z and $\|R_x\| \leq 1/x$ for all $x > 0$, then Z is the generator of a one-parameter contraction semigroup, and conversely. (Theorem 8.3.2)

(iii) For every $\varepsilon > 0$ there exists a densely defined operator Z acting in a reflexive Banach space such that $\|R_x\| \leq (1 + \varepsilon)/x$ for all $x > 0$, but Z is not the generator of a one-parameter semigroup . (Theorem 8.3.10)

(iv) T_t is a bounded holomorphic semigroup if and only if there exist $\alpha > 0$ and $N < \infty$ such that the associated resolvents satisfy $\|R_z\| \leq N|z|^{-1}$ for all z such that $|\mathrm{Arg}(z)| \leq \alpha + \pi/2$. (Theorems 8.4.1 and 8.4.2)

(v) If $\lambda \notin \mathrm{Spec}(Z)$, then $z \in \mathrm{Spec}(Z)$ if and only if $(\lambda - z)^{-1} \in \mathrm{Spec}(R_\lambda)$, and both imply that $e^{zt} \in \mathrm{Spec}(T_t)$ for all $t \geq 0$. (Lemma 8.1.9 and Theorem 8.2.7)

(vi) There exists a one-parameter group T_t acting on a Hilbert space $\mathcal{H}$ such that $\mathrm{Spec}(Z) \subseteq i\mathbf{R}$ but $e^{|t|} \in \mathrm{Spec}(T_t)$ for all $t \in \mathbf{R}$. (Theorem 8.2.9)

This chapter describes a number of more advanced results about one-parameter semigroup s, mostly concerned with growth bounds. We recall from Theorem

6.1.23 that every one-parameter semigroup satisfies a growth bound of the form

$$\|T_t\| \leq M e^{\omega t} \tag{10.1}$$

for all $t \geq 0$. In this section we introduce various constants related to ω above, and find inequalities between them.

The exponential growth rate ω_0, of the semigroup T_t was defined in Section 6.2 as the infimum of all permissible constants ω in (10.1). Finding values of M and ω for which (10.1) holds provides very limited information about the behaviour of the semigroup norms for several reasons. In Example 10.2.9 we show that it is possible for $\|T_t\|$ to be highly oscillatory as a function of t. It is also easy to produce examples in which $\|T_t\|$ is close to 1 until t is quite large, and then starts to decrease at an exponential rate. Nevertheless we need to understand the role that ω_0 plays before moving on.

Problem 10.1.1 Given $\alpha \in \mathbf{R}$, find the exact value of $\|T_t\|$ for the one-parameter group defined on $L^2(\mathbf{R}, (1+x^2)^{\alpha/2}\mathrm{d}x)$ by

$$(T_t f)(x) := f(x-t).$$

<u>Note</u> The answer differs for $\alpha < 0$ and $\alpha > 0$. $\square$

Problem 10.1.2 By changing the weight in Problem 10.1.1, find an example of a one-parameter semigroup T_t acting on a Hilbert space $\mathcal{H}$ such that $\|T_t\| = 1$ for all $t \geq 0$ but

$$\lim_{t \to \infty} \|T_t f\| = 0$$

for all $f \in \mathcal{H}$. $\square$

Problem 10.1.3 Prove that the exponential growth rate of a one-parameter semigroup is a similarity invariant, i.e. it is unaffected by changing from the given norm on $\mathcal{B}$ to an equivalent norm. $\square$

A function $p : [0, \infty) \to [-\infty, \infty)$ is said to be subadditive if

$$p(x+y) \leq p(x) + p(y) \tag{10.2}$$

for all $x, y \geq 0$.

Problem 10.1.4 Prove that if $p : [0, \infty) \to \mathbf{R}$ is concave with $p(0) \geq 0$ then it is subadditive. Give an example of a non-concave subadditive function. $\square$

Lemma 10.1.5 *If p is subadditive on $[0, \infty)$ and bounded above on $[0, 1]$ then*

$$-\infty \le \inf_{t>0} t^{-1} p(t) = \lim_{t\to\infty} t^{-1} p(t) < \infty. \tag{10.3}$$

Proof. If $a > 0$ and $p(a) = -\infty$ then $p(t) = -\infty$ for all $t > a$ by (10.2), and the lemma is trivial, so let us assume that p is finite everywhere. Since it is bounded above on $[0, 1]$ it is bounded above on every finite interval, again by (10.2).

If $a^{-1} p(a) < \gamma$ and $na \le t < (n+1)a$ for some positive integer n then

$$t^{-1} p(t) \le t^{-1} \{ np(a) + p(t - na) \}$$

$$\le a^{-1} p(a) + t^{-1} \sup\{ p(s) : 0 \le s \le a \},$$

which is less than γ for all large enough t. This implies the stated result. $\square$

Theorem 10.1.6 *If T_t is a one-parameter semigroup on a Banach space $\mathcal{B}$ then*

$$\omega_0 = \inf_{0 < t < \infty} t^{-1} \log \|T_t\| = \lim_{t\to\infty} t^{-1} \log \|T_t\| \tag{10.4}$$

satisfies $-\infty \le \omega_0 < \infty$ and

$$\mathrm{Rad}(T_t) = e^{\omega_0 t}$$

for all $t > 0$.

Proof. The function

$$p(t) := \log \|T_t\|$$

satisfies the conditions of Lemma 10.1.5, so

$$\lim_{t\to\infty} t^{-1} p(t) = \omega_0$$

exists. If $t > 0$ then Theorem 4.1.3 implies that

$$\mathrm{Rad}(T_t) = \lim_{n\to\infty} \|T_{nt}\|^{1/n}$$

$$= \lim_{n\to\infty} \exp\{ n^{-1} p(nt) \}$$

$$= e^{\omega_0 t}. \qquad \square$$

We have already seen that the rate of decay (asymptotic stability) of $\|T_t f\|$ as $t \to \infty$ may depend on f: in Problem 8.4.7 we proved that if T_t is a bounded holomorphic semigroup and $f \in \mathrm{Ran}(Z^n)$ then $\|T_t f\| = O(t^{-n})$ as $t \to \infty$; in Example 6.3.5 we saw that the Gaussian semigroup T_t on $L^2(\mathbf{R}^N)$ satisfies $\|T_t f\| = O(t^{-N/4})$ as $t \to \infty$ for all $f \in L^1(\mathbf{R}^N) \cap L^2(\mathbf{R}^N)$.[1]

The following general result is of some interest. Define S to be the set of all $\omega \in \mathbf{R}$ such that

$$\|T_t x\| \leq M_\omega e^{\omega t} \|x\| \tag{10.5}$$

for some M_ω, all $t \geq 0$ and all $x \in \mathrm{Dom}(Z)$, where $\|x\| := \|x\| + \|Zx\|$. We then put

$$\omega_1 := \inf\{S\}.$$

We also put

$$s := \sup\{\mathrm{Re}(z) : z \in \mathrm{Spec}(Z)\}.$$

It was shown by Wrobel that for any $0 < \sigma < 1$ there exists a one-parameter semigroup T_t acting on a Hilbert space $\mathcal{H}$ such that $s = 0$, $\omega_0 = 1$ and $\omega_1 = \sigma$.[2]

Theorem 10.1.7 *We always have the inequalities* $s \leq \omega_1 \leq \omega_0$. *If there exists* $a \geq 0$ *such that* T_t *is norm continuous for* $t \geq a$, *then they are all equal.*

Proof. It follows from Problem 6.1.4 that if $a \notin \mathrm{Spec}(Z)$ then there exists a constant $c > 0$ such that

$$c^{-1}\|AR(a, Z)\| \leq \|A\| \leq c\|AR(a, Z)\|$$

for all operators A on $\mathcal{B}$, where

$$\|A\| := \sup\{\|Af\| : \|f\| \leq 1\}.$$

This implies that $\omega \in S$ if and only if

$$\|T_t R(a, Z)\| \leq M'_\omega e^{\omega t}$$

for some M'_ω and all $t \geq 0$. Theorem 8.2.10 states that if $z \in \mathrm{Spec}(Z)$ then

$$e^{zt}(a - z)^{-1} \in \mathrm{Spec}(T_t R(a, Z)).$$

If $\omega \in S$, we deduce that

$$|e^{zt}(a - z)^{-1}| \leq \|T_t R(a, Z)\| \leq M'_\omega e^{\omega t}.$$

---

[1] See [Arendt et al. 2001, Meyn and Tweedie 1996, Aldous and Fill] for references to the substantial literature on asymptotic stability, particularly in the stochastic context.
[2] See [Wrobel 1989, Ex. 4.1].

Since $t > 0$ is arbitrary, this implies that $\mathrm{Re}(z) \leq \omega$ whenever $z \in \mathrm{Spec}(Z)$ and $\omega \in S$. Hence $s \leq \omega_1$. The inequality $\omega_1 \leq \omega_0$ follows directly from the definitions of the two quantities.

The final statement of the theorem is proved by combining Theorems 8.2.12 and 10.1.6. □

10.2 Short time growth bounds

Although the constant ω_0 defined in (10.4) controls the long time asymptotics of $\|T_t\|$, it has little influence on the short time (i.e. transient) behaviour of the norm. The reason is that if ω is close to ω_0 then the constant M in the bound $\|T_t\| \leq M\mathrm{e}^{\omega t}$ may be very large, even in examples of real importance. We start by giving a very simple example of this phenomenon.

Example 10.2.1 Let Z_n be the $n \times n$ matrix

$$(Z_n)_{r,s} := \begin{cases} -1 & \text{if } r = s, \\ 2 & \text{if } r + 1 = s, \\ 0 & \text{otherwise,} \end{cases}$$

for which $\mathrm{Spec}(Z_n) = \{-1\}$. Put $T_{n,t} := \mathrm{e}^{Z_n t}$ for all $t \geq 0$, regarded as acting on $\mathbf{C}^n$ with the Euclidean norm. On writing down the (upper triangular) matrix of $T_{n,t}$ one sees that

$$(2t)^n \mathrm{e}^{-t}/n! \leq \|T_{n,t}\| \leq \{1 + 2t + (2t)^2/2! + \cdots + (2t)^n/n!\}\mathrm{e}^{-t}$$

for all $t > 0$. Therefore $\|T_{n,t}\| \to 0$ as $t \to \infty$ at an exponential rate. However, for smaller t the norm grows rapidly. Figure 10.1 plots $\|T_{n,t}\|$ as a function of $t \geq 0$ for $n := 12$. For larger n the short time growth of the norm is even more dramatic. It is easy to construct diagonalizable matrices Z_n exhibiting the same phenomenon. □

We proved in Theorem 8.3.10, item (iii) in the last section, that it may not be easy to prove that an operator Z is the generator of a one-parameter semigroup from *numerical* information about its resolvent norms. On the other hand Theorem 10.2.5 below states that one can use the pseudospectra to obtain *lower* bounds, not on the semigroup norms, but on certain regularizations of these norms, defined below. This demonstrates that the pseudospectra have a much greater effect on the short time (i.e. transient) behaviour of the

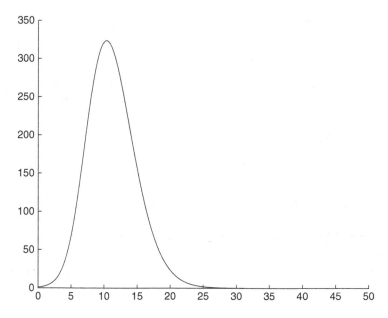

Figure 10.1: Plot of $\| T_{n,t} \|$ as a function of t for $n = 12$

semigroup than the spectrum does.[3] Regularization of the semigroup norms $\| T_t \|$ is necessary because the norms themselves may be highly oscillatory as functions of t; see Problem 10.2.10 below.

Although our main application is to one-parameter semigroup s, we assume below only that $\mathcal{B}$ is a Banach space and that $T_t : \mathcal{B} \to \mathcal{B}$ is a strongly continuous family of operators defined for $t \geq 0$, satisfying $\| T_0 \| = 1$ and $\| T_t \| \leq Me^{\omega t}$ for some M, ω and all $t \geq 0$. We define $N(t)$ to be the upper log-concave envelope of $\| T_t \|$. In other words $\nu(t) := \log(N(t))$ is defined to be the smallest concave function satisfying $\nu(t) \geq \log(\| T_t \|)$ for all $t \geq 0$. It is immediate that $N(t)$ is continuous for $t > 0$, and that

$$1 = N(0) \leq \lim_{t \to 0+} N(t).$$

In many cases one may have $N(t) = \| T_t \|$, but we do not study this question, asking only for lower bounds on $N(t)$ which are based on pseudospectral information.

---

[3] The theorems in this section are almost all taken from [Davies 2005A], but Trefethen has been emphasizing the importance of this point for a number of years. See [Trefethen and Embree 2005].

We assume throughout this section that

$$\limsup_{t \to +\infty} t^{-1} \log(\|T_t\|) = 0. \tag{10.6}$$

This identity can often be achieved by replacing T_t by $e^{kt}T_t$ for a suitable value of $k \in \mathbf{R}$. In the semigroup context it may be rewritten in the form $\omega_0 = 0$, and implies that $\|T_t\| \geq 1$ for all $t \geq 0$ by Theorem 10.1.6. If we define the operators R_z on $\mathcal{B}$ by

$$R_z f := \int_0^\infty (T_t f) e^{-zt} \, dt$$

then $\|R_z\|$ is uniformly bounded on $\{z : \mathrm{Re}(z) \geq \gamma\}$ for any $\gamma > 0$, and the norm converges to 0 as $\mathrm{Re}(z) \to +\infty$. In the semigroup context R_z is the resolvent of the generator Z of the semigroup, but in the general context it is simply a norm analytic function of z defined for $\mathrm{Re}(z) > 0$.

The following lemma compares $N(t)$ with the alternative regularization[4]

$$L(t) := \sup\{\|T_s\| : 0 \leq s \leq t\}.$$

Lemma 10.2.2 *If $t > 0$ then*

$$\|T_t\| \leq L(t) \leq N(t).$$

If T_t is a one-parameter semigroup then we also have

$$N(t) \leq L(t/n)^{n+1} \tag{10.7}$$

for all positive integers n and $t \geq 0$.

Proof. The log-concavity of $N(t)$ and the assumption (10.6) imply that $N(t)$ is a non-decreasing function of t. We deduce that $\|T_t\| \leq L(t) \leq N(t)$. If T_t is a one-parameter semigroup we claim that $L(t/n)^{1+ns/t}$ is a log-concave function of s which dominates $\|T_s\|$ for all $s \geq 0$. This implies

$$N(s) \leq L(t/n)^{1+ns/t}$$

for all $s \geq 0$. Putting $s = t$ yields (10.7).

Since the log-concavity is immediate, our claim depends on proving that

$$\|T_s\| \leq L(t/n)^{1+ns/t} \tag{10.8}$$

---

[4] The idea for studying $N(t)$ arose after corresponding results for $L(t)$ had been obtained by Trefethen. See [Trefethen and Embree 2005, p. 140].

for all $s \geq 0$. If $0 \leq s \leq t/n$ then $\|T_s\| \leq L(t/n)$ by the definition of the RHS, and this implies (10.8) because $L(u) \geq 1$ for all $u \geq 0$. If $s > t/n$ there exists a positive integer r such that $rt/n < s \leq (r+1)t/n$. Putting $u := s/(r+1)$ we see that $0 < u \leq t/n$, so

$$\|T_s\| = \|T_{(1+r)u}\| \leq \|T_u\|^{1+r} \leq L(t/n)^{1+r} \leq L(t/n)^{1+ns/t}$$

as required. $\square$

In the following lemma we put

$$N'(0+) := \lim_{\varepsilon \to 0+} \varepsilon^{-1}\{N(\varepsilon) - N(0)\} \in [0, +\infty]$$

and

$$\rho := \min\{\omega : \|T_t\| \leq e^{\omega t} \text{ for all } t \geq 0\}.$$

Lemma 10.2.3 *The constant ρ satisfies*

$$\rho = N'(0+) \geq \limsup_{t \to 0} t^{-1}\left\{\|T_t\| - 1\right\}.$$

If T_t is a one-parameter semigroup and $\mathcal{B}$ is a Hilbert space then

$$\rho = \sup\{\mathrm{Re}(z) : z \in \mathrm{Num}(Z)\} \tag{10.9}$$

where $\mathrm{Num}(Z)$ is the numerical range of the generator Z.

Proof. If $N'(0+) \leq \omega$ then, since $N(t)$ is log-concave,

$$\|T_t\| \leq N(t) \leq e^{\omega t}$$

for all $t \geq 0$. The converse is similar. The second statement follows from the fact that, assuming Z to be the generator of a one-parameter semi-group, $(Z - \omega I)$ is the generator of a contraction semigroup if and only if $\mathrm{Num}(Z - \omega I)$ is contained in $\{z : \mathrm{Re}(z) \leq 0\}$; see Theorem 8.3.5. $\square$

We study the function $N(t)$ via a transform, defined for all $\omega > 0$ by

$$M(\omega) := \sup\{N(t)e^{-\omega t} : t \geq 0\}.$$

Putting $\nu(t) := \log(N(t))$ and $\mu(\omega) := \log(M(\omega))$ we obtain

$$\mu(\omega) = \sup\{\nu(t) - \omega t : t \geq 0\}.$$

Up to a sign, the function μ is the Legendre transform of ν (also called the conjugate function), and must be convex. It follows directly from the definition that $M(\omega)$ is a monotonic decreasing function of ω which converges as $\omega \to +\infty$ to $\limsup_{t \to 0} \|T_t\|$. Hence $M(\omega) \geq 1$ for all $\omega > 0$. We also have

$$N(t) = \inf\{M(\omega)e^{\omega t} : 0 < \omega < \infty\} \tag{10.10}$$

for all $t > 0$ by the properties of the Legendre transform (i.e. simple convexity arguments).

In the semigroup context the constant c introduced below measures the deviation of the generator Z from being the generator of a contraction semigroup. The lemma is most useful when c is much larger than 1. If $c = 1$ it provides no useful information.

Lemma 10.2.4 *If $a > 0$, $b \in \mathbf{R}$ and*

$$c := a\|R_{a+ib}\| \geq 1,$$

then

$$M(\omega) \geq \tilde{M}(\omega) := \max\{(a - \omega)c/a, 1\}. \tag{10.11}$$

Proof. The formula

$$R_{a+ib} = \int_0^\infty T_t \mathrm{e}^{-(a+ib)t} \, \mathrm{d}t \tag{10.12}$$

implies that

$$\frac{c}{a} = \|R_{a+ib}\| \leq \int_0^\infty N(t)\mathrm{e}^{-at} \, \mathrm{d}t \leq \int_0^\infty M(\omega)\mathrm{e}^{\omega t-at} \, \mathrm{d}t = \frac{M(\omega)}{a - \omega}$$

for all ω such that $0 < \omega < a$. The estimate follows easily. $\square$

Theorem 10.2.5 *(Davies)*[5] *If $a > 0$, $b \in \mathbf{R}$,*

$$c := a\|R_{a+ib}\| \geq 1$$

and $r := a - a/c$, then

$$N(t) \geq \min\{\mathrm{e}^{rt}, c\}$$

for all $t \geq 0$.

Proof. This uses

$$N(t) \geq \inf\{\tilde{M}(\omega)\mathrm{e}^{\omega t} : \omega > 0\},$$

which is proved by using (10.10) and (10.11). $\square$

The above theorem provides a lower bound on $N(t)$ from a single value of the resolvent norm. The constants $c(a)$, defined for $a > 0$ by

$$c(a) := a \sup\{\|R_{a+ib}\| : b \in \mathbf{R}\}, \tag{10.13}$$

[5] See [Davies 2005A].

are immediately calculable from the pseudospectra. It follows from (10.12) and $\omega_0 = 0$ that $c(a)$ remains bounded as $a \to +\infty$.

Corollary 10.2.6 *Under the above assumptions one has*

$$N(t) \geq \sup_{\{a:c(a)\geq 1\}} \left\{ \min\{e^{r(a)t}, c(a)\} \right\} \tag{10.14}$$

where

$$r(a) := a - a/c(a).$$

The constant s_0 in the following theorem is defined in Section 10.6; we prove that it equals ω_0 if $\mathcal{B}$ is a Hilbert space in Theorem 10.6.4

Theorem 10.2.7 *If T_t is a one-parameter semigroup and $s_0 = \omega_0 = 0$ then $c(a) \geq 1$ for all $a > 0$.*

Proof. If $c(a) < 1$ then by using the resolvent expansion (8.3) one obtains

$$\|R_{a+ib+z}\| \leq \frac{c(a)}{a} (1 - |z|c(a)/a)^{-1}$$

for all z such that $|z| < a/c(a)$. This implies that $s_0 < 0$. $\square$

The quantities $c(a)$ defined by (10.13) are simpler to evaluate if the one-parameter semigroup is positivity-preserving in the sense that $f \geq 0$ implies $T_t f \geq 0$ for all $t \geq 0$. The following is only one of many special properties of positivity-preserving semigroups to be found in Chapter 12.

Lemma 10.2.8 *Let T_t be a positivity-preserving one-parameter semigroup acting in $L^p(X, dx)$ for some $1 \leq p < \infty$. If $\omega_0 = 0$ then*

$$\|R_{a+ib}\| \leq \|R_a\|$$

for all $a > 0$ and $b \in \mathbf{R}$. Hence $c(a) = a\|R_a\|$.

Proof. Let $f \in L^p$ and $g \in L^q = (L^p)^*$, where $1/p + 1/q = 1$. Then

$$|\langle R_{a+ib}f, g \rangle| = \left| \int_0^\infty \langle T_t f, g \rangle e^{-(a+ib)t} \, dt \right|$$

$$\leq \int_0^\infty |\langle T_t f, g \rangle| e^{-at} \, dt$$

$$\leq \int_0^\infty \langle T_t |f|, |g| \rangle e^{-at} \, dt$$

$$= \langle R_a |f|, |g| \rangle$$

$$\leq \|R_a\| \, \|f\|_p \, \|g\|_q.$$

By letting f and g vary we obtain the statement of the lemma. (See Problem 13.1.4 for the crucial inequality $|\langle Xf, g \rangle| \leq \langle X|f|, |g| \rangle$, which holds for all positivity-preserving operators X. It is elementary if X has a non-negative integral kernel.) $\square$

Example 10.2.9 Let T_t be the positivity-preserving, one-parameter semigroup acting on $L^2(\mathbf{R}^+)$ with generator

$$(Zf)(x) := f'(x) + v(x)f(x)$$

where v is a real-valued, bounded measurable function on $\mathbf{R}^+$. One may regard Z as a bounded perturbation of the generator Y of the semigroup $(S_t f)(x) := f(x+t)$. One has the explicit formula

$$(T_t f)(x) = \frac{a(x+t)}{a(x)} f(x+t) \tag{10.15}$$

for all $f \in L^2$ and all $t \geq 0$, where

$$a(x) := \exp\left\{ \int_0^x v(s)\,ds \right\}.$$

The function a is continuous and satisfies

$$e^{-\|v\|_\infty t} a(x) \leq a(x+t) \leq e^{\|v\|_\infty t} a(x)$$

for all x, t. Hence $\|T_t\| \leq e^{\|v\|_\infty t}$ for all $t \geq 0$.

The detailed behaviour of $\|T_t\|$ depends on the choice of v, or equivalently of a. If $c > 0$ and $0 < \gamma < 1$ then the unbounded potential $v(x) := c(1-\gamma)x^{-\gamma}$ corresponds to the choice

$$a(x) := \exp\{cx^{1-\gamma}\}.$$

Instead of deciding the precise domain of the generator Z, we define the one-parameter semigroup T_t directly by (10.15), and observe that

$$N(t) = \|T_t\| = \exp\{ct^{1-\gamma}\}$$

for all $t \geq 0$. This implies that

$$M(\omega) = \exp\{c'\omega^{1-1/\gamma}\}$$

for all $\omega > 0$. If c is large and γ is close to 1, the semigroup norm grows rapidly for small t, before becoming almost stationary. The behaviour of $\|T_t f\|$ as $t \to \infty$ depends upon the choice of f, but one has $\lim_{t\to\infty} \|T_t f\| = 0$ for a dense set of f, including all f with compact support.

For this unbounded potential v, every z with $\mathrm{Re}(z) < 0$ is an eigenvalue, the corresponding eigenvector being

$$f(x) := \exp\left\{zx - c(1-\gamma)x^{1-\gamma}\right\}.$$

Hence

$$\mathrm{Spec}(Z) = \{z : \mathrm{Re}(z) \leq 0\}. \qquad \square$$

Problem 10.2.10 The above example may be used to show that $\|T_t\|$ can be highly oscillatory. If $k > 1$ prove that the choice

$$a(x) := 1 + (k-1)\sin^2(\pi x/2) \qquad (10.16)$$

in (10.15) leads to $\|T_{2n}\| = 1$ and $\|T_{(2n+1)}\| = k$ for all positive integers n. In this case the regularizations $N(t)$ and $L(t)$ are not equal, but both are equal to k for $t \geq 1$. $\quad\square$

10.3 Contractions and dilations

In this section we consider a single contraction A on $\mathcal{H}$, i.e. an operator A such that $\|A\| \leq 1$. Our main result is a dilation theorem for contractions.[6]

Theorem 10.3.1 *(Sz.-Nagy) If A is a contraction on $\mathcal{H}$ then there exists a Hilbert space $\mathcal{K}$ containing $\mathcal{H}$ and a unitary operator U on $\mathcal{K}$ such that*

$$A^n = PU^nP\big|_{\mathcal{H}}$$

for all non-negative integers n, where P is the orthogonal projection of $\mathcal{K}$ onto $\mathcal{H}$.

Proof. We put $\mathcal{K} := l^2(\mathbf{Z}, \mathcal{H})$ and identify $\phi \in \mathcal{H}$ with the sequence $\tilde{f}_r := \delta_{r,0}\phi$ in $\mathcal{K}$. The theorem follows immediately provided there exists a unitary operator $U := D + N$, where N has a strictly upper triangular block matrix and $D_{r,s} := \delta_{r,0}\delta_{s,0}A$.

---

[6] Much fuller treatments of the theorem may be found in [Davies 1980B, Sz.-Nagy and Foias 1970]. These texts prove the uniqueness of the dilation under further conditions, but we do not need that for the applications that we make of the theorem.

We construct a unitary operator whose block matrix is of the form

$$
U := \begin{pmatrix}
\ddots & \ddots & & & & & & \\
& 0 & 1 & & & & & \\
& & 0 & 1 & & & & \\
& & & 0 & C & D & & \\
& & & A & B & & & \\
& & & & 0 & 1 & & \\
& & & & & 0 & 1 & \\
& & & & & & 0 & \ddots \\
& & & & & & & \ddots
\end{pmatrix}
$$

where the blank entries all vanish. More precisely we put

$$
(Uf)_r := \begin{cases}
Af_0 + (I - |A^*|^2)^{1/2} f_1 & \text{if } r = 0, \\
(I - |A|^2)^{1/2} f_0 - A^* f_1 & \text{if } r = -1, \\
f_{r+1} & \text{otherwise.}
\end{cases}
$$

The unitarity of U follows by Lemma 9.5.1. $\quad\square$

Problem 10.3.2 Given a complex constant z such that $|z| < 1$, let $\mathcal{L}$ be the m-dimensional subspace of $\mathcal{H} := l^2(\mathbf{Z})$ consisting of all sequences of the form $f_r := \chi(r)p(r)z^r$, where χ is the characteristic function of $[0, +\infty)$ and p is a polynomial of degree at most $(m - 1)$. Let P be the orthogonal projection of $\mathcal{H}$ onto $\mathcal{L}$ and let U denote the unitary operator on $\mathcal{H}$ given by $(Uf)_r := f_{r+1}$. Prove that if $A := PUP|_{\mathcal{L}}$ then

$$
A^n = PU^n P|_{\mathcal{L}}
$$

for all non-negative integers n. Find the spectrum of the operator A. $\quad\square$

In Theorem 10.3.1 we constructed a unitary dilation of a given contraction, but one can also start with a unitary operator and go in the reverse direction.

Theorem 10.3.3 *Let U be a unitary operator on the Hilbert space $\mathcal{H}$ and let $\mathcal{L}_0$, $\mathcal{L}_1$ be two closed subspaces such that $\mathcal{L}_1 \subseteq \mathcal{L}_0$, $U\mathcal{L}_0 \subseteq \mathcal{L}_0$ and $U\mathcal{L}_1 \subseteq \mathcal{L}_1$. Let P denote the orthogonal projection of $\mathcal{H}$ onto $\mathcal{L} := \mathcal{L}_0 \cap \mathcal{L}_1^{\perp}$. Then the contraction $A := PUP|_{\mathcal{L}}$ satisfies*

$$
A^n = PU^n P|_{\mathcal{L}}
$$

for all non-negative integers n.

Proof. Let P_i denote the orthogonal projections of $\mathcal{H}$ onto $\mathcal{L}_i$ for $i = 0, 1$. The hypotheses imply that $P_0 P_1 = P_1 P_0 = P_1$ and $P_i U P_i = U P_i$ for $i = 0, 1$. Moreover $P = (I - P_1) P_0$.

We start by proving that $P U^n P_1 = 0$ for all non-negative integers n. This is obvious for $n = 0$ and the inductive step is

$$PU^{n+1}P_1 = PU^n.UP_1 = PU^n(P_1 UP_1) = (PU^n P_1)UP_1 = 0.$$

We also prove the theorem inductively, noting that it is elementary for $n = 0$. The inductive step is

$$PU^n P.PUP = PU^n(P_0 - P_1 P_0)UP_0(I - P_1)$$
$$= PU^n P_0 UP_0(I - P_1)$$
$$= PU^{n+1}P_0(I - P_1)$$
$$= PU^{n+1}P. \qquad \square$$

Problem 10.3.4 Find the subspaces $\mathcal{L}_0$ and $\mathcal{L}_1$ that recast Problem 10.3.2 into the setting of Theorem 10.3.3. $\square$

One may use the dilation theorem to give another proof of von Neumann's theorem, previously discussed in Section 9.5.

Theorem 10.3.5 *(von Neumann) If A is a contraction on a Hilbert space $\mathcal{H}$ then $\|f(A)\| \le \|f\|_\infty$ for all analytic functions f on the closure of the unit disc D.*

Proof. We first note that it is sufficient to prove the result for polynomials, by approximation. In this case Theorem 10.3.1 yields

$$\|p(A)\| = \|Pp(U)P|_{\mathcal{H}}\| \le \|p(U)\| \le \|p\|_\infty.$$

The final inequality uses the spectral theorem for unitary operators. $\square$

The main dilation theorem of this section can be extended to one-parameter contraction semigroups as follows.[7]

Theorem 10.3.6 *(semigroup dilation theorem) If A_t is a one-parameter contraction semigroup on $\mathcal{H}$ then there exists a Hilbert space $\mathcal{K}$ containing $\mathcal{H}$*

---

[7] See [Davies 1980B] or [Sz.-Nagy and Foias 1970] for the proof.

and a one-parameter unitary group U_t on $\mathcal{K}$ such that

$$A_t = PU_tP|_{\mathcal{H}}$$

for all $t \geq 0$, where P is the orthogonal projection of $\mathcal{K}$ onto $\mathcal{H}$.

10.4 The Cayley transform

We have already shown (Theorem 8.3.5) that an operator Z with dense domain $\mathcal{D}$ in a Hilbert space $\mathcal{H}$ is the generator of a one-parameter semigroup of contractions if and only if it is dissipative, i.e.

$$\operatorname{Re}\langle Zf, f \rangle \leq 0$$

for all $f \in \mathcal{D}$, and also

$$(\lambda I - Z)\mathcal{D} = \mathcal{H}$$

for some, or equivalently all, $\lambda > 0$. Our goal is to replace the second condition by one that is sometimes easier to verify.

If Z is a densely defined dissipative operator and $0 \neq f \in \mathcal{D}$ then

$$\operatorname{Re}\langle (Z - I)f, f \rangle \leq -\langle f, f \rangle < 0,$$

so $(Z - I)f \neq 0$. In other words $(Z - I)$ is one-one on its domain. We define the Cayley transform of a dissipative operator Z with dense domain $\mathcal{D}$ by

$$Cf := (Z + I)(Z - I)^{-1}f,$$

its domain being $\mathcal{E} = (Z - I)\mathcal{D}$, which need not be a dense subspace of $\mathcal{H}$.

Lemma 10.4.1 *The Cayley transform C is a contraction from $\mathcal{E}$ into $\mathcal{H}$, and $(C - I)$ has dense range. Moreover every contraction C with this property is the Cayley transform of a unique densely defined dissipative operator Z.*

Proof. Given Z, we start by showing that C has the stated properties. If $f \in \mathcal{D}$ then

$$\begin{aligned}
\|(Z + I)f\|^2 &= \|f\|^2 + 2\operatorname{Re}\langle Zf, f \rangle + \|Zf\|^2 \\
&\leq \|f\|^2 - 2\operatorname{Re}\langle Zf, f \rangle + \|Zf\|^2 \\
&= \|(Z - I)f\|^2.
\end{aligned}$$

Putting $g := (Z - I)f$, this establishes that $\|Cg\| \le \|g\|$ for all $g \in \mathcal{E}$. It follows from the definition of C that

$$(C - I)g = 2(Z - I)^{-1}g \tag{10.17}$$

for all $g \in \mathcal{E}$. Therefore the range of $(C - I)$ equals the domain $\mathcal{D}$ of $(Z - I)$, which is dense by hypothesis. The equation (10.17) also establishes that the relationship between Z and C is one-one.

We next prove that if C is a contraction with domain $\mathcal{E}$ and $(C - I)$ has dense range $\mathcal{D}$ then $(C - I)$ is one-one: equivalently if $f \in \mathcal{E}$ and $Cf = f$ then $f = 0$. We start by writing $\mathcal{E} := \mathcal{E}_0 \oplus \mathcal{E}_1$ where $\mathcal{E}_0 := \mathbf{C}f$ and $\mathcal{E}_1 := \{h \in \mathcal{E} : \langle h, f \rangle = 0\}$. If $h \in \mathcal{E}_1$, $\varepsilon > 0$ and $\langle f, Ch \rangle \ne 0$ we put $\delta := \varepsilon \langle f, Ch \rangle$. We then have

$$\begin{aligned}
\|C(\delta h + f)\|^2 &= \|\delta Ch + f\|^2 \\
&= |\delta|^2 \|Ch\|^2 + 2\varepsilon |\langle f, Ch \rangle|^2 + \|f\|^2 \\
&> |\delta|^2 \|h\|^2 + \|f\|^2 \\
&= \|\delta h + f\|^2
\end{aligned}$$

for all sufficiently small $\varepsilon > 0$, because a term of size $O(\varepsilon)$ is larger than terms of size $O(\varepsilon^2)$ as $\varepsilon \to 0$. This contradicts the fact that C is a contraction, so we deduce that $\langle f, Ch \rangle = 0$, and hence $C\mathcal{E}_1 \subseteq \mathcal{E}_1$. This implies that

$$\mathcal{D} = (C - I)\mathcal{E} = (C - I)\mathcal{E}_0 + (C - I)\mathcal{E}_1 \subseteq \mathcal{E}_1.$$

The density of $\mathcal{D}$ and the definition of $\mathcal{E}_1$ finally imply that $f = 0$.

We now prove that a contraction C such that $\mathcal{D} := \mathrm{Ran}(C - I)$ is dense is the Cayley transform of a densely defined dissipative operator Z. Given C, we use the fact that $(C - I)$ is one-one to define Z by (10.17). Since

$$\mathrm{Dom}(Z) = \mathrm{Ran}((Z - I)^{-1}) = \mathrm{Ran}(C - I) = \mathcal{D},$$

we see that Z is densely defined. Since C is a contraction and

$$C = (Z + I)(Z - I)^{-1}$$

we deduce that

$$\|(Z + I)f\|^2 \le \|(Z - I)f\|^2$$

for all $f \in \mathcal{D}$. By expanding both sides and simplifying we see that this is equivalent to Z being dissipative. $\quad\square$

We define a maximal dissipative operator to be a (densely defined) dissipative operator that has no dissipative extensions. The following theorem does not extend to general Banach spaces.

Theorem 10.4.2 *The operator Z is the generator of a one-parameter contraction semigroup on the Hilbert space $\mathcal{H}$ if and only if it is a maximal dissipative operator.*

Proof. It is immediate from the proof of Lemma 10.4.1, in particular (10.17), that there is a one-one correspondence between contractive extensions C' of C and dissipative extensions Z' of Z. Therefore Z is maximal if and only if C is maximal.

If Z is the generator of a one-parameter contraction semigroup then $\text{Ran}(I - Z) = \mathcal{H}$ so $\text{Dom}(C) = \mathcal{H}$ and C is maximal.

Conversely suppose that C is a contraction with domain $\mathcal{E}$, that $(C - I)$ has dense range and that C has no proper extension with these properties. By taking limits we deduce that $\mathcal{E}$ is a closed subspace of $\mathcal{H}$. If it were a proper closed subspace then we would be able to put $C' = CP$, where P is the orthogonal projection of $\mathcal{H}$ onto $\mathcal{E}$, to get a proper extension of C. Therefore $\text{Ran}(I - Z) = \mathcal{E} = \mathcal{H}$. Theorem 8.3.5 now implies that Z is the generator of a one-parameter contraction semigroup. $\square$

In Lemma 5.4.4 we proved that a closed symmetric operator H is self-adjoint if and only if its deficiency indices are both zero. In the following theorem we deal with the case in which this is not necessarily so. We first clarify the relationship with Lemma 5.4.4, in which the dimensions of the following two subspaces are called the deficiency indices of H.

Lemma 10.4.3 *If C is the Cayley transform of $Z := iH$, where H is a densely defined symmetric operator, then*

$$\text{Dom}(C)^{\perp} = \{f \in \text{Dom}(H^*) : H^*f = if\},$$
$$\text{Ran}(C)^{\perp} = \{f \in \text{Dom}(H^*) : H^*f = -if\}.$$

Proof. By definition $\text{Dom}(C) = (iH - I)\mathcal{D}$. Therefore the following statements are equivalent.

$$f \perp \text{Dom}(C),$$
$$\langle (iH - I)g, f \rangle = 0 \qquad \text{for all } g \in \mathcal{D},$$
$$\langle Hg, f \rangle = \langle g, if \rangle \qquad \text{for all } g \in \mathcal{D}.$$

The third of these statements is equivalent to $f \in \text{Dom}(H^*)$ and $H^*f = if$. The proof that $f \perp \text{Ran}(C)$ iff $f \in \text{Dom}(H^*)$ and $H^*f = -if$ is similar. $\square$

Theorem 10.4.4 *If H is a densely defined symmetric operator then the Cayley transform C of $Z = iH$ is isometric. If H is maximal symmetric then either iH or $-iH$ is the generator of a one-parameter semigroup of isometries. If $H = H^*$ then H is the generator of a one-parameter unitary group.*

Proof. The hypotheses imply that $\text{Re}\langle Zf, f \rangle = 0$ for all $f \in \text{Dom}(Z)$. Therefore

$$\|(iH + 1)f\|^2 = \|Hf\|^2 + \|f\|^2 = \|(iH - 1)f\|^2$$

for all such f, which implies that $C := (iH + I)(iH - I)^{-1}$ is an isometry.

If $\mathcal{L} := \text{Dom}(C)^\perp$ and $\mathcal{M} := \text{Ran}(C)^\perp$ are both non-zero then they contain subspaces $\mathcal{L}'$ and $\mathcal{M}'$ which have the same positive dimension, and there exists a unitary operator V mapping $\mathcal{L}'$ onto $\mathcal{M}'$. We define the isometric extension $C' : \text{Dom}(C) \oplus \mathcal{L}' \to \text{Ran}(C) \oplus \mathcal{M}'$ by

$$C'(f \oplus g) := Cf \oplus Vg.$$

Since there is a one-one correspondence between symmetric extensions of H and isometric extensions of C, we conclude that H is not maximal symmetric.

If H is maximal symmetric there are two cases to consider. If $\mathcal{L} = \{0\}$ then $\text{Ran}(I - Z) = \mathcal{H}$, so Z is the generator of a one-parameter contraction semigroup T_t on $\mathcal{H}$ by Theorem 8.3.5. If $f \in \text{Dom}(Z)$ and $f_t = T_t f$ then

$$\frac{\text{d}}{\text{d}t}\|f_t\|^2 = \langle Zf_t, f_t \rangle + \langle f_t, Zf_t \rangle = 2\text{Re}\langle Zf_t, f_t \rangle = 0.$$

Therefore T_t is a one-parameter semigroup of isometries.

A similar argument shows that if $\mathcal{M} = \{0\}$ then $(-iH)$ is the generator of a one-parameter semigroup of isometries.

If $H = H^*$ then Lemma 10.4.3 implies that the deficiency indices of H both vanish, because a symmetric operator cannot have complex eigenvalues. Therefore $\mathcal{L} = \mathcal{M} = \{0\}$, and $\pm iH$ are both generators of one-parameter semigroup s of isometries. This is sufficient to establish that iH is the generator of a one-parameter unitary group by using Theorem 6.1.23. $\square$

Problem 10.4.5 The formula

$$(T_t f)(x) := \begin{cases} f(x - t) & \text{if } 0 \le t \le x, \\ 0 & \text{otherwise,} \end{cases}$$

defines a one-parameter semigroup of isometries on $L^2(0, \infty)$ whose generator Z is given formally by $(Zf)(x) := -f'(x)$. Prove that $C_c^1(0, \infty)$ is a core for the generator Z and that every $f \in \text{Dom}(Z)$ is a continuous function satisfying $f(0) = 0$. Prove also that $\text{Re}\langle Zf, f \rangle = 0$ for all $f \in \text{Dom}(Z)$. □

If T_t is a one-parameter contraction semigroup on $\mathcal{H}$, we say that it is unitary on the closed subspace $\mathcal{L}$ if T_t maps $\mathcal{L}$ isometrically onto $\mathcal{L}$ for every $t \geq 0$. We say that T_t is completely non-unitary if the only such subspace is $\{0\}$.

Theorem 10.4.6 *If T_t is a one-parameter contraction semigroup on $\mathcal{H}$ then there is a unique orthogonal decomposition $\mathcal{H} := \mathcal{H}_1 \oplus \mathcal{H}_2$ such that $\mathcal{H}_1$ and $\mathcal{H}_2$ are both invariant, T_t is unitary on $\mathcal{H}_1$ and completely non-unitary on $\mathcal{H}_2$.*

Proof. We start by identifying a closed linear subspace on which T_t is isometric. Let $\mathcal{L}$ denote the set of all $f \in \mathcal{H}$ such that $\|T_t f\| = \|f\|$ for all $t \geq 0$. If $f, g \in \mathcal{L}$ then

$$
\begin{aligned}
2\|f\|^2 + 2\|g\|^2 &= 2\|T_t f\|^2 + 2\|T_t g\|^2 \\
&= \|T_t(f + g)\|^2 + \|T_t(f - g)\|^2 \\
&\leq \|f + g\|^2 + \|f - g\|^2 \\
&= 2\|f\|^2 + 2\|g\|^2.
\end{aligned}
$$

Therefore $\|T_t(f + g)\| = \|f + g\|$. It is easy to show that $\mathcal{L}$ is closed under scalar multiples and norm limits, so we conclude that it is a closed linear subspace of $\mathcal{H}$.

If T_t maps a closed linear subspace $\mathcal{M}$ isometrically onto $\mathcal{M}$ for all $t \geq 0$ then it is immediate that $\mathcal{M} \subseteq \mathcal{L}$. Therefore $\mathcal{M} = T_t(\mathcal{M}) \subseteq T_t(\mathcal{L})$ for all $t \geq 0$, and we conclude that

$$
\mathcal{M} \subseteq \mathcal{H}_1 := \bigcap_{t \geq 0} T_t(\mathcal{L}).
$$

If we prove that T_t is unitary when restricted to $\mathcal{H}_1$ then it follows that $\mathcal{H}_1$ is the largest subspace with this property.

We first prove that $\mathcal{H}_1$ is an invariant subspace for T_t. It follows from its definition that $T_s(\mathcal{L}) \subseteq \mathcal{L}$ for all $s \geq 0$. Hence

$$
T_s(\mathcal{H}_1) \subseteq T_s(T_t \mathcal{L}) = T_t(T_s \mathcal{L}) \subseteq T_t \mathcal{L}
$$

for all $s, t \geq 0$. This implies that $T_s \mathcal{H}_1 \subseteq \mathcal{H}_1$ for all $s \geq 0$.

We next prove that T_t maps $\mathcal{H}_1$ onto $\mathcal{H}_1$ for all $t \geq 0$. If $f \in \mathcal{H}_1$ and $t \geq 0$ then there exists $g \in \mathcal{L}$ such that $f = T_t g$; moreover g is unique because T_t is isometric when restricted to $\mathcal{L}$. If $s \geq 0$ then there also exists $g_s \in \mathcal{L}$ such that $f = T_{t+s} g_s$. It follows by the uniqueness property that $g = T_s g_s$. Hence $g \in T_s(\mathcal{L})$ for all $s \geq 0$, so $g \in \mathcal{H}_1$.

We finally have to prove that $\mathcal{H}_2 := \mathcal{H}_1^\perp$ is an invariant subspace for T_t. The direct sum decomposition $\mathcal{H} = \mathcal{H}_1 \oplus \mathcal{H}_2$ allows us to write T_t in the block form

$$T_t := \begin{pmatrix} U_t & A_t \\ 0 & B_t \end{pmatrix}$$

where $U_t : \mathcal{H}_1 \to \mathcal{H}_1$ is unitary, $A_t : \mathcal{H}_2 \to \mathcal{H}_1$ and $B_t : \mathcal{H}_2 \to \mathcal{H}_2$. Since T_t^* is a contraction, we see that

$$\|f\|^2 + \|A_t^* f\|^2 = \|U_t^* f\|^2 + \|A_t^* f\|^2$$
$$= \|T_t^*(f \oplus 0)\|^2$$
$$\leq \|f \oplus 0\|^2$$
$$= \|f\|^2$$

for all $f \in \mathcal{H}_1$. Hence $A_t^* = 0$, so $A_t = 0$ and $T_t(\mathcal{H}_2) \subseteq \mathcal{H}_2$. $\square$

Problem 10.4.7 If e^{Zt} is a one-parameter contraction semigroup on a finite-dimensional space $\mathcal{H}$, use the Jordan canonical form for Z to prove that T_t is completely non-unitary if and only if every eigenvalue λ of Z satisfies $\mathrm{Re}(\lambda) < 0$. $\square$

10.5 One-parameter groups

We start with a classical result due to Sz.-Nagy.[8]

Theorem 10.5.1 *(Sz.-Nagy) Let A be a bounded invertible operator acting on a Hilbert space $\mathcal{H}$ and suppose that $\|A^n\| \leq c$ for all $n \in \mathbf{Z}$. Then there exists a bounded invertible operator S such that SAS^{-1} is unitary. Moreover $\|S^{\pm 1}\| \leq c$.*

Proof. The bound

$$1 = \|A^{-n} A^n\| \leq \|A^{-n}\| \|A^n\| \leq c \|A^n\|$$

[8] See [Sz.-Nagy 1947].

implies that $c^{-1} \leq \|A^n\| \leq c$ for all $n \in \mathbf{Z}$. We now define new inner products

$$\langle f, g \rangle_n := n^{-1} \sum_{r=0}^{n-1} \langle A^r f, A^r g \rangle$$

for all positive integers n, and use the above inequalities to deduce that

$$c^{-2} \|f\|^2 \leq \|f\|_n^2 \leq c^2 \|f\|^2$$

for all $f \in \mathcal{H}$. This implies the existence of positive self-adjoint operators B_n such that $c^{-2}I \leq B_n \leq c^2 I$ and

$$\langle f, g \rangle_n = \langle B_n f, g \rangle$$

for all $f, g \in \mathcal{H}$.

We next show that $A^{\pm 1}$ are close to being contractions with respect to the norms $\| \cdot \|_n$, provided n is large enough. If $f \in \mathcal{H}$ then

$$\|Af\|_n^2 = n^{-1} \sum_{r=1}^{n-1} \|Af\|^2 + n^{-1} \|A^n f\|^2$$

$$\leq \|f\|_n^2 + n^{-2} \sum_{r=0}^{n-1} \|A^{n-r} A^r f\|^2$$

$$\leq \|f\|_n^2 + c^2 n^{-2} \sum_{r=0}^{n-1} \|A^r f\|^2$$

$$= (1 + c^2/n) \|f\|_n^2.$$

Therefore

$$\|A\|_n^2 \leq 1 + c^2/n.$$

A similar calculation applies to A^{-1}.

The final step is to let $n \to \infty$. Since B_n need not converge we use the weak operator compactness (see Problem 10.5.2 below) of the set

$$\mathcal{B} := \{ B \in \mathcal{L}(\mathcal{H}) : c^{-2}I \leq B \leq c^2 I \}, \tag{10.18}$$

to pass to a subsequence $B_{n(r)}$ that converges in the weak operator topology as $r \to \infty$ to a limit B_∞ that also lies in $\mathcal{B}$. Letting $r \to \infty$ in

$$0 \leq \langle B_{n(r)} A^{\pm 1} f, A^{\pm 1} f \rangle \leq (1 + c^2/n(r)) \langle B_{n(r)} f, f \rangle$$

yields

$$0 \leq \langle B_\infty A^{\pm 1} f, A^{\pm 1} f \rangle \leq \langle B_\infty f, f \rangle.$$

We now put $S := B_\infty^{1/2}$ and $g := Sf$ to get

$$\|SA^{\pm 1}S^{-1}g\| \le \|g\|$$

for all $g \in \mathcal{H}$. In other words $SA^{\pm 1}S^{-1}$ are both contractions. The final bound $\|S^{\pm 1}\| \le c$ follows directly from (10.18) and the definition of S. $\quad\square$

Problem 10.5.2 Let

$$K := \prod_{f,g \in \mathcal{H}} K_{f,g}$$

where

$$K_{f,g} := \{z \in \mathbf{C} : |z| \le c^2 \|f\|\,\|g\|\},$$

so that K is compact for the product topology. Let M be the subset of K consisting of all functions $z : \mathcal{H} \times \mathcal{H} \to \mathbf{C}$ for which

$$z_{f+h,g} = z_{f,g} + z_{h,g}$$

$$z_{f,g} = \overline{z_{g,f}}$$

$$c^{-2}\|f\|^2 \le z_{f,f} = \overline{z_{f,f}} \le c^2 \|f\|^2$$

$$z_{\alpha f,g} = \alpha z_{f,g}$$

for all f, g, $h \in \mathcal{H}$ and $\alpha \in \mathbf{C}$. Prove that M is a compact subset of K and that it is homeomorphic to the set $\mathcal{B}$ defined in (10.18), provided the latter is given its weak operator topology. $\quad\square$

Note The one-sided analogue of Theorem 10.5.1 is false: Foguel has constructed a power-bounded operator A (i.e. an operator such that $\|A^n\|$ are uniformly bounded for $n \ge 1$) which is not similar to a contraction.[9]

Our proof of Theorem 10.5.1 used a compactness argument but was otherwise rather explicit and computational. One can also base the proof on the following fixed-point theorem.

Problem 10.5.3 (Markov-Kakutani theorem) Let X be a compact convex set in a locally convex Hausdorff topological vector space. Let $S : X \to X$ be continuous and affine in the sense that

$$S(\lambda x + (1-\lambda)y) = \lambda S(x) + (1-\lambda)S(y)$$

[9] See [Foguel 1964] and [Lebow 1968]. [Chernoff 1976] discusses the corresponding problem for one-parameter semigroup s. For a complete analysis of the one-sided case sees [Paulsen 1984] and [Pisier 1997]

for all x, $y \in S$ and all $\lambda \in (0, 1)$. By adapting the proof of Theorem 10.5.1 prove that there exists $a \in S$ such that $S(a) = a$. □

Theorem 10.5.4 *Let T_t be a one-parameter group acting on a Hilbert space $\mathcal{H}$ and suppose that $\|T_t\| \leq c$ for all $t \in \mathbf{R}$. Then there exists a bounded invertible operator S such that $U_t = ST_tS^{-1}$ are unitary for all $t \in \mathbf{R}$.*

Proof. This is a routine modification of the proof of Theorem 10.5.1. One puts

$$\langle f, g \rangle_n := \frac{1}{n} \int_0^n \langle T_t f, T_t g \rangle \, dt$$

to obtain

$$c^{-1}\|f\| \leq \|f\|_n \leq c\|f\|$$

for all $f \in \mathcal{H}$. One then rewrites the bound

$$\|T_s\|_n^2 \leq 1 + c^2 |s|/n$$

in the form

$$0 \leq \langle B_n T_s f, T_s f \rangle \leq (1 + c^2 |s|/n)\langle B_n f, f \rangle$$

and passes to a subsequence $B_{n(r)}$ as before. □

The remainder of this section discusses generalizations of the above theorems. We will need to use the following standard result below.

Problem 10.5.5 *Let $\mathcal{D}$ be a dense linear subspace of the Hilbert space $\mathcal{H}$, and let $Q : \mathcal{D} \times \mathcal{D} \to \mathbf{C}$ be a map which is linear in the first variable and conjugate linear in the second. Suppose also that*

$$Q(f, g) = \overline{Q(g, f)}$$

for all $f, g \in \mathcal{D}$ and there exist α, $\beta \in \mathbf{R}$ such that

$$\alpha\|f\|^2 \leq Q(f, f) \leq \beta\|f\|^2$$

for all $f \in \mathcal{D}$. Prove that there exists a bounded self-adjoint operator $A : \mathcal{H} \to \mathcal{H}$ such that $\alpha I \leq A \leq \beta I$ and

$$Q(f, g) = \langle Af, g \rangle$$

for all $f, g, \in \mathcal{D}$. □

Theorem 10.5.6 *If Z is the generator of the one-parameter group T_t on the Hilbert space $\mathcal{H}$ then the following conditions are equivalent.*

(i) *There exists a constant $a \geq 0$ such that*

$$|\mathrm{Re}\langle Zf, f\rangle| \leq a\|f\|^2$$

for all $f \in \mathrm{Dom}(Z)$.

(ii) *There exists a constant $a \geq 0$ such that*

$$\|T_t\| \leq e^{a|t|}$$

for all $t \in \mathbf{R}$.

(iii) *One has $Z := iH + A$ where H is self-adjoint and A is bounded and self-adjoint.*

Proof. (i)$\Rightarrow$(ii) If $f \in \mathrm{Dom}(Z)$ and $f_t := T_t f$ then

$$\left|\frac{\mathrm{d}}{\mathrm{d}t}\|f_t\|^2\right| = |\langle Zf_t, f_t\rangle + \langle f_t, Zf_t\rangle| \leq 2a\|f_t\|^2$$

for all $t \in \mathbf{R}$. Hence $\|T_t f\| \leq e^{a|t|}\|f\|$ for all $f \in \mathrm{Dom}(Z)$; the same bound holds for all $f \in \mathcal{H}$ by continuity.

(ii)$\Rightarrow$(iii) The first step is to construct the bounded operator A of the theorem. If $f \in \mathrm{Dom}(Z)$ then

$$-a\|f\|^2 \leq \mathcal{Q}(f, f) \leq a\|f\|^2$$

where

$$\mathcal{Q}(f, g) := \frac{1}{2}\left(\langle Zf, g\rangle + \langle f, Zg\rangle\right).$$

It follows by Problem 10.5.5 that there exists a bounded self-adjoint operator A on $\mathcal{H}$ and an unbounded symmetric operator H with $\mathrm{Dom}(H) = \mathrm{Dom}(Z)$ such that $Z = iH + A$. Since iH is a bounded perturbation of Z, Theorem 11.4.1 implies that it is the generator of a one-parameter group U_t. Since H is symmetric U_t is a one-parameter group of isometries. Differentiating $U_t^* = U_{-t}$ at $t = 0$ yields $H = H^*$ by Theorem 7.3.3.

(iii)$\Rightarrow$(i) The definition of Z yields

$$|\mathrm{Re}\langle Zf, f\rangle| \leq \|A\| \|f\|^2$$

for all $f \in \mathrm{Dom}(Z)$ immediately. Since H is self-adjoint its deficiency indices vanish and iH is the generator of a one-parameter group of isometries by Theorem 10.4.4. Since Z is a bounded perturbation of iH, it is the generator of a one-parameter group by Theorem 11.4.1. $\square$

Theorem 10.5.7 *(Haase)*[10] *If $T_t := e^{Zt}$ is a one-parameter group on the Hilbert space $\mathcal{H}$ there exists an equivalent inner product $\langle \cdot, \cdot \rangle_Q$ on $\mathcal{H}$ such that $Z := iH + A$ where H is Q-self-adjoint and A is bounded and Q-self-adjoint.*

Proof. We use the inner product of

$$\langle f, g \rangle_Q := \int_{-\infty}^{\infty} \langle T_t f, T_t g \rangle e^{-2b|t|} \, dt$$

where $b > a$ and $\|T_t\| \le M e^{a|t|}$ for all $t \in \mathbf{R}$. We start by proving that this is equivalent to the given inner product. If $f \in \mathcal{H}$ then

$$\|f\|_Q^2 \le \int_{-\infty}^{\infty} M^2 e^{2a|t| - 2b|t|} \|f\|^2 \, dt = \frac{M^2}{b - a} \|f\|^2.$$

In the reverse direction we use the fact that

$$1 = \|T_0\| \le \|T_t\| \|T_{-t}\|$$

to deduce that

$$\|T_t\| \ge M^{-1} e^{-a|t|}$$

for all $t \in \mathbf{R}$. Therefore

$$\|f\|_Q^2 \ge \int_{-\infty}^{\infty} M^{-2} e^{-2a|t| - 2b|t|} \|f\|^2 \, dt = \frac{\|f\|^2}{M^2 (b + a)}$$

for all $f \in \mathcal{H}$.

We next prove that $\|T_s\|_Q \le e^{b|s|}$ for all $s \in \mathbf{R}$. If $f \in \mathcal{H}$ then

$$\|T_s f\|_Q^2 = \int_{-\infty}^{\infty} e^{-2b|t|} \|T_{s+t} f\|^2 \, dt$$

$$= \int_{-\infty}^{\infty} e^{-2b|t-s|} \|T_t f\|^2 \, dt$$

$$\le \int_{-\infty}^{\infty} e^{2b|s| - 2b|t|} \|T_t f\|^2 \, dt$$

$$= e^{2b|s|} \|f\|_Q^2.$$

The proof is completed by applying Theorem 10.5.6. $\square$

---

[10] See [Haase 2004].

Problem 10.5.8 The conditions of Theorem 10.5.6 imply that $\mathrm{Dom}(Z) = \mathrm{Dom}(Z^*)$. Construct a one-parameter group acting on the Hilbert space $\mathcal{H} = L^2(\mathbf{R})$ such that

$$\mathrm{Dom}(Z) \cap \mathrm{Dom}(Z^*) = \{0\}. \qquad \square$$

10.6 Resolvent bounds in Hilbert space

If $T_t := \mathrm{e}^{Zt}$ is a one-parameter semigroup acting on a Banach space $\mathcal{H}$ then we have defined ω_0 to be the infimum of all constants a such that

$$\|T_t\| \leq M(a)\mathrm{e}^{at}$$

for some $M(a)$ and all $t \geq 0$. We define s_0 to be the infimum of all constants b such that

$$\sup\{\|R(z, Z)\| : \mathrm{Re}(z) \geq b\} < \infty.$$

It follows directly from the identity

$$R(z, Z) = \int_0^\infty T_t \mathrm{e}^{-zt} \, \mathrm{d}t$$

of Theorem 8.2.1 that $s_0 \leq \omega_0$. In a finite-dimensional context these constants are equal but in general they may differ.

In Theorem 10.6.4 we prove the surprising fact that they are always equal in a Hilbert space. The proof depends on three preliminary lemmas.

Lemma 10.6.1 *Let A be a closed operator acting in the Banach space $\mathcal{B}$ and satisfying $\mathrm{Spec}(A) \cap S = \emptyset$, where*

$$S := \{x + iy : \alpha < x < \beta \text{ and } y \in \mathbf{R}\}.$$

Suppose also that $\|R(z, A)\| \leq k$ for all $z \in S$. If $0 < p < \infty$ and $f \in \mathcal{B}$ and

$$\int_{-\infty}^\infty \|R(u + iy, A)f\|^p \, \mathrm{d}y \leq c$$

for some $u \in S$ then

$$\int_{-\infty}^\infty \|R(v + iy, A)f\|^p \, \mathrm{d}y \leq c(1 + (\beta - \alpha)k)^p$$

for all $v \in S$.

Proof. By using the translation invariance of the integrals with respect to y we can reduce to the case in which u and v are both real.

The resolvent identity

$$R(v+iy, A)f = R(u+iy, A)f + (u-v)R(v+iy, A)R(u+iy, A)f$$

yields the bound

$$\|R(v+iy, A)f\| \le (1 + (\beta - \alpha)k)\|R(u+iy, A)f\|,$$

from which the statement of the lemma follows directly. □

Lemma 10.6.2 *Let* $T_t := e^{Zt}$ *be a one-parameter semigroup on the Banach space $\mathcal{B}$ and let* $\mathrm{Re}(z) > \omega_0$. *Then*

$$\int_0^\infty te^{-zt}\langle T_t f, \phi\rangle\, \mathrm{d}t = \langle R(z, Z)^2 f, \phi\rangle$$

for all $f \in \mathcal{B}$ and $\phi \in \mathcal{B}^*$.

Proof. This relies upon the formula

$$\frac{\mathrm{d}}{\mathrm{d}t}\{te^{-zt}\langle T_t R(z, Z)f, \phi\rangle\} = e^{-zt}\langle T_t R(z, Z)f, \phi\rangle - te^{-zt}\langle T_t f, \phi\rangle.$$

The condition $\mathrm{Re}(z) > \omega_0$ allows us to integrate both sides with respect to t to obtain the stated result. □

Lemma 10.6.3 *Let* $T_t := e^{Zt}$ *be a one-parameter semigroup on the Hilbert space $\mathcal{H}$ and let* $a > \omega_0$. *Then there exists a constant c such that*

$$\int_{-\infty}^\infty \|R(a+iy, Z)f\|^2\, \mathrm{d}y = 2\pi \int_0^\infty e^{-2at}\|T_t f\|^2\, \mathrm{d}t$$

$$\le c\|f\|^2 \tag{10.19}$$

for all $f \in \mathcal{H}$.

Proof. If $\{e_n\}_{n=1}^\infty$ is a complete orthonormal set in $\mathcal{H}$ and $a > \omega_0$ then

$$\int_0^\infty e^{-at}\langle T_t f, e_n\rangle e^{-iyt}\, \mathrm{d}t = \langle R(a+iy, Z)f, e_n\rangle.$$

Note that the integrand lies in $L^1(0, \infty) \cap L^2(0, \infty)$. Using the unitarity of the Fourier transform we deduce that

$$\int_0^\infty |e^{-at}\langle T_t f, e_n\rangle|^2\, \mathrm{d}t = \frac{1}{2\pi}\int_{-\infty}^\infty |\langle R(a+iy, Z)f, e_n\rangle|^2\, \mathrm{d}y.$$

Summing this over n yields the statement of the lemma. An upper bound to the second integral in (10.19) is obtained by using the hypothesis that $a > \omega_0$. □

Theorem 10.6.4 *If* $T_t := e^{Zt}$ *is a one-parameter semigroup acting on a Hilbert space* $\mathcal{H}$ *then* $s_0 = \omega_0$.

Proof. By adding a suitable multiple of the identity operator to Z this reduces to proving that if $s_0 < 0$ then $\omega_0 < 0$, and hence to the proposition that if there is a constant c such that $\|R(z, Z)\| \leq c$ for all z such that $\mathrm{Re}(z) \geq 0$ then $\omega_0 < 0$.

Given $f, g \in \mathcal{H}$ and z satisfying $\mathrm{Re}(z) > 0$ we consider the function

$$\phi_z(t) := \begin{cases} te^{-zt}\langle T_t f, g \rangle & \text{if } t \geq 0, \\ 0 & \text{otherwise.} \end{cases}$$

If $\mathrm{Re}(z) > \omega_0$ then $\phi_z \in L^1(\mathbf{R})$ and its Fourier transform is

$$\psi_z(y) := \langle R(z + iy, Z)^2 f, g \rangle$$

by Lemma 10.6.2. We consider this latter function of y for all values of the parameter z in

$$S := \{x + iv : 0 < x < \omega_0^+ + 2 \text{ and } v \in \mathbf{R}\}$$

where $\omega_0^+ := \max\{\omega_0, 0\}$. Since

$$|\psi_z(y)| \leq \|R(z + iy, Z)f\| \, \|R(\bar{z} - iy, Z^*)g\|$$

we deduce that

$$\|\psi_z\|_1 \leq \left\{ \int_{-\infty}^{\infty} \|R(z + iy, Z)f\|^2 \, dy \right\}^{1/2} \left\{ \int_{-\infty}^{\infty} \|R(\bar{z} - iy, Z^*)g\|^2 \, dy \right\}^{1/2}$$

whenever the RHS is finite.

There exists a constant k such that $\|R(z, Z)\| \leq k$ for all $z \in S$ by hypothesis, and there exists a constant c such that

$$\int_{-\infty}^{\infty} \|R((\omega_0^+ + 1) + iy, Z)f\|^2 \, dy \leq c\|f\|^2$$

by Lemma 10.6.3. It follows by Lemma 10.6.1 that there exists a constant b such that

$$\int_{-\infty}^{\infty} \|R(z + iy, Z)f\|^2 \, dy \leq b\|f\|^2$$

for all $z \in S$. A similar argument implies that

$$\int_{-\infty}^{\infty} \|R(\bar{z} - iy, Z^*)g\|^2 \, dy \leq b\|g\|^2$$

for all $z \in S$. We deduce that

$$\|\psi_z\|_1 \leq b\|f\| \, \|g\|$$

for all f, $g \in \mathcal{H}$ and all $z \in S$. We now take the inverse Fourier transform of ψ_z to obtain

$$|t e^{-zt} \langle T_t f, g \rangle| \leq \frac{b}{2\pi} \|f\| \|g\|$$

for all $t \geq 0$, $z \in S$ and f, $g \in \mathcal{H}$. Letting $z \to 0$ we deduce that $\|T_t\| < 1$ for sufficiently large $t > 0$. This implies that $\omega_0 < 0$ by Theorem 10.1.6.　□

The proof of the above theorem is entirely constructive in the sense that all of the constants are controlled. This suggests, correctly, that one can obtain a variety of related bounds by modifying the assumptions.

Theorem 10.6.5 *(Eisner-Zwart)*[11] *Let $T_t := e^{Zt}$ be a one-parameter semigroup acting on the Hilbert space $\mathcal{H}$. Then one has the relations $(i) \Rightarrow (ii) \Rightarrow (iii)$ between the conditions*

(i) $\|T_t\| \leq M$ for some $M > 0$ and all $t \geq 0$;
(ii) $\|R(\lambda, Z)\| \leq M/\mathrm{Re}(z)$ for all z such that $\mathrm{Re}(z) > 0$;
(iii) $\|T_t\| \leq K(1+t)$ for some $K > 0$ and all $t \geq 0$.

However (ii) does not imply a bound of the form $\|T_t\| \leq K(1+t^\alpha)$ for any $\alpha < 1$.

---

[11] For the proof see [Zwart 2003, Eisner and Zwart 2005].

11

Perturbation theory

11.1 Perturbations of unbounded operators

Very few differential equations can be solved in closed form, and mathematicians have developed a variety of techniques for understanding the general properties of solutions of many of the others. One of the earliest is by means of perturbation theory. As well as providing a method of evaluating solutions by means of series expansions, it provides valuable theoretical insights. The latter are the focus of attention in this chapter.

If Z is an unbounded operator in a Banach space $\mathcal{B}$, one can define several types of perturbation of Z. The simplest case arises when $W := Z + A$ where $\text{Dom}(W) := \text{Dom}(Z)$ and A is a bounded operator on $\mathcal{B}$. Whatever the technical assumptions the goal is to determine spectral and other properties of W, assuming that Z is an operator which can be analyzed in great detail.

If A is an unbounded operator with $\text{Dom}(A) \supseteq \text{Dom}(Z)$ we say that A is relatively bounded with respect to Z if there exist constants c, d such that

$$\|Af\| \leq c\|Zf\| + d\|f\| \tag{11.1}$$

for all $f \in \text{Dom}(Z)$. The infimum of all possible constants c is called the relative bound of A with respect to Z. If $a \notin \text{Spec}(Z)$ then Problem 6.1.4 implies that A is relatively bounded with respect to Z if and only if $AR(a, Z)$ is a bounded operator on $\mathcal{B}$.

Many of the calculations in the chapter are related in some way to the following result.

Problem 11.1.1 Let Z be a closed operator acting in $\mathcal{B}$ and let A be relatively bounded with respect to Z. If z does not lie in $\text{Spec}(Z)$ or $\text{Spec}(Z + A)$ then

$$R(z, Z + A) - R(z, Z) = R(z, Z + A)AR(z, Z) \qquad \square$$

Lemma 11.1.2 *If Z is closed and A has relative bound less than one with respect to Z then $W := Z + A$ is also closed.*

Proof. Suppose that $0 < c < 1$, $0 \leq d < \infty$ and

$$\|Af\| \leq c\|Zf\| + d\|f\|$$

for all $f \in \mathrm{Dom}(Z)$. Consider the two norms

$$\|f\|_1 := \|Zf\| + \|f\|,$$
$$\|f\|_2 := \|Wf\| + \|f\|,$$

on $\mathrm{Dom}(Z)$. It follows from Problem 6.1.1 that $\mathrm{Dom}(Z)$ is complete with respect to the first norm and that it is sufficient to prove that it is complete with respect to the second norm. We shall prove that the two norms are equivalent.

If $f \in \mathrm{Dom}(Z)$ then

$$\|Wf\| + \|f\| \leq \|Zf\| + \|Af\| + \|f\|$$
$$\leq (1+c)\|Zf\| + (1+d)\|f\|$$
$$\leq (1+c+d)(\|Zf\| + \|f\|).$$

Conversely

$$(1+d)(\|Wf\| + \|f\|) \geq \|Wf\| + (1+d)\|f\|$$
$$\geq \|Zf\| - \|Af\| + (1+d)\|f\|$$
$$\geq (1-c)\|Zf\| + \|f\|$$
$$\geq (1-c)(\|Zf\| + \|f\|). \qquad \square$$

By considering the case $A := -Z$, one sees that the conclusion of Lemma 11.1.2 need not hold if $c = 1$ in (11.1).

Theorem 11.1.3 *Let Z be a closed operator acting in the Banach space $\mathcal{B}$ and suppose that $a \notin \mathrm{Spec}(Z)$. If A is relatively bounded with respect to Z then $a \notin \mathrm{Spec}(Z + cA)$ provided $c \in \mathbf{C}$ satisfies $|c| \, \|AR(a, Z)\| < 1$. Moreover*

$$\lim_{c \to 0} \|R(a, Z + cA) - R(a, Z)\| = 0.$$

Proof. We need to prove that if c satisfies the stated condition then $a \notin \text{Spec}(Z + cA)$ and

$$R(a, Z + cA) = R(a, Z)(1 - cAR(a, Z))^{-1}. \tag{11.2}$$

If we denote the right-hand side of (11.2) by B then it is immediate that B is bounded and one-one with range equal to $\text{Dom}(Z)$. The first statement of the theorem is completed by observing that

$$(aI - (Z + cA))B = ((aI - Z) - cA)R(a, Z)(1 - cAR(a, Z))^{-1}$$

$$= (I - cAR(a, Z))(1 - cAR(a, Z))^{-1}$$

$$= I.$$

The norm convergence of the resolvent as $c \to 0$ follows directly from (11.2). $\square$

Theorem 11.1.3 is a semicontinuity result for the spectrum under small perturbations. The following example shows that full continuity cannot be proved under such conditions.

Problem 11.1.4 Consider the bounded operator A_c defined on $l^2(\mathbf{Z})$ by

$$(A_c f)(n) := \begin{cases} f(n+1) & \text{if } n \neq 0, \\ cf(n+1) & \text{if } n = 0. \end{cases}$$

Prove that $\text{Spec}(A_c)$ equals $\{z \in \mathbf{C} : |z| = 1\}$ unless $c = 0$, in which case it equals $\{z \in \mathbf{C} : |z| \leq 1\}$. $\square$

Theorem 11.1.5 *(Riesz) Let γ be a closed contour enclosing the compact component S of the spectrum of the closed operator A acting in $\mathcal{B}$, and suppose that $T = \text{Spec}(A) \backslash S$ is outside γ. Then*

$$P := \frac{1}{2\pi i} \int_\gamma R(z, A) \, dz$$

is a bounded projection commuting with A. The restriction of A to $P\mathcal{B}$ has spectrum S and the restriction of A to $(I - P)\mathcal{B}$ has spectrum T.

Proof. The proof of Theorem 1.5.4 is not directly applicable because it uses the boundedness of A, but we can use that theorem together with Lemma 8.1.9 to prove this one.

If a is just outside γ then $a \notin \text{Spec}(A)$ and $B := R(a, A)$ is a bounded operator. Suppose that γ is parametrized by $s \in [0, 1]$ with $\gamma(0) = \gamma(1)$. If we put $\sigma(s) := (a - \gamma(s))^{-1}$ then the map $z \to w := (a - z)^{-1}$ maps the part of the spectrum of A inside γ one-one onto the part of the spectrum of B inside σ by Lemma 8.1.9. The following identities establish that the associated spectral projections are equal.

$$
\begin{aligned}
\int_\sigma R(w, B)\, dw &= \int_0^1 R(\sigma(s), B)\sigma'(s)\, ds \\
&= \int_0^1 \left\{ (a - \gamma(s))^{-1} - (aI - A)^{-1} \right\}^{-1} (a - \gamma(s))^{-2} \gamma'(s)\, ds \\
&= \int_\gamma \left\{ R(z, A) - (z - a)^{-1} I \right\}\, dz \\
&= \int_\gamma R(z, A)\, dz. \qquad \qquad \square
\end{aligned}
$$

The following theorem generalizes Theorem 1.5.6 to unbounded operators and also to spectral projections with rank greater than 1.[1] For simplicity we restrict attention to bounded perturbations, but the theorem can be modified so as to apply to relatively bounded perturbations.

Theorem 11.1.6 *(Rellich) Let γ be a simple closed curve enclosing a non-empty compact subset S of the spectrum of a closed operator A acting in the Banach space $\mathcal{B}$. Suppose that $\|R(z, A)\| \leq c$ for all $z \in \gamma$ and that B is a bounded operator on $\mathcal{B}$. Then $\mathrm{Spec}(A + tB) \cap U \neq \emptyset$ for all $t \in \mathbf{C}$ such that $|t|\, \|B\|\, c < 1$, where U is the region inside γ. The spectral projections of $A + tB$ associated with the region U depend analytically on t and the spectral subspaces all have the same dimension.*

Proof. If t satisfies the stated bound and $A_t := A + tB$ then $\mathrm{Spec}(A_t) \cap \gamma = \emptyset$ and

$$
R(z, A_t) = R(z, A)(1 - tBR(z, A))^{-1}
$$

for all $z \in \gamma$ by (11.2). Moreover $R(z, A_t)$ is a jointly analytic function of (z, t) and

$$
\|R(z, A_t)\| \leq \frac{c}{1 - |t|\, \|B\|\, c}
$$

for all relevant (z, t).

The spectral projection of A_t associated with the region U is given by

$$
P_t := \frac{1}{2\pi i} \int_\gamma R(z, A_t)\, dz
$$

[1] As before it is a special case of results of Rellich which are treated systematically in [Kato 1966A].

by Theorem 11.1.5. These projections depend analytically on t. Lemma 1.5.5 now implies that the rank of P_t does not depend on t. □

Example 11.1.7 Complex scaling is an important technique in quantum mechanics, particularly when determining resonances numerically. It is a large subject in its own right, so we only attempt to give a flavour of the method. By starting with the eigenfunction rather than the potential and confining ourselves to the one-dimensional case, we make the calculations entirely elementary.

Let f be an analytic function which does not vanish anywhere in the sector $S_\alpha := \{z : |\mathrm{Arg}(z)| < \alpha\}$. Suppose also that

$$-f''(x) + V(x)f(x) = \lambda f(x)$$

for all $x > 0$. Then V is the restriction to the positive real line of the analytic function

$$V(z) := \lambda + \frac{f''(z)}{f(z)}$$

defined on the sector S_α. If $|\theta| < \alpha$ and we define $f_\theta(x) := f(e^{i\theta}x)$ for all $x > 0$ then

$$-f_\theta''(x) + V_\theta(x)f_\theta(x) = e^{2i\theta}\lambda f(x)$$

for all $x > 0$, where

$$V_\theta(x) := e^{2i\theta}V(e^{i\theta}x).$$

Starting from $f(z) := e^{-z^2/2}$ one discovers that if $|\theta| < \pi/4$ then $e^{2i\theta}$ is an eigenvalue for the NSA harmonic oscillator

$$(A_\theta g)(x) := -g''(x) + e^{4i\theta}x^2 g(x)$$

acting in $L^2(0, \infty)$ subject to Neumann boundary conditions at 0, or in $L^2(\mathbf{R})$. Starting from $f(z) := ze^{-z}$ one finds that $-e^{2i\theta}$ is an eigenvalue for the hydrogen atom with a complex coupling constant

$$(B_\theta g)(x) := -g''(x) - 2e^{i\theta}x^{-1}g(x)$$

acting in $L^2(0, \infty)$ subject to Dirichlet boundary conditions at 0.

Conversely, if one is given a potential V that is analytic in the sector S_α then one may consider the analytic family of operators H_z acting in $L^2(0, \infty)$ according to the formula

$$(H_z f)(x) := -f''(x) + V(zx)f(x).$$

Under suitable technical assumptions, which include the specification of boundary conditions, an obvious modification of Theorem 11.1.6 allows one to conclude that the eigenvalues and eigenfunctions of H_z depend analytically on z in a manner that can be analyzed in detail. □

11.2 Relatively compact perturbations

In this section we consider relatively compact perturbations of an operator Z and their effect on the essential spectrum. We assume throughout the section that Z is a closed, densely defined operator acting in a Banach space $\mathcal{B}$ and that $\text{Spec}(Z) \neq \mathbf{C}$. We define $\mathcal{D}$ to be the vector space $\text{Dom}(Z)$ provided with the Banach space norm

$$\||f\|| := \|Zf\| + \|f\|. \tag{11.3}$$

We say that Z is a Fredholm operator if it is closed and Fredholm considered as a bounded operator from $\mathcal{D}$ to $\mathcal{B}$. We define the essential spectrum $\text{EssSpec}(Z)$ of Z to be the set of all $z \in \mathbf{C}$ such that $zI - Z$ is not Fredholm in this sense. Evidently the essential spectrum is a closed subset of the spectrum of Z.

Lemma 11.2.1 *Let Z be a closed operator on $\mathcal{B}$ and let $z \in \mathbf{C}$. If the sequence of vectors $f_n \in \mathcal{D}$ converges weakly in $\mathcal{D}$ to 0 as $n \to \infty$ and satisfies*

$$\lim_{n \to \infty} \||f_n\|| = 1, \qquad \lim_{n \to \infty} \|Zf_n - zf_n\| = 0,$$

then $z \in \text{EssSpec}(Z)$.

Proof. This is a minor modification of the proof of Lemma 4.3.15, in which the projection P acts in the Banach space $\mathcal{D}$. □

The essential spectrum of an unbounded operator Z can be determined from any of its resolvent operators, and this has the advantage of avoiding explicit reference to its domain.

Theorem 11.2.2 *Let Z be a closed unbounded operator acting in $\mathcal{B}$ and let* $\lambda \notin \mathrm{Spec}(Z)$. *Then* $z \in \mathrm{EssSpec}(Z)$ *if and only if* $z \neq \lambda$ *and*

$$(\lambda - z)^{-1} \in \mathrm{EssSpec}\{(\lambda I - Z)^{-1}\}.$$

Proof. We start from the identity

$$(\lambda - z)^{-1}I - (\lambda I - Z)^{-1} = (\lambda - z)^{-1} \{(\lambda I - Z) - (\lambda - z)I\} (\lambda I - Z)^{-1}$$
$$= (\lambda - z)^{-1}(zI - Z)(\lambda I - Z)^{-1}.$$

Since $(\lambda I - Z)^{-1}$ is a bounded invertible map from $\mathcal{B}$ onto $\mathrm{Dom}(Z)$, the latter being given its natural norm, we see that $((\lambda - z)^{-1}I - (\lambda I - Z)^{-1})$ is Fredholm if and only if $zI - Z$ is Fredholm. This is equivalent to the statement of the theorem. $\square$

Corollary 11.2.3 *If Z_1, Z_2 are two closed unbounded operators acting in $\mathcal{B}$ and there exists $\lambda \notin \mathrm{Spec}(Z_1) \cup \mathrm{Spec}(Z_2)$ for which*

$$(\lambda I - Z_1)^{-1} - (\lambda I - Z_2)^{-1}$$

is compact, then Z_1 and Z_2 have the same essential spectrum.

If Z is a closed operator on $\mathcal{B}$ and A is a perturbation satisfying

$$\mathrm{Dom}(Z) \subseteq \mathrm{Dom}(A)$$

then we say that A is relatively compact with respect to Z if A is compact considered as an operator from $\mathcal{D}$ to $\mathcal{B}$. If $a \notin \mathrm{Spec}(Z)$ then Problem 6.1.4 implies that A is relatively compact with respect to Z if and only if $AR(a, Z)$ is a compact operator. [2]

Lemma 11.2.4 *If A is relatively compact with respect to Z then it is relatively bounded. If $\mathcal{B}$ is reflexive and satisfies the approximation property of Section 4.2 then the relative bound is 0.*

Proof. The first statement is elementary. If $\mathcal{B}$ has the approximation property and $a \notin \mathrm{Spec}(Z)$ then $AR(a, Z)$ may be approximated arbitrarily closely by

---

[2] Our theorems on relatively compact and rank 1 perturbations have several variants and may be extended in a number of directions. See [Desch and Schappacher 1988, Arendt and Batty 2005A, Arendt and Batty 2005B].

finite rank operators by Theorem 4.2.4. That is, given $\varepsilon > 0$ there exist $f_1, \ldots, f_n \in \mathcal{B}$ and $\phi_1, \ldots, \phi_n \in \mathcal{B}^*$ such that

$$\left\| AR(a, Z)f - \sum_{r=1}^{n} f_r \langle f, \phi_r \rangle \right\| < \varepsilon \|f\| \tag{11.4}$$

for all $f \in \mathcal{B}$. We next observe that $R(a, Z)^*$ has dense range in $\mathcal{B}^*$: if this were not true the Hahn-Banach theorem together with reflexivity would imply that there exists a non-zero $f \in \mathcal{B}$ such that $R(a, Z)f = 0$. It follows that by slightly changing ϕ_r one can achieve (11.4) as well as $\phi_r = R(a, Z)^* \psi_r$, where $\psi_r \in \mathcal{B}^*$. Putting $g := R(a, Z)f$ we see that (11.4) is equivalent to

$$\left\| Ag - \sum_{r=1}^{n} f_r \langle g, \psi_r \rangle \right\| < \varepsilon \|(aI - Z)g\|$$

for all $g \in \mathrm{Dom}(Z)$. Therefore

$$\|Ag\| < \varepsilon \|Zg\| + \left(\varepsilon |a| + \sum_{r=1}^{n} \|f_r\| \, \|\psi_r\| \right) \|g\|$$

for all such g. $\quad\square$

Problem 11.2.5 Let Z be the closed operator on $\mathcal{B} := L^1(\mathbf{R})$ defined by $(Zf)(x) := xf(x)$ with the maximal domain. Defining $\mathcal{D}$ as usual, find $\phi \in \mathcal{D}^*$ and $g \in \mathcal{B}$ such that the rank one operator $Af := \phi(f)g$ does not have relative bound 0 with respect to Z. $\quad\square$

The fact that the following theorem does not require $\mathcal{B}$ to be reflexive is crucial for its application in Theorem 14.3.5.

Theorem 11.2.6 *Let Z be a closed operator acting in $\mathcal{B}$ with $\mathrm{Spec}(Z) \neq \mathbf{C}$. If A is a relatively compact perturbation of Z then Z and $Z + A$ have the same essential spectrum. Moreover $Z + A$ is closed on the same domain as Z.*

Proof. The hypotheses of the theorem imply that $Z : \mathcal{D} \to \mathcal{B}$ is bounded and $A : \mathcal{D} \to \mathcal{B}$ is compact. Therefore $zI - Z - A$ is Fredholm regarded as a bounded operator from $\mathcal{D}$ to $\mathcal{B}$ if and only if $zI - Z$ is Fredholm.

We prove that $Z + A$ is closed on $\mathrm{Dom}(Z)$ without using Lemma 11.1.2. By adding a suitable constant to Z we reduce to the case in which $0 \notin \mathrm{Spec}(Z)$, so that $Z : \mathcal{D} \to \mathcal{B}$ is bounded and invertible. Since $Y := Z + A$ is Fredholm, there exist closed subspaces $\mathcal{D}_1$ and $\mathcal{D}_2$ of $\mathcal{D}$ such that $\mathcal{D} = \mathcal{D}_1 \oplus \mathcal{D}_2$, $Y|_{\mathcal{D}_1} = 0$, $\mathcal{D}_1$ is finite-dimensional, and

$$\|Yf\| \geq c \|\|f\|\| \tag{11.5}$$

for some $c > 0$ and all $f \in \mathcal{D}_2$. See Theorem 4.3.5.

Now suppose that $f_n \in \mathcal{D}$ and that $\|f_n - f\| \to 0$ and $\|Yf_n - k\| \to 0$ as $n \to \infty$ for some $f, k \in \mathcal{B}$. We may write f_n in the form $f_n := g_n + h_n$ where $g_n \in \mathcal{D}_1$, $h_n \in \mathcal{D}_2$ and $Yg_n = 0$ for all n. Since $\|Yh_n - k\| \to 0$, (11.5) implies that h_n is a Cauchy sequence in $\mathcal{D}_2$. Therefore there exists $h \in \mathcal{D}_2$ such that $\|\|h_n - h\|\| \to 0$ as $n \to \infty$. Since $g_n = f_n - h_n$ we deduce that g_n converges in $\mathcal{D}_1$, the two norms being equivalent on this space because it is finite-dimensional. Therefore f_n converges in $\mathcal{D}$. The limit must coincide with f, so $f \in \mathcal{D}$. Since $Y : \mathcal{D} \to \mathcal{B}$ is bounded $Yf = k$ and Y is closed. $\square$

Theorem 11.2.7 *Let* $H_0 f := -f''$ *in* $L^2(0, \pi)$ *subject to Dirichlet boundary conditions at* $0, \pi$. *If* V *is a possibly complex-valued potential and* $V \in L^2(0, \pi)$ *then* $H := H_0 + V$ *has empty essential spectrum and compact resolvent operators.*

Proof. The normalized eigenfunctions of H_0 are $\phi_n(x) := (2/\pi)^{1/2} \sin(nx)$, where $n \in \mathbf{N}$, and the corresponding eigenvalues are n^2. The resolvent operator H_0^{-1} has the integral kernel

$$G(x, y) := \sum_{n=1}^{\infty} n^{-2} \phi_n(x)\phi_n(y). \tag{11.6}$$

Because this series converges uniformly, G is a continuous function on $[0, \pi]^2$ which vanishes on the boundary of the square. The operator VH_0^{-1} has the Hilbert-Schmidt kernel $V(x)G(x, y)$. Theorem 11.2.6 implies that H has empty essential spectrum. A further calculation of the same type shows that the Hilbert-Schmidt norm of $V(aI - H_0)^{-1}$ converges to 0 as $a \to -\infty$. Hence $a \notin \mathrm{Spec}(H)$ for all large enough negative a by Theorem 11.1.3. The formula (11.2) becomes

$$R(a, H) = R(a, H_0)(I - AR(a, H))^{-1}$$

and implies that $R(a, H)$ is compact. $\square$

Example 11.2.8 *Let* H act in $L^2(a, b)$ subject to Dirichlet boundary conditions according to the formula

$$(Hf)(x) := (H_0 f + Vf)(x) := -f''(x) + V(x)f(x)$$

where V is a complex-valued, continuous function on $[a, b]$. (The results of this example have been extended to much more general Sturm-Liouville operators and other boundary conditions.) We infer as in Theorem 11.2.7 that H has empty essential spectrum and compact resolvent.

If $a := 0$ and $b := \pi$ then $\mathrm{Spec}(H_0) = \{n^2 : n \in \mathbf{N}\}$ and $\|V(zI - H_0)^{-1}\| < 1$ provided $\mathrm{dist}(z, \mathrm{Spec}(H_0)) > \|V\|_\infty$. Theorem 11.1.3 implies that

$$\mathrm{Spec}(H) \subseteq \bigcup_{n=1}^{\infty} B(n^2, \|V\|_\infty).$$

For large enough n these balls are disjoint and there is exactly one eigenvalue in each ball by Theorem 11.1.6. The precise asymptotics of the eigenvalues is well understood in the self-adjoint case, but much less so for complex-valued V.[3]

Assuming that 0 is not an eigenvalue of H there are linearly independent solutions ϕ and ψ of $Hf = 0$ satisfying $\phi(a) = 0$, $\phi'(a) = 1$, $\psi(b) = 0$, $\psi'(b) = -1$. By differentiating the RHS one sees that the Wronskian $w := \phi'(x)\psi(x) - \phi(x)\psi'(x)$ is independent of x. Putting $x = a$ and using the assumption that 0 is not an eigenvalue of H we deduce that $w \neq 0$. Direct calculations show that $Hf = g$ if and only if

$$f(x) = \int_a^b G(x, y)g(y)\,dy$$

where

$$G(x, y) := \begin{cases} w^{-1}\phi(x)\psi(y) & \text{if } x \leq y, \\ w^{-1}\psi(x)\phi(y) & \text{if } y \leq x. \end{cases}$$

Hence G is the integral kernel for the resolvent operator H^{-1}. The Green function for the particular case $V = 0$, $a = 0$ and $b = \pi$ is written down in Example 5.6.10. □

Lemma 11.2.9 *Let Z be a closed operator on $\mathcal{B}$ and let $A : \mathcal{D} \to \mathcal{B}$ be defined by $Af := \phi(f)g$ where $\phi \in \mathcal{D}^*$ and $g \in \mathcal{B}$. Then $z \notin \mathrm{Spec}(Z)$ is an eigenvalue of $Z + A$ if and only if*

$$\phi(R(z, Z)g) = 1. \tag{11.7}$$

The LHS is an analytic function of z, so the solutions of (11.7) form a discrete subset of $\mathbf{C} \backslash \mathrm{Spec}(Z)$.

Proof. The eigenvalue equation $Zf + \phi(f)g = zf$ may be rewritten in the form $(zI - Z)f = \phi(f)g$. This is in turn equivalent to $f = \phi(f)R(z, Z)g$. The assumption that $z \notin \mathrm{Spec}(Z)$ implies that $\phi(f) \neq 0$. Normalizing to the case $\phi(f) = 1$ leads to the stated conclusion. □

---

[3] See, however, [Tkachenko 2002]. The paper of [Albeverio et al. 2006] provides the solution of the inverse spectral problem for a class of complex-valued distributional potentials.

Problem 11.2.10 Let H be defined in $L^2(\mathbf{R})$ by

$$(Hf)(x) := xf(x) + \phi(x) \int_{\mathbf{R}} f(s)\phi(s)\,\mathrm{d}s$$

where $\phi(x) := c(x+i)^{-n}$ for some $c \in \mathbf{C}$ and $n \in \mathbf{N}$. Find all the eigenvalues of H and describe how they move as c varies. □

We conclude the section with an application of the above theorem to Schrödinger operators with possibly complex potentials. Example 11.4.10 provides further information about the following operator.

Theorem 11.2.11 *Let $Z := \Delta$ on $L^2(\mathbf{R}^N)$. Let A be the operator of multiplication by a function $a \in L^p(\mathbf{R}^N)$ where $p = 2$ if $N \leq 3$, and $p > N/2$ if $N \geq 4$. Also let B be the operator of multiplication by a bounded measurable function b which vanishes as $|x| \to \infty$, i.e. $b \in L_0^\infty(\mathbf{R}^N)$. Then $(A+B)(\lambda I - Z)^{-1}$ is compact for all $\lambda > 0$. Therefore the essential spectrum of $Z + A + B$, regarded as acting in $L^2(\mathbf{R}^N)$, is $(-\infty, 0]$.*

Proof. We rely upon the results in Section 5.7. In the notation of that section we have

$$(A+B)(\lambda I - Z)^{-1} = (a(Q) + b(Q))g(P) \qquad (11.8)$$

where $g(\xi) := (|\xi|^2 + \lambda)^{-1}$. We note that $g \in L^q \cap L_0^\infty$ provided $q > N/2$. Problem 5.7.4 implies that (11.8) is a compact operator on $L^2(\mathbf{R}^N)$. Theorem 11.2.6 now implies that the essential spectrum of $Z + A + B$ is the same as that of Z. The latter equals $(-\infty, 0]$ by Theorem 8.1.1. □

This theorem is far from the sharpest that can be proved,[4] but the most important case excluded concerns Schrödinger operators with locally L^1 potentials in one dimension, for which we refer to Section 14.3.

11.3 Constant coefficient differential operators on the half-line

The spectral properties of ordinary differential operators depend heavily on the boundary conditions. However, we shall see that their essential spectrum does not. We establish this first at an abstract level.

---

[4] See [Simon 1982] for a comprehensive survey.

Lemma 11.3.1 *Let $\mathcal{B}_0$ be a closed subspace of finite codimension in the Banach space $\mathcal{B}_1$. Let A_1 be a bounded operator from $\mathcal{B}_1$ to $\mathcal{B}$ and let A_0 be its restriction to $\mathcal{B}_0$. Then A_1 is Fredholm if and only if A_0 is Fredholm.*

Proof. We start by writing $\mathcal{B}_1 := \mathcal{B}_0 \oplus \mathcal{L}$ where $\mathcal{L}$ is finite-dimensional. We define $D : \mathcal{B}_1 \to \mathcal{B}$ by $D(f \oplus g) := (A_0 f) \oplus 0$. Since $(A_1 - D)$ is of finite rank either both A_1 and D are Fredholm or neither is. Since $\mathrm{Ran}(D) = \mathrm{Ran}(A_0)$ and $\mathrm{Ker}(D) = \mathrm{Ker}(A_0) \oplus \mathcal{L}$ either both A_0 and D are Fredholm or neither is. This completes the proof. $\square$

Lemma 11.3.2 *Let A_1 and A_2 be two closed operators acting in the Banach space $\mathcal{B}$ and suppose that they coincide on $\mathcal{L} = \mathrm{Dom}(A_1) \cap \mathrm{Dom}(A_2)$. If $\mathcal{L}$ has finite codimension in both $\mathrm{Dom}(A_1)$ and $\mathrm{Dom}(A_2)$ then*

$$\mathrm{EssSpec}(A_1) = \mathrm{EssSpec}(A_2).$$

Proof. Given $z \in \mathbf{C}$ and $j = 1, 2$ let B denote the restriction of either of $(zI - A_j)$ to $\mathcal{L}$. It follows from Lemma 11.3.1 that $(zI - A_j)$ is Fredholm if and only if B is Fredholm. This completes the proof. $\square$

The spectral properties of the constant coefficient differential operator

$$(Af)(x) := \sum_{r=0}^{n} a_r f^{(r)}(x), \tag{11.9}$$

with $a_n \neq 0$, depend heavily on the interval chosen and on the boundary conditions. Our next theorem shows that the essential spectrum is much more stable in this respect. The first half of the theorem can easily be extended to a wide class of variable coefficient differential operators.

Let $\mathcal{D}$ be the space of n times continuously differentiable functions on $[0, \infty)$ all of whose derivatives lie in L^2. Given any linear subspace L of $\mathbf{C}^n$ let $\mathcal{D}_L$ denote the subspace consisting of all $f \in \mathcal{D}$ such that $(f(0), f'(0), ..., f^{(n-1)}(0)) \in L$.

The operator A is closable on $\mathcal{D}_L$ by an argument very similar to that of Example 6.1.9. We denote the closure by A_L, and refer to it as the operator A acting in $L^2(0, \infty)$ subject to the imposition of the boundary conditions L.

Theorem 11.3.3 *The essential spectrum of the operator A_L is independent of the choice of L. Indeed*

$$\mathrm{EssSpec}(A_L) = \{\sigma(\xi) : \xi \in \mathbf{R}\},$$

where the symbol σ of the operator is given by

$$\sigma(\xi) := \sum_{r=0}^{n} a_r i^r \xi^r.$$

Proof. Each of the stated operators is an extension of the operator A_M corresponding to the choice $M = \{0\}$. Moreover $\mathcal{D}_M$ has finite codimension in $\mathcal{D}_L$ for any choice of L, so the first statement follows from Lemma 11.3.2. Let B_M denote the 'same' differential operator but acting in $L^2(-\infty, 0)$ and subject to the boundary conditions $f(0) = f'(0) = \cdots = f^{(n-1)}(0) = 0$. Finally let T denote the 'same' operator acting in $L^2(\mathbf{R})$. Since $A_M \oplus B_M$ is the restriction of T to a subdomain of finite codimension, it follows by Lemma 11.3.2 that

$$\operatorname{EssSpec}(A_M) \subseteq \operatorname{EssSpec}(A_M) \cup \operatorname{EssSpec}(B_M)$$

$$= \operatorname{EssSpec}(A_M \oplus B_M)$$

$$= \operatorname{EssSpec}(T)$$

$$= \{\sigma(\xi) : \xi \in \mathbf{R}\}.$$

The last equality follows by using the Fourier transform to prove that T is unitarily equivalent to the operator of multiplication by $\sigma(\cdot)$ acting on its maximal domain.

Conversely suppose that $z \in \mathbf{C}$ and there exists $\xi \in \mathbf{R}$ such that $\sigma(\xi) = z$. Let $\phi \in C_c^\infty(0, \infty) \subseteq \operatorname{Dom}(A_L)$ satisfy $\phi(x) = 0$ if $x \leq 1$ or $x \geq 4$, and $\phi(x) = 1$ if $2 \leq x \leq 3$. Then define $f_n \in C_c^\infty(0, \infty)$ for positive n by

$$f_n(x) := e^{i\xi x} \phi(x/n).$$

A direct computation establishes that

$$\lim_{n \to \infty} \|A_L f_n - z f_n\| / \|\|f_n\|\| = 0.$$

(The denominator diverges more rapidly than the numerator as $n \to \infty$.) An application of Lemma 11.2.1 proves that $z \in \operatorname{EssSpec}(A_L)$. $\square$

In order to determine the full spectrum of such operators acting in $L^2(0, \infty)$ we must specify the boundary conditions. We suppose from now on that A is of even order, i.e.

$$(Af)(x) := \sum_{r=0}^{2n} a_r f^{(r)}(x),$$

where $a_{2n} = 0$. We impose the 'Dirichlet' boundary conditions

$$f(0) = f'(0) = \cdots = f^{(n-1)}(0) = 0,$$

but the theorem below may easily be extended to other boundary conditions. The symbol of A is given by

$$\sigma(\xi) := \sum_{r=0}^{2n} a_r i^r \xi^r$$

and satisfies $|\text{Re}(\sigma(\xi))| \to +\infty$ as $\xi \to \pm\infty$.

Theorem 11.3.4[5] *The spectrum of the operator A is the union of $S = \{\sigma(\xi) : \xi \in \mathbf{R}\}$, the set of eigenvalues of A and the set of eigenvalues of A^*. If $z \notin S$ then z cannot be an eigenvalue of both A and A^*. It is an eigenvalue of one of these operators unless the winding number of γ around z equals n.*

Proof. The first statement of the theorem follows directly from Theorem 11.3.3. We next identify the eigenvalues of A.

If we disregard the boundary conditions then $Af = zf$ has exactly $2n$ linearly independent solutions. We present an explicit basis for the intersection of the solution space with $L^2(0, \infty)$. If $z \notin \text{EssSpec}(A)$ then none of the roots of $\sigma(\xi) = z$ lies on the real axis. Suppose first that the roots $\xi_1, \ldots, \xi_{2n}$ are distinct. After re-ordering them we may assume that $\xi_1, \ldots, \xi_k$ have negative imaginary parts while the remainder have positive imaginary parts – we include 0 as a possible value of k. If we disregard the boundary conditions the space of L^2 solutions of $Af = zf$ is k-dimensional and consists precisely of the functions of the form

$$f(x) := \sum_{r=1}^{k} \alpha_r e^{\xi_r x}.$$

These only lie in the domain of A if they satisfy the boundary conditions at $x = 0$. These boundary conditions involve the derivatives

$$f^{(m)}(0) = \sum_{r=1}^{k} \alpha_r \xi_r^m$$

for $m = 0, 1, 2, \ldots$. The non-vanishing of the Vandermonde determinant implies that if $k \leq n$ the only such solution is $f = 0$ while if $k > n$ a non-zero solution exists.

---

[5] See [Edmunds and Evans 1987, Theorem IX.7.3]. The reformulation in terms of winding numbers may be found in [Reddy 1993, Davies 2000B].

If the equation $\sigma(\xi) = z$ has some repeated roots then the set of solutions of $Af = zf$ are generated by functions of the form $x^s e^{\xi_r x}$ where the possible values of $s \geq 0$ depend upon the multiplicity of ξ_r. The same argument applies, since the relevant generalized Vandermonde determinants are still non-zero.[6]

We see that z is an eigenvalue of A if and only if the number of roots of $\sigma(\xi) = z$ that have negative real parts is greater than n. By a similar argument one sees that z is an eigenvalue of A^* if and only if the number of roots of $\sigma(\xi) = z$ that have positive real parts is greater than n. These two facts imply the second statement of the theorem.

We have now completed the proof, except for the fact that the result is expressed in terms of the number of solutions of $\sigma(\xi) = z$ that have positive or negative real parts. The final statement of the theorem follows by applying Rouche's theorem to the integral

$$\frac{1}{2\pi i} \int_{-\infty}^{\infty} \frac{\sigma'(\xi)}{\sigma(\xi) - z} \, d\xi. \qquad \square$$

Problem 11.3.5 Let A be the convection-diffusion operator defined on $L^2(0, \infty)$ by

$$(Af)(x) := f''(x) + f'(x)$$

subject to the Dirichlet boundary conditions $f(0) = 0$. Find the spectrum of A. Compare your result with Example 9.3.20 and Theorem 9.3.21. $\square$

11.4 Perturbations: semigroup based methods

One may often show that if Z is the generator of a one-parameter semigroup and A is a perturbation which is 'small' in some sense then $Z + A$ is also the generator of a one-parameter semigroup . We start by considering the easiest case, in which A is a bounded operator.

Theorem 11.4.1 *Let Z be the generator of a one-parameter semigroup S_t on the Banach space $\mathcal{B}$ and suppose that*

$$\|S_t\| \leq M e^{at}$$

---

[6] See [Muir 1923].

for all $t \geq 0$. *If* A *is a bounded operator on* $\mathcal{B}$ *then* $(Z + A)$ *is the generator of a one-parameter semigroup* T_t *on* $\mathcal{B}$ *such that*

$$\|T_t\| \leq M e^{(a+M\|A\|)t}$$

for all $t \geq 0$.

Proof. We define the operators T_t by

$$T_t f := S_t f + \int_{s=0}^{t} S_{t-s} A S_s f \, ds$$

$$+ \int_{s=0}^{t} \int_{u=0}^{s} S_{t-s} A S_{s-u} A S_u f \, du \, ds$$

$$+ \int_{s=0}^{t} \int_{u=0}^{s} \int_{v=0}^{u} S_{t-s} A S_{s-u} A S_{u-v} A S_v f \, dv \, du \, ds + \cdots . \quad (11.10)$$

The nth term is an n-fold integral whose integrand is a norm continuous function of the variables. It is easy to verify that the series is norm convergent and that

$$\|T_t f\| \leq M e^{at} \|f\| \sum_{n=0}^{\infty} (tM\|A\|)^n/n!$$

$$= M e^{(a+M\|A\|)t}$$

for all $f \in \mathcal{B}$.

The proof that $T_s T_t = T_{s+t}$ for all $s,\, t \geq 0$ is a straightforward but lengthy exercise in multiplying together series term by term and rearranging integrals, which we leave to the reader. If $f \in \mathcal{B}$ then

$$\lim_{t \to 0} \|T_t f - f\| \leq \lim_{t \to 0} \left\{ \|S_t f - f\| + \sum_{n=1}^{\infty} M e^{at} \|f\| (tM\|A\|)^n/n! \right\}$$

$$= 0$$

so T_t is a one-parameter semigroup .

If $f \in \mathcal{B}$ then

$$\lim_{t \to 0} \|t^{-1}(T_t f - f) - t^{-1}(S_t f - f) - Af\|$$

$$\leq \lim_{t \to 0} \left\| t^{-1} \int_0^t S_{t-s} A S_s f \, ds - Af \right\|$$

$$+ \lim_{t \to 0} t^{-1} M e^{at} \|f\| \sum_{n=2}^{\infty} (tM\|A\|)^n/n!$$

$$= 0.$$

It follows that f lies in the domain of the generator Y of T_t if and only if it lies in the domain of Z, and that

$$Yf := Zf + Af$$

for all such f. $\square$

As well as being illuminating in its own right (11.10) easily leads to the identities

$$T_t f = S_t f + \int_{s=0}^{t} T_{t-s} A S_s f \, ds \tag{11.11}$$

$$= S_t f + \int_{s=0}^{t} S_{t-s} A T_s f \, ds. \tag{11.12}$$

Corollary 11.4.2 *Let T_t^λ be the one-parameter semigroup on $\mathcal{B}$ with generator $Z + \lambda A$, where Z is the generator of a one-parameter semigroup, A is a bounded operator and $\lambda \in \mathbf{C}$. Then for every $t \geq 0$, T_t^λ is an entire function of the coupling constant λ.*

Proof. This is an immediate consequence of (11.10). $\square$

We mention in passing that the analytic dependence of an $n \times n$ matrix $A(z)$ on a complex parameter z does not imply that the eigenvalues of $A(z)$ depend analytically on z: branch points may occur even for $n = 2$.

Problem 11.4.3 *Let $S_t := e^{Zt}$ be a one-parameter semigroup acting on $L^2(X, dx)$ and suppose that*

$$(S_t f)(x) := \int_X K_t(x, y) f(y) \, dy$$

for all $f \in L^2(X)$, where K is a non-negative integral kernel which depends continuously on $t > 0$ and on $x, y \in X$. Suppose also that $T_t = e^{(Z+A)t}$ where A is multiplication by a (possibly complex-valued) bounded function a on X. Use the expansion (11.10) to prove that T_t has an integral kernel L satisfying

$$|L_t(x, y)| \leq e^{\|a\|_\infty t} K_t(x, y)$$

for all $t > 0$ and $x, y \in X$. $\square$

We next extend the method of Theorem 11.4.1 to what we call class $\mathcal{P}$ perturbations, after Phillips. We do not intend to imply that this class of unbounded perturbations is well adapted to all applications, but it is simple and provides a prototype for more sophisticated results. We discuss some alternatives on page 348.

We say that the operator A is a class $\mathcal{P}$ perturbation of the generator Z of the one-parameter semigroup S_t if

> (i) A is a closed operator,
> (ii) $\mathrm{Dom}(A) \supseteq \bigcup_{t>0} S_t(\mathcal{B})$,
> (iii) $\int_0^1 \|AS_t\| \, dt < \infty$. $\qquad\qquad\qquad\qquad\qquad$ (11.13)

Note that AS_t is bounded for all $t > 0$ under conditions (i) and (ii) by the closed graph theorem.

Lemma 11.4.4 *If A is a class $\mathcal{P}$ perturbation of the generator Z then*

$$\mathrm{Dom}(A) \supseteq \mathrm{Dom}(Z).$$

If $\varepsilon > 0$ then

$$\|AR(\lambda, Z)\| \le \varepsilon \qquad\qquad\qquad (11.14)$$

for all large enough $\lambda > 0$. Hence A has relative bound 0 with respect to Z.

Proof. Combining (11.13) with the bound

$$\|AS_t\| \le \|AS_1\| M e^{a(t-1)}$$

valid for all $t \ge 1$, we see that

$$\int_0^\infty \|AS_t\| e^{-\lambda t} \, dt < \infty$$

for all $\lambda > a$. If $\varepsilon > 0$ then for all large enough λ we have

$$\int_0^\infty \|AS_t\| e^{-\lambda t} \, dt \le \varepsilon.$$

Now

$$\int_0^\infty S_t e^{-\lambda t} f \, dt = R(\lambda, Z) f$$

for all $f \in \mathcal{B}$, so by the closedness of A we see that $R(\lambda, Z)f \in \mathrm{Dom}(A)$ and

$$\|AR(\lambda, Z)f\| \le \varepsilon \|f\|$$

as required to prove (11.14).

If $g \in \mathrm{Dom}(Z)$ and we put $f := (\lambda I - Z)g$ then we deduce from (11.14) that

$$\|Ag\| \leq \varepsilon \|(\lambda I - Z)g\|$$
$$\leq \varepsilon \|Zg\| + \varepsilon \lambda \|g\|$$

for all large enough $\lambda > 0$. This implies the last statement of the theorem. $\square$

Theorem 11.4.5 *If A is a class $\mathcal{P}$ perturbation of the generator Z of the one-parameter semigroup S_t on $\mathcal{B}$ then $Z + A$ is the generator of a one-parameter semigroup T_t on $\mathcal{B}$.*

Proof. Let a be small enough that

$$c := \int_0^{2a} \|AS_t\| \, \mathrm{d}t < 1.$$

We may define T_t by the convergent series (11.10) for $0 \leq t \leq 2a$, and verify as in the proof of Theorem 11.4.1 that $T_s T_t = T_{s+t}$ for all s, $t \geq 0$ such that $s + t \leq 2a$. We now extend the definition of T_t inductively for $t \geq 2a$ by putting

$$T_t := (T_a)^n T_{t-na}$$

if $n \in \mathbf{N}$ and $na < t \leq (n+1)a$. It is straightforward to verify that T_t is a semigroup.

Now suppose that $\|S_t\| \leq N$ for $0 \leq t \leq a$. If $f \in \mathcal{B}$ then

$$\|T_t f - f\| \leq \|S_t f - f\| + \sum_{n=1}^{\infty} N \left(\int_0^t \|AS_s\| \, \mathrm{d}s \right)^n \|f\|,$$

so

$$\lim_{t \to 0} \|T_t f - f\| = 0$$

and T_t is a one-parameter semigroup on $\mathcal{B}$.

It is an immediate consequence of the definition that

$$T_t f = S_t f + \int_0^t T_{t-s} A S_s f \, \mathrm{d}s \tag{11.15}$$

for all $f \in \mathcal{B}$ and all $0 \leq t \leq a$. Suppose that this holds for all t such that $0 < t \leq na$. If $na < u \leq (n+1)a$ then

$$
\begin{aligned}
T_u f &= T_a T_{u-a} f \\
&= T_a \left\{ S_{u-a} f + \int_0^{u-a} T_{u-a-s} A S_s f \, \mathrm{d}s \right\} \\
&= S_a S_{u-a} f + \int_0^a T_{a-s} A S_s (S_{u-a} f) \, \mathrm{d}s \\
&\quad + \int_0^{u-a} T_{u-s} A S_s f \, \mathrm{d}s \\
&= S_u f + \int_0^u T_{u-s} A S_s f \, \mathrm{d}s.
\end{aligned}
$$

By induction (11.15) holds for all $t \geq 0$.

We finally have to identify the generator Y of T_t. The subspace

$$
\mathcal{D} := \bigcup_{t>0} S_t \{ \mathrm{Dom}(Z) \}
$$

is contained in $\mathrm{Dom}(Z)$ and is invariant under S_t and so is a core for Z by Theorem 6.1.18. If $f \in \mathcal{D}$ then there exist $g \in \mathrm{Dom}(Z)$ and $\varepsilon > 0$ such that $f = S_\varepsilon g$. Hence

$$
\begin{aligned}
\lim_{t \to 0} t^{-1}(T_t f - f) &= \lim_{t \to 0} t^{-1}(S_t f - f) + \lim_{t \to 0} t^{-1} \int_0^t T_{t-s}(A S_\varepsilon) S_s g \, \mathrm{d}s \\
&= Z f + (A S_\varepsilon) g \\
&= (Z + A) f.
\end{aligned}
$$

Therefore $\mathrm{Dom}(Y)$ contains $\mathcal{D}$ and $Y f = (Z+A) f$ for all $f \in \mathcal{D}$. If $f \in \mathrm{Dom}(Z)$ then there exists a sequence $f_n \in \mathcal{D}$ such that $\| f_n - f \| \to 0$ and $\| Z f_n - Z f \| \to 0$ as $n \to \infty$. It follows by Lemma 11.4.4 that $\| A f_n - A f \| \to 0$ and hence that $Y f_n$ converges. Since Y is a generator it is closed, and we deduce that

$$
Y f = (Z + A) f
$$

for all $f \in \mathrm{Dom}(Z)$.

Multiplying (11.15) by $\mathrm{e}^{-\lambda t}$ and integrating over $(0, \infty)$ we see as in the proof of Lemma 11.4.4 that if $\lambda > 0$ is large enough then

$$
R(\lambda, Y) f = R(\lambda, Z) f + R(\lambda, Y) A R(\lambda, Z) f
$$

for all $f \in \mathcal{B}$. If λ is also large enough that

$$
\| A R(\lambda, Z) \| < 1
$$

we deduce that

$$R(\lambda, Y) = R(\lambda, Z)(I - AR(\lambda, Z))^{-1}.$$

Hence

$$\mathrm{Dom}(Y) = \mathrm{Ran}(R(\lambda, Y)) = \mathrm{Ran}(R(\lambda, Z)) = \mathrm{Dom}(Z)$$

and $Y = Z + A$. $\quad\square$

Problem 11.4.6 Prove that if A is closed, $0 < \alpha < 1$, $c_1 > 0$ and

$$\|AS_t\| \leq c_1 t^{-\alpha}$$

for all $0 < t \leq 1$, then

$$\|AR(\lambda, Z)\| = O(\lambda^{\alpha - 1})$$

as $\lambda \to +\infty$. Deduce that there exists a constant c_2 such that for all small enough $\varepsilon > 0$ and all $f \in \mathrm{Dom}(Z)$ one has

$$\|Af\| \leq \varepsilon \|Zf\| + c_2 \varepsilon^{-\alpha/(1-\alpha)} \|f\|. \qquad \square$$

Theorem 11.4.7 provides a converse to Problem 11.4.6 for holomorphic semigroups. A generalization of this theorem is presented in Theorem 11.5.7.

Theorem 11.4.7 *Suppose that the holomorphic semigroup S_t has generator Z and that*

$$\|S_t\| \leq c_1, \qquad \|ZS_t\| \leq c_2/t$$

for all t such that $0 < t \leq 1$. Suppose also that the operator A has domain containing $\mathrm{Dom}(Z)$ and that there exists $\alpha \in (0, 1)$ such that

$$\|Af\| \leq \varepsilon \|Zf\| + c_3 \varepsilon^{-\alpha/(1-\alpha)} \|f\| \qquad (11.16)$$

for all $f \in \mathrm{Dom}(Z)$ and $0 < \varepsilon \leq 1$. Then

$$\|AS_t\| \leq (c_2 + c_1 c_3) t^{-\alpha} \qquad (11.17)$$

for all t such that $0 < t \leq 1$. Hence A is a class $\mathcal{P}$ perturbation of Z and Theorem 11.4.5 is applicable.

Proof. Under the stated conditions on t and ε we have

$$\|AS_t f\| \leq \varepsilon \|ZS_t f\| + c_3 \varepsilon^{-\alpha/(1-\alpha)} \|S_t f\|$$
$$\leq (\varepsilon c_2 t^{-1} + c_1 c_3 \varepsilon^{-\alpha/(1-\alpha)}) \|f\|.$$

If we put $\varepsilon := t^{1-\alpha}$ we obtain (11.17). $\quad\square$

Problem 11.4.8 Given $0 < \alpha < 1$, prove that (11.16) holds for all $\varepsilon > 0$ if and only if there is a constant c_4 such that

$$\|Af\| \le c_4 \|Zf\|^{\alpha} \|f\|^{1-\alpha}$$

for all $f \in \mathrm{Dom}(Z)$.   □

Our next result can be adapted to holomorphic semigroups, but our more limited version has a simpler proof.[7]

Problem 11.4.9 Suppose that $0 < \alpha < 1$, H is a non-negative self-adjoint operator on $\mathcal{H}$ and A is a linear operator with $\mathrm{Dom}(A) \supseteq \mathrm{Dom}(H)$. Use Problem 11.4.8 and the spectral theorem to prove that if

$$\|Af\| \le c\|(H+I)^{\alpha}f\| \tag{11.18}$$

for all $f \in \mathrm{Dom}(H)$ then there exists c_3 such that

$$\|Af\| \le \varepsilon\|Hf\| + c_3\varepsilon^{-\alpha/(1-\alpha)}\|f\|$$

for all $f \in \mathrm{Dom}(H)$ and all ε satisfying $0 < \varepsilon \le 1$. This implies that Theorem 11.4.7 can be applied to perturbations satisfying (11.18).   □

The application of the above results to partial differential operators could take up an entire chapter, because of the variety of different conditions which might be imposed on the coefficients. We can do no more than indicate some of the standard applications. We start with the case in which $(Z + A)$ is a Schrödinger operator.

Example 11.4.10 The following develops the example in Theorem 11.2.11 using semigroup techniques. Let $Z := \Delta$ and let $S_t := e^{Zt}$ be the Gaussian one-parameter semigroup acting in $L^2(\mathbf{R}^N)$ and given for $t > 0$ by $S_t f := k_t * f$ where

$$k_t(x) := (4\pi t)^{-N/2} e^{-|x|^2/4t}.$$

(See Example 6.3.5 and Theorem 6.3.2.) Let $A \in L^p(\mathbf{R}^N)$ where $2 \le p < \infty$. One may use (5.10) to prove that

$$\|AS_t\| \le c_N \|A\|_p t^{-N/2p}$$

for all t satisfying $0 < t \le 1$. This implies that the conditions of Theorem 11.4.5 are satisfied provided $p \ge 2$ and $p > N/2$. Hence $Z + A$ is the generator of a one-parameter semigroup T_t acting on $L^2(\mathbf{R}^N)$.   □

---

[7] See [Cachia and Zagrebnov 2001] and [Kato 1966A, Chap. 9, Cor. 2.5].

One may also apply the methods described above to higher order differential operators.

Theorem 11.4.11 *Let* $Z := -H$ *where* $H = H^* := (-\Delta)^n \geq 0$ *acts in* $L^2(\mathbf{R}^N)$. *Also let* A *be a lower order perturbation of the form*

$$(Af)(x) := \sum_{|\alpha| < 2n} a_\alpha(x)(D^\alpha f)(x).$$

If $a_\alpha \in L^{p_\alpha}(\mathbf{R}^N) + L^\infty(\mathbf{R}^N)$ *for each* α, *where* $p_\alpha \geq 2$ *and* $p_\alpha > N/(2n - |\alpha|)$, *then* $Z + A$ *is the generator of a one-parameter semigroup and* A *has relative bound* 0 *with respect to* Z.

Proof. It follows by applying Problem 11.4.9, Theorem 11.4.7 and then Theorem 11.4.5 that it is sufficient to prove that for each α there exists $\beta < 1$ for which

$$X_\alpha := a_\alpha(\cdot)D^\alpha(H+1)^{-\beta}$$

is bounded. Following the notation of Theorem 5.7.3, we may put $X_\alpha := a_\alpha(Q)b_\alpha(P)$ where

$$b_\alpha(\xi) := \frac{i^{|\alpha|}\xi^\alpha}{(|\xi|^{2n}+1)^\beta}.$$

If $a_\alpha \in L^\infty(\mathbf{R}^N)$ then $\|X\| \leq \|a_\alpha\|_\infty \|b_\alpha\|_\infty < \infty$ provided $|\alpha|/2n < \beta < 1$. On the other hand if $a_\alpha \in L^p(\mathbf{R}^N)$ where $p \geq 2$ and $p > N/(2n - |\alpha|)$ then there exists β such that

$$\frac{N + |\alpha|p}{2np} < \beta < 1.$$

This implies that $(|\alpha| - 2n\beta)p + N < 0$ and hence $b_\alpha \in L^p(\mathbf{R}^N)$. The boundedness of X may now be deduced from Theorem 5.7.3. $\square$

Corollary 11.4.12 *If* $a_\alpha \in L^{p_\alpha}(\mathbf{R}^N) + L_0^\infty(\mathbf{R}^N)$ *for each* α, *where* $p_\alpha \geq 2$ *and* $p_\alpha > N/(2n - |\alpha|)$, *then*

$$\text{EssSpec}(H + A) = [0, \infty).$$

Proof. By examining the proof, and in particular Theorems 5.7.1 and 5.7.3, in more detail we see that A is a relatively compact perturbation of H. We can therefore apply Theorem 11.2.6. $\square$

The following theorem enables us to construct a one-parameter semigroup acting on $L^p(\mathbf{R})$ for a Schrödinger operator with a singular potential, *without specifying* the L^p domain of its generator.[8]

Theorem 11.4.13 *Let* $S_t := \mathrm{e}^{\Delta t}$ *be the Gaussian one-parameter semigroup acting in* $L^1(\mathbf{R})$ *and given by* $S_t f := k_t * f$ *where*

$$k_t(x) := (4\pi t)^{-1/2}\mathrm{e}^{-x^2/4t}.$$

(See Example 6.3.5 and Theorem 6.3.2.) If $V \in L^1(\mathbf{R})$ *then* $Z := \Delta + V$ *is the generator of a one-parameter semigroup* T_t *acting on* $L^1(\mathbf{R})$. *This semigroup may be extended consistently to one-parameter semigroup s on* $L^p(\mathbf{R})$ *for all* $1 \le p < \infty$.

Proof. We first consider the perturbed semigroup as acting in L^1. The formula

$$(V\mathrm{e}^{\Delta t}f)(x) = \int_{\mathbf{R}} V(x)(4\pi t)^{-1/2}\mathrm{e}^{-(x-y)^2/4t}f(y)\,\mathrm{d}y$$

implies

$$\|V\mathrm{e}^{\Delta t}\| \le (4\pi t)^{-1/2}\|V\|_1$$

for all $t > 0$ by Theorem 2.2.5. This implies that $T_t := \mathrm{e}^{(\Delta+V)t}$ is a one-parameter semigroup acting on L^1 by Theorem 11.4.5. Let $\|T_t\| \le M\mathrm{e}^{at}$ for all $t \ge 0$.

By comparing the perturbation expansions on the two sides we see that

$$\langle T_t f, g \rangle = \langle f, T_t g \rangle$$

for all $f, g \in L^1 \cap L^\infty$, where the inner product is complex linear in both terms. Therefore

$$|\langle T_t f, g \rangle| = |\langle f, T_t g \rangle| \le \|f\|_\infty \|T_t g\|_1 \le \|f\|_\infty M\mathrm{e}^{at}\|g\|_1.$$

Since g is arbitrary this implies that

$$\|T_t f\|_\infty \le M\mathrm{e}^{at}\|f\|_\infty$$

for all $f \in L^1 \cap L^\infty$ and all $t \ge 0$. The proof is now completed by applying Lemma 6.1.30. □

We finally discuss modifications of condition (11.13) in the definition of class $\mathcal{P}$ perturbations. One can modify or weaken it in several ways. We emphasize

---

[8] A systematic investigation of the same idea in higher dimensions is provided in [Simon 1982].

that in particular contexts, such as those relating to Schrödinger operators, further refinements are needed to get the optimal results.[9]

(i) The proof and conclusion of Theorem 11.4.5 remain valid if we replace condition (i) in the definition of a class $\mathcal{P}$ perturbation on page 342 by the (weaker) assumption that $S_t \mathcal{B} \subseteq \mathrm{Dom}(Z)$ for all $t > 0$ and A is bounded from $\mathrm{Dom}(Z)$ to $\mathcal{B}$ with respect to the natural norms of the two spaces.

(ii) One may assume the Miyadera-Voigt condition[10]

$$\int_0^\delta \|AS_t x\| \, dt \leq c\|x\|$$

for some $\delta > 0$, some $c < 1$ and all x in a dense linear subspace of $\mathcal{B}$. This has the advantage of being applicable in some cases in which S_t is a one-parameter group of isometries; note that if S_t is a one-parameter group and A is a class $\mathcal{P}$ perturbation then A must be bounded.

(iii) If A is a more singular perturbation then both definitions may fail. One alternative is to assume that $A = BC$ where B and C lie in some class of unbounded perturbations for which

$$\int_0^1 \|CS_t B\| \, dt < \infty.$$

It is possible to show that the perturbation expansion (11.10) is still convergent for small enough t under this condition.[11]

(iv) In extreme cases A is not an operator in any obvious sense but one can make the assumption[12]

$$\|S_s A S_t\| \leq c(st)^{-1/2+\varepsilon}$$

for some $\varepsilon > 0$ and all $s, t \in (0, 1)$.

The following example indicates that (iii) may be used to define the one-parameter group $e^{(i\Delta+V)t}$ acting on $L^2(\mathbf{R})$ for $t \in \mathbf{R}$ and complex-valued $V \in L^1(\mathbf{R})$.

Problem 11.4.14 Prove that if $A, B \in L^2(\mathbf{R})$ then the operator

$$C_t := Ae^{i\Delta t}B$$

---

[9] See [Simon 1982] for a detailed and systematic survey.
[10] See [Miyadera 1966, Voigt 1977].
[11] See [Kato 1966B, Davies 1974] for the details.
[12] See [Davies 1977].

satisfies

$$\|C_t\| \le (4\pi|t|)^{-1/2}\|A\|_2\|B\|_2$$

for all $t \in \mathbf{R}\backslash\{0\}$. □

Problem 11.4.15 Prove that if $S_t := \mathrm{e}^{-Ht}$ where H is a non-negative, self-adjoint operator acting in the Hilbert space $\mathcal{H}$ then condition (iv) above implies that $(H+I)^{-1/2}A(H+I)^{-1/2}$ is a bounded operator. □

11.5 Perturbations: resolvent based methods

In the last section we considered situations in which the perturbed semigroups could be constructed directly. In this section we adopt a more indirect method, based on estimates of the resolvent operators instead of the semigroups. Some of the hypotheses in this section are weaker than they were for the analogous results in the last section, but it needs to be noted that the conclusions are also weaker – the existence of the perturbed semigroups is proved, but the validity of the perturbation expansion (11.10) is not proved.

Let Z and A be operators acting in the Banach space $\mathcal{B}$ and satisfying

$$\mathrm{Dom}(Z) \subseteq \mathrm{Dom}(A).$$

We need the concept of relative bound defined in Section 5.1 and the concept of dissipativity defined in Section 8.3.

Theorem 11.5.1 *Suppose that Z is the generator of a one-parameter contraction semigroup S_t on $\mathcal{B}$ and that A is a perturbation of Z with relative bound less than $1/2$. If $Z + A$ is also dissipative (as happens if A is dissipative) then $Z + A$ is the generator of a one-parameter contraction semigroup.*

Proof. Since $Z + A$ is dissipative it is sufficient by Theorem 8.3.5 to show that

$$\mathrm{Ran}(\lambda I - Z - A) = \mathcal{B}$$

for some λ satisfying $\lambda > 0$. Since

$$\mathrm{Ran}(\lambda I - Z - A) = (\lambda I - Z - A)(\lambda I - Z)^{-1}\mathcal{B}$$
$$= (I - A(\lambda I - Z)^{-1})\mathcal{B}$$

it suffices to show that

$$\|A(\lambda I - Z)^{-1}\| < 1 \tag{11.19}$$

for all large enough $\lambda > 0$. If

$$\|Ag\| \le a\|Zg\| + b\|g\| \tag{11.20}$$

for all $g \in \mathrm{Dom}(Z)$ then

$$\|A(\lambda I - Z)^{-1}f\| \le a\|Z(\lambda I - Z)^{-1}f\| + b\|(\lambda I - Z)^{-1}f\|$$

$$\le a\lambda\|(\lambda I - Z)^{-1}f\| + a\|f\| + b\|(\lambda I - Z)^{-1}f\|$$

$$\le (2a + b\lambda^{-1})\|f\|$$

for all $f \in \mathcal{B}$; in the final inequality we used the Hille-Yosida Theorem 8.3.2. If $a < 1/2$ then $(2a + b\lambda^{-1}) < 1$ for all large enough $\lambda > 0$, so (11.19) is valid. $\square$

Corollary 11.5.2 *It is sufficient in Theorem 11.5.1 that the relative bound of A with respect to Z is less than 1.*

Proof. We first note that Z and $Z + A$ are dissipative, so $Z + \lambda A$ is dissipative for all $\lambda \in (0, 1)$. Suppose that (11.20) holds for some a satisfying $0 < a < 1$. Let n be a positive integer satisfying $0 < 1/n \le (1 - a)/2$ and let $\varepsilon := 1/n$. We prove inductively that $Z + m\varepsilon A$ is the generator of a one-parameter contraction semigroup on $\mathcal{B}$ for all m such that $0 \le m \le n$.

Suppose that this holds for some integer m satisfying $0 \le m < n$. Then

$$\|\varepsilon A f\| \le \frac{1}{2}(1 - a)\|Af\|$$

$$\le \frac{1}{2}(1 - am\varepsilon)\|Af\|$$

$$\le \frac{1}{2}\{a\|Zf\| + b\|f\| - am\varepsilon\|Af\|\}$$

$$\le \frac{a}{2}\|(Z + m\varepsilon A)f\| + \frac{b}{2}\|f\|$$

for all $f \in \mathrm{Dom}(Z) = \mathrm{Dom}(Z + m\varepsilon A)$. Therefore $Z + (m + 1)\varepsilon A$ is the generator of a one-parameter contraction semigroup by Theorem 11.5.1. This calculation allows us to carry out a finite induction from $m = 0$ to $m = n$. $\square$

In applications the operators Z and A are often only specified on a core of Z.

Lemma 11.5.3 *Let $\mathcal{D}$ be a core for the generator Z of a one-parameter contraction semigroup on the Banach space $\mathcal{B}$. If the operator A has domain $\mathcal{D}$ and satisfies*

$$\|Af\| \le a\|Zf\| + b\|f\| \tag{11.21}$$

for all $f \in \mathcal{D}$, then A may be extended uniquely to $\mathrm{Dom}(Z)$ *so as to satisfy (11.21) for all $f \in \mathrm{Dom}(Z)$.*

Proof. Since $\mathcal{D}$ is a core for Z the subspace $\mathcal{E} := (I - Z)\mathcal{D}$ is dense in $\mathcal{B}$. The operator B defined on $\mathcal{E}$ by

$$Bf := A(I - Z)^{-1}f$$

is bounded by Problem 6.1.4 and so may be extended uniquely to a bounded linear operator on $\mathcal{B}$, which we also denote by B. The operator A is then extended to $\mathrm{Dom}(Z)$ by putting

$$Af := B(I - Z)f$$

for all $f \in \mathrm{Dom}(Z)$. If $f \in \mathrm{Dom}(Z)$ then since $\mathcal{D}$ is a core, there exist $f_n \in \mathcal{D}$ such that $\|f_n - f\| \to 0$ and $\|Zf_n - Zf\| \to 0$ as $n \to \infty$. The bounds

$$\|Af_n\| \le a\|Zf_n\| + b\|f_n\|,$$

valid for all n, imply (11.21) by continuity. $\square$

Problem 11.5.4 Suppose that L acts in $L^2(\mathbf{R}^N)$ and is given by

$$(Lf)(x) := \nabla \cdot (a(x)\nabla f(x)), \tag{11.22}$$

where $0 < \alpha \le a(x) \le \beta < \infty$ for all $x \in \mathbf{R}^N$ and $\nabla a(x)$ is bounded on $\mathbf{R}^N$. Then one may write $L = Z + A$ where $Z := \beta\Delta$ and

$$(Af)(x) := (a(x) - \beta)(\Delta f)(x) + \nabla a(x) \cdot \nabla f(x).$$

Prove that A satisfies the conditions of Corollary 11.5.2. $\square$

Problem 11.5.5 Formulate and prove an analogous result for the operator

$$(Lf)(x) := -\Delta\{a(x)\Delta f(x)\}$$

acting in $L^2(\mathbf{R}^N)$. $\square$

This approach is not capable of treating operators such as (11.22) in which $a(\cdot)$ is matrix-valued or not differentiable. In such cases one needs to use quadratic form techniques.[13]

Problem 11.5.6 Prove that the domain of the operator defined formally in $L^2(\mathbf{R})$ by

$$(Lf)(x) := \frac{\mathrm{d}}{\mathrm{d}x}\left(a(x)\frac{\mathrm{d}f}{\mathrm{d}x}\right)$$

---

[13] See, for example, [Davies 1989].

cannot contain $C_c^\infty(\mathbf{R})$ if

$$a(x) := \frac{1 + |x|^\alpha}{2 + |x|^\alpha}$$

and $0 < \alpha \le 1/2$. By combining the ideas in Theorem 11.4.11 and Problem 11.5.4 show that if $\alpha > 1/2$ then L, defined on a suitable domain containing $C_c^\infty(\mathbf{R})$, is the generator of a one-parameter contraction semigroup. $\square$

We conclude with a more general version of Theorem 11.4.7. The main hypothesis is weaker than (11.16), but the perturbed semigroup has eventually to be constructed from its resolvent operators rather than directly from the perturbation expansion (11.10).

Theorem 11.5.7 *(Hille) Let Z be the generator of a bounded holomorphic semigroup S_t on the Banach space $\mathcal{B}$ and let A be a perturbation with $\mathrm{Dom}(A) \supseteq \mathrm{Dom}(Z)$. If A has a sufficiently small relative bound with respect to Z then $Z + A - cI$ is also the generator of a bounded holomorphic semigroup T_t for all large enough c.*

Proof. The constants in the proof are all explicit, but we have suppressed reference to their values in the statement of the theorem.

Theorem 8.4.1 implies that there exist constants N and $\alpha \in (0, \pi/2)$ such that $\|R(w, Z)\| \le N|w|^{-1}$ for all $w \in S$, where $S := \{w : |\mathrm{Arg}(w)| < \alpha + \pi/2\}$. Our main task is to prove that there exists a constant $c > 0$ such that

$$\|AR(w+c, Z)\| \le 1/2 \qquad (11.23)$$

for all $w \in S$. We assume that

$$\|Ag\| \le \varepsilon\|Zg\| + b\|g\|$$

for all $g \in \mathrm{Dom}(Z)$, where $0 \le \varepsilon(N+1) \le 1/4$. If $f \in \mathcal{B}$ and $g := R(w+c, Z)f$ then

$$\|AR(w+c, Z)f\| = \|Ag\|$$
$$\le \varepsilon\|Zg\| + b\|g\|$$
$$\le \varepsilon\|Zg - (c+w)g\| + (b + |c+w|)\|g\|$$
$$= \varepsilon\|f\| + (b + |c+w|)\|R(w+c, Z)f\|$$
$$\le \varepsilon\|f\| + \frac{(b + |c+w|)N}{|c+w|}\|f\|$$

$$\leq \{\varepsilon(N+1) + b/|c+w|\}\|f\|$$
$$\leq \{1/4 + b/|c+w|\}\|f\|.$$

We obtain (11.23) by choosing c large enough to ensure that $b/|c+w| \leq 1/4$ for all $w \in S$.

Armed with (11.23), we now observe that the resolvent identity

$$R(w, Z+A-cI) = R(w+c, Z) + R(w, Z+A-cI)AR(w+c, Z)$$

implies

$$R(w, Z+A-cI) = R(w+c, Z)\{I - AR(w+c, Z)\}^{-1}$$

and hence

$$\|R(w, Z+A-cI)\| \leq \frac{2N}{|w+c|} \leq \frac{M}{|w|}$$

for all $w \in S$. This assumes the existence of the resolvent $R(w, Z+A-cI)$, a matter which is dealt with as in Theorem 11.1.3. One completes the proof by applying Theorem 8.4.2. $\square$

12

Markov chains and graphs

12.1 Definition of Markov operators

This chapter and the next are concerned with the spectral theory of positive (i.e. positivity preserving) operators. This theory can be developed at many levels. The space $\mathcal{B}$ on which the operator acts may be an ordered Banach space, a Banach lattice, or $L^p(X, \mathrm{d}x)$ where $1 \leq p \leq \infty$. In this chapter we usually put $\mathcal{B} := l^1(X)$, where X is a (finite or) countable set. As well as providing the simplest context for the theorems, this case has a wide variety of important applications to probability theory and graph theory.

We start with some comments about the significance of the differences between the l^1 and l^2 norms. In the context of Markov semigroups the relevant norm is the l^1 norm. All of the probabilistic properties of Markov operators are naturally formulated in terms of this norm, and they need not even be bounded with respect to the l^2 norm. Nevertheless spectral theory has traditionally been developed in a Hilbert space context, largely because this is technically much easier. In many situations one can prove that the spectrum of an operator is the same whether this operator is regarded as acting in $l^1(X)$ or $l^2(X)$, but this is not always the case; see Theorem 12.6.2.

Given $u \in X$ we define $\delta_u \in l^1(X)$ by

$$\delta_u(x) := \begin{cases} 1 & \text{if } x = u, \\ 0 & \text{otherwise.} \end{cases}$$

Note that

$$f = \sum_{x \in X} f_x \delta_x$$

is an l^1 norm convergent expansion for all $f \in l^1(X)$. If $A : l^1(X) \to l^1(X)$ is a bounded linear operator then we define its matrix $A_{x,y}$ by the formula

$$A_{x,y} := (A\delta_y)(x),$$

so that

$$(Af)(x) = \sum_{y \in X} A_{x,y} f(y)$$

for all $f \in l^1(X)$, the series being absolutely convergent. In our present context Theorems 2.2.5 and 2.2.8 state that the norm of every bounded linear operator A on $l^1(X)$ is given by

$$\|A\| = \sup_{y \in X} \left\{ \sum_{x \in X} |A_{x,y}| \right\}. \tag{12.1}$$

Conversely an infinite matrix $A_{x,y}$ determines a bounded operator A on $l^1(X)$ if and only if the RHS of (12.1) is finite.

The analogous statement for $l^\infty(X)$ is not true, unless X is finite, because the subspace consisting of functions of finite support is not dense in $l^\infty(X)$. However, if $A_{x,y}$ is a matrix with subscripts $x, y \in X$ then the formula

$$(Af)(x) = \sum_{y \in X} A_{x,y} f(y)$$

defines a bounded linear operator A on $l^\infty(X)$ if and only if the RHS of (12.2) is finite. In that case

$$\|A\| = \sup_{x \in X} \left\{ \sum_{y \in X} |A_{x,y}| \right\}. \tag{12.2}$$

We say that a linear operator $P : l^1(X) \to l^1(X)$ is a Markov operator if its matrix satisfies $P_{x,y} \geq 0$ and $\sum_{x \in X} P_{x,y} = 1$ for all $y \in X$. One may equivalently require that

$$f \geq 0 \quad \text{implies} \quad Pf \geq 0, \qquad \langle Pf, 1 \rangle = \langle f, 1 \rangle \tag{12.3}$$

for all $f \in l^1(X)$. Here the angular brackets refer to the natural pairing between the Banach space $l^1(X)$ and its dual space $l^\infty(X)$. A third definition requires that $P(K) \subseteq K$, where

$$K := \{ f : f \geq 0 \text{ and } \sum_{x \in X} f(x) = 1 \} \tag{12.4}$$

is the set of all probability distributions on X.

The matrix $P_{x,y}$ describes a situation in which a particle or other entity at the site y jumps randomly to another site x with 'transition probability'

$P_{x,y}$. The diagonal entry $P_{y,y}$ gives the probability that no jump occurs, and the two assumptions on the matrix entries are then simply the conditions that probabilities are always non-negative and that the sum of the probabilities of all possible outcomes must equal 1. One may regard the jumps as taking place between times t and $t+1$, but this makes two important assumptions, the first one being that the transition probabilities do not vary with time. The second, Markov, assumption is that the probability of jumping from y to x between times t and $t+1$ does not depend upon how the particle got to the site y at time t. In other words the history of the particle is irrelevant. This assumption is not always valid, and in such situations one must use much more sophisticated stochastic ideas than are to be found here.

If a particle starts at the site $x_0 = a$ it may then move successively to $x_1, x_2, \ldots, x_n$. The Markov laws state that the probability of each jump is independent of the previous one, so the probability of this particular path, which we call ω, is

$$\mathcal{P}(\omega) := \prod_{r=1}^{n} P(y_r, y_{r-1}).$$

If we denote by $\Omega(n, a)$ the sample space of all paths of length n starting at $a \in X$ then it is easy to see that

$$\sum_{\omega \in \Omega(n,a)} \mathcal{P}(\omega) = 1.$$

One may associate a directed graph $(X, \mathcal{E})$ with a Markov operator P by putting $(y, x) \in \mathcal{E}$ if $P(x, y) > 0$. One then says that $\omega := (x_0, x_2, \ldots, x_n)$ is a permitted path if $P(x_r, x_{r-1}) > 0$ for all relevant r. Equivalently $P(\omega) > 0$. We will see that the graph of a Markov operator provides valuable insights into its behaviour.

Under the standing assumptions of time independence and the Markov property, if $f \in K$ is the distribution of some system at time 0 then the induced distribution at time $t > 0$ is $P^t f$. The long time behaviour of the system depends on the existence and nature of the limit of $P^t f$ as $t \to \infty$.

Although Markov operators are naturally defined on $l_{\mathbf{R}}^1(X)$, if one wants to ask questions about their spectral properties one has to pass to $l_{\mathbf{C}}^1(X)$. If $1 \le p < \infty$ the complexification of a real operator $A_{\mathbf{R}} : l_{\mathbf{R}}^p(X) \to l_{\mathbf{R}}^p(X)$ is defined by

$$A_{\mathbf{C}}(f + ig) := (A_{\mathbf{R}}f) + i(A_{\mathbf{R}}g)$$

where f, g are arbitrary functions in $l_{\mathbf{R}}^p(X)$. One may readily check that $A_{\mathbf{C}}$ is a complex-linear operator acting on the complex linear space $l_{\mathbf{C}}^p(X)$. An alternative proof of our next theorem is given in Theorem 13.1.2.

Theorem 12.1.1 *Let $A_{\mathbf{R}} : l_{\mathbf{R}}^p(X) \to l_{\mathbf{R}}^p(X)$ be real and let $A_{\mathbf{C}}$ be its complexification, where $1 \leq p < \infty$. Then $A_{\mathbf{C}}$ has the same norm as $A_{\mathbf{R}}$.*

Proof. For $p = 1$ the statement follows immediately from the fact that (12.1) holds whether one works in the real or the complex space. If $p > 1$ then the inequality

$$\|A_{\mathbf{R}}\| \leq \|A_{\mathbf{C}}\|$$

follows directly from the definition of the norm of an operator. To prove the converse we use the fact that

$$|a + ib|^p = c^{-1} \int_{-\pi}^{\pi} |a \cos \theta + b \sin \theta|^p \, d\theta$$

for all $a, b \in \mathbf{R}$, where

$$c := \int_{-\pi}^{\pi} |\cos \theta|^p \, d\theta.$$

If $f, g \in l_{\mathbf{R}}^p(X)$ then

$$\|A_{\mathbf{C}}(f + ig)\|^p = \sum_{x \in X} |(A_{\mathbf{R}}f)(x) + i(A_{\mathbf{R}}g)(x)|^p$$

$$= c^{-1} \sum_{x \in X} \int_{-\pi}^{\pi} |(A_{\mathbf{R}}f)(x) \cos(\theta) + (A_{\mathbf{R}}g)(x) \sin(\theta)|^p \, d\theta$$

$$= c^{-1} \int_{-\pi}^{\pi} \sum_{x \in X} |(A_{\mathbf{R}}f)(x) \cos(\theta) + (A_{\mathbf{R}}g)(x) \sin(\theta)|^p \, d\theta$$

$$= c^{-1} \int_{-\pi}^{\pi} \|A_{\mathbf{R}}(f \cos(\theta) + g \sin(\theta)))\|^p \, d\theta$$

$$\leq \|A_{\mathbf{R}}\|^p c^{-1} \int_{-\pi}^{\pi} \|f \cos(\theta) + g \sin(\theta)\|^p \, d\theta$$

$$= \|A_{\mathbf{R}}\|^p c^{-1} \int_{-\pi}^{\pi} \sum_{x \in X} |f(x) \cos(\theta) + g(x) \sin(\theta)|^p \, d\theta$$

$$= \|A_{\mathbf{R}}\|^p c^{-1} \sum_{x \in X} \int_{-\pi}^{\pi} |f(x) \cos(\theta) + g(x) \sin(\theta)|^p \, d\theta$$

$$= \|A_{\mathbf{R}}\|^p \sum_{x \in X} |f(x) + ig(x)|^p$$

$$= \|A_{\mathbf{R}}\|^p \|f + ig\|^p. \qquad \square$$

From now on we often do not specify whether an operator acts on the real or complex Banach space, because of the freedom afforded by the above theorem.

12.2 Irreducibility and spectrum

Let $\mathcal{B} := l_{\mathbf{R}}^1(X)$. Given $f, g \in \mathcal{B}$ we can define several new functions, such as

$$|f|(x) := |f(x)|,$$

$$f_+(x) := \max\{f(x), 0\},$$

$$f_-(x) := -\min\{f(x), 0\},$$

$$(f \vee g)(x) := \max\{f(x), g(x)\},$$

$$(f \wedge g)(x) := \min\{f(x), g(x)\}.$$

Each of the operations can be defined in terms of the map $f \to |f|$. In particular

$$f \vee g = \frac{f+g}{2} + \left|\frac{f-g}{2}\right|,$$

$$f \wedge g = \frac{f+g}{2} - \left|\frac{f-g}{2}\right|.$$

We define the positive part of $\mathcal{B}$ by

$$\mathcal{B}_+ := \{f \in \mathcal{B} : f \geq 0\},$$

where $f \geq 0$ means $f(x) \geq 0$ for all $x \in X$.

Problem 12.2.1 Prove that $\mathcal{B}_+$ is a closed convex cone and that the map $f \to |f|$ is a non-linear contraction from $\mathcal{B}$ to $\mathcal{B}_+$, i.e.

$$\| \,|f| - |g|\, \| \leq \|f - g\|$$

for all $f, g \in \mathcal{B}$. $\square$

We will need a number of lemmas.

Lemma 12.2.2 *If $0 \leq f \leq g \in \mathcal{B}_+$ then $\|f\| \leq \|g\|$, and $\|f\| = \|g\|$ implies $f = g$.*

We define a linear sublattice $\mathcal{L}$ of $\mathcal{B}$ to be a linear subspace such that $f \in \mathcal{L}$ implies $|f| \in \mathcal{L}$.

Problem 12.2.3 Given a linear sublattice $\mathcal{L}$ in $\mathcal{B}$, prove that if $f, g \in \mathcal{L}$ then $f_+, f_-, f \vee g, f \wedge g \in \mathcal{L}$. Use Problem 12.2.1 to prove that the norm closure of $\mathcal{L}$ is also a linear sublattice. ☐

Problem 12.2.4 If S is any subset of $\mathcal{B}$, prove that the subspace

$$\{f \in \mathcal{B} : f(x) = 0 \text{ for all } x \in S\}$$

is a linear sublattice. If $a, b \in X$ and $\gamma > 0$, prove that

$$\{f \in \mathcal{B} : f(b) = \gamma f(a)\}$$

is a linear sublattice, but that if $\gamma < 0$ it is not. ☐

Lemma 12.2.5 *If $A : \mathcal{B} \to \mathcal{B}$ is a positive operator then*

$$\|A\| = \sup \left\{ \frac{\|Af\|}{\|f\|} : 0 \neq f \in \mathcal{B}_+ \right\}. \tag{12.5}$$

If A is positive and has norm 1 then

$$\mathcal{L} := \{f \in \mathcal{B} : Af = f\},$$

is a linear sublattice of $\mathcal{B}$.

Proof. If $f \in \mathcal{B}$ then $-|f| \leq f \leq |f|$ and $A \geq 0$ together imply $-A(|f|) \leq A(f) \leq A(|f|)$. Hence

$$|A(f)| \leq A(|f|).$$

If c denotes the RHS of (12.5) then one sees immediately that $c \leq \|A\|$. Conversely

$$\|Af\| = \| \, |A(f)| \, \| \leq \|A(|f|)\| \leq c \| \, |f| \, \| = \|f\|$$

for all $f \in \mathcal{B}$. Therefore $\|A\| \leq c$.

If $\|A\| = 1$ and $Af = f$ then

$$\| \, |f| \, \| = \|f\| = \|A(f)\| = \| \, |A(f)| \, \| \leq \|A(|f|)\| \leq \| \, |f| \, \|.$$

Since the two extreme quantities are equal, Lemma 12.2.2 implies that

$$A(|f|) = |A(f)| = |f|,$$

so $\mathcal{L}$ is a linear sublattice. ☐

Given $f \in \mathcal{B}$ we define

$$\operatorname{supp}(f) := \{x \in X : f(x) \neq 0\}.$$

Let A be a positive operator acting on $\mathcal{B}$. We say that $E \subseteq X$ is an invariant set if for every $f \in \mathcal{B}_+$ such that $\mathrm{supp}(f) \subseteq E$ one has $\mathrm{supp}(Af) \subseteq E$. This is equivalent to the condition that $x \in E$ and $(x, y) \in \mathcal{E}$ implies $y \in E$. We say that A is irreducible if the only invariant sets are X and $\emptyset$. This is equivalent to the operator-theoretic condition that for all $x, y \in X$ there exists $n > 0$ such that $(A^n)_{x,y} > 0$. From a graph-theoretic perspective irreducibility demands that for all $x, y \in X$ there exists a path $\omega := (y = x_0, x_1, \ldots, x_n = x)$ such that $(x_{r-1}, x_r) \in \mathcal{E}$ for all relevant r.

Theorem 12.2.6 *If $A : \mathcal{B} \to \mathcal{B}$ is a positive, irreducible operator and $\|A\| = 1$ then the subspace $\mathcal{L} := \{f : Af = f\}$ is of dimension at most 1. If $\mathcal{L}$ is one-dimensional then the associated eigenfunction satisfies $f(x) > 0$ for all $x \in X$ (possibly after replacing f by $-f$).*

Proof. If $f \in \mathcal{L}_+ := \mathcal{L} \cap \mathcal{B}_+$ and $f(y) > 0$ and $A_{x,y} > 0$ then

$$f(x) = (Af)(x) = \sum_{u \in X} A_{x,u} f(u) \geq A_{x,y} f(y) > 0.$$

This implies that the set $E := \mathrm{supp}(f)$ is invariant with respect to A.

Using the irreducibility assumption we deduce that $f \in \mathcal{L}_+$ implies $\mathrm{supp}(f) = X$, unless f vanishes identically. If $f \in \mathcal{L}$ then $f_\pm \in \mathcal{L}$ because $\mathcal{L}$ is a sub-lattice. It follows that either $f_+ = 0$ or $f_- = 0$. This establishes that every non-zero $f \in \mathcal{L}$ is strictly positive, possibly after multiplying it by -1.

If $f, g \in \mathcal{L}_+$ and $\lambda = f(a)/g(a)$ for some choice of $a \in X$ then $h = f - \lambda g$ lies in $\mathcal{L}$ and vanishes at a. Hence h is identically zero, and f, g are linearly dependent. We conclude that $\dim(\mathcal{L}) = 1$. $\square$

If P is a Markov operator then $P^* 1 = 1$, and this implies that $1 \in \mathrm{Spec}(P)$ by Problem 1.2.14. However, it does not imply that 1 is an eigenvalue of P without further hypotheses. If X is finite, then 1 is indeed always an eigenvalue of P.

Corollary 12.2.7 *Let X be a finite set and let $P : l^1(X) \to l^1(X)$ be an irreducible Markov operator. Then $\mathcal{L}$ is one-dimensional and there exists a unique vector $\mu \in \mathbf{R}^X$ such that $\mu(x) > 0$ for all $x \in X$, $\sum_{x \in X} \mu(x) = 1$ and $P\mu = \mu$.*

Theorem 12.2.8 *Suppose that $p(x) \geq 0$ for all $x \in \mathbf{Z}$, $p(x) = 0$ if $|x| \geq k$ and $\sum_{x \in \mathbf{Z}} p(x) = 1$. Then the Markov operator $P : l^1(\mathbf{Z}) \to l^1(\mathbf{Z})$ defined by $Pf = p * f$ has no eigenvalues, and in particular $\mathcal{L} = \{0\}$.*

Proof. If $p * f = \lambda f$ for some $\lambda \in \mathbf{C}$ then, on putting

$$\hat{p}(\theta) := \sum_{n \in \mathbf{Z}} p(n) e^{-in\theta},$$

$$\hat{f}(\theta) := \sum_{n \in \mathbf{Z}} f(n) e^{-in\theta},$$

a direct calculation yields $\hat{p}(\theta)\hat{f}(\theta) = \lambda \hat{f}(\theta)$ for all $\theta \in [-\pi, \pi]$. Note that the first series is finite and the second converges absolutely and uniformly. Now $\hat{p}(\theta)$ is an entire function of θ, so there are only a finite number of solutions of $\hat{p}(\theta) = \lambda$ in $[-\pi, \pi]$. We conclude that $\hat{f}(\theta) = 0$ for all except a finite number of points in $[-\pi, \pi]$. But $\hat{f}$ is continuous so $\hat{f}$ must vanish identically. Therefore $f = 0$. $\square$

12.3 Continuous time Markov chains

Markov operators and their integer powers describe the evolution of a random system whose state changes at integer times, or whose state is only inspected at integer times. It is clearly also of interest to ask the same questions for random systems which change continuously in time.

We again assume that $\mathcal{B} := l^1(X)$ where X is a finite or countable set. We say that P_t is a Markov semigroup on $l^1(X)$ if it is a one-parameter semigroup and each operator P_t is a Markov operator. The condition $t \geq 0$ is important in the following arguments.

Theorem 12.3.1 *Let $A : l^1(X) \to l^1(X)$ be a bounded linear operator. Then e^{At} is a positive operator for all $t \geq 0$ if and only if $A(x, y) \geq 0$ for all $x \neq y$. It is a Markov operator for all $t \geq 0$ if and only if in addition to the above condition*

$$A(y, y) = - \sum_{\{x : x \neq y\}} A(x, y)$$

for all $y \in X$.

Proof. Suppose that e^{At} is positive and $x \neq y$. Then

$$A(x, y) = \lim_{t \to 0+} t^{-1} \langle e^{At} \delta_y, \delta_x \rangle$$

and the RHS is non-negative. Conversely suppose that $A(x, y) \geq 0$ for all $x \neq y$. We may write $A := B + cI$ where $B \geq 0$ and $c := \inf\{A(x, x) : x \in X\}$. Note that $|c| \leq \|A\|$. It follows that

$$e^{At} = e^{ct} \sum_{n=0}^{\infty} \frac{t^n B^n}{n!} \geq 0$$

for all $t \geq 0$. If e^{At} is positive for all $t \geq 0$ then it is a Markov semigroup if and only if

$$\langle e^{At} f, 1 \rangle = \langle f, 1 \rangle$$

for all $f \in l^1(X)$ and $t \geq 0$. Differentiating this at $t = 0$ implies that $\langle Af, 1 \rangle = 0$ for all $f \in l^1(X)$, or equivalently that $\sum_{x \in X} A(x, y) = 0$ for all $y \in X$. Conversely if this holds then

$$\langle e^{At} f, 1 \rangle = \sum_{n=0}^{\infty} \frac{t^n}{n!} \langle A^n f, 1 \rangle = \langle f, 1 \rangle$$

for all $t \in \mathbf{R}$. $\quad\square$

Note A continuous time Markov semigroup $P_t := e^{At}$ acting on $l^1(X)$ with a bounded generator A is actually defined for all $t \in \mathbf{C}$, not just for $t \geq 0$. The Markov property implies that $\|P_t\| = 1$ for all $t \geq 0$. However for $t < 0$ the operators P_t are generally not positive. If X is finite then all of the eigenvalues of A must satisfy $\mathrm{Re}(\lambda) \leq 0$ because $|e^{\lambda t}| \leq \|e^{At}\| = 1$ for all $t \geq 0$. By evaluating the trace of A one sees that at least one eigenvalue λ must satisfy $\mathrm{Re}(\lambda) < 0$, unless A is identically zero. Since $\|e^{At}\| \geq |e^{\lambda t}|$, it follows that the norm grows exponentially as $t \to -\infty$.

Theorem 12.3.1 has an analogue for subMarkov semigroups.

Theorem 12.3.2 *Let* $A : l^1(X) \to l^1(X)$ *be a bounded linear operator such that* $A(x, y) \geq 0$ *for all* $x \neq y$, *so that* $P_t := e^{At} \geq 0$ *for all* $t \geq 0$. *Then the following are equivalent.*

(i) P_t *is a subMarkov operator for all* $t \geq 0$ *in the sense that*

$$0 \leq \langle P_t f, 1 \rangle \leq \langle f, 1 \rangle$$

for all $f \in l^1(X)_+$ *and all* $t \geq 0$;

(ii) $\sum_{x \in X} A(x, y) \leq 0$ *for all* $y \in X$;

(iii) $\langle Af, 1 \rangle \leq 0$ *for all* $f \in l^1(X)_+$.

Proof. This depends upon using the formula

$$\langle P_t f, 1 \rangle = \langle f, 1 \rangle + \int_0^t \langle A P_s f, 1 \rangle \, ds. \qquad \square$$

Example 12.3.3 Consider the discrete Laplacian A defined by

$$Af(n) := f(n-1) - 2f(n) + f(n+1)$$

for all $f \in l^1(\mathbf{Z})$ and all $n \in \mathbf{Z}$. It is evident that this satisfies the conditions for the generator of a Markov semigroup P_t. The semigroup is most easily written down by using Fourier analysis. One has

$$\{Af\}\hat{}\,(\theta) = (2\cos(\theta) - 2)\hat{f}(\theta)$$

where

$$\hat{f}(\theta) := \sum_{n \in \mathbf{Z}} f_n e^{-in\theta}.$$

Hence

$$(P_t f)\hat{}\,(\theta) = e^{-2t(1-\cos(\theta))}\hat{f}(\theta)$$

for all $t > 0$. It follows that

$$P_t f = k_t * f \qquad (12.6)$$

where $k_t \geq 0$ is determined by the identity

$$\hat{k}_t(\theta) := e^{-2t(1-\cos(\theta))}.$$

The Fourier coefficients of this function are

$$k_t(n) := \frac{1}{2\pi} \int_{-\pi}^{\pi} e^{-2t(1-\cos(\theta))+in\theta} \, d\theta.$$

One sees that $t \to k_t(n)$ is a Bessel function. $\quad\square$

The following method of constructing continuous time Markov chains is fairly general, and is said to represent the chain as a Poisson process over a discrete time chain.

Lemma 12.3.4 *If Q is a Markov operator acting on $l^1(X)$ and c is a positive constant then the operator*

$$A := c(Q - I)$$

is the generator of a Markov semigroup P_t.

Proof. We only have to check that $A(x, y) \geq 0$ for all $x \neq y$ and that $\langle Af, 1 \rangle = 0$ for all $f \in l^1(X)$. $\quad \square$

The semigroup operators are given by

$$P_t := e^{ctQ - ctI} = e^{-ct} \sum_{n=0}^{\infty} \frac{c^n t^n Q^n}{n!}.$$

This may be rewritten in the form

$$P_t = \sum_{n=0}^{\infty} a(t, n) Q^n$$

where $a(t, n) > 0$ for all $t, n \geq 0$ and $\sum_{n=0}^{\infty} a(t, n) = 1$. This equation can be interpreted as describing a particle which makes jumps at random real times according to the Poisson law $a(t, n)$, and when it jumps it does so from one point of X to another according to the law of Q.

The above lemma has a converse.

Lemma 12.3.5 *If A is the bounded infinitesimal generator of a continuous time Markov semigroup P_t acting on $l^1(X)$, then there exists a positive constant c and a Markov operator Q such that*

$$A = c(Q - I).$$

Proof. If $c := \sup\{-A(x, x) : x \in X\}$ then $0 \leq c \leq \|A\|$. The case $c = 0$ implies $A = 0$, for which we can make any choice of Q. Now suppose that $c > 0$. It is immediate that $B := A + cI$ is a positive operator and that

$$\langle Bf, 1 \rangle = \langle Af, 1 \rangle + c \langle f, 1 \rangle = c \langle f, 1 \rangle$$

for all $f \in l^1(X)$. Therefore $Q := c^{-1}B$ is a Markov operator and $A = c(Q - I)$. $\quad \square$

If $P_t(x, y) > 0$ for all $x, y \in X$ and all $t > 0$ we say that P_t is irreducible.

Theorem 12.3.6 *Let the continuous time Markov semigroup P_t acting on $l^1(X)$ for $t \geq 0$ have the bounded infinitesimal generator $A := c(Q - I)$. If $x, y \in X$ then either $P_t(x, y) > 0$ for all $t > 0$ or $P_t(x, y) = 0$ for all $t > 0$. The Markov semigroup P_t is irreducible if and only if the Markov operator Q is irreducible.*

Proof. The first statement follows directly from the formula

$$P_t(x, y) = e^{-ct} \sum_{n=0}^{\infty} \frac{c^n t^n}{n!} Q^n(x, y)$$

in which every term is non-negative. If $x \neq y$ then $P_t(x, y) > 0$ if and only if $Q^n(x, y) > 0$ for some $n \geq 1$. On the other hand $P_t(x, x) > 0$ for all $x \in X$ and all $t > 0$, whatever Q may be.

The second statement of the theorem depends on the fact that Q is irreducible if and only if for all $x, y \in X$ there exists $n \geq 1$ for which $Q^n(x, y) > 0$.   □

12.4 Reversible Markov semigroups

A bounded operator $A : l^1(X) \to l^1(X)$ is said to be reversible (or to satisfy detailed balance) with respect to a positive weight $\rho : X \to (0, \infty)$ if

$$A(x, y)\rho(y) = A(y, x)\rho(x)$$

for all $x, y \in X$. If $A \geq 0$ then reversibility implies that the associated graph $(X, \mathcal{E})$ is undirected in the sense that $(x, y) \in \mathcal{E}$ if and only if $(y, x) \in \mathcal{E}$.

Lemma 12.4.1 *If the Markov operator P is reversible with respect to ρ and $\rho \in l^1(X)_+$ then $P\rho = \rho$.*

Proof. We have

$$(P\rho)(x) = \sum_{y \in X} P(x, y)\rho(y) = \sum_{y \in X} P(y, x)\rho(x) = \rho(x)$$

for all $x \in X$.   □

Lemma 12.4.2 *If P is a reversible Markov operator and X is finite then $\mathrm{Spec}(P) \subseteq [-1, 1]$.*

Proof. Since P is a contraction on $l^1(X)$, its spectrum is contained in $\{z : |z| \leq 1\}$. It remains to prove that the spectrum is real. The operator $B := \rho^{-1/2} P \rho^{1/2}$ is similar to P and therefore has the same spectrum. Its matrix satisfies

$$B(x, y) = \rho(x)^{-1/2} \left[P(x, y)\rho(y) \right] \rho(y)^{-1/2}$$
$$= \rho(x)^{-1/2} \left[P(y, x)\rho(x) \right] \rho(y)^{-1/2}$$
$$= B(y, x).$$

Since B is real and symmetric, its spectrum is real.   □

The above proof does not extend to infinite sets X, because the multiplication operators $\rho^{\pm 1/2}$ need not be bounded, and in the important applications they are not.

Lemma 12.4.3 *Let* A *be the bounded generator of a Markov semigroup* P_t *acting on* $l^1(X)$, *and let* $\rho : X \to (0, \infty)$. *Then* P_t *is reversible with respect to* ρ *for every* $t \geq 0$ *if and only if* A *is reversible with respect to* ρ.

Proof. We use the reversibility condition in the form $A\rho = \rho A'$, where A' is the transpose of A. We use this to deduce that $A^n \rho = \rho(A')^n$. This implies the reversibility of the semigroup by using the power series expansion of e^{At}. In the converse direction we differentiate the reversibility condition for P_t at $t = 0$. $\square$

The remainder of this section will be concerned with the construction of reversible Markov operators, which are of importance in statistical dynamics.[1] The goal is to describe the relaxation to equilibrium of a system of interacting particles. We allow X to be infinite and assume that $\sim$ is a symmetric relation such that

$$0 < n(y) := \#\{x : x \sim y\} \leq k$$

for some $k < \infty$ and all $y \in X$. We also assume that $x \sim x$ is false for all $x \in X$.

We leave the proof of the next theorem to the reader, since it is a routine verification of the necessary conditions. The semigroup e^{At} is said to define the Glauber dynamics of the statistical system.

Theorem 12.4.4 *Let* $\rho : X \to (0, +\infty)$ *be a weight such that*

$$c^{-1} \leq \rho(x)/\rho(y) \leq c \tag{12.7}$$

whenever $x \sim y$. *Then*

$$A(x, y) := \begin{cases} \rho(x)^{1/2}\rho(y)^{-1/2} & \text{if } x \sim y, \\ -\sum_{\{u : u \sim x\}} \rho(u)^{1/2}\rho(y)^{-1/2} & \text{if } x = y, \\ 0 & \text{otherwise,} \end{cases}$$

is the matrix of a bounded linear operator A *on* $l^1(X)$, *and* A *satisfies the detailed balance condition with respect to* ρ. *Moreover* A *is the generator of a one-parameter Markov semigroup on* $l^1(X)$ *that satisfies the detailed balance condition with respect to* ρ.

---

[1] See [Streater 1995, pp. 104 ff.] for a comprehensive account of this approach to non-equilibrium statistical mechanics.

In applications X might describe (a discrete approximation to) the possible configurations of a large molecule and one puts

$$\rho(x) := Z^{-1} e^{-\beta h(x)} \tag{12.8}$$

where $h(x)$ is the ground state energy of a particular configuration. The function $h : X \to \mathbf{R}$ is called the Hamiltonian of the system, $\beta > 0$ is the inverse temperature of the environment, the partition function $Z := \sum_{x \in X} e^{-\beta h(x)}$ is assumed to be finite, and ρ is called the equilibrium state or Gibbs state.

Example 12.4.5 We consider a model in statistical dynamics for which S is a large finite subset of $\mathbf{Z}^n$ and $X := 2^S$, so that every state $x \in X$ is a function $x : S \to \{-1, 1\}$. Every state can be represented by a diagram consisting of $\pm$'s on the plane region S. A typical state is given below; it assumes the choice $S := \{1, \ldots, 6\}^2 \subseteq \mathbf{Z}^2$. Even in this case $\#(X) = 2^{36}$.

$$
\begin{array}{cccccc}
+ & + & - & + & + & + \\
+ & - & - & - & + & + \\
- & + & - & - & + & - \\
+ & + & + & - & - & - \\
- & + & + & + & + & - \\
- & + & + & - & + & +
\end{array}
$$

We define the Hamiltonian or energy function $h(x)$ of a state x to be the sum of the energies associated with its 'bonds' as follows. We start by specifying a function $J : \{-1, 1\}^2 \to \mathbf{R}$ and assume that $J(u, v) = J(v, u)$ for all u, v. We interpret $J(u, v)$ as the energy of the bond (u, v). We then put

$$h(x) := \sum_{\{s, t \in S : |s-t|=1\}} J(x(s), x(t))$$

where $|s - t|$ is the Euclidean distance between s and t in S; other measures of closeness of s and t could be used.

We say that $x \sim y$ if the states differ at only one site s. Every $x \in X$ has exactly $\#(S)$ neighbours. If $x \sim y$ then the sums defining $h(x)$ and $h(y)$ differ only for those bonds which start or end at the relevant site $s \in S$. If n is the dimension, then there are $4n$ such bonds if s is an interior point of S, but fewer if it is a boundary point. Hence $x \sim y$ implies

$$|h(x) - h(y)| \leq 4n\|J\|_\infty. \tag{12.9}$$

Given the function h and $\beta > 0$, we define the Gibbs state ρ by (12.8) and then the reversible Markov semigroup on $l^1(X)$ as described in Theorem 12.4.4. The bound (12.9) provides the necessary upper and lower bounds on $\rho(x)/\rho(y)$ for such x, y. $\square$

Problem 12.4.6 Prove that the Markov semigroup defined in Example 12.4.5 is irreducible. □

Example 12.4.7 We say that the bond interactions are ferromagnetic if $J(u, v) := (u - v)^2$, which implies that

$$h(x) := \sum_{\{(s,t):|s-t|=1\}} (x(s) - x(t))^2.$$

Assuming that S is connected, the minimum value of $h(x)$ is taken at two sites: when $x(s) = 1$ for all $s \in S$ and also when $x(s) = -1$ for all $s \in S$. If the temperature is very small, i.e. $\beta > 0$ is very large, then the equilibrium state

$$\rho_\beta(x) = Z^{-1} e^{-\beta h(x)}$$

is highly concentrated around these two minima, with equal probabilities of being close to each. □

Problem 12.4.8 Find the minimum energy configurations x if the bonds are anti-ferromagnetic, i.e. $J(u, v) := -(u - v)^2$. □

12.5 Recurrence and transience

In this section we study the long time asymptotics of irreducible Markov operators and semigroups, obtaining results that are more specific than those in Section 10.1. We start with the discrete time case. Recall that a Markov operator P acting on $l^1(X)$ is said to be irreducible if for all $x, y \in X$ there exists $n > 0$ such that $P^n(x, y) > 0$; in other words it is possible to get from x to y if one waits a suitable length of time. We also say that P is aperiodic if $P^n(x, x) > 0$ for *all* large enough n.

Problem 12.5.1 Let $X := \{1, 2, \ldots, n\}$ where we identify n with 0 to get a periodic set. Define the Markov operator $P : l^1(X) \to l^1(X)$ by

$$(Pf)(x) := pf(x+1) + qf(x-1)$$

where $p > 0$, $q > 0$ and $p + q = 1$. Prove that P is aperiodic if and only if n is odd. Also find the spectrum of P and the number of eigenvalues of modulus 1.[2] □

---

[2] The matrix of P is an example of a circulant matrix. Every eigenvector f is of the form $f_j = w^j$ for all $j \in X$ where $w \in \mathbf{C}$ satisfies $w^n = 1$.

If P is an irreducible Markov operator, we say that it is recurrent (at x) if

$$\sum_{n=1}^{\infty} P^n(x, x) = +\infty \tag{12.10}$$

and if this sum is finite we say that it is transient. The reason for using these names will become clear in Theorem 12.5.5.

Lemma 12.5.2 *If P is an irreducible Markov operator then the notions of recurrence, transience and aperiodicity are independent of the choice of the point $x \in X$.*

Proof. Let $c(x)$ stand for the sum in (12.10). Since P is irreducible, given $x \neq y \in X$ there exist $a, b > 0$ such that $P^a(x, y) > 0$ and $P^b(y, x) > 0$. This implies

$$0 \leq P^a(x, y)P^n(y, y)P^b(y, x)$$
$$\leq \sum_{u, v \in X} P^a(x, u)P^t(u, v)P^b(v, x)$$
$$= P^{n+a+b}(x, x) \tag{12.11}$$

for all $n \geq 0$. Hence

$$0 \leq P^a(x, y)c(y)P^b(y, x) \leq c(x).$$

Combining this with a similar inequality in the reverse direction we see that $c(y) < \infty$ if and only if $c(x) < \infty$.

The inequality (12.11) also establishes that if $P^n(y, y) > 0$ for all large enough n then $P^{n+a+b}(x, x) > 0$ for all large enough n. $\quad\square$

Problem 12.5.3 If X is finite prove that every irreducible Markov operator on $l^1(X)$ is recurrent. $\quad\square$

Problem 12.5.4 Let P be an irreducible Markov operator on $l^1(X)$ and suppose that

$$P^n(x, x) \sim e^{na(x)}$$

as $n \to +\infty$ in the sense that

$$\lim_{n \to +\infty} n^{-1} \log(P^n(x, x)) = a(x).$$

Prove that $a(x)$ is independent of x. Note that if $a(x) < 0$ for some (every) x then P is transient. $\quad\square$

If P is an irreducible Markov operator and $x \in X$, we put $p_n := P^n(x, x)$. Alternatively

$$p_n := \sum_{n,x} \mathcal{P}(\omega)$$

where $\omega := \{x_0, x_1, \ldots, x_n\}$ and

$$\mathcal{P}(\omega) := P(x_n, x_{n-1})P(x_{n-1}, x_{n-2})\ldots P(x_1, x_0).$$

The notation $\sum_{n,x}$ indicates that one sums over all paths ω for which $x = x_0 = x_n$.

We also define the first return probability f_n at x for the time n by

$$f_n := \begin{cases} 0 & \text{if } n = 0, \\ p_1 & \text{if } n = 1, \\ \sum'_{n,x} \mathcal{P}(\omega) & \text{if } n > 1, \end{cases}$$

where $\sum'_{n,x}$ denotes the sum over all paths ω which start at x at time zero and return to x again *for the first time* at time n.

The following theorem states that the irreducible Markov operator P is recurrent if and only if every path which starts at x eventually returns to x with probability 1.

Theorem 12.5.5 *One always has*

$$\sum_{n=0}^{\infty} f_n \le 1. \tag{12.12}$$

Moreover the sum equals 1 *if and only if*

$$\sum_{n=0}^{\infty} p_n = \infty.$$

Proof. The first statement is the consequence of the fact that one is summing the probabilities of disjoint events. The second relies on the formula

$$p_n = \delta_{n,0} + f_n p_0 + f_{n-1} p_1 + f_{n-2} p_2 + \cdots + f_1 p_{n-1},$$

which is proved by dividing all relevant paths into subclasses. If $0 < s < 1$ and we define

$$p(s) := \sum_{n=0}^{\infty} p_n s^n, \qquad f(s) := \sum_{n=0}^{\infty} f_n s^n,$$

then

$$p(s) = 1 + f(s)p(s).$$

Taking the limit $s \to 1-$ in

$$p(s) = 1/(1 - f(s))$$

yields the stated result. □

Problem 12.5.6 Let P be the irreducible Markov operator on $l^1(\mathbf{Z}^+)$ associated with the infinite tridiagonal matrix

$$P := \begin{pmatrix} 0 \, \alpha & & & & \\ 1 \, 0 \, \alpha & & & \\ & \beta \, 0 \, \alpha & & \\ & & \beta \, 0 \, \alpha & \\ & & & \beta \, 0 \, \ddots \\ & & & & \ddots \, \ddots \end{pmatrix}$$

where $\alpha > 0$, $\beta > 0$ and $\alpha + \beta = 1$. Prove that P is transient if and only if $\alpha < 1/2$. □

We now progress to the analogous questions for continuous time Markov semigroups.

Theorem 12.5.7 *Let the Markov semigroup P_t on $l^1(X)$ have a bounded infinitesimal generator. If P_t is irreducible then the condition*

$$\int_0^\infty P_t(x, x) \, \mathrm{d}t < \infty \tag{12.13}$$

holds for all $x \in X$ or for no $x \in X$.

Proof. This is essentially the same as the discrete time case (Lemma 12.5.2) with the sum replaced by an integral. □

If the condition (12.13) holds for all $x \in X$ we say that P_t is transient and otherwise we say it is recurrent.

Problem 12.5.8 In Example 12.3.3

$$P_t(0, 0) = \frac{1}{2\pi} \int_{-\pi}^{\pi} \mathrm{e}^{-2(1 - \cos(\theta))t} \, \mathrm{d}\theta.$$

By evaluating the integrals involved prove that this semigroup is recurrent. □

Example 12.5.9 The discrete Laplacian on $l^1(\mathbf{Z}^2)$ is of the form $A := A_1 + A_2$ where

$$(A_1 f)(m, n) := f(m-1, n) - 2f(m, n) + f(m+1, n)$$

and

$$(A_2 f)(m, n) := f(m, n-1) - 2f(m, n) + f(m, n+1).$$

Since $\|A_1\| = \|A_2\| \leq 4$, we see that $\|A\| \leq 8$. It is also clear that A generates a Markov semigroup P_t. One can show that

$$P_t f = h_t * f$$

for all $f \in l^1(\mathbf{Z}^2)$ and for a suitable function h_t by copying the procedure in Example 12.3.3. The main modification is that one needs to use the two-dimensional version of Fourier series, in which the transform of a function on $\mathbf{Z}^2$ is a periodic function on $[-\pi, \pi]^2$.

The following observation makes the computation of P_t easy. A direct calculation shows that $A_1 A_2 = A_2 A_1$. Therefore

$$\mathrm{e}^{At} = \mathrm{e}^{(A_1 + A_2)t} = \mathrm{e}^{A_1 t} \mathrm{e}^{A_2 t}$$

for all $t \geq 0$. Since we have already solved the one-dimensional problem, we can immediately write

$$h_t(u) = k_t(u_1) k_t(u_2)$$

where k_t is the sequence defined in Example 12.3.3.   □

Problem 12.5.10 Write down the definition of the discrete Laplacian acting on $\mathbf{Z}^3$ and find the expression for $P_t(0, 0)$. Prove that the Markov semigroup is transient.   □

Lemma 12.5.11 *Let $P_t := \mathrm{e}^{At}$ for all $t \geq 0$, where $A := c(Q - I)$, $c > 0$ and Q is an irreducible Markov operator. Then P_t is recurrent if and only if Q is recurrent.*

Proof. This depends upon integrating both sides of the formula

$$P_t(x, x) = \sum_{n=0}^{\infty} a(t, n) Q^n(x, x)$$

with respect to t, where

$$a(t, n) := \mathrm{e}^{-ct} c^n t^n / n!$$ □

12.6 Spectral theory of graphs

Recall that an undirected graph $(X, \mathcal{E})$ is a graph such that $(x, y) \in \mathcal{E}$ implies $(y, x) \in \mathcal{E}$. The graph is said to be connected if there is a path $(x = x_0, x_1, \ldots, x_n = y)$ joining every pair of points $x, y \in X$, such that $(x_{r-1}, x_r) \in \mathcal{E}$ for all $r \in \{1, \ldots, n\}$. The length of the shortest such path is called the graph distance $d(x, y)$ between x and y.

Every graph, directed or not, has an associated incidence matrix J, defined by

$$J(x, y) := \begin{cases} 1 & \text{if } (y, x) \in \mathcal{E}, \\ 0 & \text{otherwise.} \end{cases}$$

The spectral properties of J provide invariants for the graph. If X is finite there is no ambiguity in talking about the spectrum of J, but for infinite X one has to state what Banach space J acts on and impose conditions which imply that it is bounded on that space. The spectrum of J is also called the spectrum of the graph itself.[3]

The degree of a point a in a graph $(X, \mathcal{E})$ is defined to be $\#\{x : (a, x) \in \mathcal{E}\}$. We say $(X, \mathcal{E})$ is locally finite if every point has finite degree and that $(X, \mathcal{E})$ has constant degree k if every point has the same degree k.

A path $(x_0, x_1, \ldots, x_n)$ in a graph $(X, \mathcal{E})$ is said to be closed if $x_0 = x_n$ and it is said to be minimal if all other points in the path are different from each other and from x_0. A tree is defined to be an undirected, connected graph $(X, \mathcal{E})$ which contains no minimal closed paths of length greater than 2. In a tree any pair of points can be joined in exactly one way by a path whose length is $d(x, y)$. We say that a tree $(X, \mathcal{E})$ is a k-tree if it has constant degree $k < \infty$. There is only one k-tree up to isomorphism, and it is infinite with no free ends. A k-tree is said to be hyperbolic if $k \geq 3$. This implies that the number $s(r)$ of points whose graph distance from $a \in X$ equals r is given by

$$s(r) = k(k-1)^{r-1}. \tag{12.14}$$

This grows exponentially as $r \to \infty$ provided $k \geq 3$, as happens for the surface areas of spheres in hyperbolic geometry. Figure 12.1 shows the 4 vertices at distance 1 and the 12 vertices at distance 2 from a chosen vertex of a 4-tree.

We show that the l^2 spectrum of a k-tree is different from its l^1 spectrum provided $k \geq 3$. This is false for $k = 2$ because a 2-tree is isomorphic as a graph to $\mathbf{Z}$. We actually prove that k does not lie in the l^2 spectrum of the

---

[3] The spectral theory of graphs is a well-developed subject to which hundreds of papers and several books have been devoted. See, for example, [Chung 1997, Woess 2000].

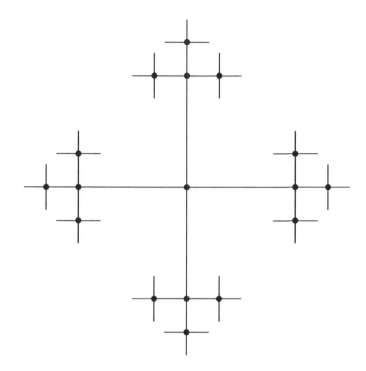

Figure 12.1: 17 neighbouring vertices in a 4-tree

graph; it does lie in the l^1 spectrum because $k^{-1}J$ is a Markov operator on $l^1(X)$.

Theorem 12.6.2 below is not an isolated result. The L^p spectrum of (the Laplacian on) hyperbolic space also depends on p. In particular, the L^2 spectrum of two-dimensional hyperbolic space is $[1/4, \infty)$ while its L^1 spectrum is the whole of the region on or inside the parabola $y^2 = x$; this contains zero, as it should for probabilistic reasons.[4]

Lemma 12.6.1 *There exists $\psi > 0$ on X with*

$$0 < (J\psi)(x) \le \lambda\psi(x) \tag{12.15}$$

for all $x \in X$, where $\lambda := 2(k-1)^{1/2}$. The l^2 spectrum of J satisfies

$$\mathrm{Spec}(J) \subseteq [-\lambda, \lambda].$$

---

[4] See [Lohoué and Rychener 1982]. A short survey of further results of this type may be found in [Davies 1989, Sect. 5.7].

Proof. We start by choosing some base point $a \in X$ arbitrarily, and then define $\psi(x) := \mu^{d(x,a)}$ where $0 < \mu < 1$. Direct calculations establish that

$$(J\psi)(x) = \begin{cases} k\mu\psi(x) & \text{if } x = a, \\ \{(k-1)\mu + \mu^{-1}\}\psi(x) & \text{if } x \neq a. \end{cases}$$

This implies that $0 \leq J\psi \leq \lambda\psi$ where $\lambda := (k-1)\mu + \mu^{-1}$. The quantity $(k-1)\mu + \mu^{-1}$ is minimized by putting $\mu := (k-1)^{-1/2}$, and this leads to (12.15).

Since $J = J^*$, the second statement of the proof now follows by Theorem 13.1.6, but we provide a more direct proof. We emphasize that neither proof requires $\psi \in l^2(X)$, which does not hold in this context. If $f \in l^2(X)$ then using the Schwarz inequality we have

$$\|Jf\|_2^2 = \sum_{x \in X} \left| \sum_{y \in X} J(x,y)f(y) \right|^2$$

$$\leq \sum_{x \in X} \left(\sum_{y \in X} J(x,y)\psi(y) \right) \left(\sum_{y \in X} J(x,y)\psi(y)^{-1}|f(y)|^2 \right)$$

$$\leq \sum_{x \in X} \left(\sum_{y \in X} J(x,y)\lambda\psi(x)\psi(y)^{-1}|f(y)|^2 \right)$$

$$= \sum_{y \in X} \left(\sum_{x \in X} J(y,x)\lambda\psi(x)\psi(y)^{-1}|f(y)|^2 \right)$$

$$\leq \sum_{y \in X} \lambda^2 |f(y)|^2$$

$$= \lambda^2 \|f\|_2^2. \qquad \square$$

Theorem 12.6.2 *(Kesten)*[5] *The l^2 spectrum of the incidence matrix J of a k-tree equals $[-\lambda, \lambda]$, where $\lambda := 2(k-1)^{1/2}$ satisfies $0 < \lambda < k$ if $k \geq 3$.*

Proof. It follows from Lemma 12.6.1 that we only have to prove that $s \in$ Spec(J) for all $s \in [-\lambda, \lambda]$.

First choose an arbitrary base point $a \in X$. Given $\theta \in \mathbf{R}$ and $n \in \mathbf{N}$ define $f_n \in l^2(X)$ by

---

[5] See [Kesten 1959].

$$f_n(x) := \begin{cases} (\mu e^{i\theta})^{d(x,a)} & \text{if } d(x,a) \le n, \\ 0 & \text{otherwise,} \end{cases}$$

where $\mu := (k-1)^{-1/2}$ as before. An easy calculation shows that

$$\|f_n\|_2^2 = 1 + \sum_{r=1}^{n} \mu^{2r} k(k-1)^{r-1} = 1 + nk/(k-1).$$

Moreover

$$(Jf_n)(x) = \begin{cases} k\mu e^{i\theta} & \text{if } d(x,a) = 0, \\ \lambda \cos(\theta) f_n(x) & \text{if } 1 \le d(x,a) \le n-1, \\ (\mu e^{i\theta})^{n-1} & \text{if } d(x,a) = n, \\ (\mu e^{i\theta})^{n} & \text{if } d(x,a) = n+1, \\ 0 & \text{if } d(x,a) > n+1. \end{cases}$$

Putting $s := \lambda \cos(\theta)$ we obtain

$$\|Jf_n - sf_n\|_2^2 = |k\mu e^{i\theta} - s|^2$$

$$+ k(k-1)^{n-1} \left| (\mu e^{i\theta})^{n-1} - s(\mu e^{i\theta})^{n} \right|^2$$

$$+ k(k-1)^{n} \left| (\mu e^{i\theta})^{n} \right|^2$$

$$= O(1)$$

as $n \to \infty$. Therefore

$$\|Jf_n - sf_n\|_2 / \|f_n\|_2 = O(n^{-1/2})$$

as $n \to \infty$. This establishes that $s \in \mathrm{Spec}(J)$ for all $s \in [-\lambda, \lambda]$. $\square$

Example 12.6.3 The above theorem again illustrates the dangers of uncritical numerical approximation. If one chooses a base point $a \in X$ and puts $X_n := \{x \in X : \mathrm{dist}(x,a) \le n\}$ then one may determine the spectrum of the operator J_n obtained by restricting J to $l^p(X_n)$; this is independent of p because the space is finite-dimensional. However, X_n has a large number of free ends, and one has to decide how to define J_n near these; this is analogous to choosing boundary conditions for a differential operator. The fact that the limit of $\mathrm{Spec}(J_n)$ depends on the choice of boundary conditions ultimately explains the p-dependence of the spectrum of J in $l^p(X)$. $\square$

Another difference between the random walk on a k-tree and that on $\mathbf{Z}^N$ is that in the former case a random path moves away from its starting point at a <u>linear rate</u> as the time increases.

Problem 12.6.4 Prove that if $d(x)$ denotes the distance of x from a fixed centre a in the k-tree $(X, \mathcal{E})$ then

$$\sum_{x \in X} d(x)(P^t \delta_a)(x) \sim t(k-2)/k$$

as $t \to +\infty$, provided $k \geq 3$. $\square$

We now turn from trees to more general graphs with constant degree $k < \infty$. We show that whether k lies in the l^2 spectrum of the incidence matrix J on X depends on the rate of growth of the volume of balls in the graph as their radius increases. We say that an undirected, connected graph $(X, \mathcal{E})$ has polynomial volume growth if

$$v(n) \leq cn^s$$

for some choice of the base point $a \in X$, some $c, s > 0$ and all positive integers n, where $v(n) := \#\{x : d(x, a) \leq n\}$. The infimum of all values of s for which such an inequality holds is called the asymptotic dimension of the graph at infinity. This condition is not satisfied by k-trees. On the other hand $\mathbf{Z}^N$ is a graph with polynomial growth if we put $(x, y) \in \mathcal{E}$ when

$$\sum_{r=1}^{N} |x_r - y_r| = 1.$$

Moreover $\mathbf{Z}^N$ has asymptotic dimension N. We finally say that a graph has subexponential volume growth if

$$\limsup_{n \to \infty} v(n)^{1/n} = 1.$$

Problem 12.6.5 Prove that if $(X, \mathcal{E})$ is an undirected, locally finite, connected graph then its asymptotic dimension does not depend on the choice of the base point. Prove the same for the concept of subexponential volume growth. $\square$

Problem 12.6.6 Give an example of an undirected, locally finite, connected graph whose asymptotic dimension s satisfies $1 < s < 2$. Which positive real numbers are possible asymptotic dimensions of graphs? $\square$

The following theorem does not establish that the l^1 spectrum of the graph equals the l^2 spectrum, but it provides a first step in that direction.

Theorem 12.6.7 *Let J be the incidence matrix of an undirected, connected graph $(X, \mathcal{E})$ with constant degree $k < \infty$ and subexponential volume growth. Then k lies in the l^2 spectrum of J.*

Proof. Choose a base point $a \in X$ and put

$$v(n) := \#\{x : d(x, a) \le n\},$$

$$s(n) := \#\{x : d(x, a) = n\}.$$

On defining

$$f_n(x) := \begin{cases} 1 \text{ if } d(x, a) \le n, \\ 0 \text{ otherwise,} \end{cases}$$

one sees immediately that $\|f_n\|_2^2 = v(n)$ for all n. Using the fact that $(Jf_n)(x) = kf_n(x)$ for all x such that $d(x, a) \le n - 1$, one obtains

$$\|Jf_n - kf_n\|_2^2 / \|f_n\|_2^2 \le k^2 \frac{s(n) + s(n+1)}{v(n)}$$

$$\le k^2 \frac{s(n) + s(n+1)}{v(n-1)}$$

$$= k^2 \left\{ \frac{v(n+1)}{v(n-1)} - 1 \right\}.$$

If we show that the $\liminf$ of the final expression is 0, then it follows that k lies in the l^2 spectrum of J.

If $\liminf_{n \to \infty} v(n+1)/v(n-1) = 1 + 2\varepsilon$ for some $\varepsilon > 0$ then there exists N such that $v(n+1)/v(n-1) \ge 1 + \varepsilon$ for all $n \ge N$. This implies that

$$v(N + 2r) \ge v(N)(1 + \varepsilon)^r$$

for all $r \ge 1$. Hence $v(r)$ grows at an exponential rate as $r \to \infty$, contrary to the hypothesis of the theorem. $\quad\square$

Problem 12.6.8 Let X be the subgraph of $\mathbf{Z}^n$ obtained by removing a set S of vertices and all of the edges that have one of their ends in S. Suppose that for every positive integer n there exists $a_n \in \mathbf{Z}^n$ such that the Euclidean ball with centre a and radius n does not meet S. Prove that the spectrum of the incidence matrix of X contains $2n$, which is the maximum value of the degree of all points in X. $\quad\square$

13
Positive semigroups

13.1 Aspects of positivity

In this chapter we extend some of the ideas in Chapter 12 to a more general context and describe some of the special spectral properties of positive operators. These were first discovered for $n \times n$ matrices with non-negative entries by Perron and Frobenius, but many aspects of the theory can be extended to a much more general level. [1]

When we write $\mathcal{B} := L^p(X, dx)$ in this chapter, we usually refer to the space of real-valued functions. We assume throughout that the measure space satisfies the assumptions listed on page 35. Sometimes we will consider the corresponding complex space, and when we need to distinguish between these we do so by adding subscripts, as in $\mathcal{B}_R$ and $\mathcal{B}_C$.

If X is a countable set and dx is the counting measure we write $l^p(X)$ in place of $L^p(X, dx)$. A number of the theorems have slightly less technical proofs in the discrete case, because one does not have to worry about null sets and can use pointwise evaluation of functions.

Later in the chapter we assume that X is a compact metric space, and consider certain positive one-parameter semigroup s acting on $C(X)$.

If $f \in \mathcal{B}$, the positive and negative parts of f are defined by

$$f_+ := \max\{f, 0\} = \tfrac{1}{2}(|f| + f),$$
$$f_- := \max\{-f, 0\} = \tfrac{1}{2}(|f| - f).$$

Note that $|f| \leq |g|$ implies $\|f\| \leq \|g\|$. The set $\mathcal{B}_+$ of all non-negative $f \in \mathcal{B}$ is a convex cone, and is closed with respect to the norm and weak topologies

---

[1] This chapter is only an introduction to a large and important subject. Systematic accounts may be found in [Schaefer 1974, Engel and Nagel 1999]. In this book we concentrate on the infinite-dimensional theory, but [Minc 1988] reveals a wealth of more detailed results for finite matrices with non-negative entries.

of $\mathcal{B}$. An operator $A : \mathcal{B} \to \mathcal{B}$ is said to be positive, symbolically $A \geq 0$, if $Af \geq 0$ for all $f \geq 0$. We say that T_t is a positive one-parameter semigroup on $\mathcal{B}$ if $T_t \geq 0$ for all $t \geq 0$.

Lemma 13.1.1 *If A is a positive operator acting on $\mathcal{B} = L_{\mathbf{R}}^p(X, \mathrm{d}x)$, then A is bounded and*

$$\|A\| = \sup\{\|Af\|/\|f\| : f \geq 0 \text{ and } f \neq 0\}.$$

Proof. Suppose first that for all $n \in \mathbf{Z}^+$ there exists $f_n \geq 0$ such that $\|f_n\| = 1$ and $\|Af_n\| \geq 4^n$. If we put

$$f := \sum_{n=1}^{\infty} 2^{-n} f_n$$

then $f \geq 0$, $\|f\| \leq 1$ and $0 \leq 2^{-n} f_n \leq f$ for all n. Hence $0 \leq 2^{-n} Af_n \leq Af$, and

$$2^n \leq 2^{-n} \|Af_n\| \leq \|Af\|$$

for all n. The contradiction implies that there exists c such that $\|Af\| \leq c$ whenever $f \geq 0$ and $\|f\| = 1$. If c is the smallest such constant then $c \leq \|A\| \leq +\infty$.

Given $f \in \mathcal{B}$, the inequality $-|f| \leq f \leq |f|$ implies $-A|f| \leq Af \leq A|f|$ and hence $|Af| \leq A|f|$. Therefore

$$\|Af\| = \| |Af| \| \leq \| A|f| \| \leq c \| |f| \| = c\|f\|.$$

This implies that $\|A\| \leq c$. $\square$

In order to study the spectrum of an operator $A_{\mathbf{R}}$ acting on $\mathcal{B}_{\mathbf{R}} = L_{\mathbf{R}}^p(X, \mathrm{d}x)$, one must pass to the complexification $\mathcal{B}_{\mathbf{C}} = L_{\mathbf{C}}^p(X, \mathrm{d}x)$. The complex-linear operator $A_{\mathbf{C}}$ is defined in the natural way by $A_{\mathbf{C}}(f + ig) := A_{\mathbf{R}}f + iA_{\mathbf{R}}g$. The proof in Theorem 13.1.2 that $\|A_{\mathbf{C}}\| = \|A_{\mathbf{R}}\|$ is only valid for positive operators. One may also adapt the proof of Theorem 12.1.1 to $L^p(X, \mathrm{d}x)$; this does not require A to be positive, but Problem 13.1.3 shows that it does require $p = q$.

Theorem 13.1.2 *Let $1 \leq p, q \leq \infty$ and let $A_{\mathbf{R}} : L_{\mathbf{R}}^p(X, \mathrm{d}x) \to L_{\mathbf{R}}^q(X, \mathrm{d}x)$ be a positive linear operator. Then*

$$|A_{\mathbf{C}}(f + ig)| \leq A_{\mathbf{R}}(|f + ig|)$$

for all $f, g \in L_{\mathbf{R}}^p(X, \mathrm{d}x)$. Hence $\|A_{\mathbf{C}}\| = \|A_{\mathbf{R}}\|$.

Proof. Given $\theta \in \mathbf{R}$ we have

$$|(A_\mathbf{R} f) \cos(\theta) + (A_\mathbf{R} g) \sin(\theta)| = |A_\mathbf{R} (f \cos(\theta) + g \sin(\theta))|$$
$$\leq A_\mathbf{R} (|f \cos(\theta) + g \sin(\theta)|)$$
$$\leq A_\mathbf{R} (|f + ig|).$$

Let $u, v, w : X \to \mathbf{R}$ be functions in the classes of $A_\mathbf{R} f$, $A_\mathbf{R} g$, $A_\mathbf{R}(|f + ig|)$. Then we have shown that

$$|u(x) \cos(\theta) + v(x) \sin(\theta)| \leq w(x)$$

for all x not in some null set $N(\theta)$. If $\{\theta_n\}_{n=1}^\infty$ is a countable dense subset of $[-\pi, \pi]$ then

$$|u(x) + iv(x)| = \sup_{1 \leq n < \infty} |u(x) \cos(\theta_n) + v(x) \sin(\theta_n)| \leq w(x)$$

for all x not in the null set $\bigcup_{n=1}^\infty N(\theta_n)$. This implies the first statement of the theorem, from which the second follows immediately. $\quad\square$

Problem 13.1.3 The following shows that the positivity condition in Theorem 13.1.2 is necessary if $p \neq q$. Consider the matrix

$$A := \begin{pmatrix} 1 & 1 \\ 1 & -1 \end{pmatrix}$$

as a bounded operator from l^∞ to l^1. Show that $\|A_\mathbf{R}\| = 2$ but $\|A_\mathbf{C}\| = 2^{3/2}$. $\quad\square$

Problem 13.1.4 Let A be a positive linear operator on $L^p(X, \mathrm{d}x)$ where $1 \leq p < \infty$, and let $1/p + 1/q = 1$. Use Theorem 13.1.2 to prove that

$$|\langle A\phi, \psi \rangle| \leq \langle A|\phi|, |\psi| \rangle$$

for all complex-valued $\phi \in L^p(X, \mathrm{d}x)$ and $\psi \in L^q(X, \mathrm{d}x)$. Also give the much more elementary proof available when A has a non-negative integral kernel. $\quad\square$

Our next lemma might be regarded as an operator version of the Schwarz inequality. An operator version of the Hölder inequality (2.1) may be proved by the same method.

Lemma 13.1.5 *Let $1 \leq p, q \leq \infty$ and let $A_\mathbf{R} : L_\mathbf{R}^p(X, \mathrm{d}x) \to L_\mathbf{R}^q(X, \mathrm{d}x)$ be a positive linear operator. Then*

$$|A_\mathbf{C}(fg)(x)|^2 \leq \{A_\mathbf{R}(|f|^2)(x)\}\{A_\mathbf{R}(|g|^2)(x)\}$$

almost everywhere, for all $f, g \in L_\mathbf{C}^{2p}(X, \mathrm{d}x)$.

Proof. If A has a non-negative integral kernel K then

$$|A_{\mathbf{C}}(fg)(x)|^2 = \left| \int_X \{K(x,y)^{1/2} f(y)\}\{K(x,y)^{1/2} g(y)\} \, \mathrm{d}y \right|^2$$

$$\leq \int_X K(x,y)|f(y)|^2 \, \mathrm{d}y \int_X K(x,y)|g(y)|^2 \, \mathrm{d}y$$

$$= \{A_{\mathbf{R}}(|f|^2)(x)\}\{A_{\mathbf{R}}(|g|^2)(x)\}.$$

This finishes the proof if X is finite or countable. We deal with the general case by using an approximation procedure.

If $\mathcal{E} := \{E_1, \ldots, E_n\}$ is a sequence of disjoint Borel sets with finite measures $|E_r|$, we define the orthogonal projection $P_{\mathcal{E}}$ by

$$P_{\mathcal{E}} f := \sum_{r=1}^n |E_r|^{-1} \chi_{E_r} \langle f, \chi_{E_r} \rangle.$$

We then note that the operator $P_{\mathcal{E}} A$ has the non-negative integral kernel

$$K(x,y) := \sum_{r=1}^n |E_r|^{-1} \chi_{E_r}(x) \{A^*(\chi_{E_r})(y)\}.$$

Hence

$$|P_{\mathcal{E}}(p)(x)|^2 \leq \{P_{\mathcal{E}}(q)(x)\}\{P_{\mathcal{E}}(r)(x)\}$$

almost everywhere, where $p := A_{\mathbf{C}}(fg)$, $q := A_{\mathbf{R}}(|f|^2)$ and $r := A_{\mathbf{R}}(|g|^2)$. The proof is completed by choosing a sequence of increasingly fine partitions $\mathcal{E}(n)$ for which $P_{\mathcal{E}(n)}(p)$, $P_{\mathcal{E}(n)}(q)$ and $P_{\mathcal{E}(n)}(r)$ converge to p, q and r respectively not only in norm but also almost everywhere (see Theorem 2.1.7). $\square$

The following theorem has a wider scope than is apparent at first sight, because it is not required that $\phi \in L^2(X, \mathrm{d}x)$.

Theorem 13.1.6 *Let A be a positive linear operator on $L^2(X, \mathrm{d}x)$ and let ϕ be a measurable function on X. If $\phi(x) > 0$ almost everywhere, $0 \leq A\phi \leq \lambda\phi$ and $0 \leq A^*\phi \leq \mu\phi$ then*

$$\|A\| \leq (\lambda\mu)^{1/2}.$$

Proof. Assume first that $\phi \in L^2(X, \mathrm{d}x)$, so that $\phi(x)^2 \mathrm{d}x$ is a finite measure and $L^\infty(X) \subseteq L^2(X, \phi^2\mathrm{d}x)$. We define the unitary operator $U: L^2(X, \phi^2\mathrm{d}x) \to L^2(X, \mathrm{d}x)$ by $Uf := \phi f$. We then observe that $B := U^{-1}AU$ is positive and satisfies $0 \leq B1 \leq \lambda 1$ and $0 \leq B^*1 \leq \mu 1$. If $f \in L^\infty(X)$ then

$$|(Bf)(x)|^2 \leq B(|f|^2)(x)B(1)(x) \leq \lambda B(|f|^2)(x)$$

almost everywhere by Lemma 13.1.5. Therefore

$$\|Bf\|_2^2 \leq \lambda \langle B(|f|^2), 1 \rangle = \lambda \langle |f|^2, B^*(1) \rangle \leq \lambda\mu \langle |f|^2, 1 \rangle = \lambda\mu \|f\|_2^2.$$

Since $L^\infty(X)$ is dense in $L^2(X, \phi^2 dx)$ we deduce that $\|A\| = \|B\| \leq (\lambda\mu)^{1/2}$.

If $\phi \notin L^2(X, dx)$ then the assumptions of the theorem have to be interpreted appropriately. We assume that $0 \leq A\tilde{\phi} \leq \lambda\phi$ for all $\tilde{\phi} \in L^2(X, dx)$ that satisfy $0 \leq \tilde{\phi} \leq \phi$, and similarly for A^*. We then define B as before, and observe that $0 \leq Bf \leq \lambda 1$ and $0 \leq B^*f \leq \mu 1$ for all $f \in L^\infty(X) \cap L^2(X, \phi^2 dx)$ such that $0 \leq f \leq 1$.

From this point on we work in the weighted L^2 space. Let $\mathcal{D}$ denote the set of all bounded functions on X whose supports have finite measure with respect to the measure $\phi(x)^2 dx$. If $f \in \mathcal{D}$ and $\text{supp}(f) = E$ then

$$|(Bf)(x)|^2 = |(B(f\chi_E))(x)|^2 \leq B(|f|^2)(x)B(\chi_E^2)(x) \leq \lambda B(|f|^2)(x)$$

almost everywhere, by Lemma 13.1.5. If the set F has finite measure then

$$\int_F |(Bf)|^2 \phi^2 \, dx \leq \lambda \int_F B(|f|^2)\phi^2 \, dx$$
$$= \lambda \langle B(|f|^2), \chi_F \rangle$$
$$= \lambda \langle |f|^2, B^*\chi_F \rangle$$
$$\leq \lambda\mu \langle |f|^2, 1 \rangle$$
$$= \lambda\mu \|f\|_2^2.$$

Since F is arbitrary subject to having finite measure we deduce that

$$\|Bf\|_2^2 \leq \lambda\mu \|f\|_2^2$$

for all $f \in \mathcal{D}$, and since $\mathcal{D}$ is a dense subspace of L^2 we obtain the same bound for all $f \in L^2$. Therefore $\|A\| = \|B\| \leq (\lambda\mu)^{1/2}$. $\square$

Corollary 13.1.7 *Let A be a positivity preserving*[2] *self-adjoint linear operator on $L_{\mathbf{R}}^2(X, dx)$ and let $\phi \in L_{\mathbf{R}}^2(X, dx)$. If $\phi(x) > 0$ almost everywhere and $A\phi = \lambda\phi$ then*

$$\|A\| = \lambda.$$

Our next lemma states that the singularity closest to the origin of certain operator-valued analytic functions lies on the positive real axis.

---

[2] Because A is self-adjoint, we use this term instead of 'positive' to distinguish it from the condition that $\langle Af, f \rangle \geq 0$ for all f.

Lemma 13.1.8 *Suppose that* A_n *are positive operators on* $\mathcal{B} := L^p(X, dx)$ *and that for all* z *such that* $|z| < R$ *the series*

$$A(z) := \sum_{n=0}^{\infty} A_n z^n \qquad (13.1)$$

converges in norm to an operator $A(z)$. *Suppose also that* $A(z)$ *may be analytically continued to the region* $\{z : |z - R| < S\}$. *Then the series (13.1) is convergent for all* z *such that* $|z| < R + S$.

Proof. If $0 \leq f \in \mathcal{B}$ and $0 \leq g \in \mathcal{B}^*$ then the function

$$F(z) := \langle A(z)f, g \rangle$$

is analytic in

$$D := \{z : |z| < R\} \cup \{z : |z - R| < S\}.$$

We have

$$F^{(n)}(R) = \lim_{r \to R-} F^{(n)}(r)$$

$$= \sum_{m=n}^{\infty} \frac{m!}{(m-n)!} \langle A_m f, g \rangle R^{m-n}$$

by a monotone convergence argument that uses the non-negativity of the coefficients. Moreover

$$0 \leq \sum_{n=0}^{\infty} F^{(n)}(R) x^n / n! < \infty$$

for all x such that $0 \leq x < S$ by the analyticity of F in $\{z : |z - R| < S\}$. Therefore the series

$$\sum_{m=0}^{\infty} \langle A_m f, g \rangle (R + x)^m = \sum_{n=0}^{\infty} \sum_{m=n}^{\infty} \frac{m!}{(m-n)!n!} \langle A_m f, g \rangle R^{m-n} x^n$$

of non-negative terms is convergent for $0 \leq x < S$, and the series

$$\sum_{m=0}^{\infty} \langle A_m f, g \rangle z^m$$

has radius of convergence at least $(R + S)$. The same holds for all $f \in \mathcal{B}$ (resp. $g \in \mathcal{B}^*$) since every element of $\mathcal{B}$ (resp. $\mathcal{B}^*$) is a linear combination of four elements of $\mathcal{B}_+$ (resp. $\mathcal{B}_+^*$). The proof that (13.1) is norm convergent for all z such that $|z| < R + S$ is similar to that of Lemma 1.4.10. $\square$

Theorem 13.1.9 *Let A be a positive operator on $\mathcal{B}$ and let*

$$r := \max\{|z| : z \in \mathrm{Spec}(A)\}$$

be its spectral radius. Then $r \in \mathrm{Spec}(A)$.

Proof. If $|z| > r$ then the series

$$(zI - A)^{-1} = \sum_{n=0}^{\infty} z^{-n-1} A^n$$

is norm convergent. Since the analytic function $z \to (zI - A)^{-1}$ cannot be analytically continued to any set $\{z : |z| > r - \varepsilon\}$, the function must have a singularity at $z = r$ by Lemma 13.1.8. Therefore $r \in \mathrm{Spec}(A)$. $\square$

All of the above ideas can be adapted to the context of one-parameter semigroup s.

Semigroups of the following type occur in population growth models and the neutron diffusion equation. These models are unstable if $\|T_t\|$ increases indefinitely with t. In such cases one either has a population explosion, or it is prevented by some non-linear effect.

Lemma 13.1.10 *Let the operator $Z := -M + A$ act in $\mathcal{B} := L^p(X, \mathrm{d}x)$, where M denotes the operator of multiplication by a measurable function m that is bounded below and A is a bounded, positive operator on $\mathcal{B}$. Then $T_t := \mathrm{e}^{Zt}$ is a positive one-parameter semigroup.*

Proof. Putting $S_t := \mathrm{e}^{-Mt}$, we note that Z is a bounded perturbation of $-M$, so Theorem 11.4.1 is applicable. T_t is a positive operator for all $t \geq 0$ because every term in the perturbation expansion (11.10) is positive. $\square$

Given $f \in \mathcal{B}_+$ one might interpret $f(t, y) := (T_t f)(y)$ as the local density of some entities at a site y, which can increase (if $m(y) < 0$) or decrease (if $m(y) > 0$) as time passes without moving from y. Entities at the position y can cause new entities to appear at the position x at the rate $A(x, y)$ if A has an integral kernel $A(x, y)$. The nth term on the right-hand side of (11.10) describes the part of the state at time t for which exactly $(n-1)$ such creations have taken place.

13.2 Invariant subsets

Given any Borel set E, the linear subspace

$$\mathcal{L}_E := \{f : \mathrm{supp}(f) \subseteq E\}$$

is called a closed order ideal in $\mathcal{B}$ by virtue of possessing the following properties. It is a closed linear subspace $\mathcal{L}$ of $\mathcal{B}$ such that if $|f| \leq |g|$ and $g \in \mathcal{L}$ then $f \in \mathcal{L}$. We see that if f lies in some ideal, then so do $f_{\pm}$ and $|f|$.

Theorem 13.2.1 *Every ideal $\mathcal{L}$ in $\mathcal{B} := L^p(X, \mathrm{d}x)$ is of the form $\mathcal{L}_E$ for some Borel set E, which is uniquely determined up to a null set.*

Proof. We start by observing that since we always assume that the σ-field of Borel subsets of X is countably generated, the space $\mathcal{B}$ is separable. Let f_n be a countable dense subset of $\mathcal{L}$ and let E be the support of

$$k := \sum_{n=1}^{\infty} 2^{-n} |f_n| / \|f_n\|.$$

It is immediate that $\mathrm{supp}(f_n) \subseteq E$ for all n, and this implies by a limiting argument that $\mathrm{supp}(f) \subseteq E$ (up to a null set) for every $f \in \mathcal{L}$. Therefore $\mathcal{L} \subseteq \mathcal{L}_E$.

Conversely let $0 \leq g \in \mathcal{L}_E$ and define $g_n := \min\{g, nk\}$. Since $0 \leq g_n \leq nk$, $k \in \mathcal{L}$ and $\mathcal{L}$ is an ideal, $g_n \in \mathcal{L}$. Since $\|g_n - g\| \to 0$ as $n \to \infty$ by the dominated convergence theorem it follows that $g \in \mathcal{L}$. If $h \in \mathcal{L}_E$ we conclude from the decomposition $h = h_+ - h_-$ and the above argument that $h \in \mathcal{L}$. Therefore $\mathcal{L} = \mathcal{L}_E$. $\square$

Now let A be a positive operator on $\mathcal{B}$. We say that the Borel set E is invariant with respect to A if $A(\mathcal{L}_E) \subseteq \mathcal{L}_E$, or equivalently if $\mathrm{supp}(f) \subseteq E$ implies $\mathrm{supp}(Af) \subseteq E$ (up to null sets).

Problem 13.2.2 Prove that the class of all invariant sets with respect to a given positive operator A is closed under countable unions and countable intersections. If $p = 2$ and A is self-adjoint, prove that it is also closed under complements. $\square$

Theorem 13.2.3 *Let $1 \leq p < \infty$ and let A be a positive operator on $L^p(X, \mathrm{d}x)$. If E is a measurable subset of X then the following are equivalent.*

(i) E is invariant with respect to A;
(ii) there exists $f \geq 0$ in L^p such that $\mathrm{supp}(f) = E$ and $\mathrm{supp}(Af) \subseteq E$;
(iii) there exists $g \geq 0$ in L^p such that $\mathrm{supp}(g) = E$ and $0 \leq Ag \leq \mu g$ for some $\mu > 0$.

If X is countable and $\mathrm{d}x$ is the counting measure then they are also equivalent to

(iv) if $y \in E$, $x \notin E$ and $n \in \mathbf{N}$ then $A^n(x, y) = 0$;

(v) the set E is invariant in the directed graph $(X, \mathcal{E})$ defined by specifying that $(y, x) \in \mathcal{E}$ if $A(x, y) > 0$.

Proof. (i) $\Rightarrow$ (ii) If E_n is a sequence of sets of finite measure with union equal to E then $f_n := \chi_{E_n} / |E_n|^{1/p}$ satisfy $\|f_n\|_p = 1$. The function $f = \sum_{n=1}^{\infty} 2^{-n} f_n$ is non-negative, lies in L^p and has support equal to E. Since E is invariant $\operatorname{supp}(Af) \subseteq E$.

(ii) $\Rightarrow$ (iii) If $\mu > \|A\|$ then the series

$$g := \sum_{n=0}^{\infty} \mu^{-n} A^n f$$

is norm convergent in L^p. We have $0 \le g \in L^p$, $\operatorname{supp}(g) = E$ and

$$0 \le Ag = \sum_{n=0}^{\infty} \mu^{-n} A^{n+1} f = \mu \sum_{n=1}^{\infty} \mu^{-n} A^n f \le \mu g.$$

(iii) $\Rightarrow$ (i) If $0 \le h \in L^p$ and $\operatorname{supp}(h) \subseteq E$ we define h_n for all $n \in \mathbf{N}$ by $h_n := h \wedge (ng)$. It is immediate by the dominated convergence theorem that $\|h - h_n\| \to 0$ as $n \to \infty$. Also $0 \le h_n \le ng$ implies $0 \le Ah_n \le nAg \le n\mu g$ so $\operatorname{supp}(Ah_n) \subseteq E$. On letting $n \to \infty$ we obtain $\operatorname{supp}(Ah) \subseteq E$, so E is invariant.

(i) $\Rightarrow$ (iv) If $y \in E$ then $\delta_y \in \mathcal{L}_E$ so $A^n \delta_y \in \mathcal{L}_E$ for all $n \in \mathbf{N}$. If also $x \notin E$ then

$$0 = (A^n \delta_y)(x) = A^n(x, y).$$

(iv) $\Rightarrow$ (v) This depends on the fact that $A^n(x, y) := (A^n \delta_y)(x)$ is the sum of all (necessarily non-negative) expressions of the form

$$A(x_n, x_{n-1}) \dots A(x_2, x_1) A(x_1, x_0)$$

such that $x_0 = y$ and $x_n = x$. If $A^n(x, y) = 0$ it follows that there cannot exist a path $(y = x_0, x_1, \dots, x_n = x)$ such that $x_{r-1} \to x_r$ for all $r \in \{1, \dots, n\}$. If $x \in E$ and $y \notin E$ no such path can exist for any $n \in \mathbf{N}$. Hence E is invariant in the directed graph on X.

(v) $\Rightarrow$ (i) By using the decomposition $f = f_+ - f_-$ one sees that it is sufficient to prove that if $0 \le f \in L^p$, $\operatorname{supp}(f) \subseteq E$ and E is invariant in the graph-theoretic sense then $\operatorname{supp}(Af) \subseteq E$. We assume that E is countable, the finite case being easier. We have $\|f_n - f\| \to 0$ as $n \to \infty$ where

$$f_n := \sum_{r=1}^{n} f(x_r) \delta_{x_r}$$

and $\{x_r\}_{r=1}^\infty$ is any enumeration of the points in E. We have

$$\text{supp}(Af_n) \subseteq \bigcup_{r=1}^n \text{supp}(A\delta_{x_r})$$

and this is contained in E because (v) implies $\text{supp}(A\delta_x) \subseteq E$ for all $x \in E$. Letting $n \to \infty$, $\text{supp}(f_n) \subseteq E$ for all n implies $\text{supp}(f) \subseteq E$. $\quad\square$

We adapt these ideas to semigroups in the obvious way; namely we say that E is an invariant set with respect to the positive one-parameter semigroup T_t acting on $\mathcal{B}$ if $T_t(\mathcal{L}_E) \subseteq \mathcal{L}_E$ for all $t \geq 0$.

Theorem 13.2.4 *Let $1 \leq p < \infty$ and $\mathcal{B} := l^p(X)$, where X is a countable set. Let $T_t := e^{Zt}$ where $Z := -M + A$, M is the operator of multiplication by a measurable function that is bounded below and A is a positive bounded operator. If E is a measurable subset of X then the following are equivalent.*

(i) E is invariant under the semigroup T_t;
(ii) if $y \in E$ and $x \notin E$ then $T_t(x, y) = 0$ for all $t > 0$;
(iii) if $y \in E$ and $x \notin E$ then $T_t(x, y) = 0$ for some $t > 0$;
(iv) E is invariant under A.

Proof. (i) $\Rightarrow$ (ii) This follows by applying Theorem 13.2.3 to each T_t separately.

(ii) $\Rightarrow$ (iii) is trivial.

(iii) $\Rightarrow$ (iv) This depends upon (11.11), all the terms of which are non-negative. If $T_{n,t}$ denotes the term in (11.11) involving an n-dimensional integral, then (iii) implies that $T_{n,t}(x, y) = 0$ and hence that

$$A(x_n, x_{n-1})...A(x_2, x_1)A(x_1, x_0) = 0$$

for every path such that $x_0 = y$ and $x_n = x$. This implies that $A^n(x, y) = 0$ for all $n \in \mathbf{N}$. We deduce (iv) by applying Theorem 13.2.3.

(iv) $\Rightarrow$ (i) On inspecting (11.11) one sees that all of the operators concerned leave $\mathcal{L}_E$ invariant. $\quad\square$

Problem 13.2.5 Find all the invariant sets for the positive one-parameter semigroup T_t acting on $L^p(\mathbf{R})$ for all $t \geq 0$ according to the formula

$$(T_t f)(x) := f(x - t). \qquad\qquad \square$$

13.3 Irreducibility

A positive operator A acting on $\mathcal{B} = L^p(X, dx)$ is said to be irreducible, or to act irreducibly on X, if its only invariant sets are $\emptyset$ and X, and other sets that differ from these by a null set.

Theorem 13.3.1 *Let* $1 \leq p < \infty$ *and let* A *be a positive operator on* $L^p(X, dx)$. *The following are equivalent.*

(i) A acts irreducibly on X;
(ii) if $f \geq 0$ *in* L^p *and* $\operatorname{supp}(Af) \subseteq \operatorname{supp}(f)$ *then either* $f = 0$ *or* $\operatorname{supp}(f) = X$;
(iii) if $g \geq 0$ *in* L^p *and* $0 \leq Ag \leq \mu g$ *for some* $\mu > 0$ *then either* $g = 0$ *or* $\operatorname{supp}(g) = X$.

If X is countable and dx *is the counting measure then these are also equivalent to*

(iv) for all $x, y \in X$, *there exists* $n \in \mathbf{N}$ *for which* $A^n(x, y) > 0$;
(v) the directed graph associated with A is irreducible.

Proof. Each of the statements is the translation of the corresponding statement of Theorem 13.2.3 when the only possible sets E are $\emptyset$ and X. □

The following lemma is needed in the proof of Theorem 13.3.3.

Lemma 13.3.2 *Let* $S \subseteq \mathbf{N}$ *be a semigroup in the sense that* $(m + n) \in S$ *for all* $m, n \in S$. *Then there exists* $p \in \mathbf{N}$ *such that* $S \subseteq \mathbf{N}p$ *and also* $kp \in S$ *for all sufficiently large* $k \in \mathbf{N}$. *We call* p *the period of S.*

Proof. The set

$$G := \{m - n : m, n \in S\}$$

is a subgroup of $\mathbf{Z}$ so there exists $p \in \mathbf{N}$ such that $G = \mathbf{Z}p$. If $a, b, c \in S$ then p is a factor of $((a + b) - c)$ and of $(b - c)$ since both of these lie in G. Hence p is a factor of a, and $S \subseteq \mathbf{N}p$.

It follows from the definition of G that there exist $m, n \in S$ such that $m - n = p$. We claim that $(n^2 + rp) \in S$ for all $r \in \mathbf{N}$. To prove this put $r = sn + t$ where $s \geq 0$ and $1 \leq t \leq n$ to get

$$n^2 + rp = n^2 + (sn + t)p = n^2 + snp + t(m - n) = tm + (n + sp - t)n,$$

which lies in S because the coefficients of m and n are both non-negative. □

The number p identified in the following theorem is called the period of the irreducible operator A. If $p = 1$ we say that A is aperiodic.

Theorem 13.3.3 *Let* $1 \leq p < \infty$ *and let* A *be a positive operator on* $l^p(X)$. *Given* $x \in X$ *let* p_x *be the period of the semigroup*

$$S_x := \{n \in \mathbf{N} : A^n(x, x) > 0\}.$$

If A *is irreducible then* p_x *is independent of* x.

Proof. We first establish that S_x is a semigroup. If $m, n \in S_x$ then

$$A^{m+n}(x, x) = \sum_{y \in X} A^m(x, y) A^n(y, x)$$

$$\geq A^m(x, x) A^n(x, x)$$

$$> 0,$$

so $(m + n) \in S_x$.

Since A is irreducible, for any $x, y \in X$ there exist m, n such that

$$A^m(x, y) > 0, \quad A^n(y, x) > 0.$$

Therefore

$$A^{m+n+kp_x}(y, y) \geq A^n(y, x) A^{kp_x}(x, x) A^m(x, y) > 0$$

for all large enough $k \in \mathbf{N}$. This implies that p_x is a multiple of p_y. One proves that p_y is a multiple of p_x similarly, so $p_x = p_y$. $\square$

Problem 13.3.4 Construct an irreducible positive operator A acting on $l^2(X)$ for a suitable choice of the finite set X, with the following properties. A has period 1, but for every $x \in X$ and every $n \leq 10^{10}$ one has $A^n(x, x) > 0$ if and only if n is even. $\square$

We say that a positive one-parameter semigroup T_t is irreducible on $\mathcal{B} = L^p(X, dx)$ if the only invariant sets of the semigroup are $\emptyset$ and X, and other sets that differ from these by a null set.

Theorem 13.3.5 *Let* $T_t := e^{Zt}$ *be a one-parameter semigroup on* $l^p(X)$ *where* $1 \leq p < \infty$, $Z := -M + A$, M *is the operator of multiplication by a measurable function that is bounded below, and* A *is a positive, bounded operator on* $l^p(X)$. *Then the following are equivalent.*

(i) T_t *is irreducible;*
(ii) if $x, y \in X$ *then* $T_t(x, y) > 0$ *for some* $t > 0$;

(iii) if $x, y \in X$ then $T_t(x, y) > 0$ for all $t > 0$;

(iv) A is irreducible.

Proof. Each statement is an immediate consequence of Theorem 13.2.4 using the fact that the only possible sets E are $\emptyset$ and X. □

We next turn to the properties of eigenfunctions of positive operators. We say that $\mathcal{L}$ is a linear sublattice of $\mathcal{B} = L^p(X, \mathrm{d}x)$ if it is a linear subspace and $f \in \mathcal{B}$ implies $|f| \in \mathcal{B}$. Given $f, g \in \mathcal{B}$ we put

$$(f \vee g)(x) := \max\{f(x), g(x)\},$$

$$(f \wedge g)(x) := \min\{f(x), g(x)\}.$$

If $\mathcal{L}$ is a linear sublattice of $\mathcal{B}$ then $f, g \in \mathcal{L}$ implies that $f_\pm \in \mathcal{L}$, $f \vee g \in \mathcal{L}$ and $f \wedge g \in \mathcal{L}$. Moreover the closure $\overline{\mathcal{L}}$ of $\mathcal{L}$ is also a linear sublattice of $\mathcal{B}$, for the same reason as in Problem 12.2.3.

Theorem 13.3.6 *Let A be a positive contraction acting on $\mathcal{B} := L^p(X, \mathrm{d}x)$, where $1 \leq p < \infty$. Then*

$$\mathcal{L} = \{f : Af = f\}$$

is a closed linear sublattice of $\mathcal{B}$. If A is irreducible then $\mathcal{L}$ has dimension at most 1 and $f \in \mathcal{L}$ implies $f(x) > 0$ except on a null set (after multiplying f by -1 if necessary).

Proof. The proof that $\mathcal{L}$ is a closed linear sublattice follows Theorem 12.2.6 closely, as does the proof that if $f \in \mathcal{L}$ then either $f(x) > 0$ almost everywhere or $f(x) < 0$ almost everywhere.

We have to prove that $\dim(\mathcal{L}) \leq 1$ by a different method. If f and g are two positive functions in $\mathcal{L}$, put $h_s := sf - (1 - s)g$ for all $s \in [0, 1]$. We note that $I_+ := \{s : h_s \geq 0\}$ is a closed interval containing 1 while $I_- := \{s : h_s \leq 0\}$ is a closed interval containing 0. These intervals must intersect at some point c, at which $h_c = 0$. This implies that f and g are linearly dependent. □

Note that $1 \in \mathrm{Spec}(A)$ does not imply that 1 is an eigenvalue of A without some further assumption, such as the compactness of A or the finiteness of X. A similar comment applies to our next theorem.

Theorem 13.3.7 *Let $T_t := \mathrm{e}^{Zt}$ be a positive one-parameter contraction semigroup acting on $\mathcal{B} := L^p(X, \mathrm{d}x)$, where $1 \leq p < \infty$. Then*

$$\mathcal{L} := \{f \in \mathrm{Dom}(Z) : Zf = 0\}$$

is a closed linear sublattice of $\mathcal{B}$. If T_t is irreducible then $\mathcal{L}$ has dimension at most 1 and $f \in \mathcal{L}$ implies $f(x) > 0$ except on a null set (after replacing f by $-f$ if necessary).

Proof. An elementary calculation shows that

$$\mathcal{L} = \{f : T_t f = f \text{ for all } t > 0\},$$

after which one follows the same argument as in Theorem 13.3.6. $\square$

13.4 Renormalization

Let A be a positive operator on $L^p(X, \mathrm{d}x)$, where $1 \leq p < \infty$. Suppose also that $A\phi = \lambda\phi$, where $\phi(x) > 0$ for all x outside some null set and $\|\phi\|_p = 1$.

We can transfer the operator to a new L^p space as follows. Define a new, finite measure on X by $\tilde{\mathrm{d}}x = \phi(x)^p \mathrm{d}x$ and define the isometry $T: L^p(X, \tilde{\mathrm{d}}x) \to L^p(X, \mathrm{d}x)$ by

$$(Tf)(x) = f(x)\phi(x).$$

Then $\tilde{A} = T^{-1}AT: L^p(X, \tilde{\mathrm{d}}x) \to L^p(X, \tilde{\mathrm{d}}x)$ is positive and has the same norm and spectrum as A. The maps from A to $\tilde{A}$ and from $\mathrm{d}x$ to $\tilde{\mathrm{d}}x$ are called renormalizations.

Lemma 13.4.1 *Under the above conditions the operator $\tilde{A}$ restricts to a bounded operator $\tilde{A}_\infty$ on $L^\infty(X, \tilde{\mathrm{d}}x)$ whose norm is λ. If $p = 2$ and A is self-adjoint then $\tilde{A}$ extends (or restricts) compatibly to bounded operators $\tilde{A}_q$ on $L^q(X, \tilde{\mathrm{d}}x)$ for all $1 \leq q \leq \infty$.*

Proof. We first observe that

$$1 \in L^\infty(X, \tilde{\mathrm{d}}x) \subseteq L^p(X, \tilde{\mathrm{d}}x),$$

the inclusion being a contraction, and that $\tilde{A}1 = \lambda 1$. If $\|f\|_\infty \leq 1$ then $-1 \leq f \leq 1$ and by the positivity of $\tilde{A}$ we have

$$-\lambda 1 = -\tilde{A}1 \leq \tilde{A}f \leq \tilde{A}1 = \lambda 1.$$

Therefore $\|\tilde{A}\|_\infty \leq \lambda$. The identity $\tilde{A}1 = \lambda 1$ now implies $\|\tilde{A}\|_\infty = \lambda$.

The fact that $\tilde{A}$ extends or restricts to a bounded operator on $L^p(X, \tilde{\mathrm{d}}x)$ for all $p \in [1, \infty]$ uses self-adjointness and interpolation; see Problem 2.2.18. $\square$

It is known that the spectra of $\tilde{A}$ and $\tilde{A}_\infty$ need not coincide.[3] The following theorem provides conditions under which this is true.

Theorem 13.4.2 *Let A be a bounded self-adjoint operator on $L^2(X, dx)$, where X has finite measure. If A is ultracontractive in the sense that it is bounded considered as an operator from $L^2(X, dx)$ to $L^\infty(X, dx)$ then A extends (or restricts) to a bounded operator A_p on $L^p(X, dx)$ for $1 \le p \le \infty$. These operators are compact for $1 < p < \infty$ and have the same spectrum for $1 \le p \le \infty$.*

Proof. We omit the subscript p on A_p. It follows from Theorem 4.2.17 that A is compact as an operator on L^2. Since A is bounded from L^2 to L^∞ and L^∞ is continuously embedded in L^2, A is bounded on L^∞. By a duality argument it is bounded on L^1. Theorem 4.2.14 now implies that A is compact considered as an operator on L^p for all $p \in (1, \infty)$.

Since A is bounded from L^1 to L^2 and L^2 is continuously embedded in L^1 it follows that A^2 is a compact operator on L^1. By taking adjoints it is also compact on L^∞. Hence it is compact as an operator on L^p for all $p \in [1, \infty]$ by Theorem 4.2.14, and its spectrum is independent of p by Theorem 4.2.15. By considering the case $p = 2$ we deduce that $\mathrm{Spec}(A_p^2) \subseteq [0, \infty)$ for all $p \in [1, \infty]$. Theorem 1.2.18 now implies that

$$\mathrm{Spec}(A_p) \subseteq \mathbf{R} \tag{13.2}$$

for all $p \in [1, \infty]$.

Since L^∞ is continuously embedded in L^1 a similar argument implies that A^3 is a compact operator on L^1. By taking adjoints it is also compact on L^∞. Hence it is compact as an operator on L^p for all $p \in [1, \infty]$ by Theorem 4.2.14, and its spectrum is independent of p by Theorem 4.2.15. By combining (13.2) and Theorem 1.2.18 with

$$\mathrm{Spec}(A_p^3) = \mathrm{Spec}(A_2^3) \subseteq \mathbf{R}$$

we deduce that

$$\mathrm{Spec}(A_p) = \mathrm{Spec}(A_2) \subseteq \mathbf{R}$$

for all $p \in [1, \infty]$. $\square$

Problem 13.4.3 Formulate and prove a weaker version of Theorem 13.4.2 when the self-adjointness assumption is omitted. $\square$

---

[3] See [Davies 1989, Theorem 4.3.5] for an example which arose in the study of the harmonic oscillator, and for which the fact that the two spectra differed was quite a surprise.

Example 13.4.4 The ideas above can be used to study certain convection-diffusion operators. We give a partial account of the simplest case.[4] Let $a :$ $\mathbf{R} \to \mathbf{R}$ be a smooth function and define the operator $L : C_c^\infty(\mathbf{R}) \to C_c^\infty(\mathbf{R})$ by

$$(Lf)(x) := -f''(x) + 2a'(x)f'(x).$$

A routine calculation shows that

$$\langle Lf, g \rangle = \langle f', g' \rangle$$

for all $f, g \in C_c^\infty(\mathbf{R})$, where the inner products are calculated in $\mathcal{H} := L^2$ $(\mathbf{R}, \phi(x)^2 dx)$ and $\phi(x) := e^{-a(x)}$. It follows that L is symmetric and non-negative in the sense that $\langle Lf, f \rangle \geq 0$ for all $f \in C_c^\infty(\mathbf{R})$.

We may transfer L to $\tilde{\mathcal{H}} := L^2(\mathbf{R}, dx)$ by means of the unitary map $U :$ $\mathcal{H} \to \tilde{\mathcal{H}}$ defined by $(Uf)(x) := \phi(x)^{-1}f(x)$. Putting $H := U^{-1}LU$ we obtain

$$(Hf)(x) := -f''(x) + V(x)f(x)$$

for all $f \in C_c^\infty(\mathbf{R})$, where $V(x) := a'(x)^2 - a''(x)$. It follows that the Schrödinger operator H is symmetric and non-negative in $\mathcal{H}$ in the same sense as before.

Formally one sees that $L1 = 0$ and $H\phi = 0$, but the two functions are not in the domains of the respective operators. We assume that H is essentially self-adjoint on $C_c^\infty(\mathbf{R})$, and denote its self-adjoint closure by the same symbol. We then assume that $\phi \in \mathrm{Dom}(H)$ and that the equation $H\phi = 0$ is valid. It may be proved that e^{-Ht} is a positivity preserving one-parameter semigroup on $\mathcal{H}$. By applying U we deduce that e^{-Lt} is a positivity preserving one-parameter semigroup on $\tilde{\mathcal{H}}$. One may add a constant to a to ensure that $\|\phi\|_2 = 1$; this implies that $\phi(x)^2 dx$ is a probability measure and $e^{-Lt}1 = 1$ for all $t \geq 0$. It follows by using Lemma 13.4.1 that e^{-Lt} restricts to a one-parameter contraction semigroup on $L^p(\mathbf{R}, \phi(x)^2 dx)$ for all $1 \leq p < \infty$.

A more detailed analysis of the properties of these semigroups turns out to depend on the precise rate at which $a(x) \to +\infty$ as $|x| \to \infty$. The cases $a(x) := (1 + x^2)^s$ are quite different in certain key respects depending on whether $s > 1$ or $0 < s \leq 1$. In particular Theorem 13.4.2 is only applicable to e^{-Lt} if $s > 1$. See the indicated references. $\square$

13.5 Ergodic theory

A measure space (X, Σ, dx) is said to be a probability space if the measure of X is 1. One also says that dx is a probability measure or distribution. We

---

[4] Extensions and technical details may be found in [Davies and Simon 1984] and [Davies 1989, Chap. 4].

start by considering a measure-preserving and invertible map $\tau : X \to X$ on such a probability space. Then the formula

$$(Af)(x) = f(\tau x)$$

defines an invertible isometry on $L^p(X)$ for all $1 \leq p \leq \infty$. We restrict attention to the case $p = 2$, when A is a unitary operator.

The map τ is said to be ergodic if the only invariant sets of X are $\emptyset$ and X. Note that $\tau(E) \subseteq E$ implies $\tau(E) = E$ and $\tau(X \backslash E) = X \backslash E$ because τ is measure preserving and invertible. The following theorem is often summarized by saying that for an ergodic system the space average of any quantity equals its time average. We discuss this at the end of the section.

Theorem 13.5.1 *If τ is a measure preserving and invertible map on the probability space (X, dx) then*

$$\lim_{n \to \infty} \frac{1}{n+1}(1 + A + A^2 + \cdots + A^n)f = Pf \qquad (13.3)$$

for all $f \in L^2(X)$, where P is the spectral projection of A associated with the eigenvalue 1. If also τ is ergodic then

$$Pf = \langle f, 1 \rangle 1$$

for all $f \in L^2(X)$.

Proof. The first part is a consequence of the Spectral Theorem 5.4.1. Since A is unitary it is equivalent to the operator of multiplication by $m(s)$ in some space $L^2(S \, ds)$, where $|m(s)| = 1$ almost everywhere. The equation (13.3) follows directly from

$$\lim_{n \to \infty} \frac{1}{n+1}(1 + m(s) + m(s)^2 + \cdots + m(s)^n) = \chi_E(s)$$

almost everywhere, where $E := \{s : m(s) = 1\}$.

The second part follows from the fact that ergodicity is equivalent to the subspace $\mathcal{L} := \{f : Af = f\}$ being of dimension 1. $\square$

It is not always easy to determine whether a map τ is ergodic, just as it is not always easy to determine whether a number is irrational, even if it can be computed with arbitrary accuracy by an explicit algorithm.

Theorem 13.5.2 *Let $X := [0, 1]$ with 0 and 1 identified, and put $\tau(x) := x + t$ where all additions are carried out mod 1. Then τ is ergodic if and only if t is irrational.*

Proof. Once again τ is ergodic if and only if $\mathcal{L} := \{f : Af = f\}$ is one-dimensional. The eigenfunctions of A are given by

$$f_n(x) := e^{2\pi i n x}$$

where $n \in \mathbf{Z}$, and the corresponding eigenvalues are

$$\lambda_n := e^{2\pi i n t}.$$

Clearly $\lambda_0 = 1$. This is the only eigenvalue equal to 1 if and only if t is irrational. $\square$

If μ is a probability measure on a compact metric space K, the support of μ is defined to be $K \backslash U$ where U is the largest open set for which $\mu(U) = 0$. Let K be a compact metric space and μ a probability measure with support equal to K, so that $\mu(U) > 0$ for all non-empty open sets U. The compact space $X := K^{\mathbf{Z}}$ of all maps $x : \mathbf{Z} \to K$ is a probability space if we provide it with the infinite product measure $dx := \mu^{\mathbf{Z}}$. The bilateral shift τ on X is defined by

$$(\tau(x))_n := x_{n-1}.$$

Theorem 13.5.3 *The bilateral shift is an ergodic map on X.*

Proof. Let $\mathcal{F}_n$ denote the subspace of functions $f \in L^2(X)$ that depend only on those coordinates x_m of $x \in X$ for which $-n \leq m \leq n$. If K is finite then $\dim(\mathcal{F}_n) = (\#(K))^{2n+1}$. It may be seen that $\mathcal{F} := \bigcup_{n=1}^{\infty} \mathcal{F}_n$ is dense in $L^2(X)$.

If $f, g \in \mathcal{F}_n$ then $A^m f$ and g depend on entirely different coordinates if $m > 2n + 1$. Therefore

$$\langle A^m f, g \rangle = \langle f, 1 \rangle \langle 1, g \rangle$$

for all such m. A density argument now implies that

$$\lim_{m \to \infty} \langle A^m f, g \rangle = \langle f, 1 \rangle \langle 1, g \rangle$$

for all $f, g \in L^2(X)$. It is immediate from this that 1 is the only eigenvalue of A, and that its multiplicity is 1. The ergodicity of τ follows immediately. $\square$

Problem 13.5.4 Write out the analogous definitions and theorems for a one-parameter group of measure preserving maps on a probability space. Let X be the torus $[0, 1]^2$ subject to periodic boundary conditions, and let $\tau_t(x, y) := (x + t, y + ut)$ for all $t \in \mathbf{R}$, where all additions are evaluated mod 1. Find the precise conditions on u for this group to be ergodic. $\square$

The above theorems do no more than indicate some of the issues that may arise in more complicated dynamical systems. Given a Hamiltonian system with phase space X, one can solve Hamilton's equations to obtain volume-preserving dynamics on X. The operators on $L^2(X)$ associated with integer times cannot be ergodic because the Hamiltonian function is invariant under the evolution. However, the evolution may well be ergodic on some or all of the energy surfaces. Another example is the billiards problem. Here one considers a region in $\mathbf{R}^2$ with a sufficiently regular boundary and a particle moving in straight lines with perfect reflection at the boundary. Once again one might hope that generically the motion is ergodic.

Intensive research into such problems has shown that the proof of ergodicity is extremely hard in those cases in which it has been proved. The KAM theory shows that certain types of Hamiltonian systems cannot be ergodic. Some systems are believed to be ergodic on the basis of numerical and other evidence, but there is no proof. Nevertheless, the issue is of great importance in equilibrium statistical mechanics, and research is still active.

A much greater range of phenomena arise for dynamical systems. One considers a continuous map S on a topological space X, which we assume to be a compact metric space. This induces a so-called Koopman operator

$$(Vf)(x) := f(S(x))$$

which we consider as acting on $C(X)$; one can consider V as acting on a variety of other function spaces. We say that a probability measure μ on X is invariant with respect to S if $\mu(E) = \mu(S^{-1}(E))$ for all Borel subsets of X. This is equivalent to

$$\int_X f(S(x)) \, \mu(\mathrm{d}x) = \int_X f(x) \, \mu(\mathrm{d}x)$$

holding for all non-negative measurable functions f on X. One may then define the Perron-Frobenius operator U to be the restriction of V^* to L^1 $(X, \mu(\mathrm{d}x))$, which is an invariant subspace for V^*; once again one may consider the 'same' operator acting on other function spaces. These caveats are important because in examples the spectrum of the operator often depends on the Banach space being considered.[5]

Problem 13.5.5 Let X denote the interval $[0, 1]_{\mathrm{per}}$ and let $S : X \to X$ be defined by $S(x) = 2x \bmod 1$. Prove that the Lebesgue measure is invariant

---

[5] We refer the reader to [Bedford et al. 1991, Baladi 2000] for reviews of this subject and [Antoniou et al. 2002] for the analysis of a surprisingly difficult example, with references to a number of earlier studies.

under S. Deduce that V is an isometric operator on $L^2([0, 1], dx)$ and find its exact range. Write down an explicit formula for the Perron-Frobenius operator V^*, also regarded as an operator on $L^2([0, 1], dx)$. If $0 \neq f \in \{\mathrm{Ran}(V)\}^\perp$ and $|z| < 1$ prove that $f_z := \sum_{n=0}^\infty z^n V^n f$ is a (non-zero) eigenvector of V^*. Deduce that $\mathrm{Spec}(V) = \{z : |z| \leq 1\}$. See also Problem 5.3.4. $\square$

13.6 Positive semigroups on $C(X)$

In this section we consider the spectral properties of a positive one-parameter semigroup $T_t : C(X) \to C(X)$ such that $T_t 1 = 1$ for all $t \geq 0$, where X is a compact metric space.[6] We will call T_t a Markov semigroup although this is essentially the dual of the earlier meaning that we gave to this term. Note, however, that no special measure on X is specified. All the results in this section, except Theorem 13.6.14 and its corollary, are classical.

The concepts in this section may be regarded as an introduction to the much deeper theory of stochastic differential equations, in which X is replaced by an infinite-dimensional set, normally some class of functions. One of the main tasks is to replace the uniform norm on $C(X)$ by another norm defined on a subspace of $C(X)$ by a formula that is well-adapted to the problem in hand. This is a well-developed but highly non-trivial subject.[7] Our goal here is to explain how to walk, leaving those interested to find out for themselves how much harder running is.

$C(X)$ is an ordered Banach space in which quantities such as $|f|$, f^+, f^- can be defined, and many of the results proved for $L^p(X, dx)$ may be adapted to $C(X)$. In particular the proof of Theorem 13.1.2 shows that if A is a positive operator on $C(X)$ then

$$|Af| \leq A(|f|) \tag{13.4}$$

for all $f \in C(X)$. However, $C(X)$ has two differences from the L^p spaces. The first is that one has to replace the phrase 'measurable set' by 'open set' in various places; this is not quite as harmless as it appears, because complements of open sets are not open. The second and more important difference is that $0 \leq f \leq g$ and $\|f\| = \|g\|$ do not together imply that $f = g$. The following problem shows that this is not a mere technical detail.

---

[6] We always take the norm in $C(X)$ to be the supremum norm and define $f \geq 0$ to mean that $f(x) \geq 0$ for all $x \in X$. We make no assumption about the existence of some favoured measure on X.

[7] See [Hairer and Mattingly 2005] for one of the latest and deepest contributions to the stochastic Navier-Stokes equation, as well as references to earlier literature.

Problem 13.6.1 Let $Q : C[0, 1] \to C[0, 1]$ be defined by

$$(Qf)(x) := (1 - x)f(0) + xf(1).$$

Prove that Q is positive, $Q1 = 1$, and that $\mathcal{L} := \{f : Qf = f\}$ is a two-dimensional subspace of $C[0, 1]$ but not a closed sublattice. $\square$

Theorem 13.6.2 *A bounded operator Z on $C(X)$ is the generator of a norm continuous semigroup of positive operators if and only if*

$$Z + \|Z\|I \geq 0. \tag{13.5}$$

Proof. If (13.5) holds then

$$T_t = \mathrm{e}^{-\|Z\|t} \sum_{n=0}^{\infty} (Z + \|Z\|I)^n t^n / n!$$

so T_t is a positive operator for all $t \geq 0$.

Conversely suppose that $T_t \geq 0$ for all $t \geq 0$. If δ_x denotes the unit measure concentrated at x then

$$T_t^* \delta_x := \lambda(x, t)\delta_x + \mu_{x,t}$$

where $\lambda(x, t) \geq 0$ and $\mu_{x,t}$ is a non-negative, countably additive measure satisfying $\mu_{x,t}(\{x\}) = 0$. We have

$$
\begin{aligned}
|1 - \lambda(x, t)| &\leq |1 - \lambda(x, t)| + \|\mu_{x,t}\| \\
&= \|T_t^* \delta_x - \delta_x\| \\
&\leq \|\mathrm{e}^{Z^*t} - 1\| \\
&\leq \sum_{n=1}^{\infty} \|Z\|^n t^n / n! \\
&= \mathrm{e}^{\|Z\|t} - 1
\end{aligned}
$$

for all $t \geq 0$. Therefore

$$\lambda(x, t) \geq 2 - \mathrm{e}^{\|Z\|t}$$

for all such t. If $\alpha > \|Z\|$ we deduce that there exists $\delta > 0$ such that $\lambda(x, t) \geq \mathrm{e}^{-\alpha t}$ for all $t \in (0, \delta)$.

If $f \in C(X)^+$ and $x \in X$ then

$$(e^{(Z+\alpha I)t} f)(x) = e^{\alpha t} \langle f, e^{Z^* t} \delta_x \rangle$$
$$\geq e^{\alpha t} \langle f, \lambda(x, t) \delta_x \rangle$$
$$\geq \langle f, \delta_x \rangle$$
$$= f(x)$$

provided $0 < t < \delta$. Therefore

$$(Z + \alpha I)f = \lim_{t \to 0} t^{-1} \{ e^{(Z+\alpha I)t} f - f \} \geq 0$$

for all $f \in \mathcal{B}^+$. Letting $\alpha \to \|Z\|$ we deduce that $Z + \|Z\|I \geq 0$. $\quad\square$

Theorem 13.6.3 *If Z is the generator of a norm continuous Markov semigroup T_t on $C(X)$ then $0 \in \mathrm{Spec}(Z)$ and*

$$\mathrm{Spec}(Z) \subseteq \{ z : |z + \|Z\| | \leq \|Z\| \}. \tag{13.6}$$

Proof. If we differentiate $T_t 1 = 1$ at $t = 0$ we obtain $Z(1) = 0$, which implies that $0 \in \mathrm{Spec}(Z)$. Since $W := Z + \|Z\|I$ is a positive operator, we see that

$$\|W\| = \|W(1)\| = \|Z\|.$$

Therefore

$$\mathrm{Spec}(W) \subseteq \{ z : |z| \leq \|Z\| \}.$$

This implies (13.6). $\quad\square$

Example 13.6.4 Suppose that X is a compact metric space and $Q(x, E) \in \mathbf{R}$ for all $x \in X$ and all Borel subsets E of X. We say that Q is a Markov kernel on X if

(i) $Q(x, X) = 1$ for all $x \in X$.
(ii) $E \to Q(x, E)$ is a non-negative, countably additive measure for all $x \in X$.
(iii) $x \to \int_X f(y) Q(x, dy)$ is continuous for all $f \in C(X)$.

Given a continuous function $\sigma : X \to \mathbf{R}^+$, one may consider the evolution equation

$$\frac{\partial}{\partial t} f(x, t) = -\sigma(x) f(x) + \int_X \sigma(x) f(y) Q(x, dy). \tag{13.7}$$

The associated Markov semigroup describes a particle which jumps from x to some other position randomly, the rate being $\sigma(x)$, and the new location being controlled by the kernel Q.

If X is not compact, but σ is still bounded, then a Markov semigroup with a bounded generator may still be associated with the evolution equation. However, if σ is unbounded the subject becomes much more difficult and interesting. Probabilistically, a particle may jump from one position to another more and more rapidly as it moves away from its starting point, and it may have a finite first passage time to infinity. This corresponds to the technical possibility that the natural minimal solution to the evolution equation does not satisfy $T_t 1 = 1$. In such cases one needs to specify a re-entry law at ∞ in order to associate a Markov semigroup with (13.7). $\square$

Theorem 13.6.5 *A densely defined operator Z on $C(X)$ with $1 \in \mathrm{Dom}(Z)$ is the generator of a Markov semigroup if and only if the following conditions are all satisfied.*

(i) *If $f \in \mathrm{Dom}(Z)$ then $\overline{f} \in \mathrm{Dom}(Z)$ and $Z\overline{f} = \overline{Zf}$.*
(ii) *If $g \in C(X)$ then there exists $f \in \mathrm{Dom}(Z)$ such that $f - Zf = g$.*
(iii) *If $a \in X$, $f \in \mathrm{Dom}(Z)$ and $f(x) \le f(a)$ for all $x \in X$ then $(Zf)(a) \le 0$.*

Proof. If Z is the generator of a Markov semigroup then (i) is elementary and (ii) is a consequence of Theorem 8.3.2. Since $T_t 1 = 1$ for all $t \ge 0$ it follows that $1 \in \mathrm{Dom}(Z)$ and that $Z1 = 0$. If $f \in \mathrm{Dom}(Z)$ and $f(x) \le f(a)$ for all $x \in X$ then we put $g := f + \|f\|1$, so that $g \ge 0$ and $\|g\| = \langle g, \delta_a \rangle$. Therefore

$$
\begin{aligned}
(Zf)(a) &= \langle Zf, \delta_a \rangle \\
&= \langle Zg, \delta_a \rangle \\
&= \lim_{t \to 0} t^{-1} \{ \langle T_t g, \delta_a \rangle - \langle g, \delta_a \rangle \} \\
&\le \lim_{t \to 0} t^{-1} \{ \|T_t\| \, \|g\| \, \|\delta_a\| - \|g\| \} \\
&\le 0.
\end{aligned}
$$

Conversely suppose that Z satisfies (i)–(iii), and let $Z_{\mathbf{R}}$ denote the restriction of Z to $\mathrm{Dom}(Z) \cap C_{\mathbf{R}}(X)$. The hypothesis (iii) implies that $Z_{\mathbf{R}}(1) = 0$. If $f \in \mathrm{Dom}(Z_R)$ then one of $\pm f(a) = \|f\|$ holds and (iii) then implies that one of $\langle Z_R f, \pm \delta_a \rangle \le 0$ holds. Therefore $Z_{\mathbf{R}}$ satisfies the weak dissipativity condition of Theorem 8.3.5.

Condition (ii) states that $\mathrm{Ran}(I - Z_{\mathbf{R}}) = C_{\mathbf{R}}(X)$, so by (the real version of) Theorem 8.3.5 $Z_{\mathbf{R}}$ is the generator of a one-parameter contraction semigroup on $C_{\mathbf{R}}(X)$. The proof that the complexification of T_t is a Markov semigroup is straightforward. $\square$

If U is an open set in the compact metric space X we define the closed subspace $\mathcal{J}_U \subseteq C(X)$ by

$$\mathcal{J}_U := \{f \in C(X) : \overline{\{x : f(x) \neq 0\}} \subseteq U\}.$$

It is easy to show that $\mathcal{J}_U = \overline{C_c(U)}$, where $C_c(U)$ is the space of $f \in C(X)$ whose supports

$$\mathrm{supp}(f) := \overline{\{x : f(x) \neq 0\}}$$

are contained in U.[8] We say that U is invariant under the Markov semigroup T_t on $C(X)$ if $T_t(\mathcal{J}_U) \subseteq \mathcal{J}_U$ for all $t \geq 0$, and that T_t is irreducible if $\emptyset$ and X are the only invariant sets.

Problem 13.6.6 Prove that $\mathcal{J}_U$ is an order ideal. Prove also that every order ideal in $C(X)$ is of the form $\mathcal{J}_U$ for some open set U in X. $\square$

Lemma 13.6.7 *If T_t is a Markov semigroup on $C(X)$ and $0 \leq f \in C(X)$ then*

$$U := \bigcup_{t \geq 0} \{x : (T_t f)(x) > 0\}$$

is an invariant set under T_t.

Proof. If $\mathcal{J}$ is the set of all $g \in C(X)$ such that

$$|g| \leq T_{t_1} f + T_{t_2} f + \cdots + T_{t_n} f$$

for some finite sequence $t_r \geq 0$, then an application of (13.4) implies that $\mathcal{J}$ is a linear subspace invariant under T_t. Since $\mathcal{J}_U$ is the norm closure of $\mathcal{J}$ it is also invariant under T_t. $\square$

Theorem 13.6.8 *If $T_t := e^{Zt}$ is an irreducible norm continuous Markov semigroup on $C(X)$, and $f \in C(X)^+$ is not identically zero, then*

$$(T_t f)(x) > 0$$

for all $t > 0$ and all $x \in X$.

Proof. Given f as specified and $t > 0$ put

$$U_n := \{x : \{(Z + \|Z\| I)^n f\}(x) > 0\}$$

and

$$W_t := \{x : (T_t f)(x) > 0\}.$$

---

[8] This is different from our previous definition of support, but in the present context no special measure is identified, and the previous definition is inapplicable.

Since $Z + \|Z\| I \geq 0$, the identity

$$(T_t f)(x) = e^{-\|Z\| t} \sum_{n=0}^{\infty} \{(Z + \|Z\| I)^n f\}(x) t^n / n! \qquad (13.8)$$

implies that $W_t = \bigcup_{n \geq 1} U_n$. Therefore W_t does not depend on t, and we may drop the subscript.

If $g \in C_c(W)$ then there exists α such that $|g| \leq \alpha T_1 f$. Therefore $|T_t g| \leq \alpha T_{t+1} f \in \mathcal{J}_W$. This implies that $T_t(\mathcal{J}_W) \subseteq \mathcal{J}_W$ and hence, by irreducibility, that $W = X$. The theorem now follows immediately. $\quad\square$

Example 13.6.9 Theorem 13.6.8 depends on the norm continuity of the semigroup. If $X := [0, 1]$ with periodic boundary conditions and $(T_t f)(x) := f(x + t)$ then T_t is irreducible but the conclusion of Theorem 13.6.8 is false. $\quad\square$

Our next theorem assumes the conclusion of Theorem 13.6.8, but does not require the semigroup T_t to be norm continuous.

Theorem 13.6.10 *Let T_t be a Markov semigroup on $C(X)$ such that if $f \in C(X)^+$ is not identically zero then $(T_t f)(x) > 0$ for all $x \in X$ and all $t > 0$. If $\{T_t f : t \geq 0\}$ has norm compact closure for all $f \in C(X)$ then it is mixing in the sense that there exists a (necessarily unique) probability measure μ with support X such that*

$$\lim_{t \to \infty} \|T_t f - \langle f, \mu \rangle 1\| = 0 \qquad (13.9)$$

for all $f \in C(X)$.

Proof. If $f \in C(X)^+$ then

$$m(t) := \min\{(T_t f)(x) : x \in X\}$$

is continuous and monotonically increasing with $0 \leq m(t) \leq \|f\|$ for all $t \geq 0$. Let $m := \lim_{t \to \infty} m(t)$ and let g be the norm limit of some sequence $T_{t_n} f$ such that $t_n \to \infty$ as $n \to \infty$. If $t \geq 0$ then

$$\min\{(T_t g)(x) : x \in X\} = \lim_{n \to \infty} \min\{(T_{t+t_n} f)(x) : x \in X\}$$

$$= \lim_{n \to \infty} m(t_n + t)$$

$$= m. \qquad (13.10)$$

We show that this implies that $g(x) = m$ for all $x \in X$. Putting $t = 0$ in (13.10) we see that $g = m1 + h$ for some $h \in C(X)^+$. If h is not identically zero then our hypotheses imply that $(T_t h)(x) > 0$ for all $x \in X$ and hence that $\min\{(T_t g)(x) : x \in X\} > m$ for all $t > 0$. This contradicts (13.10).

We have now proved that the only possible norm limit of any sequence $T_{t_n} f$ is $m1$. The compactness hypothesis implies that $T_t f$ converges in norm to $m1$ as $t \to \infty$. We deduce immediately m depends linearly on f. Indeed

$$m = \int_X f(x)\mu(\mathrm{d}x)$$

where μ is a probability measure on X. Since $m > 0$ for every $f \in C(X)^+$ that is not identically zero, we finally see that the support of μ equals X. $\square$

Example 13.6.11 The compactness condition of the above theorem is satisfied if T_t is compact for all $t > 0$, but it is much weaker than that. Put $X := [0, 1]_{\mathrm{per}}$ and define the Markov semigroup T_t on $C(X)$ by

$$(T_t f)(x) := \mathrm{e}^{-\alpha t} f(x + t) + (1 - \mathrm{e}^{-\alpha t})\langle f, 1 \rangle 1$$

where $\alpha > 0$. One sees immediately that T_t satisfies all the conditions of the theorem. However, the operators T_t are not compact and do not depend norm continuously on t. The generator

$$(Zf)(x) := -\alpha f(x) + f'(x) + \alpha \langle f, 1 \rangle 1$$

of the semigroup is unbounded. $\square$

The remainder of this section is devoted to the study of the peripheral point spectrum. This is defined as the set of all purely imaginary eigenvalues of the generator Z of a Markov semigroup T_t acting on $C(X)$, where X is a compact metric space.[9]

Theorem 13.6.12 *If* $T_t := \mathrm{e}^{Zt}$ *is an irreducible Markov semigroup acting on* $C(X)$, *then the peripheral point spectrum of* Z *is a subgroup of* $i\mathbf{R}$. *Moreover each such eigenvalue has multiplicity* 1.

Proof. Let $Zf = i\alpha f$ where $\alpha \in \mathbf{R}$ and $\|f\| = 1$. Then $T_t f = \mathrm{e}^{i\alpha t} f$ for all $t \geq 0$, so (13.4) implies

$$0 \leq |f| = |T_t f| \leq T_t(|f|) \leq 1.$$

---

[9] The theorem below is a part of the classical Perron-Frobenius theory when X is finite and T_t is replaced by the powers of a single Markov operator. A Banach lattice version may be found in [Schaefer 1974, pp. 329 ff.].

Putting $g := 1 - |f|$ we deduce that $0 \leq T_t g \leq g$ for all $t \geq 0$, so $U := \{x : g(x) > 0\}$ is an invariant set by Lemma 13.6.7. Since $\|f\| = 1$, $U \neq X$, so using the irreducibility hypothesis, we see that $U = \emptyset$; hence $|f(x)| = 1$ for all $x \in X$.

If $h \in C(X)$, $\|h\| = 1$ and $Zh = i\alpha h$ for the same α as above then for any choice of $a \in X$ the eigenfunction $f(a)h - h(a)f$ vanishes at a. By the above argument it must vanish everywhere. Hence h is a multiple of f, and the eigenvalue $i\alpha$ must Have multiplicity 1.

For each $x \in X$ and $t \geq 0$ let $\mu_{x,t}$ be the probability measure on X such that

$$(T_t k)(x) = \int_X k(u) \mu_{x,t}(\mathrm{d}u)$$

for all $k \in C(X)$. If f, α are as above then the identities $|f(x)| = \|f\| = 1$ and

$$\mathrm{e}^{i\alpha t} f(x) = \int_X f(u) \mu_{x,t}(\mathrm{d}u)$$

together imply that $\mu_{x,t}(E) = 1$, where

$$E := \{u \in X : f(u) = \mathrm{e}^{i\alpha t} f(x)\}.$$

For every $h \in C(X)$ we have

$$(T_t(\overline{f}h))(x) = \int_E \overline{f(u)} h(u) \mu_{x,t}(\mathrm{d}u)$$

$$= \int_E \overline{\mathrm{e}^{i\alpha t} f(x)} h(u) \mu_{x,t}(\mathrm{d}u)$$

$$= \mathrm{e}^{-i\alpha t} \overline{f(x)} (T_t h)(x).$$

Now suppose that $h \in \mathrm{Dom}(Z)$ and $Zh = i\beta h$ for some $\beta \in \mathbf{R}$. Since $T_t h = \mathrm{e}^{i\beta t} h$ for all $t \geq 0$ we see that

$$T_t(\overline{f}h) = \mathrm{e}^{i(\beta - \alpha)t} \overline{f} h$$

for all $t \geq 0$. Therefore $\overline{f}h \in \mathrm{Dom}(Z)$ and

$$Z(\overline{f}h) = i(\beta - \alpha)\overline{f}h.$$

This concludes the proof that the peripheral point spectrum of Z is a subgroup of $i\mathbf{R}$. $\square$

Those familiar with Fourier analysis on locally compact abelian groups will see that the following example can be extended to any subgroup Γ of $\mathbf{R}$ provided S^n is replaced by the compact dual group of Γ.

Example 13.6.13 Given $\alpha \in \mathbf{R}^n$ we define the subgroup Γ of $\mathbf{R}$ by

$$\Gamma := \{m \cdot \alpha : m \in \mathbf{Z}^n\}.$$

Putting $S := \{z \in \mathbf{C} : |z| = 1\}$, we also define the one-parameter group T_t on $C(S^n)$ by

$$(T_t f)(z) := f(e^{i\alpha_1 t} z_1, ..., e^{i\alpha_n t} z_n)$$

where $z := (z_1, ..., z_n)$. We claim that the generator Z of T_t has peripheral point spectrum $i\Gamma$.

If $f_m(z) := z_1^{m_1} ... z_n^{m_n}$ for some $m \in \mathbf{Z}^n$ and all $z \in S^n$ then

$$T_t f_m = e^{itm \cdot \alpha} f_m$$

for all $t \in \mathbf{R}$. Therefore $f_m \in \mathrm{Dom}(Z)$ and

$$Z f_m = im \cdot \alpha f_m.$$

This proves that $i\Gamma$ is contained in the peripheral point spectrum of Z.

Conversely suppose that $T_t f = e^{i\sigma t} f$ for all $t \in \mathbf{R}$, where $f \in C(S^n)$ is not identically zero. Since T_t may be extended to a one-parameter unitary group on $L^2(S^n)$, we have

$$e^{i\sigma t} \langle f, f_m \rangle = \langle T_t f, f_m \rangle = \langle f, T_{-t} f_m \rangle = \langle f, e^{-itm \cdot \alpha} f_m \rangle = e^{itm \cdot \alpha} \langle f, f_m \rangle$$

for all $m \in \mathbf{Z}^n$. Since $\{f_m : m \in \mathbf{Z}^n\}$ is a complete orthonormal set in $L^2(S^n)$, we deduce that $\sigma = m \cdot \alpha$ for some $m \in \mathbf{Z}^n$. Note also that $\langle f, f_r \rangle = 0$ unless $r \cdot \alpha = m \cdot \alpha$. If $\alpha_1, ..., \alpha_n$ are rationally independent this implies that $\langle f, f_r \rangle = 0$ unless $r = m$. $\square$

We conclude the section with two recent results whose technical assumptions differ slightly from those elsewhere in this section.[10]

Theorem 13.6.14 *(Davies) Let X be a locally compact, metrizable space, and let $\mathrm{d}x$ be a Borel measure with support equal to X. Let $T_t := e^{Zt}$ be a positive one-parameter semigroup acting on $L^p(X, \mathrm{d}x)$ for some $p \in [1, \infty)$ and assume the Feller property $T_t(L^p(X)) \subseteq C(X)$ for all $t > 0$. Then the peripheral point spectrum of Z cannot contain any non-zero points.*

Corollary 13.6.15 *Let X be a countable set and let $T_t := e^{Zt}$ be a positive one-parameter semigroup acting on $l^p(X)$ for some $p \in [1, \infty)$. Then the peripheral point spectrum of Z cannot contain any non-zero points.*

---

[10] We refer to [Davies 2005C] for the proofs and to [Keicher 2006] for a generalization of the results to atomic Banach lattices.

14

NSA Schrödinger operators

14.1 Introduction

There is an extensive literature on the spectral theory of self-adjoint Schrödinger operators, motivated by their applications in quantum theory and other areas of mathematical physics. The subject has been dominated by three techniques: the spectral theorem, the use of variational methods for estimating eigenvalues, and theorems related to scattering theory.

By comparison, the non-self-adjoint (NSA) theory is in its infancy. Attempts to carry over techniques from the self-adjoint theory have had a limited success, but numerical experiments have shown that the NSA theory has crucial differences. Natural NSA analogues of self-adjoint theorems often turn out to be false, and recent studies have revealed new and unexpected phenomena. This chapter reveals some of the results in this field. Because the subject is so new, we mostly confine attention to the one-dimensional theory, in which special techniques allow some progress to be made. Many of the results in this chapter were discovered after 1990, and we make no claim that they have reached their final form. In some cases we do not give complete proofs.

By a NSA Schrödinger operator we will mean an operator of the form

$$(H_p f)(x) := (H_0 f)(x) + V(x) f(x) \tag{14.1}$$

acting in $L^p(\mathbf{R}^N)$, where $1 \leq p < \infty$, $H_0 f := -\Delta f$ and V is a complex-valued potential. The precise domain of the operators will be specified in each section.

One can also study Schrödinger operators with NSA boundary conditions.

Example 14.1.1 Let $Hf(x) := -f''(x)$ act in $L^2(0, \infty)$ subject to the boundary condition $f'(0) + cf(0) = 0$, where c is a complex constant. The only possible eigenvalue of H is $-c^2$, the corresponding eigenfunction being $f(x) := -e^{-cx}$. This lies in $L^2(0, \infty)$ if and only if $\text{Re}(c) > 0$. Theorem 11.3.3 implies that

$$\text{Spec}(H) = \begin{cases} [0, \infty) & \text{if } \text{Re}(c) \leq 0, \\ [0, \infty) \cup \{-c^2\} & \text{if } \text{Re}(c) > 0. \end{cases}$$

Note that the single complex eigenvalue $-c^2$ is absorbed into the positive real axis as c approaches the imaginary axis from the right.

If $\text{Re}(\lambda) > 0$ and $\lambda \neq c$ then the method of Example 11.2.8 yields

$$((\lambda^2 I + H)^{-1} f)(x) = \int_0^\infty G(x, y) f(y) \, dy$$

where

$$G(x, y) := \begin{cases} w^{-1} \phi(x) \psi(y) & \text{if } x \leq y, \\ w^{-1} \psi(x) \phi(y) & \text{if } y \leq x, \end{cases}$$

and

$$\phi(x) := (\lambda - c) e^{\lambda x} + (\lambda + c) e^{-\lambda x},$$

$$\psi(x) := e^{-\lambda x},$$

$$w := 2\lambda(\lambda - c).$$

The Green function G satisfies the conditions of Corollary 2.2.19 and hence determines a bounded operator on $L^2(0, \infty)$. $\square$

14.2 Bounds on the numerical range

One way of controlling the spectrum of a NSA operator is by using the fact that under fairly weak conditions it is contained in the closure of the numerical range. For bounded operators this is proved in Theorem 9.3.1, while for unbounded operators it is a consequence of Lemma 9.3.14. The main hypothesis of this lemma can often be proved by using Theorem 11.5.1 or Corollary 11.5.2. In this section we concentrate on bounding the numerical range itself.

We assume that H is a non-negative self-adjoint operator acting in a Hilbert space $\mathcal{H}$ and that $\mathcal{V}_R$ is a real vector space of symmetric operators on $\mathcal{H}$, each of which has domain containing $\text{Dom}(H)$. We also suppose that every $V \in \mathcal{V}_R$ satisfies a bound of the form

$$-c(V)\langle f, f \rangle \leq \langle Vf, f \rangle + \langle Hf, f \rangle$$

for all $f \in \mathrm{Dom}(H)$. If $V \in \mathcal{V}_\mathbf{C} := \mathcal{V}_\mathbf{R} + i\mathcal{V}_\mathbf{R}$ then it follows immediately (see below) that the numerical range of $H + V$, which includes all of its eigenvalues, satisfies

$$\mathrm{Num}(H + V) \subseteq \{z : \mathrm{Re}(z) \geq -c(\mathrm{Re}(V))\}.$$

Our goal is to obtain sharper bounds on the numerical range of $H + V$.[1]

Theorem 14.2.1 *If $-\pi/2 < \theta < \pi/2$ and $x + iy \in \mathrm{Num}(H + V)$ then*

$$x\cos(\theta) - y\sin(\theta) \geq -\cos(\theta)c(V_\theta)$$

where

$$V_\theta := \mathrm{Re}(e^{i\theta}V/\cos(\theta))$$
$$= \mathrm{Re}(V) - \tan(\theta)\mathrm{Im}(V).$$

Proof. If $\|f\| = 1$ and $\langle (H + V)f, f \rangle = x + iy$ then

$$\mathrm{Re}\{e^{i\theta}(x + iy)\} = \mathrm{Re}\{e^{i\theta}\langle Vf, f \rangle\} + \mathrm{Re}\{e^{i\theta}\langle Hf, f \rangle\}$$
$$= \cos(\theta)\{\langle V_\theta f, f \rangle + \langle Hf, f \rangle\}$$
$$\geq -\cos(\theta)c(V_\theta). \qquad \square$$

Corollary 14.2.2 $\mathrm{Num}(H + V)$ *is contained in the region on or inside the envelope of the family of lines*

$$x\cos(\theta) - y\sin(\theta) = -\cos(\theta)c(V_\theta) \qquad (14.2)$$

where $-\pi/2 < \theta < \pi/2$.

Proof. The intersection of the half planes is the region on or inside the envelope of the lines. $\quad \square$

Example 14.2.3 One can determine the envelope explicitly if $\mathcal{H} := L^2(X, \mathrm{d}x)$ and $\mathcal{V}_\mathbf{R}$ is a vector space of real-valued functions on X, regarded as multiplication operators (i.e. potentials). We also assume that $c(W/s) := k(|W|)/s^\gamma$

---

[1] An early version of the ideas in this section, for bounded potentials only, appeared in [Abramov et al. 2001]. The theorems here are abstracted from [Frank et al. 2006], where the authors also prove a Lieb-Thirring bound for Schrödinger operators with complex potentials. The application of Lieb-Thirring bounds to self-adjoint Schrödinger operators is an important subject, discussed in [Weidl 1996, Laptev and Weidl 2000].

for some $\gamma > 1$, all $W \in \mathcal{V}_{\mathbf{R}}$ and all $s > 0$. We assume that $0 \leq k(W_1) \leq k(W_2)$ if $0 \leq W_1 \leq W_2 \in \mathcal{V}_{\mathbf{R}}$. These conditions hold if $X := \mathbf{R}^N$ and

$$c(V) = \left\{ \int_{\mathbf{R}^N} |V(x)|^p \, \mathrm{d}x \right\}^\alpha$$

for suitable positive p, α, a property that is commonplace in the theory of Schrödinger operators. $\quad\square$

Theorem 14.2.4 *Under the assumptions of Example 14.2.3,* $\mathrm{Num}(H + V)$ *is contained on or inside the curve given parametrically by*

$$x = k(|V|) \cos(\theta)^{-\gamma} \left\{ (\gamma - 1) \sin(\theta)^2 - \cos(\theta)^2 \right\},$$

$$y = k(|V|) \gamma \cos(\theta)^{1-\gamma} \sin(\theta),$$

where $-\pi/2 < \theta < \pi/2$.

Proof. It follows by Corollary 14.2.2 that we need to find the envelope of the lines

$$x \cos(\theta) - y \sin(\theta) = -\cos(\theta)^{1-\gamma} k(|V|) \tag{14.3}$$

where $-\pi/2 < \theta < \pi/2$. This is obtained by solving the simultaneous equations

$$x \cos(\theta) - y \sin(\theta) = -\cos(\theta)^{1-\gamma} k(|V|),$$

$$x \sin(\theta) + y \cos(\theta) = (\gamma - 1) \cos(\theta)^{-\gamma} \sin(\theta) k(|V|). \qquad\square$$

<u>Note 1</u> The envelope of Theorem 14.2.4 crosses the x-axis at $x = -k(|V|)$ and the y-axis at

$$y = \pm k(|V|) \gamma^{\gamma - 1/2} (\gamma - 1)^{1-\gamma}.$$

<u>Note 2</u> By putting $\cos(\theta) = \delta > 0$ and letting $\delta \to 0$, we obtain the asymptotic form

$$y \sim \pm a x^{1 - 1/\gamma}$$

of the envelope as $x \to +\infty$, where $a > 0$ may be computed explicitly.

<u>Note 3</u> The case $\gamma = 1$ may be treated similarly. The envelope is the semicircle $|x + iy| = k(|V|)$, $x \leq 0$ together with the two lines $x \geq 0$, $y = \pm k(|V|)$. This case is applicable when dealing with bounded functions V, with $c(V) := \|V\|_\infty$.

<u>Note 4</u> If $Hf := -f''$ acting in $L^2(\mathbf{R})$ and $\mathcal{V}_{\mathbf{R}} := L^1(\mathbf{R})$ then the method of proof of Theorem 14.3.1 implies that

$$c(V) = \|V\|_1^2 / 4.$$

Numerical range methods cannot lead to the optimal bound

$$\text{Spec}(H+V) \subseteq [0, +\infty) \cup \{z : |z| \leq \|V\|_1^2/4\}$$

of Theorem 14.3.1 and Corollary 14.3.11. The numerical range is always convex, and so arguments using it cannot imply the non-existence of eigenvalues with large real parts.

Problem 14.2.5 Test the sharpness of the bounds on the numerical range proved in Note 2 by evaluating

$$z(\delta, \alpha) := \frac{\langle (H+V)f, f \rangle}{\langle f, f \rangle}$$

for all $\delta \in (0, 1)$ and all $\alpha > 0$, where

$$V(x) := i(2\delta)^{-1}|x|^{\delta-1}e^{-x^2}$$

satisfies $\|V\|_1 \leq 1$ for all $\delta \in (0, 1)$ and

$$f(x) := e^{-\alpha x^2/2}$$

lies in $\text{Dom}(H)$ for all $\alpha > 0$.   □

14.3 Bounds in one space dimension

The spectral bounds of the last section can be improved dramatically for Schrödinger operators in one dimension. We start by assuming that

$$(H_2 f)(x) := (H_0 f)(x) + V(x)f(x) \tag{14.4}$$

acts in $L^2(\mathbf{R})$, where $H_0 f := -f''$ and the complex potential V lies in $L^1(\mathbf{R}) \cap L^2(\mathbf{R})$. The condition $V \in L^2(\mathbf{R})$ implies that V is a relatively compact perturbation of H_0 by Theorem 11.2.11. The same theorem implies that the spectrum of H_2 is equal to $[0, \infty)$ together with eigenvalues which can only accumulate on the non-negative real axis or at infinity. Moreover the domain of H_2 is equal to the Sobolev space $W^{2,2}(\mathbf{R})$ by Example 3.2.2, and this is contained in $C_0(\mathbf{R})$ by Theorem 3.2.1.

Theorem 14.3.1[2] *If $V \in L^1(\mathbf{R}) \cap L^2(\mathbf{R})$ then every eigenvalue λ of the*

---

[2] This theorem was proved in the stated form in [Abramov et al. 2001]. It was subsequently extended by two very different arguments to all $V \in L^1(\mathbf{R})$ in [Brown and Eastham 2002] and [Davies and Nath 2002]. See Corollary 14.3.11 below.

Schrödinger operator H which does not lie on the positive real axis satisfies

$$|\lambda| \le \|V\|_1^2/4. \tag{14.5}$$

Hence

$$\mathrm{Spec}(H) \subseteq [0, \infty) \cup \{z \in \mathbf{C} : |z| \le \|V\|_1^2/4\}.$$

Proof. Let $\lambda := -z^2$ be an eigenvalue of H_2 where $\mathrm{Re}(z) > 0$, and let f be the corresponding eigenfunction, so that $f \in W^{2,2}(\mathbf{R}) \subseteq C_0(\mathbf{R})$. Then

$$(H_0 + z^2)f = -Vf$$

so

$$-f = (H_0 + z^2)^{-1}Vf.$$

Putting $X := |V|^{1/2}$, $W := V/X$ and $g := Wf \in L^2$, we deduce that

$$-g = W(H_0 + z^2)^{-1}Xg$$

so

$$-1 \in \mathrm{Spec}(W(H_0 + z^2)^{-1}X).$$

We complete the proof by estimating the Hilbert-Schmidt norm of this operator, whose kernel is

$$W(x)\frac{e^{-z|x-y|}}{2z}X(y).$$

We have

$$1 \le \|W(H_0 + z^2)^{-1}X\|_2^2$$

$$= (4|z|^2)^{-1} \int_{\mathbf{R}^2} |W(x)|^2 e^{-2\mathrm{Re}(z)|x-y|}|X(y)|^2 \, dxdy$$

$$\le (4|z|^2)^{-1}\|V\|_1^2.$$

Therefore $|z| \le \|V\|_1^2/4$. □

Note that the same bounds can be proved for the corresponding operator on the half-line subject to either Neumann or Dirichlet boundary conditions, by minor adjustments to the proof.

Example 14.3.2 One may show that the constant $1/4$ in (14.5) is sharp by evaluating an eigenvalue of a well-known Schrödinger operator.[3]

---

[3] Depending on the choice of a and b, the differential operator may have other eigenvalues. See [Langmann et al. 2006] for a complete analysis in the self-adjoint case.

We consider the differential equation

$$-f''(x) + V(x)f(x) = \lambda f(x)$$

on **R** where V is the Pöschl-Teller potential

$$V(x) := -\frac{b(b+1)a^2}{\{\cosh(ax)\}^2}$$

with $a > 0$ and $\mathrm{Re}(b) > 0$. A direct calculation shows that the function

$$f(x) := \{\cosh(ax)\}^{-b}$$

lies in $\mathcal{S} \subseteq \mathrm{Dom}(H)$ and that it satisfies the equation with $\lambda := -b^2 a^2$. Moreover

$$|\lambda| = \frac{\|V\|_1^2}{4|b+1|^2}.$$

Letting $b \to 0$ subject to the constraint $\mathrm{Re}(b) > 0$, one sees that the constant $1/4$ is sharp. Further examination of the calculation shows that the eigenvalue λ may be as close as one likes to any point on the circle $\{z : |z| = \|V\|_1^2/4\}$. $\square$

Those who are familiar with the theory of self-adjoint Schrödinger operators might expect that H has only a finite number of isolated eigenvalues under the hypotheses, but in the NSA case this need not be true. Relative compactness of the potential only implies that any accumulation point of the eigenvalues must lie on the non-negative real axis. Another difference with the self-adjoint theory is that eigenvalues do not only appear or disappear at the origin as the potential varies. They may emerge from or be absorbed into points on the positive real axis.

In Corollary 14.3.11 we obtain the bound (14.5) assuming only that $V \in L^1(\mathbf{R})$. Our strategy is to regard H as acting in $L^1(\mathbf{R})$, and then to obtain the L^2 result by interpolation at the end of the main theory. We do not specify the domain of H acting as an operator in $L^2(\mathbf{R})$ or use the theory of quadratic forms. Our main goal is not to remove an unnecessary technical condition but to obtain explicit bounds on the eigenvalues for a much larger class of potentials.

We say that the complex-valued potential V lies in $\mathcal{V}$ if V has a decomposition $V = W + X$ where $W \in L^1(\mathbf{R})$ and X lies in the space $L_0^\infty(\mathbf{R})$ of all bounded measurable functions on **R** which vanish at infinity. For every $V \in \mathcal{V}$ many such decompositions exist. Roughly speaking $V \in \mathcal{V}$ if V is locally L^1 and it decays to zero at infinity.

We consider the operator $H_1 := H_0 + V$ acting in $L^1(\mathbf{R})$, where $V \in \mathcal{V}$. It follows from the bounds below that H_1 is a densely defined operator with the same domain as H_0 in $L^1(\mathbf{R})$. We will use the notation $\|A\|_p$ to denote the norm of any operator A acting on $L^p(\mathbf{R})$.

The next two lemmas will be used to determine the essential spectrum of the operator H_1.

Lemma 14.3.3 *If K is a uniformly continuous, bounded function on $\mathbf{R}$ then the operator S defined by*

$$(Sf)(x) := \int_{\mathbf{R}} K(x-y) f(y) \, dy$$

is compact from $L^1(\mathbf{R})$ to $C[a, b]$ for any finite a, b.

Proof. We note that for $f \in L^1(\mathbf{R})$ and $x \in [a, b]$

$$|(Sf)(x)| \leq \int_{\mathbf{R}} |K(x-y)| |f(y)| \, dy$$

$$\leq \|K\|_\infty \|f\|_1.$$

This implies that $\|S\| \leq \|K\|_\infty < \infty$. If we show that $\{Sf : \|f\|_1 \leq 1\}$ is an equicontinuous family of functions then by using the Arzela-Ascoli Theorem 4.2.7 we may conclude that the operator S is compact.

Let $\varepsilon > 0$. By the uniform continuity of K there exists a $\delta > 0$ such that $|K(u) - K(v)| < \varepsilon$ whenever $|u - v| < \delta$. For such u, v we conclude that

$$|(Sf)(u) - (Sf)(v)| \leq \int_{\mathbf{R}} |K(u-y) - K(v-y)| |f(y)| \, dy$$

$$< \varepsilon. \qquad \square$$

Lemma 14.3.4 *If $V \in L^1(\mathbf{R})$ and K is uniformly continuous and bounded on $\mathbf{R}$ then the operator T defined by*

$$(Tf)(x) := \int_{\mathbf{R}} V(x) K(x-y) f(y) \, dy$$

is compact from $L^1(\mathbf{R})$ to $L^1(\mathbf{R})$.

Proof. We regard $C[-n, n]$ as a subspace of $L^1(\mathbf{R})$ by putting every function in the former space equal to 0 outside $[-n, n]$. We put

$$V_n(x) := \begin{cases} V(x) & \text{if } |x| \leq n, \\ 0 & \text{otherwise.} \end{cases}$$

We put $T_n := V_n S_n$, where $S_n : L^1(\mathbf{R}) \to C[-n, n]$ is given by

$$(S_n f)(x) := \int_{\mathbf{R}} K(x - y) f(y) \, \mathrm{d}y.$$

An application of Lemma 14.3.3 implies that T_n is a compact operator from $L^1(\mathbf{R})$ to $L^1(\mathbf{R})$. Since $\|T_n - T\|_1 \to 0$ as $n \to \infty$ we conclude that T is a compact operator on $L^1(\mathbf{R})$. $\square$

Theorem 14.3.5 *The essential spectrum of the operator H_1 is $[0, \infty)$ for every $V \in \mathcal{V}$.*

Proof. Let $\lambda > 0$. Given $\varepsilon > 0$ we may write $V := W_\varepsilon + X_\varepsilon$ where $W_\varepsilon \in L^1(\mathbf{R})$ and $\|X_\varepsilon\|_\infty < \varepsilon$. This implies that

$$\|V(H_0 + \lambda^2)^{-1} - W_\varepsilon(H_0 + \lambda^2)^{-1}\|_1 \leq \varepsilon \|(H_0 + \lambda^2)^{-1}\|_1.$$

Letting $\varepsilon \to 0$ we deduce by approximation that $V(H_0 + \lambda^2)^{-1}$ is compact. The proof is completed by applying Theorem 11.2.6 and Example 8.4.5. $\square$

To determine the spectrum of the operator H_1 completely we need only find its discrete eigenvalues. This is a numerical problem, but useful bounds are provided by Theorem 14.3.9. These bounds are expressed in terms of a function F on $(0, \infty)$, which is defined by

$$F(s) := \sup_{y \in \mathbf{R}} \left\{ \int_{\mathbf{R}} |V(x)| \, \mathrm{e}^{-s|x-y|} \, \mathrm{d}x \right\}. \tag{14.6}$$

Lemma 14.3.6 *$F(s)$ is a positive, decreasing, convex function of s for $s > 0$. It is bounded if and only if $V \in L^1(\mathbf{R})$, in which case*

$$\lim_{s \to 0+} F(s) = \|V\|_1.$$

The equation $F(s) = 2s$ has a unique solution.

Proof. It is clear that $F(s)$ is a positive decreasing function of s for $s > 0$. We next observe that for a fixed value of y and compactly supported $V \in L^1(\mathbf{R})$ the integral

$$\int_{\mathbf{R}} |V(x)| \, \mathrm{e}^{-s|x-y|} \, \mathrm{d}x$$

is a convex function of s; this can be proved by differentiating twice under the integral sign with respect to s. Convexity with respect to s for each fixed y then follows for all $V \in \mathcal{V}$ by an approximation argument. Finally convexity is preserved on taking the supremum with respect to y.

The behaviour of $F(s)$ as $s \to 0+$ follows directly from its definition. The last statement of the theorem may be seen by inspecting the graphs of $F(s)$ and $2s$. $\square$

Problem 14.3.7 Prove that if $|V(x)|$ is a decreasing function of $|x|$ then

$$F(s) = 2 \int_0^\infty |V(x)| \, e^{-xs} \, dx. \qquad \square$$

Lemma 14.3.8 *If* $\mathrm{Re}(\lambda) > 0$ *then*

$$\| V(H_0 + \lambda^2)^{-1} \|_1 = \frac{F(\mathrm{Re}(\lambda))}{2|\lambda|}.$$

Proof. We have

$$V(H_0 + \lambda^2)^{-1} f(x) = \int_{\mathbf{R}} K(x, y) f(y) \, dy$$

where

$$K(x, y) := V(x) \frac{e^{-\lambda|x-y|}}{2\lambda}.$$

Theorem 2.2.5 now implies that

$$\| V(H_0 + \lambda^2)^{-1} \|_1 = \sup_{y \in \mathbf{R}} \left\{ \int_{\mathbf{R}} |K(x, y)| \, dx \right\}$$

$$= \frac{F(\mathrm{Re}(\lambda))}{2|\lambda|}. \qquad \square$$

Theorem 14.3.9 *Let* $z := -\lambda^2$, *where* $\lambda := \lambda_1 + i\lambda_2$, $\lambda_1 > 0$ *and* $\lambda_2 \in \mathbf{R}$. *If* z *is an eigenvalue of the Schrödinger operator* H_1 *then*

$$|\lambda_2| \le \sqrt{F(\lambda_1)^2/4 - \lambda_1^2} \qquad (14.7)$$

and $0 < \lambda_1 \le \mu$, *where* $\mu > 0$ *is determined by* $F(\mu) = 2\mu$. *Therefore*

$$\mathrm{Spec}(H_1) \subseteq [0, \infty) \cup$$

$$\left\{ -(\lambda_1 + i\lambda_2)^2 \in \mathbf{C} : 0 < \lambda_1 \le \mu \text{ and } |\lambda_2| \le \sqrt{F(\lambda_1)^2/4 - \lambda_1^2} \right\}.$$

In this estimate one may replace F *by any upper bound of* F.

Proof. We proceed by contradiction. If (14.7) is false then $F(\lambda_1) < 2|\lambda|$, so $\|V(H_0 + \lambda^2 I)^{-1}\|_1 < 1$ by Lemma 14.3.8. The resolvent formula

$$(H_1 + \lambda^2 I)^{-1} = (H_0 + \lambda^2 I)^{-1} \left(I + V(H_0 + \lambda^2 I)^{-1}\right)^{-1}$$

now implies that $-\lambda^2 \notin \mathrm{Spec}(H_1)$.   □

We now extend the above results from $L^1(\mathbf{R})$ to $L^p(\mathbf{R})$ for $1 < p < \infty$. Theorem 11.4.13 implies that there is a one-parameter semigroup T_t on $L^1(\mathbf{R})$ whose generator is $-H_1$. Moreover this semigroup may be extended compatibly to $L^p(\mathbf{R})$ for all $1 \leq p < \infty$; we denote the generators of the corresponding semigroups by $-H_p$. We regard H_0 as acting in any of the L^p spaces without explicitly indicating this.

Theorem 14.3.10 *The essential spectrum of H_p equals $[0, \infty)$ for all $1 \leq p < \infty$. In addition the spectrum of H_p does not depend on p.*

Proof. We first observe that for all large enough $\lambda > 0$ the formula

$$C_1 := (H_0 + \lambda^2 I)^{-1} - (H_1 + \lambda^2 I)^{-1}$$
$$= (H_1 + \lambda^2 I)^{-1} V(H_0 + \lambda^2 I)^{-1}$$

defines a compact operator on $L^1(\mathbf{R})$. The formula

$$C_p := (H_0 + \lambda^2 I)^{-1} - (H_p + \lambda^2 I)^{-1}$$

defines a family of consistent bounded operators on $L^p(\mathbf{R})$, and Theorem 4.2.14 implies that they are all compact. It follows by Theorem 11.2.6 or Corollary 11.2.3 that the essential spectrum of H_p equals that of H_0. Example 8.4.5 implies that this equals $[0, \infty)$.

The proof that the non-essential parts of the spectra of H_p do not depend on p uses the same argument as Theorem 4.2.15.   □

Theorem 14.3.10 allows us to apply any spectral results proved in $L^1(\mathbf{R})$ to other L^p spaces, and from now on we do this freely; we also drop the subscript on H_p.

Corollary 14.3.11[4] *If $V \in L^1(\mathbf{R})$ and z is an eigenvalue of $H := H_0 + V$ then either $z \geq 0$ or*

$$|z| \leq \frac{\|V\|_1^2}{4}.$$

---

[4] See the historical comments on Theorem 14.3.1 above.

Proof. Lemma 14.3.6 implies that

$$F(s) \leq \|V\|_1$$

and we substitute this into (14.7). $\quad\square$

Corollary 14.3.12 *Let* $V \in L^p(\mathbf{R})$ *where* $1 < p < \infty$, *and put* $k := (2/q)^{1/q}\|V\|_p$ *where* $1/p + 1/q = 1$. *If* $z := -\lambda^2$ *is an eigenvalue of* H_1, *where* $\lambda := \lambda_1 + i\lambda_2$, $\lambda_1 > 0$ *and* $\lambda_2 \in \mathbf{R}$, *then* $\lambda_1 \leq (k/2)^{q/(q+1)}$ *and*

$$|\lambda_2| \leq \sqrt{\frac{k^2}{4}\lambda_1^{-2/q} - \lambda_1^2}.$$

Proof. We note that

$$
\begin{aligned}
F(s) &= \sup_{y \in \mathbf{R}} \int_{\mathbf{R}} |V(x)| e^{-s|x-y|} \, \mathrm{d}x \\
&\leq \|V\|_p \left(\int_{\mathbf{R}} e^{-s|t|q} \, \mathrm{d}t \right)^{1/q} \\
&= \|V\|_p \left(\frac{2}{qs} \right)^{1/q} \\
&= ks^{-1/q}.
\end{aligned}
$$

We insert this estimate into Theorem 14.3.9 to obtain the result. To obtain the value of μ we simply solve $\frac{k^2}{4}\mu^{-2/q} - \mu^2 = 0$. $\quad\square$

Problem 14.3.13 Compute the function F when $V(x) := ce^{-|x|}$ and c is a complex constant. Compare the conclusions of Theorem 14.3.9 and Corollary 14.3.11 for this example. $\quad\square$

Example 14.3.14 If $V(x) := c|x|^{a-1}$ where $0 < a < 1$ and c is a complex constant such that $|c| = 1$ then

$$F(s) = 12\Gamma(a)s^{-a}$$

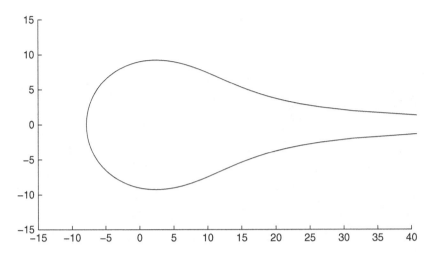

Figure 14.1: Bounds on the spectrum of $H_0 + V$ in Example 14.3.14

for all $s > 0$. The curve (14.7) bounding the spectrum of $H_0 + V$ is given in polar coordinates by

$$r = \Gamma(a)^{2/(1+a)} \{\sin(\theta/2)\}^{-2a/(1+a)}$$

where $0 < \theta < 2\pi$. The curve is depicted in Figure 14.1 for $a = 1/4$. $\square$

14.4 The essential spectrum of Schrödinger operators

In this section we consider Schrödinger operators $H := H_0 + V$ acting on $L^2(\mathbf{R}^N)$ for potentials that do not vanish at infinity. The natural domain of $H_0 := -\Delta$ is $W^{2,2}(\mathbf{R}^N)$ by Examples 3.2.2 and 6.3.5; its spectrum is $[0, \infty)$ by Examples 3.2.2 or 8.4.5. Throughout this section we assume for simplicity that V is a bounded, complex-valued potential, so that the results in Section 11.1 are applicable.

We are interested in examining the essential spectrum of H. Our first lemma states that it depends only on the asymptotic behaviour of the potential at infinity.[5]

---

[5] One may prove that the difference of the two resolvents is compact under much weaker hypotheses by using Dirichlet-Neumann bracketing, a technique discussed at length in [Reed and Simon 1979B, Sect. XIII.15], or the twisting trick of [Davies and Simon 1978], or the Enss approach to scattering theory in [Davies 1980A, Perry 1983]. These techniques also allow the invariance of the absolutely continuous and singular continuous spectrum under local perturbations to be proved.

Lemma 14.4.1 *If $H_i := H_0 + V_i$ for $i = 1, 2$ where V_i are two bounded potentials on $\mathbf{R}_N$ satisfying $\lim_{|x| \to \infty} |V_1(x) - V_2(x)| = 0$, then*

$$\mathrm{EssSpec}(H_1) = \mathrm{EssSpec}(H_2).$$

Proof. The proof is related to that of Theorem 11.2.11. We first observe that $H_2 = H_1 + W$ where W is bounded and vanishes at infinity. If $z \notin \mathrm{Spec}(H_1) \cup \mathrm{Spec}(H_0)$ then

$$W(H_1 - zI)^{-1} = AB$$

where $A := W(H_0 - zI)^{-1}$ is compact by Problem 5.7.4 and $B := (H_0 - zI)(H_1 - zI)^{-1}$ is bounded. We deduce that H_1 and H_2 have the same essential spectrum by Theorem 11.2.6. $\square$

Theorem 14.4.2 *Let*

$$\lim_{n \to \infty} V(x - a_n) = W(x)$$

for all $x \in \mathbf{R}^N$, where $a_n \in \mathbf{R}^N$ and $\lim_{n \to \infty} |a_n| = \infty$. Then

$$\mathrm{Spec}(H_0 + W) \subseteq \mathrm{EssSpec}(H_0 + V).$$

Proof. We first observe that the standard norm $\vert\!\vert\!\vert \cdot \vert\!\vert\!\vert$ on $W^{2,2}(\mathbf{R}^N)$, defined in (3.6), is equivalent to the domain norms of the operators H_0, $H_0 + V$ and $H_0 + W$ as defined in Problem 6.1.1.

If $z \in \mathrm{Spec}(H_0 + W)$ then an obvious modification of the proof of Lemma 1.2.13 implies that there exists a sequence $\phi_n \in \mathrm{Dom}(H_0)$ such that $\vert\!\vert\!\vert \phi_n \vert\!\vert\!\vert = 1$ and either $\|(H_0 + W)\phi_n - z\phi_n\| < 1/n$ for all n or $\|(H_0 + W)^*\phi_n - \bar{z}\phi_n\| < 1/n$ for all n. The two cases are treated in the same way, so we only consider the first.

If $U_r : L^2(\mathbf{R}^N) \to L^2(\mathbf{R}^N)$ are defined by $(U_r f)(x) := f(x + a_r)$ then U_r are unitary and

$$U_r^{-1}(H_0 + V)U_r = H_0 + V_r$$

where $V_r(x) := V(x - a_r)$. Our assumptions imply that

$$\lim_{r \to \infty} (H_0 + V_r)f = (H_0 + W)f$$

for all $f \in \mathrm{Dom}(H_0)$.

Let $\{e_s\}_{s=1}^\infty$ be a complete orthonormal sequence in $\mathrm{Dom}(H_0) = W^{2,2}(\mathbf{R}^N)$ for the standard inner product

$$\langle g, h \rangle_0 := \langle g, h \rangle + \langle H_0 g, H_0 h \rangle$$

on that space. For each n the sequence $U_r \phi_n$ converges weakly to 0 in $W^{2,2}(\mathbf{R}^N)$ as $r \to \infty$. Moreover

$$\lim_{r \to \infty} \|(H_0 + V)U_r \phi_n - z U_r \phi_n\| = \lim_{r \to \infty} \|(H_0 + V_r)\phi_n - z\phi_n\|$$

$$= \|(H_0 + W)\phi_n - z\phi_n\|$$

$$< 1/n.$$

Therefore there exists $r(n)$ such that $|\langle U_{r(n)}\phi_n, e_s\rangle_0| < 1/n$ for all $s = 1, \ldots, n$ and $\|(H_0 + V)U_{r(n)}\phi_n - z U_{r(n)}\phi_n\| < 1/n$. Putting $\psi_n := U_{r(n)}\phi_n$ we deduce that $\|\psi_n\| = 1$, ψ_n converges weakly to 0 in $W^{2,2}(\mathbf{R}^N)$ and $\|(H_0 + V)\psi_n - z\psi_n\| \to 0$ as $n \to \infty$. Therefore $z \in \mathrm{EssSpec}(H_0 + V)$ by Lemma 11.2.1. $\quad\square$

Example 14.4.3 Let $H := H_0 + V$, where V is bounded and periodic in the sense that $V(x + a) = V(x)$ for all $x \in \mathbf{R}^N$ and all $a \in \mathbf{Z}^N$. It follows immediately from Theorem 14.4.2 that the spectrum of H coincides with its essential spectrum. Even in the self-adjoint case (i.e. when V is real-valued) this does not imply that H has no eigenvalues: it seems possible that H might have an eigenvalue of infinite multiplicity. The proof that this cannot occur is very difficult, particularly if one includes a periodic magnetic field term in the operator. The spectral analysis of non-self-adjoint, periodic Schrödinger operators involves surprising difficulties even in one dimension.[6] $\quad\square$

Our next example has been examined in great detail in several dimensions in the self-adjoint case, with a full description of its associated scattering theory. The inclusion in (14.8) is actually an equality, as in Theorem 4.4.6.[7]

Example 14.4.4 Let $H := H_0 + V$ in $L^2(R)$, where $\lim_{n \to \infty} V(x \mp n) = W_\pm(x)$ and $W_\pm$ are complex-valued, bounded, periodic potentials. (n is supposed to be a positive integer.) Then

$$\mathrm{EssSpec}(H) \supseteq \mathrm{Spec}(H_0 + W_+) \cup \mathrm{Spec}(H_0 + W_-). \tag{14.8}$$

$\square$

---

[6] See [Kuchment 2004, Sobolev 1999] for further information about self-adjoint magnetic Schrödinger operators. Non-self-adjoint periodic Schrödinger operators in one dimension were analyzed in [Gesztesy and Tkachenko 2005], who determined the conditions under which they are of scalar type in the sense of [Dunford and Schwartz 1971, Theorem XV.6.2]; the reader might also look at the simpler Problem 4.4.11.

[7] See [Davies and Simon 1978] for a proof of this and other related results.

Problem 14.4.5 Let H be the operator with domain $W^{2,2}(\mathbf{R})$ in $L^2(\mathbf{R})$ defined by

$$(Hf)(x) = -f''(x) + i\sin(|x|^\alpha)f(x)$$

where $0 < \alpha < 1$. Use Theorems 9.3.1 and 14.4.2 to prove that

$$\mathrm{Spec}(H) = \mathrm{EssSpec}(H) = \{x + iy : x \geq 0 \text{ and } |y| \leq 1\}.$$

State and prove a generalization for bounded, complex-valued potentials that are slowly varying at infinity in the sense that $\lim_{|x| \to \infty} V'(x) = 0$. $\square$

Our final example is a huge simplification of operators that have been studied in multi-body quantum mechanics. We assume that there are two one-dimensional particles, and that they are attracted to each other and also to some external centres. The four parts of the essential spectrum that we identify correspond to both particles moving to infinity in different directions, one or other of the particles moving to infinity while the other remains in a bound state, and the two particles moving to infinity but staying bound to each other. Our potentials are complex-valued, which might seem to be non-physical, but such operators arise naturally when using complex scaling methods to identify resonances.[8]

Theorem 14.4.6 *Let* $H := H_0 + V$ *on* $L^2(\mathbf{R}^2)$, *where* $V := V_1 + V_2 + V_3$ *and* V_i *are given by* $V_1(x, y) := W_1(x)$, $V_2(x, y) := W_2(y)$, $V_3(x, y) := W_3((x - y)/\sqrt{2})$; *we assume that* W_i *are complex-valued, bounded potentials on* $\mathbf{R}$ *which also lie in* $L^1(\mathbf{R})$. *Then*

$$\mathrm{EssSpec}(H) \supseteq \bigcup_{0 \leq i \leq 3} (\mathrm{Spec}(K_i) + [0, \infty))$$

where

$$(K_i f)(x) := -f''(x) + W_i(x)f(x)$$

acting in $L^2(\mathbf{R})$, *and* $W_0 := 0$.

Proof. One applies Theorem 14.4.2 with different choices of a_n in each case. The hardest is for $i = 3$ in which case we have to put $a_n := (n, n)$, and obtain the limit operator $L_2 := H_0 + W_3((x - y)/\sqrt{2})$. This is unitarily equivalent by

---

[8] See [Reed and Simon 1979A, Cycon et al. 1987, Perry 1983] for introductions to multi-body quantum mechanics. The first of these was written prior to the geometric revolution in scattering theory of Enss, but it remains a useful account of the subject.

a rotation in $\mathbf{R}^2$ to $M_3 := H_0 + W_3(x)$. Finally separation of variables implies that

$$\text{Spec}(M_3) = \text{Spec}(K_3) + [0, \infty). \qquad \square$$

The theorem states that the essential spectrum of H contains a series of semi-infinite horizontal straight lines, starting at 0 or at any of the L^2 eigenvalues of K_i, $i = 1, 2, 3$. In fact one has equality, but this is quite hard to prove. The hypothesis $W_i \in L^1(\mathbf{R})$ implies that all of the eigenvalues, called thresholds of H, lie within a ball of finite radius and centre at 0, by Theorem 14.3.1. The operator H may also have eigenvalues corresponding to bound states of the pair of particles with each other and with the external centres.

14.5 The NSA harmonic oscillator

In this final section we summarize some results concerning the non-self-adjoint (NSA) harmonic oscillator without proofs. As well as being of mathematical interest, and in some respects exactly soluble, it arises in physics as the model for a damped or unstable laser. The fact that its eigenvectors do not form a basis is surprising and was not anticipated in the physics literature. The results in this section once again illustrate how different non-self-adjoint operators are from their self-adjoint cousins.

The harmonic oscillator is the closure of the operator

$$(H_a f)(x) := -f''(x) + ax^2 f(x) \qquad (14.9)$$

initially defined on Schwartz space $\mathcal{S}$ in $L^2(\mathbf{R})$. For $a > 0$ this is one of the most famous examples in quantum theory. Its spectrum is

$$\text{Spec}(H_a) = \{(2n+1)a^{1/2} : n = 0, 1, \ldots\}.$$

Each eigenvalue $\lambda_n := (2n+1)a^{1/2}$ is of multiplicity 1, and a corresponding eigenfunction is

$$\phi_n(x) := H_n(a^{1/4}x)e^{-a^{1/2}x^2/2}$$

where the Hermite polynomial H_n is of degree n. After normalization, the eigenfunctions provide a complete orthonormal set in $L^2(\mathbf{R})$; see Problem 3.3.14. The operator H_a is essentially self-adjoint on $\mathcal{S}$ by Problem 5.4.6 and the resolvent operators are compact.

The NSA harmonic oscillator is obtained by allowing a to be complex. At first sight this has little effect, since all of the results above except the essential self-adjointness extend to this situation with no changes. The eigenvalues are now complex, but they are given by the same formula as before. Since

$$\langle H_a f, f \rangle = \int_{\mathbf{R}} \left\{ |f'(x)|^2 + ax^2 |f(x)|^2 \right\} \, dx$$

for all $f \in \mathcal{S}$, the numerical range of H_a is contained in $\{z : 0 \leq \text{Arg}(z) \leq \text{Arg}(a)\}$.

The first evidence of a major difference between the SA and NSA harmonic oscillators comes when one tries to expand an arbitrary function $f \in L^2(\mathbf{R})$ in terms of the eigenfunctions ϕ_n. It has been proved that the sequence $\{\phi_n\}$ does not form a basis in the sense of Section 3.3 unless a is real and positive. One obtains a biorthogonal sequence by putting $\phi_n^*(x) := \overline{\phi_n(x)}$ and normalizing properly, but the norms of the spectral projections P_n grow exponentially as $n \to \infty$.

Theorem 14.5.1 *(Davies-Kuijlaars)*[9] *If $a = e^{i\theta}$ where $-\pi < \theta < \pi$ then*

$$\lim_{n \to \infty} n^{-1} \log(\|P_n\|) = 2\text{Re} \left\{ f(r(\theta)e^{i\theta/4}) \right\}$$

where

$$f(z) := \log(z + (z^2 - 1)^{1/2}) - z(z^2 - 1)^{1/2}$$

and

$$r(\theta) := (2\cos(\theta/2))^{-1/2} .$$

This theorem has profound implications for results that one would accept without thought in the self-adjoint context.

Corollary 14.5.2 *The expansion*

$$e^{-H_a t} := \sum_{n=0}^{\infty} e^{-\lambda_n t} P_n$$

is norm convergent if

$$t > t_a := \frac{\text{Re} \left\{ f(r(\theta)e^{i\theta/4}) \right\}}{\cos(\theta/2)}$$

and divergent if $0 < t < t_a$.

It follows that the sequence $\{\phi_n\}$ cannot be an Abel-Lidskii basis.

---

[9] See [Davies and Kuijlaars 2004] for the proof. The squares of the norms of the spectral projections are called the Petermann factors in the physics literature [Berry 2003].

Example 14.5.3 The above ideas may be extended to more general operators, although currently the known bounds are less sharp. If one defines the anharmonic oscillator by

$$(H_a f)(x) := -f''(x) + ax^{2m} f(x)$$

where m is a positive integer and $a > 0$, then H_a is essentially self-adjoint on $\mathcal{S}$ and it has a complete sequence $\{\phi_n\}_{n=0}^{\infty}$ of eigenfunctions such that

$$H_a \phi_n = a^{1/(m+1)} \lambda_n \phi_n$$

for all n. The eigenvalues λ_n of H_1 all have multiplicity 1 and satisfy $\lim_{n \to \infty} \lambda_n = +\infty$. All of the above statements except for the essential self-adjointness extend to complex a. The norms of the spectral projections P_n again diverge as $n \to \infty$ if a is complex, and it is known that the rate of divergence is super-polynomial.[10] $\square$

The NSA harmonic oscillator also provides an ideal example to prove the importance of pseudospectra.

Theorem 14.5.4 *(Davies) Let*

$$R_{r,\theta} := (re^{i\theta} I - H_a)^{-1}$$

where $r > 0$, $\theta \in \mathbf{R}$ and H_a is defined by (14.9). If $\operatorname{Arg}(a) < \theta < 2\pi$ then

$$\lim_{r \to \infty} \|R_{r,\theta}\| = 0.$$

On the other hand if $0 < \theta < \operatorname{Arg}(a)$ then $\|R_{r,\theta}\|$ diverges at a super-polynomial rate as $r \to \infty$.

Figure 14.2 shows level curves for the pseudospectra of the harmonic oscillator for the case $a = i$. The resolvent norm equals 1 on the outermost curve, and it increases by a factor of $\sqrt{10}$ on each successive curve moving inwards.

This theorem establishes that the norm of $(zI - H_a)^{-1}$ may increase rapidly as z moves away from the spectrum of H_a, depending on the direction in which z travels. The theorem has been extended to anharmonic oscillators in [Davies 1999A, Davies 1999B]. It has been proved that the rate of divergence of the resolvent norm is exponential if $0 < \theta < \operatorname{Arg}(a)$, but the exact exponent is not known; see [Zworski 2001, Dencker et al. 2004]. Pravda-Starov has found the precise asymptotics at infinity of the pseudospectral contours for the NSA harmonic oscillator, solving a conjecture of Boulton; see [Boulton 2002, Pravda-Starov 2005].

---

[10] See [Davies 2000A].

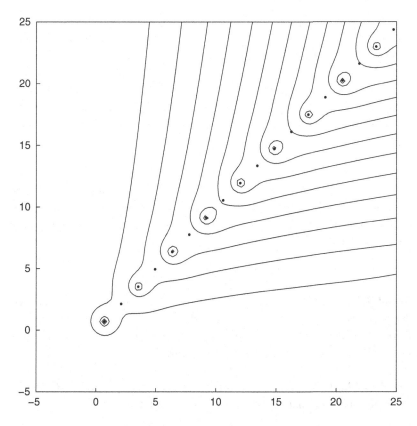

Figure 14.2: Pseudospectra of the NSA harmonic oscillator

14.6 Semi-classical analysis

Semi-classical analysis is a huge subject, and we describe only a few of the ideas in the field. Some of the material presented is half a century old, but other results have not previously been published. We start by discussing the notion of quantization at a very general level.

The goal is to associate an operator acting on $L^2(\mathbf{R}^N)$ with a function, called a symbol, on the phase space $\mathbf{R}^N \times \mathbf{R}^N$. This can be done in two directions. One might define a linear quantization map $\mathcal{Q}$ from symbols to operators or a reverse map $\mathcal{R}$ from operators to symbols. It is not realistic to expect these maps to be inverse to each other, but this might hold in the semi-classical limit $h \to 0$. There are various ways of constructing such maps, but they differ from each other by terms that vanish asymptotically as $h \to 0$. We confine attention to the coherent state quantization, because it is particularly easy to describe in operator-theoretic terms and has some nice properties. It uses the theory of reproducing kernel Hilbert spaces, which has applications ranging from the analysis of square-integrable, unitary group representations to quantum theory.[11]

In the most general formulation one starts with a Hilbert space $\mathcal{H}$ and a measurable family of vectors $\phi_x \in \mathcal{H}$, often called coherent states, parametrized by points x in a measure space (X, dx). We henceforth assume that the formula

$$(Pf)(x) := \langle f, \phi_x \rangle$$

defines an isometric embedding of $\mathcal{H}$ into $L^2(X)$. The map P allows one to regard $\mathcal{H}$ as a Hilbert space of functions in which point evaluation is continuous.[12] Equivalently we assume that $\{\phi_x\}_{x \in X}$ is a resolution of the identity in the sense that

$$\int_X |\langle f, \phi_x \rangle|^2 \, dx = \|f\|^2 \tag{14.10}$$

for all $f \in \mathcal{H}$. In almost all applications X is a topological space and ϕ_x depends norm continuously on x, but this is not needed for the general theory, provided the range of a function is understood to refer to its essential range. We will not focus on such issues below.

Problem 14.6.1 Let ρ be a positive continuous function on $U \subseteq \mathbf{C}$ and let $\mathcal{H}$ be the space of analytic functions f on U such that

$$\|f\|_2^2 := \int_U |f(x + iy)|^2 \rho(x, y) \, dx dy < \infty.$$

Prove that point evaluation $f \to f(z)$ defines a bounded linear functional ϕ_z on $\mathcal{H}$ for every $z \in U$ and that ϕ_z depends norm continuously on z. This puts such spaces into the abstract framework just described.[13]   □

The quantization procedure that we describe below can be extended to unbounded symbols and operators subject to suitable assumptions, but we confine attention to the bounded case.[14]

---

[11] The general study of reproducing kernel Hilbert spaces was initiated in [Aronsajn 1950], but the earlier work of [Gabor 1946] has also been extremely influential. Some of the early applications to physics were described in [Davies 1976, Sect. 8.5]. The method is now regularly used for studying the spectra of self-adjoint Schrödinger operators.

[12] A particular case of the operator P, often called the Gabor transform, arises in signal processing; see [Gabor 1946, Heil and Walnut 1989].

[13] The Bargmann space of analytic functions is a special case, and has had important applications in quantum field theory; see [Bargmann 1961].

[14] A much fuller account of quantization may be found in [Berezin and Shubin 1991].

Theorem 14.6.2 *The formula*

$$Af := \int_X \sigma(x)\langle f, \phi_x\rangle \phi_x \, \mathrm{d}x \tag{14.11}$$

defines a bounded linear quantization map $\mathcal{Q}(\sigma) := A$ *from* $L^\infty(X)$ *to* $\mathcal{L}(\mathcal{H})$ *with the properties*

(i) $\mathcal{Q}(\overline{\sigma}) = \{\mathcal{Q}(\sigma)\}^*$ *for all* $\sigma \in L^\infty(X)$;
(ii) if $\sigma(x) \geq 0$ *for all* $x \in X$ *then* $\mathcal{Q}(\sigma) \geq 0$;
(iii) $\|\mathcal{Q}(\sigma)\| \leq \|\sigma\|_\infty$ *for all* $\sigma \in L^\infty(X)$.

Moreover $\mathcal{Q}(\sigma) = P^*\tilde{\sigma}P$, *where* $\tilde{\sigma}$ *is the bounded multiplication operator on* $L^2(X)$ *associated with* σ.

Proof. If $f, g \in \mathcal{H}$ then

$$\begin{aligned}
\langle \mathcal{Q}(\sigma)f, g\rangle &= \int_X \sigma(x)\langle f, \phi_x\rangle\langle \phi_x, g\rangle \, \mathrm{d}x \\
&= \int_X \sigma(x)(Pf)(x)\overline{(Pg)(x)} \, \mathrm{d}x \\
&= \langle \tilde{\sigma}Pf, Pg\rangle \\
&= \langle P^*\tilde{\sigma}Pf, g\rangle.
\end{aligned}$$

Hence $\mathcal{Q}(\sigma) = P^*\tilde{\sigma}P$. The convergence of the integrals is proved by using the Schwarz inequality to obtain

$$\begin{aligned}
\int_X |\sigma(x)\langle f, \phi_x\rangle\langle \phi_x, g\rangle| \, \mathrm{d}x &\leq \|\sigma\|_\infty \int_X |\langle f, \phi_x\rangle| \, |\langle \phi_x, g\rangle| \, \mathrm{d}x \\
&\leq \|\sigma\|_\infty \left\{ \int_X |\langle f, \phi_x\rangle|^2 \, \mathrm{d}x \int_X |\langle g, \phi_x\rangle|^2 \, \mathrm{d}x \right\}^{1/2} \\
&= \|\sigma\|_\infty \|f\| \, \|g\|.
\end{aligned}$$

The other statements of the theorem follow directly. $\square$

The symbol σ above is often said to be contravariant, and γ below is then said to be a covariant symbol.

Lemma 14.6.3 *The formula*

$$\gamma(x) := \langle A\phi_x, \phi_x\rangle / \|\phi_x\|^2$$

defines a bounded linear map $\mathcal{R}(A) := \gamma$ *from* $\mathcal{L}(\mathcal{H})$ *to* $L^{\infty}(X)$ *with the properties*

(i) $\mathcal{R}(A^*) = \overline{\mathcal{R}(A)}$ *for all* $A \in \mathcal{L}(\mathcal{H})$;
(ii) *if* $A \geq 0$ *then* $\gamma(x) \geq 0$ *for all* $x \in X$;
(iii) $\|\mathcal{R}(A)\|_{\infty} \leq \|A\|$ *for all* $A \in \mathcal{L}(\mathcal{H})$.

Proof. All of the statements follow directly from the definition. □

Although $\mathcal{R}$ is not the inverse of $\mathcal{Q}$, it is approximately so if the kernel K in the next theorem is heavily concentrated near the diagonal in $X \times X$. We will give an example of this later.

Lemma 14.6.4 *If* $\sigma \in L^{\infty}(X)$ *then* $\gamma := \mathcal{R}\mathcal{Q}(\sigma)$ *is given by*

$$\gamma(x) = \int_X K(x, x')\sigma(x')\,\mathrm{d}x'$$

where

$$K(x, x') := |\langle \phi_x, \phi_{x'} \rangle|^2 / \|\phi_x\|^2$$

satisfies

(i) $K(x, x') \geq 0$ *for all* $x, x' \in X$;
(ii) $\int_X K(x, x')\,\mathrm{d}x' = 1$ *for all* $x \in X$.

Hence $\mathcal{R}\mathcal{Q}$ *is a positive linear map on* $L^{\infty}(X)$ *satisfying* $\mathcal{R}\mathcal{Q}(1) = 1$.

Proof. If we put $A := \mathcal{Q}(\sigma)$ then

$$\gamma(x) = \mathcal{R}(A)(x)$$
$$= \langle A\phi_x, \phi_x \rangle / \|\phi_x\|^2$$
$$= \int_X \sigma(x')\langle \phi_x, \phi_{x'} \rangle\langle \phi_{x'}, \phi_x \rangle / \|\phi_x\|^2\,\mathrm{d}x'$$
$$= \int_X K(x, x')\sigma(x')\,\mathrm{d}x'.$$

The other statements of the theorem follow immediately. □

Theorem 14.6.5 *Put* $w(x) := \|\phi_x\|^2$. *The quantization formula (14.11) defines a bounded linear map* $\mathcal{Q}$ *from* $L^1(X, w(x)\mathrm{d}x)$ *into the space* $\mathcal{C}_1(\mathcal{H})$ *of trace class operators on* $\mathcal{H}$. *If* $\sigma \in L^1(X, w(x)\mathrm{d}x)$ *and* $A := \mathcal{Q}(\sigma)$ *then*

$$\mathrm{tr}(A) = \int_X \sigma(x)w(x)\,\mathrm{d}x.$$

Proof. Let $\{e_n\}_{n=1}^{\infty}$ be a complete orthonormal set in $\mathcal{H}$. If $0 \leq \sigma \in L^1(X, w(x)\mathrm{d}x)$ and $A := \mathcal{Q}(\sigma)$ then

$$\mathrm{tr}(A) = \sum_{n=1}^{\infty} \int_X \sigma(x) |\langle e_n, \phi_x \rangle|^2 \, \mathrm{d}x$$

$$= \int_X \sigma(x) \sum_{n=1}^{\infty} |\langle e_n, \phi_x \rangle|^2 \, \mathrm{d}x$$

$$= \int_X \sigma(x) \|\phi_x\|^2 \, \mathrm{d}x$$

$$= \int_X \sigma(x) w(x) \, \mathrm{d}x.$$

The proof of the theorem is completed by noting that every $\sigma \in L^1(X, w(x)\mathrm{d}x)$ is a linear combination of four non-negative functions in the same space. $\quad\square$

Problem 14.6.6 Under the same hypotheses, if $L_0^{\infty}(X)$ is defined as the norm closure in $L^{\infty}(X)$ of $L^1(X, w(x)\mathrm{d}x) \cap L^{\infty}(X)$, prove that $\mathcal{Q}(\sigma)$ is compact for all $\sigma \in L_0^{\infty}(X)$. $\quad\square$

Problem 14.6.7 If $0 \leq \sigma \in L^1(X, w(x)\mathrm{d}x)$ and $\gamma := \mathcal{R}\mathcal{Q}(\sigma)$, prove that $\gamma \geq 0$ and

$$\int_X \gamma(x) w(x) \, \mathrm{d}x = \int_X \sigma(x) w(x) \, \mathrm{d}x. \qquad\square$$

The remainder of this section is devoted to determining the numerical range of a pseudodifferential operator in the semi-classical limit. The next theorem provides the key ingredient.

Theorem 14.6.8 *(Berezin) If $\sigma \in L^{\infty}(X)$, $A := \mathcal{Q}(\sigma)$ and $\gamma := \mathcal{R}(A)$ then*

$$\overline{\mathrm{Conv}}\{\gamma(x) : x \in X\} \subseteq \overline{\mathrm{Num}}(A) \subseteq \overline{\mathrm{Conv}}\{\sigma(x) : x \in X\}. \qquad (14.12)$$

Proof. The definition of $\overline{\mathrm{Num}}(A)$ yields $\{\gamma(x) : x \in X\} \subseteq \overline{\mathrm{Num}}(A)$ directly, and the LHS of (14.12) follows by the convexity of $\overline{\mathrm{Num}}(A)$, proved in Theorem 9.3.1.

The RH inclusion follows as in Theorem 9.3.4.

$$\text{Num}(A) = \{\langle P^*\tilde{\sigma}Pf, f\rangle/\|f\|^2 : 0 \neq f \in \mathcal{H}\}$$

$$= \{\langle\tilde{\sigma}g, g\rangle/\|g\|^2 : 0 \neq g \in P\mathcal{H}\}$$

$$\subseteq \{\langle\tilde{\sigma}g, g\rangle/\|g\|^2 : 0 \neq g \in L^2(X)\}$$

$$= \overline{\text{Num}}(\tilde{\sigma})$$

$$= \overline{\text{Conv}}\{\sigma(x) : x \in X\}.$$

The last line uses the normality of the multiplication operator $\tilde{\sigma}$. □

We now move to a more particular context. The definition of the coherent state quantization involves putting $\mathcal{H} := L^2(\mathbf{R}^N)$, $X := \mathbf{R}^N \times \mathbf{R}^N$, $x := (p, q)$ and $dx := d^N p\, d^N q$. Given a function $\phi \in L^2(\mathbf{R}^N)$ of norm 1 and $h > 0$ we construct functions $\phi_{p,q} \in L^2(\mathbf{R}^N)$ that are concentrated in a small neighbourhood of the point (p, q) in phase space.[15]

Lemma 14.6.9 *If*

$$\phi_{p,q}(u) := c_h\phi((u - q)/h^{1/2})e^{ip\cdot u/h} \tag{14.13}$$

where $u \in \mathbf{R}^N$, $\|\phi\|_2 = 1$, $h > 0$ *and* $c_h := (2\pi)^{-N/2}h^{-3N/4}$, *then*

$$\int_{\mathbf{R}^N \times \mathbf{R}^N} |\langle f, \phi_{p,q}\rangle|^2\, d^N p\, d^N q = \|f\|_2^2 \tag{14.14}$$

for all $f \in L^2(\mathbf{R}^N)$.

Proof. We have

$$\langle\phi_{p,q}, f\rangle = c_h \int_{\mathbf{R}^N} \phi((u - q)/h^{1/2})\overline{f(u)}e^{ip\cdot u/h}\, d^N u$$

$$= (2\pi)^{-N/2} \int_{\mathbf{R}^N} \rho_f(v, q)e^{ip\cdot v}\, d^N v$$

where

$$\rho_f(v, q) := (2\pi)^{N/2}c_h h^N \phi(h^{1/2}v - h^{-1/2}q)\overline{f(hv)}.$$

Therefore

$$\int_{\mathbf{R}^N} |\langle\phi_{p,q}, f\rangle|^2\, d^N p$$

$$= \int_{\mathbf{R}^N} |\rho_f(v, q)|^2\, d^N v$$

$$= (2\pi)^N c_h^2 h^{2N} \int_{\mathbf{R}^N} |\phi(h^{1/2}v - h^{-1/2}q)|^2|f(hv)|^2\, d^N v,$$

---

[15] The anti-Wick quantization studied in [Berezin 1971] corresponds to taking ϕ to be a Gaussian function, but this choice is not required in our context.

and

$$\int_{\mathbf{R}^N \times \mathbf{R}^N} |\langle \phi_{p,q}, f \rangle|^2 \, d^N p \, d^N q$$

$$= (2\pi)^N c_h^2 h^{5N/2} \int_{\mathbf{R}^N \times \mathbf{R}^N} |\phi(h^{1/2}v - w)|^2 |f(hv)|^2 \, d^N v \, d^N w$$

$$= (2\pi)^N c_h^2 h^{3N/2} \|\phi\|_2^2 \|f\|_2^2$$

$$= \|f\|_2^2. \qquad \qquad \square$$

Before investigating the coherent state quantization $\mathcal{Q}$ based on the family $\{\phi_{p,q}\}$ further, we describe its relationship with the classical (or Kohn-Nirenberg) quantization

$$(\mathcal{Q}_{\mathrm{cl}}(\sigma) f)(x) := (2\pi)^{-N} \int_{\mathbf{R}^N} e^{ix\cdot\xi} \sigma(x, \xi) \hat{f}(\xi) \, d\xi$$

where $\hat{f}$ is the Fourier transform of f. The integral converges absolutely for all $f \in \mathcal{S}$ provided $\sigma \in L^\infty(\mathbf{R}^N \times \mathbf{R}^N)$.

Problem 14.6.10 If ϕ, $\hat{\phi} \in L^1(\mathbf{R}^N)$, $h = 1$, $f \in \mathcal{S}$ and $\sigma \in L^\infty(\mathbf{R}^N \times \mathbf{R}^N)$, prove that

$$\mathcal{Q}(\sigma) f = \mathcal{Q}_{\mathrm{cl}}(\tilde{\sigma}) f$$

where

$$\tilde{\sigma}(x, \xi) := (2\pi)^{-N} \int_{\mathbf{R}^N \times \mathbf{R}^N} \sigma(p, q) \phi(x - q) \overline{\hat{\phi}(\xi - p)} \, e^{i(x-q)\cdot(p-\xi)} \, d^N p \, d^N q$$

is a bounded function on $\mathbf{R}^N \times \mathbf{R}^N$. Deduce that if $\sigma(p, q) := f(p) + g(q)$, where $f, g : \mathbf{R}^N \to \mathbf{C}$ are bounded functions, then

$$\mathcal{Q}(\sigma) = \tilde{f}(P) + \tilde{g}(Q)$$

for functions $\tilde{f}$, $\tilde{g}$ that you should determine. $\quad \square$

We now make the h-dependence of $\mathcal{Q}_h$, $\mathcal{R}_h$, $\phi_{h,x}$ and γ_h explicit. We write $x := (p, q)$, $X := \mathbf{R}^N \times \mathbf{R}^N$ and $dx := d^N p \, d^N q$ as convenient without explanation. We restrict the statement of Theorem 14.6.12 to bounded, continuous symbols in order to avoid technical details.

Lemma 14.6.11 *If $\sigma \in L^\infty(X)$, $A_h := \mathcal{Q}_h(\sigma)$ and $\gamma_h := \mathcal{R}_h(A_h)$ then*

$$\gamma_h(x) = \int_X h^{-N} k((x - x')/h^{1/2}) \sigma(x') \, dx' \qquad (14.15)$$

where $k : X \to [0, \infty)$ is continuous and $\int_X k(x) \, dx = 1$.

Proof. Lemma 14.6.4 states that

$$\gamma_h(x) = \int_X K_h(x, x')\sigma(x')\,dx'$$

where

$$K_h(x, x') := |\langle \phi_{h,x}, \phi_{h,x'} \rangle|^2 / \|\phi_{h,x}\|^2.$$

In our particular case

$$\|\phi_{h,x}\|_2^2 = c_h^2 \int_{\mathbf{R}^N} |\phi((u-q)/h^{1/2})e^{ip\cdot u/h}|^2\,d^N u$$

$$= c_h^2 h^{N/2} \int_{\mathbf{R}^N} |\phi(v - q/h^{1/2})|^2\,d^N v$$

$$= (2\pi)^{-N} h^{-N}$$

and

$$\langle \phi_{h,x}, \phi_{h,x'} \rangle = c_h^2 h^{N/2} e^{i(p-p')\cdot q/h} \int_{\mathbf{R}^N} \phi(w)\overline{\phi(w + (q - q')/h^{1/2})}$$

$$e^{i(p-p')\cdot w/h^{1/2}}\,d^N w$$

$$= c_h^2 h^{N/2} e^{i(p-p')\cdot q/h} \alpha((x - x')/h^{1/2})$$

for a certain function $\alpha : X \to \mathbf{C}$. The assumption $\phi \in L^2(\mathbf{R}^N)$ implies the convergence of the last integral and the continuity of the function α. Therefore

$$K(x, x') = (2\pi)^{-N} h^{-N} |\alpha((x - x')/h^{1/2})|^2.$$

This implies (14.15). The inequality $k \geq 0$ follows directly from the above calculations, and Lemma 14.6.4 implies that $\int_X k(x)\,dx = 1$. □

Theorem 14.6.12 *If σ is a bounded continuous function on $\mathbf{R}^N \times \mathbf{R}^N$ and $A_h := \mathcal{Q}_h(\sigma)$ then*

$$\lim_{h \to 0} \mathrm{Num}(A_h) = \overline{\mathrm{Conv}}\{\sigma(p, q) : p, q \in \mathbf{R}^N\}.$$

Proof. Theorem 14.6.8 implies that it is sufficient to prove that

$$\lim_{h \to 0} \gamma_h(x) = \sigma(x)$$

for all $x \in X$. This follows directly from Lemma 14.6.11 and the assumed continuity of σ. □

Note that if σ has a given modulus of continuity and enough is known about ϕ, then one may estimate the rate of convergence in the above theorem.

Theorem 14.6.12 encourages one to speculate that

$$\lim_{h \to 0} \mathrm{Spec}(A_h) = \{\sigma(p, q) : p, q \in \mathbf{R}^N\}$$

but the results in the last section demonstrate that this is surely wrong. In general one can only hope to obtain convergence of the pseudospectra in such a context.

References

[Abramov et al. 2001] Abramov, A., Aslanyan, A. and Davies, E. B.: Bounds on complex eigenvalues and resonances. *J. Phys. A, Math. Gen.* **34** (2001) 52–72.

[Adams 1975] Adams, R. A. *Sobolev Spaces*. New York: Academic Press, 1975.

[Agronovich 1996] Agranovich, M. S. Elliptic operators on closed manifolds. *Encyclopaedia of Math. Sci.*, Vol. 63, pp. 1130. Berlin: Springer, 1996.

[Albeverio et al. 2006] Albeverio, S., Hryniv, R. and Mykytyuk, Ya. On spectra of non-self-adjoint Sturm-Liouville operators. Preprint 2006.

[Aldous and Fill] Aldous, D. and Fill, J. *Reversible Markov Chains and Random Walks on Graphs*. In preparation. See
http://www.stat.berkeley.edu/users/aldous/index.html

[Anderson and Eiderman 2006] Anderson, J. M. and Eiderman, V. Ya. Cauchy transforms of point masses: The logarithmic derivative of polynomials. *Ann. Math.* **163** (2006) 1057–1076.

[Antoniou et al. 2002] Antoniou, I., Shkarin, S. A. and Yarevsky, E. The resonance spectrum of the cusp map in the space of analytic functions. *J. Math. Phys.* **43** (2002) 3746–3758.

[Arendt and Batty 2005A] Arendt, W. and Batty, C. J. K. Rank-1 perturbations of cosine functions and semigroups. Preprint 2005, *J. Funct. Anal.* to appear.

[Arendt and Batty 2005B] Arendt, W. and Batty, C. J. K. Forms, functional calculus, cosine functions and perturbations. Preprint 2005.

[Arendt et al. 2001] Arendt, W., Batty, C. J. K., Hieber, M. and Neubrander, F. *Vector-valued Laplace Transforms and Cauchy Problems*. Monographs in Mathematics, Vol. 96. Basel: Birkhäuser, 2001.

[Aronsajn 1950] Aronsajn, N. Theory of reproducing kernels. *Trans. Amer. Math. Soc.* **68** (1950) 337–404.

[Aslanyan and Davies 2001] Aslanyan, A. and Davies E. B. Separation of variables in deformed cylinders. *J. Comput. Phys.* **174** (2001) 327–344.

[Avron et al. 1994] Avron, J., Seiler, R. and Simon, B. The index of a pair of projections. *J. Funct. Anal.* **120** (1994) 220–237.

[Bargmann 1961] Bargmann, V. On a Hilbert space of analytic functions and an associated integral transform. *Commun. Pure Appl. Math.* **14** (1961) 187–214.

[Baladi 2000] Baladi, V. *Positive Transfer Operators and Decay of Correlations*. Singapore: World Scientific, 2000.

[Bañuelos and Kulczycki 2004] Bañuelos, R. and Kulczycki, T. The Cauchy process and the Steklov problem. *J. Funct. Anal.* **211** (2004) 355–423.

[Bedford et al. 1991] Bedford, T., Keane, M. and Series, C. *Ergodic Theory, Symbolic Dynamics, and Hyperbolic Spaces.* Oxford: Oxford University Press, 1991.

[Berezin 1971] Berezin, F. A. Wick and anti-Wick operator symbols. *Math. USSR Sb.* **15** (1971) 577–606.

[Berezin and Shubin 1991] Berezin, F. A. and Shubin, M. A. *The Schrödinger Equation.* Dordrecht: Kluwer Academic Publishers, 1991.

[Berry, website] `http://www.phy.bris.ac.uk/people/berry_mv/research.html`

[Berry 2003] Berry, M. V. Mode degeneracies and the Petermann excess-noise factor for unstable lasers. *J. Modern Optics* **50** (2003) 6381.

[Bollobas 1999] Bollobas, B. *Linear Analysis, an Introductory Course*, 2nd edn. Cambridge: Cambridge University Press, 1999.

[Böttcher and Silbermann 1999] Böttcher, A. and Silbermann, B. *Introduction to Large Truncated Toeplitz Matrices.* New York: Springer, 1999.

[Boulton 2002] Boulton, L. S. Non-self-adjoint harmonic oscillator semigroups and pseudospectra. *J. Oper. Theory* **47** (2002) 413–429.

[Brown and Eastham 2002] Brown, B. M. and Easham, M. S. P. Spectral instability for some Schrödinger operators. *J. Comput. Appl. Math.* **148** (2002) 49–63.

[Burke and Greenbaum 2004] Burke, J. and Greenbaum, A. Some equivalent characterizations of the polynomial numerical hull of degree k. Preprint, Nov. 2004.

[Butzer and Berens 1967] Butzer, P. L. and Berens, H. *Semigroups of Operators and Approximation.* Berlin: Springer, 1967.

[Cachia and Zagrebnov 2001] Cachia, V. and Zagrebnov, V. Operator-norm convergence of the Trotter product formula for holomorphic semigroups. *J. Oper. Theory* **46** (2001) 199–213.

[Carleson 1966] Carleson, L. On convergence and growth of partial sums of Fourier series. *Acta Math.* **116** (1966) 135–157.

[Carothers 2005] Carothers, N. L. *A Short Course on Banach Space Theory.* Cambridge: Cambridge University Press, 2005.

[Chernoff 1976] Chernoff, P. R. Two counterexamples in semigroup theory on Hilbert space. *Proc. Amer. Math. Soc.* **56** (1976) 253–255.

[Chung 1997] Chung, F. R. K. *Spectral Graph Theory.* Providence, RI: American Mathematical Society, 1997.

[Cycon et al. 1987] Cycon, H. L., Froese, R. G., Kirsch, W. and Simon, B. *Schrödinger Operators.* Texts and Monographs in Physics. Berlin: Springer, 1987.

[Daubechies 1992] Daubechies, I. *Ten Lectures on Wavelets.* CBMS-NSF Regional Conference Series in Applied Mathematics, No. 61. Philadelphia, PA: SIAM, 1992.

[Davies 1974] Davies, E. B. Time-dependent scattering theory. *Math. Ann.* **210** (1974) 149–162.

[Davies 1976] Davies, E. B. *Quantum Theory of Open Systems.* London: Academic Press, 1976.

[Davies 1977] Davies, E. B. Eigenfunction expansions for singular Schrödinger operators. *Arch. Rat. Mech. Anal.* **63** (1977) 261–272.

[Davies 1980A] Davies, E. B. On Enss' approach to scattering theory. *Duke Math. J.* **47** (1980) 171–185.

[Davies 1980B] Davies, E. B. *One-Parameter Semigroups.* London: Academic Press, 1980.

[Davies 1989] Davies, E. B. *Heat Kernels and Spectral Theory.* Cambridge Tracts in Mathematics No. 92. Cambridge: Cambridge University Press, 1989.

[Davies 1995A] Davies, E. B. The functional calculus. *J. London Math. Soc.* 2 **55** (1995) 166–176.

[Davies 1995B] Davies, E. B. L^p spectral independence and L^1 analyticity. *J. London Math. Soc.* 2 **52** (1995) 177–184.

[Davies 1995C] Davies, E. B. *Spectral Theory and Differential Operators.* Cambridge: Cambridge University Press, 1995.

[Davies 1995D] Davies, E. B. Long time asymptotics of fourth order parabolic equations. *J. d'Analyse Math.* **67** (1995) 323–345.

[Davies 1997] Davies, E. B. L^p spectral theory of higher-order elliptic differential operators. *Bull. London Math. Soc.* **29** (1997) 513–546.

[Davies 1999A] Davies, E. B. Pseudospectra, the harmonic oscillator and complex resonances. *Proc. R. Soc. London A* **455** (1999) 585–599.

[Davies 1999B] Davies, E. B. Semi-classical states for non-self-adjoint Schrödinger operators. *Commun. Math. Phys.* **200** (1999) 35–41.

[Davies 2000A] Davies, E. B. Wild spectral behaviour of anharmonic oscillators. *Bull. London Math. Soc.* **32** (2000) 432–438.

[Davies 2000B] Davies, E. B. Pseudospectra of differential operators. *J. Oper. Theory* **43** (2000) 243–262.

[Davies 2001A] Davies, E. B. Spectral properties of random non-self-adjoint matrices and operators. *Proc. R. Soc. London A* **457** (2001) 191–206.

[Davies 2001B] Davies, E. B. Spectral theory of pseudo-ergodic operators. *Commun. Math. Phys.* **216** (2001) 687–704.

[Davies 2005A] Davies, E. B. Semigroup growth bounds. *J. Oper. Theory* **53**(2) (2005) 225–249.

[Davies 2005B] Davies, E. B. Spectral bounds using higher order numerical ranges. *LMS J. Comput. Math.* **8** (2005) 17–45.

[Davies 2005C] Davies, E. B. Triviality of the peripheral point spectrum. *J. Evol. Eqns.* **5** (2005) 407–415.

[Davies 2005D] Davies, E. B. Semi-classical analysis and pseudospectra. *J. Diff. Eqns.* **216** (2005) 153–187.

[Davies 2006] Davies, E. B. Newman's proof of Wiener's theorem. See http://www.mth.kcl.ac.uk/staff/eb_davies/ newmathpapers.html

[Davies and Hager 2006] Davies, E. B. and Hager, M. Perturbations of Jordan matrices. Preprint, 2006.

[Davies and Kuijlaars 2004] Davies, E. B. and Kuijlaars, A. B. J. Spectral asymptotics of the non-self-adjoint harmonic oscillator. *J. London Math. Soc.* 2 **70** (2004) 420–426.

[Davies and Nath 2002] Davies, E. B. and Nath, J. Schrödinger operators with slowly decaying potentials. *J. Comput. Appl. Math.* **148** (2002) 1–28.

[Davies and Plum 2004] Davies, E. B. and Plum, M. Spectral pollution. *IMA J. Numer. Anal.* **24** (2004) 417–438.

[Davies and Simon 1978] Davies, E. B. and Simon, B. Scattering theory for systems with different spatial asymptotics on the left and right. *Commun. Math. Phys.* **63** (1978) 277–301.

[Davies and Simon 1984] Davies, E. B. and Simon, B. Ultracontractivity and the heat kernel for Schrödinger operators and Dirichlet Laplacians. *J. Funct. Anal.* **59** (1984) 335–395.

[Davies and Simon 2005] Davies, E. B. and Simon, B. Eigenvalue estimates for non-normal matrices and the zeros of random orthogonal polynomials on the unit circle. J. Approx. Theory 141 (2006) 189-213.

[Dencker et al. 2004] Dencker, N., Sjöstrand, J. and Zworski, M. Pseudospectra of semi-classical (pseudo)differential operators. *Commun. Pure Appl. Math.* **57** (2004) 384–415.

[Desch and Schappacher 1988] Desch, W. and Schappacher, W. Some perturbation results for analytic semigroups. *Math. Ann.* **281** (1998) 157–162.

[Diestel 1975] Diestel, J. *Geometry of Banach Spaces – Selected Topics*. Lecture Notes in Mathematics, Vol. 485. Berlin: Springer-Verlag, 1975.

[Dynkin 1965] Dynkin, E. B. *Markov Processes*, Vol. 1. Berlin: Springer, 1965.

[Dunford and Schwartz 1963] Dunford, N. and Schwartz, J. T. *Linear Operators. Part 2: Spectral Theory*. New York: Interscience, 1963.

[Dunford and Schwartz 1966] Dunford, N. and Schwartz, J. T. *Linear Operators. Part 1: General Theory*. New York: Interscience, 1966.

[Dunford and Schwartz 1971] Dunford, N. and Schwartz, J. T. *Linear Operators. Part 3: Spectral Operators*. New York: Interscience, 1971.

[Edmunds and Evans 1987] Edmunds, D. E. and Evans, W. D. *Spectral Theory and Differential Operators*. Oxford: Clarendon Press, 1987.

[Eisner and Zwart 2005] Eisner, T. and Zwart, H. Continuous-time Kreiss resolvent condition on infinite-dimensional spaces. Math. Comp. 75 (2006) 1971-1985.

[El-Fallah et al. 1999] El-Fallah, O., Nikolski, N. K. and Zarrabi, M. Resolvent estimates in Beurling-Sobolev algebras. *St. Petersburg Math. J.* **10** (1999) 901–964.

[Elst and Robinson 2006] Elst, A. F. M. ter and Robinson, D. W. Contraction semigroups on $L_\infty(\mathbf{R})$. Preprint 2006. *J. Evol. Eqns.* to appear.

[Enflo 1973] Enflo, P. A counterexample to the approximation problem in Banach spaces. *Acta Math.* **130** (1973) 309–317.

[Engel and Nagel 1999] Engel, K. J. and Nagel, R. *One-Parameter Semigroups for Linear Evolution Equations*. Graduate Texts in Mathematics 194. New York: Springer-Verlag, 1999.

[Feller 1966] Feller, W. *An Introduction to Probability Theory and its Applications*, Vol. 2. New York: Wiley, 1966.

[Foguel 1964] Foguel, S. R. A counterexample to a conjecture of Sz.-Nagy. *Proc. Amer. Math. Soc.* **15** (1964) 788–790.

[Frank et al. 2006] Frank, R. L., Laptev, A., Lieb, E. H. and Seiringer, R. Lieb-Thirring inequalities for Schrödinger operators with complex-valued potentials. Letters Math. Phys. 77 (2006) 309-316.

[Friedlander and Joshi 1998] Frielander, F. G. and Joshi, M. *Introduction to the Theory of Distributions*, 2nd edn. Cambridge: Cambridge University Press, 1998.

[Friis and Rørdam 1996] Friis, P. and Rørdam, M. Almost commuting self-adjoint matrices – a short proof of Huaxin Lin's theorem. *J. Reine Angew. Math.* **479** (1996) 121131.

[Gabor 1946] Gabor, D. Theory of communication. *J. Inst. Elec. Eng. (London)* **93** (1946) 429–457.

[Garcia-Cueva and De Francia 1985] Garcia-Cueva, J. and De Francia, J. L. *Weighted Norm Inequalities and Related Topics*. Amsterdam: North-Holland, 1985.

[Gesztesy and Tkachenko 2005] Gesztesy, F. and Tkachenko, V. When is a non-self-adjoint Hill operator a spectral operator of scalar type? C. R. Acad. Sci. Paris, Ser. I 343 (2006) 239–242.

[Gilkey 1996] Gilkey, P. B. *Invariance Theory, the Heat Equation, and the Atiyah-Singer Index Theorem*. Electronic reprint, © 1996, Peter B. Gilkey http://www.emis.de/monographs/gilkey/index.html

[Glimm and Jaffe 1981] Glimm, J. and Jaffe, A. *Quantum Physics: a Functional Integral Point of View*. New York: Springer-Verlag, 1981.

[Gohberg and Krein 1969] Gohberg, I. C. and Krein, M. G. *Introduction to the Theory of Linear Nonselfadjoint Operators in Hilbert Space*. Translations of Mathematical Monographs, Vol. 18. Providence, RI: American Mathematical Society, 1969.

[Gohberg et al. 1983] Gohberg, J., Lancaster, P. and Rodman, L. *Matrices and Indefinite Scalar Product*. Operator Theory and Applications 8. Basel: Birkhäuser, 1983.

[Goldsheid and Khoruzhenko 2000] Goldsheid, I. Ya. and Khoruzhenko, B. A. Eigenvalue curves of asymmetric tridiagonal random matrices. *Electronic J. Prob.* **5** (2000) Paper 16, 1–28.

[Goldsheid and Khoruzhenko 2003] Goldsheid, I. Ya. and Khoruzhenko, B. A. Regular spacings of complex eigenvalues in one-dimensional non-Hermitian Anderson model. *Commun. Math. Phys.* **238** (2003) 505–524.

[Goodman 1971] Goodman, R. Complex Fourier analysis on a nilpotent Lie group. *Trans. Amer. Math. Soc.* **160** (1971) 373–391.

[Grafakos 2004] Grafakos, L. *Classical and Modern Fourier Analysis*. Upper Saddle River, NJ: Pearson Educ. Inc., Prentice Hall, 2004.

[Greiner and Muller 1993] Greiner, G. and Muller, M. The spectral mapping theorem for integrated semigroups. *Semigroup Forum* **47** (1993) 115–122.

[Gustafson and Rao 1997] Gustafson, K. E. and Rao, D. K. M. *Numerical Range*. New York: Springer, 1997.

[Haase 2004] Haase, M. A decomposition theorem for generators of strongly continuous groups on Hilbert spaces. *J. Oper. Theory* **52** (2004) 21–37.

[Hairer and Mattingly 2005] Hairer, M. and Mattingley, J. Ergodicity of the 2D Navier-Stokes equations with degenerate stochastic forcing. Ann. Math. 164 no. 3 (2006) to appear.

[Halmos 1969] Halmos, P. R. Two subspaces. *Trans. Amer. Math. Soc.* **144** (1969) 381–389.

[Heil and Walnut 1989] Heil, C. E. and Walnut, D. F. Continuous and discrete wavelet transforms. *SIAM Review* **31** (1989) 628–666.

[Hille 1948] Hille, E. *Functional Analysis and Semigroups*. New York: American Mathematical Society, 1948.

[Hille 1952] Hille, E. On the generation of semigroups and the theory of conjugate functions. *Kungl. Fys. Sälls. I. Lund Förhand* **21** (1952) 1–13.

[Hille and Phillips 1957] Hille, E. and Phillips, R. S. *Functional Analysis and Semigroups*. Providence, RI: American Mathematical Society, 1957.

[Hinrichsen and Pritchard 1992] Hinrichsen, D. and Pritchard, A. J. On spectral variations under bounded real matrix perturbations. *Numer. Math.* **60** (1992) 509–524.

[Hinrichsen and Pritchard 2005] Hinrichsen, D. and Pritchard, A. J. *Mathematical Systems Theory 1. Modelling, State Space Analysis, Stability and Robustness.* Berlin: Springer-Verlag, 2005.

[Hörmander 1960] Hörmander, L. Estimates for translation invariant operators on L^p spaces. *Acta Math.* **104** (1960) 93–140.

[Hörmander 1990] Hörmander, L. *The Analysis of Linear Partial Differential Operators I*, 2nd edn. Berlin: Springer-Verlag, 1990.

[Hunt 1956] Hunt, G. A. Semigroups of measures on Lie groups. *Trans. Amer. Math. Soc.* **81** (1956) 264–293.

[Iserles and Nørsett 2006] Iserles, A. and Nørsett, S. P. From high oscillation to rapid approximation I: Modified Fourier series. Preprint 2006.

[Jensen and Cour-Harbo 2001] Jensen, A. and Cour-Harbo, A. la. *Ripples in Mathematics: The Discrete Wavelet Transform.* Berlin: Springer-Verlag, 2001.

[Jørsboe and Mejlbro 1982] Jørsboe, O. G. and Mejlbro, L. *The Carleson-Hunt Theorem on Fourier Series.* Lecture Notes in Mathematics 911. Berlin: Springer-Verlag, 1982.

[Kahane and Katznelson 1966] Kahane, J.-P. and Katznelson, Y. Sur les ensembles de divergence des sé ries trigonomtriques. *Studia Math.* **26** (1966) 305–306.

[Kato 1966A] Kato, T. *Perturbation Theory for Linear Operators.* New York: Springer, 1966.

[Kato 1966B] Kato, T. Wave operators and similarity for some non-self-adjoint operators. *Math. Ann.* **162** (1996) 258–279.

[Keicher 2006] Keicher, V. On the peripheral spectrum of bounded positive semigroups on atomic Banach lattices. Archiv Math. 87, no. 4 (2006).

[Kelley 1955] Kelley, J. L. *General Topology.* Princeton: Van Nostrand, 1955.

[Kesten 1959] Kesten, H. Symmetric random walks on groups. *Trans. Amer. Math. Soc.* **92**(2) (1959) 336–354.

[Krasnosel'skii 1960] Krasnosel'skii, M. A. On a theorem of M. Riesz. *Soviet Math. Dokl.* **1** (1960) 229–231.

[Krein 1971] Krein, S. G. *Linear Differential Equations in Banach Spaces.* American Mathematical Society Translations, Vol. 29. Providence, RI: American Mathematical Society, 1971.

[Kuchment 2004] Kuchment, P. On some spectral problems of mathematical physics. In *Partial Differential Equations and Inverse Problems*, ed. C. Conca, R. Manasevich, G. Uhlmann and M. S. Vogelius, Contemporary Mathematics, Vol. 362, 2004. Amer. Math. Soc., Providence, Rhode Island.

[Kuttler 1997] Kuttler, K. L. *Modern Analysis.* Boca Raton, FL: CRC Press, LLC, 1997.

[Langer 2001] Langer, H., Markus, A., Matsaev, V. and Tretter, C. A new concept for block operator matrices: The quadratic numerical range. *Linear Algebra Appl.* **330** (2001) 89–112.

[Langmann et al. 2006] Langmann, E., Laptev, A. and Paufler, C. Singular factorizations, self-adjoint extensions and applications to quantum many-body physics. *J. Phys. A* **39** (2006) 1057–1071.

[Laptev and Weidl 2000] Laptev, A. and Weidl, T. Recent results on Lieb-Thirring inequalities. 'Équations aux derivées partielles', La Chapelle sur Erdre, Exp. no. XX, University of Nantes, Nantes, 2000.

[Lebow 1968] Lebow, A. A power-bounded operator that is not polynomially bounded. *Mich. Math. J.* **15** (1968) 397–399.

[Lidskii 1962] Lidskii, V. B. Summability of series in the principal functions of a non-selfadjoint elliptic operator. *Trudy Moskov. Mat. Obsh.* **11** (1962) 3–35. (English transl.: *Amer. Math. Soc. Transl.*, ser. 2, **40** (1964) 193–228.)

[Lieb and Loss 1997] Lieb, E. H., Loss, M. *Analysis*. Graduate Studies in Mathematics, Vol. 14. Providence, RI: American Mathematical Society, 1997.

[Lin 1997] Lin, H. Almost commuting self-adjoint matrices and applications. *The Fields Institute Communication* **13** (1997) 193–234.

[Lohoué and Rychener 1982] Lohoué, N. and Rychener, Th. Resolvente von Δ auf symmetrischen Räumen vom nichtkompakten Typ. *Comment. Math. Helv.* **57** (1982) 445–468.

[Lorch 1939] Lorch, E. R. Bicontinuous linear transformations in certain vector spaces. *Bull. Amer. Math. Soc.* **45** (1939) 564–569.

[Lumer and Phillips 1961] Lumer, G. and Phillips, R. S. Dissipative operators in a Banach space. *Pacific J. Math.* **11** (1961) 679–698.

[Manin and Marcolli 2002] Manin, Y. I. and Marcolli, M. Continued fractions, modular symbols, and non-commutative geometry. *Selecta Math.*, new series **8** (2002) 475–521.

[Markus 1988] Markus, A. S. *Introduction to the Spectral Theory of Polynomial Operator Pencils*. Translations of Mathematical Monographs, Vol. 71. Providence, RI: American Mathematical Society, 1988.

[Martinez 2005] Martinez, C. Spectral estimates for the one-dimensional non-selfadjoint Anderson Model. *J. Oper. Theory* 56 (2006) 59-88.

[Maz'ya and Schmidt to appear] Maz'ya, V. and Schmidt, G. Approximate approximations. *Amer. Math. Soc.* to appear.

[Megginson 1998] Megginson, R. E. *An Introduction to Banach Space Theory*. New York: Springer, 1998.

[Meyn and Tweedie 1996] Meyn, S. P. and Tweedie, R. *Markov Chains and Stochastic Stability*. New York: Springer, 1996. See `http://black.csl.uiuc.edu/~meyn/pages/book.html`

[Minc 1988] Minc, H. *Nonnegative Matrices*. New York: Wiley-Interscience, 1988.

[Miyadera 1966] Miyadera, I. On perturbation theory for semi-groups of operators. *Tohoku Math. J., II. Ser.* **18** (1966) 299–310.

[Muir 1923] Muir, T. *The Theory of Determinants*, Vol. 4, pp. 177–180. New York: Dover Publ. Inc., 1923.

[Narici and Beckenstein 1985] Narici, L. and Beckenstein, E. *Topological Vector Spaces*. New York: Marcel Dekker, 1985.

[Nath 2001] Nath, J. Spectral theory of non-selfadjoint operators. PhD thesis, King's College London, 2001.

[Nelson 1959] Nelson, E. Analytic vectors. *Ann. Math.* **70** (1959) 572–614.

[Nevanlinna 1993] Nevanlinna, O. *Convergence of Iterations for Linear Equations.* Basel: Birkhäuser, 1993.

[Newman 1975] Newman, D. J. A simple proof of Wiener's $1/f$ theorem. *Proc. Amer. Math. Soc.* **48** (1975) 264–265.

[Nygaard 2005] Nygaard, O. A remark on Rainwater's theorem. *Annales Math. Inform.* **32** (2005) 125–127.

[Ouhabaz 2005] Ouhabaz, E. M. *Analysis of Heat Equations on Domains.* London Math. Soc. Mono. Series, 31. Princeton, NJ: Princeton University Press, 2005.

[Paulsen 1984] Paulsen, V. I. Every completely polynomially bounded operator is similar to a contraction. *J. Funct. Anal.* **55** (1984) 1–17.

[Pazy 1968] Pazy, A. On the differentiability and compactness of semi-groups of linear operators, *J. Math. and Mech.* **17** (1968) 1131–1141.

[Pazy 1983] Pazy, A. *Semigroups of Linear Operators and Applications to Partial Differential Equations.* New York: Springer, 1983.

[Pearcy and Shields 1979] Pearcy, C. and Shields, A. Almost commuting matrices. *J. Funct. Anal.* **33** (1979) 332–338.

[Perry 1983] Perry, P. *Scattering Theory by the Enss Method.* Mathematical Reports, Vol. 1, pp. 1–347. New York: Harwood, 1983.

[Persson 1964] Persson, A. Compact linear mappings between interpolation spaces. *Ark. Mat.* **5** (1964) 215–219.

[Pisier 1997] Pisier, G. A polynomially bounded operator on Hilbert space which is not similar to a contraction. *J. Amer. Math. Soc.* **10** (1997) 351–369.

[Pitts 1972] Pitts, C. G. C. *Introduction to Metric Spaces.* Edinburgh: Oliver and Boyd, 1972.

[Pravda-Starov 2005] Pravda-Starov, K. A complete study of the pseudospectrum for the rotated harmonic oscillator. *J. London Math. Soc.* to appear, 2005.

[Ransford 2005] Ransford, T. Eigenvalues and power growth. *Israel J. Math.* **146** (2005) 93–110.

[Ransford 2006] Ransford, T. On pseudospectra and power growth. Preprint 2006.

[Ransford and Roginskaya 2006] Ransford, T. and Roginskaya, M. Point spectra of partially power-bounded operators. *J. Funct. Anal.* **230** (2006) 432–445.

[Reddy 1993] Reddy, S. C. Pseudospectra of Wiener-Hopf integral operators and constant-coefficient differential operators. *J. Int. Eqs. Appl.* **5** (1993) 369–403.

[Reed and Simon 1979A] Reed, M. and Simon, B. *Methods of Modern Mathematical Physics,* Vol. 3. London: Academic Press, 1979.

[Reed and Simon 1979B] Reed, M. and Simon, B. *Methods of Modern Mathematical Physics,* Vol. 4. London: Academic Press, 1979.

[Rudin 1966] Rudin, W. *Real and Complex Analysis.* New York: McGraw-Hill, 1966.

[Rudin 1973] Rudin, W. *Functional Analysis.* New Delhi: Tata McGraw-Hill Publ. Co., 1973.

[Safarov 2005] Safarov, Y. Birkhoff's theorem and multidimensional numerical range. *J. Funct. Anal.* **222** (2005) 61–97.

[Schaefer 1974] Schaefer, H. H. *Banach Lattices and Positive Operators.* Berlin: Springer, 1974.

[Schmidt and Spitzer 1960] Schmidt, P. and Spitzer, F. The Toeplitz matrices of an arbitrary Laurent polynomial. *Math. Scand.* **8** (1960) 15–38.

[Shkarin 2006] Shkarin, S. Antisupercyclic operators and orbits of the Volterra operator. *J. London Math. Soc.* 2 **73** (2006) 506–528.

[Simmons 1963] Simmons, G. F. *Introduction to Topology and Modern Analysis.* New York: McGraw Hill, 1963.

[Simon 1974] Simon, B. *The $P(\phi)_2$ Euclidean (Quantum) Field Theory.* Princeton Series in Physics. Princeton, NJ: Princeton University Press, 1974.

[Simon 1982] Simon, B. Schrödinger semigroups. *Bull. Amer. Math. Soc.* **7** (1982) 445–526.

[Simon 1998] Simon, B. The classical moment problem as a self-adjoint finite difference operator. *Adv. Math.* **137** (1998) 82–203.

[Simon 2005A] Simon, B. *Trace Ideals and Their Applications*, 2nd edn. Providence, RI: American Mathematical Society, 2005.

[Simon 2005B] Simon, B. *Orthogonal Polynomials on the Unit Circle, Part 1: Classical Theory, Part 2: Spectral Theory.* AMS Colloquium Series. Providence, RI: American Mathematical Society, 2005.

[Singer 1970] Singer, I. *Bases in Banach Spaces I.* New York: Springer-Verlag, 1970.

[Singer 1981] Singer, I. *Bases in Banach Spaces II.* Berlin: Springer-Verlag, 1981.

[Sobolev 1999] Sobolev, A. V. On the spectrum of the periodic magnetic Hamiltonian. *Amer. Math. Soc. Transl.*, Series 2 **189** (1999) 233–245.

[Song and Vondraček 2003] Song, R. and Vondraček, Z. Potential theory of subordinate killed Brownian motion in a domain. *Prob. Theory Relat. Fields* **125** (2003) 578–592.

[Stein 1970] Stein, E. M. *Singular Integrals and Differentiability Properties of Functions.* Princeton, NJ: Princeton University Press, 1970.

[Streater 1995] Streater, R. F. *Statistical Dynamics.* London: Imperial College Press, 1995.

[Sutherland 1975] Sutherland, W. A. *Introduction to Metric and Topological Spaces.* Oxford: Clarendon Press, 1975.

[Sz.-Nagy 1947] Sz.-Nagy, B. On uniformly bounded linear transformations in Hilbert space. *Acta Univ. Szeged Sect. Sci. Math.* **11** (1947) 152–157.

[Sz.-Nagy and Foias 1970] Sz.-Nagy, B. and Foias, C. *Harmonic Analysis of Operators on Hilbert Space.* Amsterdam: North-Holland, 1970.

[Taylor 1996] Taylor, M. *Partial Differential Equations*, Vols. 1–3. Applied Math. Sciences, 115–117. New York: Springer-Verlag, 1996.

[Tisseur and Higham 2001] Tisseur, F. and Higham, N. J. Structured pseudospectra for polynomial eigenvalue problems, with applications. *SIAM J. Matrix Anal. Appl.* **23** (2001) 187–208.

[Tisseur and Meerbergen 2001] Tisseur, F. and Meerbergen, K. The quadratic eigenvalue problem. *SIAM Review* **43** (2001) 235–286.

[Tkachenko 2002] Tkachenko, V. Non-selfadjoint Sturm-Liouville operators with multiple spectra. In *Interpolation Theory, Systems Theory and Related Topics* (Tel Aviv/Rehovot, 1999), pp. 403–414. Oper. Theory Adv. Appl., 134. Basel: Birkhäuser, 2002.

[Trefethen et al. 2001] Trefethen, L. N., Contedini, M. and Embree, M. Spectra, pseudospectra, and localization for random bidiagonal matrices. *Commun. Pure Appl. Math.* **54** (2001) 595–623.

[Trefethen and Embree 2005] Trefethen, L. N. and Embree, M. *Spectra and Pseudospectra: The Behavior of Nonnormal Matrices and Operators.* Princeton, NJ: Princeton University. Press, 2005.

[Treves 1967] Treves, F. *Topological Vector Spaces, Distributions and Kernels.* New York: Academic Press, 1967.

[Voigt 1977] Voigt, J. On the perturbation theory for strongly continuous semigroups. *Math. Ann.* **229** (1977) 163–171.

[von Neumann 1951] von Neumann, J. Eine Spektraltheorie für allgemeine Operatoren eines unitären Raumes. *Math. Nachr.* **4** (1951) 258–281.

[Weidl 1996] Weidl, T. On the Lieb-Thirring constants $L_{\gamma,1}$ for $\gamma \geq 1/2$. *Commun. Math. Phys.* **178** (1996) 135–146.

[Weir 1973] Weir, A. J. *Lebesgue Integration and Measure.* Cambridge: Cambridge University Press, 1973.

[Wilansky 1978] Wilansky, A. *Modern Methods in Topological Vector Spaces.* New York: McGraw-Hill, 1978.

[Woess 2000] Woess, W. *Random Walks on Infinite Graphs and Groups.* Cambridge Tracts in Mathematics, Vol. 138. Cambridge: Cambridge University Press, 2000.

[Wright 2002] Wright, T. G. EigTool sofware available at http://www.comlab.ox.ac.uk/pseudospectra/eigtool

[Wrobel 1989] Wrobel, V. Asymptotic behaviour of C_0-semigroups in B-convex spaces. *Indiana Univ. Math. J.* **38** (1989) 101–114.

[Yosida 1948] Yosida, K. On the differentiability and representation of one-parameter semigroups of linear operators. *J. Math. Soc. Japan* **1** (1948) 15–21.

[Yosida 1965] Yosida, K. *Functional Analysis.* Berlin: Springer, 1965.

[Young 1988] Young, N. *An Introduction to Hilbert Space.* Cambridge: Cambridge University Press, 1988.

[Zabczyk 1975] Zabczyk, J. A note on c_0-semigroups. *Bull. Acad. Polon. Sci. sér. Sci. Math. Astr. Phys.* **23** (1975) 895–898.

[Zagrebnov 2003] Zagrebnov, V. *Topics in the Theory of Gibbs Semigroups.* Leuven Notes in Mathematical and Theoretical Physics. Leuven: Leuven University Press, 2003.

[Zwart 2003] Zwart, H. On the estimate $\|(sI - Z)^{-1}\| \leq M/\mathrm{Re}(s)$. Ulmer Seminaire, pp. 384–388, University of Ulm, 2003.

[Zworski 2001] Zworski, M. A remark on a paper of E B Davies. *Proc. Amer. Math. Soc.* **129** (2001) 2955–2957.

[Zygmund 1968] Zygmund, A. *Trigonometric Series*, Vol. 1, 2nd edn. Cambridge: Cambridge University Press, 1968.

Index

$(X, \mathcal{E})$, 357, 374
$B(0, r)$, 27
$C(K)$, 2
C^∞ vector, 201
$C^n(X)$, 6, 23
$C_0(X)$, 6
$C_c^\infty(U)$, 54
D^α, 67
$L^\infty(X, dx)$, 37
$L_0^\infty(\mathbf{R})$, 414
$L^p(X, dx)$, 36
$R(\lambda, a)$, 14
$W^{n,2}(\mathbf{R}^N)$, 77
Dom, 164
EssSpec, 118, 330
Gr, 164
Hull(p, A), 277
Hull$_n(A)$, 277
Num(A), 264
Num(p, A), 278
Num$_n(A)$, 278
Rad, 99
Spec, 14
$\mathcal{B}^*$, 6
$\mathcal{B}_+$, 359
$\mathcal{B}_c$, 5
$\mathcal{C}_2$, 151
$\mathcal{C}_p$, 153
$\mathcal{D}^\infty$, 201
$\mathcal{K}(\mathcal{B})$, 103, 118
$\mathcal{L}(\mathcal{B})$, 12
$\mathcal{S}'$, 72
dist, 2
$\langle \cdot, \cdot \rangle$, 3, 10, 36
$\| \cdot \|_2$, 45, 111, 151
$\| \cdot \|_{\mathrm{HS}}$, 45, 151
ω_0, 178, 297

$\overline{\mathcal{A}}$, 286
$\overline{\mathrm{Conv}}$, 108
s-lim, 22
supp, 36, 403
$|A|$, 138
w-lim, 19, 22
$c_0(X)$, 11
$f(H)$, 150
$f(N)$, 144
$f \vee g$ and $f \wedge g$, 359
f_+ and f_-, 359
k-tree, 374
$l^p(X)$, 3, 36

Abel-Lidskii basis, 82, 425
adjoint operator, 13
Airy operator, 250
analytic
  functions, 25
    spaces of, 428
  vector, 203
Anderson model, 134, 282, 284
anharmonic oscillator, 426
annihilator, 198
anti-ferromagnetic, 369
aperiodic, 369, 391
approximate
  eigenvalues, 247
  point spectrum, 17
approximation and regularization, 54
approximation property, 103
approximately normal, 150
Arzela-Ascoli theorem, 106
asymptotic
  dimension, 378
  stability, 298

Banach
  algebra, 13, 61, 63
  lattice, 7
  space, 1
Banach-Alaoglu theorem, 21
Banach-Mazur theorem, 108
Banach-Steinhaus theorem, 20
Bargmann, 428
base point, 376
basis, 4, 80
  unconditional, 90
Berezin, 431
Bernstein's theorem, 66
bilateral shift, 397
billiards, 398
biorthogonal, 80
Böttcher, 124
boundary conditions, 336
bounded linear functional, 6
Burke-Greenbaum theorem, 279

Calkin, 153
  algebra, 118
Carleson, 88
Cauchy problem, 163, 171
Cauchy process, 183, 186, 208
Cauchy's integral formula, 26
Cayley transform, 144, 310
Chebyshev polynomials, 274
circulant matrix, 369
Clarkson inequalities, 41
class $\mathcal{P}$, 341
closed operator, 144, 164
coherent state, 428
cokernel, 116
commutator bound, 161
commuting operators, 176
compact operator, 103, 123
compatible Banach spaces, 49, 109
complete, 80
completely non-unitary, 314
completion, 1
complex scaling, 329, 423
complexification, 357
condition number, 258
conditional basis, 80
conjugate function, 303
conjugate index, 38
connected graph, 374
consistent operators, 49, 104, 109, 177, 348
continued fractions, xii

contraction, dilation of, 307
contraction semigroup, 230
control theory, 262
convection-diffusion, 248, 274, 275, 339, 395
convex hull, 108
convexity theory, 268
convolution, 53, 63, 75, 129, 131, 274
  semigroup, 183, 205
core, 172, 210
coupling constant, 341

Davies, 277
Davies's theorem, 304, 407, 426
Davies-Kuijlaars theorem, 425
Davies-Martinez theorem, 282
Davies-Simon, 288
Davies-Simon theorem, 293
deficiency indices, 145, 312
degree, 374
detailed balance, 366
differentiability, 23, 201
differential operator, 78, 188, 211, 335, 347
dilation theorem, 309
Dini's theorem, 85
discrete Laplacian, 364
dissipative, 231, 310
distance, between subspaces, 140
distribution, 72
domain, 164, 210
du Bois Reymond, 84
dual
  operator, 13, 197
  space, 6, 72
dynamical system, 398

EigTool, 248
Eisner-Zwart theorem, 324
elliptic, 78, 79
embedding, 1
Embree, 248
Enss, 423
entire vector, 203
equicontinuous, 106
equilibrium state, 368
equivalent norms, 6
ergodic, 396
essential range, 45, 143, 148, 165
essentially bounded, 37
exponential growth rate, 297
extension, 164

Fejér's theorem, 273
Feller, 228
    property, 407
ferromagnetic, 369
finite element method, 104, 266
finite rank operator, 103
first return probability, 371
Fock space, 112
Fourier series, 4, 80
    L^2 convergence, 57
    L^p convergence, 84
    absolutely convergent, 60
    pointwise divergence, 85
Fourier transform, 67
fractional powers, 29, 208
Fréchet space, 19, 68, 203
Fréchet-Riesz theorem, 7
Fredholm operator, 116, 330
Frobenius, 380, 405
    norm, 45, 151
functional calculus, 144, 150
    holomorphic, 27

Gabor transform, 428
Gauss, xii
Gaussian function, 69, 90, 185, 206, 241, 346
Gel'fand, 61, 63, 203, 221
Gel'fand's theorem, 100
generator, 168
    classification of, 227
Gesztesy-Tkachenko, 133
Gibbs
    phenomenon, 58
    semigroup, 194
    state, 368
Glauber dynamics, 367
Gohberg, 125
Goldsheid, 134
Gram-Schmidt, 272
graph, 131, 164, 357, 374, 388
Green function, 157, 333, 408
growth bounds
    basic, 177
    long time, 369

Haar
    basis, 4
    measure, 53
Haase's theorem, 320
Hahn-Banach theorem, 6
Hamiltonian, 368

Hankel operator, 291
Hardy space, 128
harmonic oscillator, 329, 424
Hartman, 125
Hermite polynomials, 88, 274, 424
higher order hull and range, 277
Hilbert transform, 75
Hilbert-Schmidt operator, 45, 110, 151
Hille's theorem, 353
Hille-Yosida theorem, 231
Hölder
    continuous, 85
    inequality, 38
holomorphic semigroup, 237, 345, 353
homotopy invariant, 120
Hörmander, 234
hydrogen atom, 329
hyperbolic tree, 374
hypercontractive, 112

ideal, of operators, 103, 151, 155
image processing, 53
incidence matrix, 374
index, 116
interpolation, 50, 177, 183, 219, 393
invariant set, 361, 387, 389, 403
invariant subspace, 18, 168, 173, 314
inverse mapping theorem, 14
inverse temperature, 368
irreducible, 361, 365, 369, 390, 403
    isometric embedding, 1
isometry
    partial, 140
    spectrum of, 398

James's theorem, 41
Jordan decomposition, 7
Jordan matrix, 31, 34, 218, 222, 249, 269

Kakeya, 271
Keldysh, 251
kernel, 116
Kesten's theorem, 376
Khoruzhenko, 134
Koopman operator, 398
Kortweg-de Vries equation, xii
Krein's theorem, 126
Krein-Šmulian theorem, 179

Laguerre polynomials, 274
Laplacian, 69, 185

and Gaussian semigroup, 185
L^2 spectrum, 79
L^p spectrum, 241
lasers, unstable, xi, 424
Laurent operator, 53, 274
Lebesgue measure, 36
Legendre
   polynomials, 273
   transform, 303
Levy process, 76
Lieb-Thirring bound, 410
linear sublattice, 359
Liouville's theorem, 16, 26
locally finite graph, 374
log-concave envelope, 301
log-convex, 37
Lorch, 90
Lumer-Phillips theorem, 231

Markov
   operator, 356
   semigroup, 362, 399
      reversible, 366
Markov-Kakutani theorem, 317
Martinez, 282
maximal dissipative, 312
Mazur's theorem, 108
Maz'ya and Schmidt, 90
measure space, 35, 395
Mercer's theorem, 156
Millennium Bridge, xii, 252
minimal
   complete, 81
   polynomial, 272
mixing, 404
Miyadera, 228, 349
Mockenhaupt classes, 89
molecule, 368
moment problem, 74
momentum operator, 161
multi-index, 67
multiplication operator, 45, 76, 132, 143, 165,
      172, 188

Navier-Stokes equation, xii, 399
Nelson's theorem, 173
neutron diffusion equation, 386
Nevanlinna's theorem, 280
Newman's lemma, 61
non-linear, 251
non-negative operator, 135

norm continuous semigroup, 190
normal, 2, 143, 259
normed space, 1
null set, 36
numerical range, 264, 268, 272,
      409, 431

obstacle scattering, 129
one-parameter
   group, 174, 315
   semigroup, 167
      generator of, 227
      holomorphic, 237
      long time bounds, 296
      norm continuous, 190
      on dual space, 197
      short time bounds, 300
      subordinated, 205
      trace class, 194
operator
   bounded, 12
   closed, 164
   compact, 102
   Fredholm, 116
   Hilbert-Schmidt, 151
   on a Hilbert space, 135
   on an L^p space, 45
   pencil, 251
   positive, 380
   trace class, 153
order ideal, 387, 403
ordered Banach space, 355
ordering of operators, 136
orthogonal
   polynomials, 272, 288
   projection, 140
orthonormal, 4, 58, 80, 88, 93
oscillation properties, 274

partition, 35
   function, 368
   of the identity, 105
path, 357, 371
Pazy, 192
pencils of operators, 252
period of a semigroup, 390
periodic operator, 132, 282, 422
peripheral point spectrum, 287, 405
Perron, 380, 405
Perron-Frobenius operator, 398

perturbation
  bounded, 339
  of an operator, 31, 325
  of spectrum, 31, 328
  rank one, 32, 334
  relatively compact, 330
  resolvent based method, 350
  semigroup based method, 339
Phillips, 228
Plancherel theorem, 69
Poisson
  distribution, 206
  process, 364
polar decomposition, 138, 285
polarization identity, 13
polynomial
  convex hull, 280
  growth bound, 83, 378
population growth models, 386
Pöschl-Teller potential, 413
position operator, 161
positive operator, 381
power series, 26
power-bounded, 100, 317
principal part, 79
probability distribution, 356, 395
projection, 22, 140
pseudo-resolvent, 214
pseudospectra, 213, 245, 247, 426
  generalized, 255
  structured, 261

quantization, 428
quantum mechanics, 423
quotient space, 10

radiation condition, 254
radius of convergence, 26
Radon-Nikodym theorem, 43
random
  matrix, 134, 263
  walk, 206
range, 210
rank, of an operator, 31
Ransford, 100, 264
recurrence equation, 101
recurrent, 370, 372
reflexive, 10, 41
relative bound, 325
  relatively compact, 21, 331
Rellich, 31

Rellich's theorem, 328
renormalization, 393
reproducing kernel Hilbert space, 428
resolvent, 14
  bounds in Hilbert space, 321
  operators, 210
  set, 210
resonance, 254
reversible, 366
Riemann-Lebesgue lemma, 58, 71
Riesz, 7, 30
  basis, 93
Riesz's theorem, 123, 327
Riesz-Kakutani theorem, 7
Riesz-Thorin theorem, 50
Rouche's theorem, 339

sample space, 357
scalar type operator, 133, 422
scattering theory, 168, 422
Schatten, 153
Schauder basis, 80
Schauder's theorem, 108
Schrödinger operator, 163, 335, 346, 348,
  395, 408, 413, 417
Schur's theorem, 291, 293
Schwartz space, 68
second dual space, 10
sector, 147, 237
self-adjoint, 13, 143, 145
semi-classical analysis, 427
semigroup, 390
seminorm, 1, 19
separable, 36
separation theorem, 268
sequential compactness, 103
shift operator, 18, 121
Shkarin, 101
shooting method for ODEs, 253
Silbermann, 124
similarity invariant, 297
Simon, 272
singular
  integral operators, 75
  potential, 348
  values, 153
smooth, 68
Sobolev space, 77
spectral
  mapping theorem, 18, 215, 221, 223
  pollution, 266

projection, 30, 110
radius, 99
theorem, 143
spectrum
L^p dependence, 49
definition, 14, 211
essential, 118, 124, 330, 420
of a k-tree, 374
of Airy operator, 247
of consistent operators, 109, 219
of convolution operators, 65
of multiplication operator, 45
of operator pencil, 251
of Schrödinger operator, 418
of Toeplitz operator, 125
of ultracontractive operators, 394
peripheral, 287
square root lemma, 138
stability, 245
statistical dynamics, 367
Stone, 143
Stone-Weierstrass theorem, 60
strong operator limit, 22
Sturm-Liouville, 157, 273, 333, 408
subadditive, 99, 297
subexponential growth, 378
sublattice, 392
subMarkov operator, 363
subordinated semigroup, 205
subspaces, two, 140
support, 36, 397, 403
supremum norm, 2
symbol, 78, 124, 128, 211, 337
symmetric, 144
Sz.-Nagy dilation theorem, 307
Sz.-Nagy theorem, 315

three lines lemma, 49
thresholds, 424
Tietze extension theorem, 2
Toeplitz operator, 124, 128
Toeplitz-Hausdorff theorem, 265
topological vector space, 19, 220
totally bounded, 103
trace class
operator, 154

semigroup, 194
transient, 370, 372
translation invariant, 63
tree, 374
Trefethen, 248
triangle, 146, 168
tridiagonal operator, 281
trigonometric polynomial, 56
truncation, 217, 265, 274

ultracontractive, 394
unconditional basis, 90
undirected graph, 366, 374
uniform boundedness theorem, 20
uniform convexity, 41
unitary, 13, 143
unstable, 386
Urysohn's lemma, 2

Vandermonde determinant, 338
Voigt, 349
Volterra operator, 100, 249, 291
von Neumann, 143, 153
von Neumann's theorem, 286, 309

wave equation, 252
waveguides, 253
wavelets, 5, 97
weak
convergence, 165
derivative, 75
operator limit, 22
topology, 19
weak* topology, 21
weakly closed operator, 165
Weierstrass's theorem, 59
Wiener, 125
Wiener's theorem, 62
wild, 83
winding number, 126
Wintner, 125
Wright, 248
Wrobel, 299

Zabczyk's example, 222

Printed in the United States
By Bookmasters